T0304660

System Dynamics

System Dynamics: Modeling, Simulation, and Response covers linearity-based modeling techniques before delving into nonlinear systems. It compares the Bond Graph technique against traditional techniques (Newton's law, Kirchhoff's law, the law of the conservation of energy, and the heat transfer law).

Presenting transient response analyses of first- and second-order systems subjected to various inputs, the book provides a thorough discussion of computational analyses of transient responses using MATLAB®/Simulink and 20-sim software. It introduces the Lagrangian method and its application in handling mechanical and electrical systems. The book reviews the classical method for solving differential equations and includes Laplace transforms.

The book is intended for upper-level undergraduate mechanical and aerospace engineering students taking system dynamics courses.

Instructors will be able to utilize a Solutions Manual and Figure Slides for their courses.

Osama Gad is a lecturer in the Mechanical Engineering Department at Kuwait University. He received his PhD in mechanical engineering in 1997 and has almost 40 years of experience as an instructor and teaching assistant in mechanical engineering, namely in dynamics, system dynamics, theory of machines, vibrations, and control. He has written numerous articles for reputed journals and served as a peer reviewer for the *Journal of Dynamic System, Measurements and Control* (ASME), *Journal of Mechanical Design* (ASME), and *Journal of Mechanical Engineering Science* (IMechE).

System Dynamics
Modeling, Simulation, and Response

Osama Gad

CRC Press
Taylor & Francis Group
Boca Raton London New York

CRC Press is an imprint of the
Taylor & Francis Group, an **informa** business

MATLAB® and Simulink® are trademarks of The MathWorks, Inc. and are used with permission. The Math-Works does not warrant the accuracy of the text or exercises in this book. This book's use or discussion of MATLAB® or Simulink® software or related products does not constitute endorsement or sponsorship by The MathWorks of a particular pedagogical approach or particular use of the MATLAB® and Simulink® software.

First edition published 2025
by CRC Press
2385 NW Executive Center Drive, Suite 320, Boca Raton FL 33431

and by CRC Press
4 Park Square, Milton Park, Abingdon, Oxon, OX14 4RN

CRC Press is an imprint of Taylor & Francis Group, LLC

© 2025 Osama Gad

Library of Congress Control Number: 2024942781

ISBN: 978-1-032-68563-2 (hbk)
ISBN: 978-1-032-68564-9 (pbk)
ISBN: 978-1-032-68565-6 (ebk)
ISBN: 978-1-032-68566-3 (eBook+)

DOI: 10.1201/9781032685656

Typeset in Nimbus Roman
by KnowledgeWorks Global Ltd.

Access the Support Materials: www.routledge.com/9781032685632

Contents

Preface

This book is intended mainly for students majoring in mechanical engineering. The title indicates our primary objective, which is to encourage students to engage in the modeling and simulation of dynamical systems to demonstrate their dynamic responses under various input situations.

The book is primarily concerned with two topics. The first is modeling, in which the methods and approaches for translating the physics of dynamical systems into differential equation formalism are discussed, while the second is simulation and response, which is the solution of the modeling formulation of these differential equations. The material presented in this book is appropriate for use in the instruction of mechanical engineering students in their second or third year of undergraduate study and is also suitable for graduate level. It is limited to the level at which it can be taught in a one-semester course, which is an appropriate supplement to courses on vibration and control.

The book covers various engineering systems, such as mechanical, electrical, fluid, and thermal systems, making applying the ideas in engineering simple. The fundamental process for generating differential equations is based on Newton's laws for mechanical systems, Kirchhoff's rules for electrical systems, Bernoulli's equation for fluid systems, and heat transfer laws for thermal systems. Their main advantage is the capability of the modeling techniques to study both linear and nonlinear systems. These are the fundamental modeling tools that students and researchers will use. After obtaining the differential equations, it is necessary to solve them to determine the system's response to various input situations. These physical principles are expressed as relationships between variables, with no prior assignment of cause and effect. Using Bond Graphs is a consistent approach for representing systems as connections between the physical elements of dynamical systems. As an alternative to the approaches described above, it is a universal tool that can be utilized to model every kind of system independently or in combination with other systems. An exchange of power is the term that describes the interaction that takes place between these components. With this concept of power exchange, we can separate and allocate the most significant variables in a model, specifically the flows and the efforts. Understanding these Bond Graph contributions enhances one's capacity to derive models.

The Newtonian formalism of mechanical systems is explained in Chapter 2, with examples of its practical difficulties. The application of Kirchhoff's laws to developing dynamic equations for electrical systems is discussed in Chapter 4. Bernoulli's equation and heat transfer laws are considered in Chapter 5 while developing fluid and thermal dynamic equations. Chapter 3 covers the basics of the Bond Graph approach. In this chapter, the student learns about essential Bond Graph elements and how they can be categorized based on energy usage. This approach algorithms the equation-finding procedure so that computer programs can handle equation-finding and model simulation tasks after creating the Bond Graph of a system. For the student to gain a feel for Bond Graph-based system modeling without spending too much time on it, the fundamental components of this technique are introduced in detail. The 20-sim software simulates the Bond Graph models and generates their dynamic responses.

The Lagrangian method is introduced in Chapter 6, and its application in handling mechanical and electrical systems is illustrated.

Students learn Laplace transforms in Chapter 7 to prepare for the transfer function representations needed to solve differential equations. This chapter also reviews the classical method for solving differential equations.

System analyses are presented in Chapter 8. This chapter discusses transient response analyses of first- and second-order systems with different inputs. It then details how to use the MATLAB/Simulink package to do computational analyses of transient responses.

The transfer function can be used to comprehend and study a system's frequency response, as explained in Chapter 9. Students can learn about system stability, resonant frequencies, and gain and phase margins from the included Bode diagrams and how they are sketched and read.

In the last part of this book, Chapter 10, modeling and analysis of nonlinear systems are addressed. Students, particularly graduate students, will benefit from reading this chapter because it will teach them how to deal with the nonlinear aspects of dynamical systems. In this chapter, the elements of the modulated Bond Graph that represent the nonliterary aspects of the system are provided and simulated using the 20-sim software.

Osama Gad

1 Introduction

Daily, we frequently engage in exciting but complicated interactions with various systems. There will be a wide variety of situations in which these interactions occur, and it is possible to think of many potential uses for them (aircraft, cars, trains, etc.). We need to explain these systems, or a portion of them, in a form that is easy to understand for various reasons; one is the capability to model, analyze, and design.

Our objective provides a more idealized representation than the actual system attributes. We must decide on the features that should be considered instead of those that should be ignored. Modeling is an art form in which the primary objective is to select, from among a variety of available qualities, only those that are required and sufficient to portray the system accurately according to the goals of the modeler. Such a wide variety of characteristics from which to pick can make this task difficult, and this is the most fundamental idea that underlies the art of modeling; within the context of the modeling process, we have arrived at an essential stage.

To build a valuable model of a system, one must have comprehensive knowledge not only of the system that is being investigated but also of the modeling approaches that can be applied. This book provides a comprehensive review of various modeling approaches that assist us in deriving an appropriate model from our current body of information. During the modeling stage, we will demonstrate that a mathematical model can be developed based on preliminary information on the physical rules that govern the fundamental dynamic properties of the system. A model of this kind describes the connections between the various physical properties of the system elements.

Parametric models of the systems and their identification are the topics of discussion covered in this book, and we will talk about some of the more practical identification features. Simulation serves as the tool necessary for the performance of these systems, and a complete overview of simulation techniques can be applied to any dynamic system model discussed. This book serves as a primer for those unfamiliar with the fascinating fields of modeling and simulation, as illustrated in Figure 1.1.

On the other hand, it is not the goal of this book to present the reader with a comprehensive discussion of all issues, especially modeling and analysis of nonlinear dynamic systems, as doing so would have required many thousands of additional pages. A brief discussion of how to model and simulate nonlinear systems will be dealt with at the end of this book, Chapter 1. The reader of this book will come away from this experience realizing that many of the problems of linear dynamic systems have been addressed. Now, let us define the meaning of dynamic system modeling, simulation, and response.

DOI: 10.1201/9781032685656-1

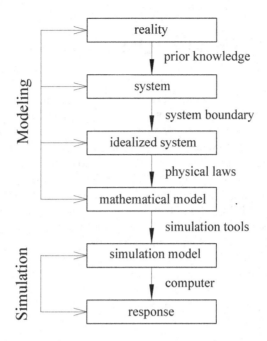

Figure 1.1 Iterative phases needed for modeling and simulation

1.1 MODELING

The practice of developing mathematical descriptions of dynamic systems, such as those that occur in mechanical, electrical, fluid, and thermal systems, is known as modeling. These descriptions should be straightforward while still including sufficient detail to fulfill the requirements of the modeler. It is essential to acknowledge the existence of a diverse collection of models, each of which describes a unique aspect of the same reality. For example, the behavior of a physical component in different systems can be investigated and explained from various viewpoints, as follows,

- The force F acting on a body in the mechanical system can be considered as a linear relationship between body mass m and its acceleration a as follows:

$$F = ma \tag{1.1}$$

- The voltage e across an electrical resistance in the electrical system can be described as a linear relationship between resistance R and current i as follows:

$$e = Ri \tag{1.2}$$

- The pressure difference Δp in the fluid system can be described as a linear relationship between mass flow rate Q that passes through a valve and its resistance R_v as follows:

$$\Delta p = R_v Q \tag{1.3}$$

- The temperature difference ΔT in the thermal system can be described as a linear relationship between heat mass flow rate q that passes through a material as follows:

$$q = C_t \Delta T \tag{1.4}$$

where C_t is the thermal capacitance.

All these physical quantities (force F, voltage e, pressure Δp, and temperature ΔT) reflect the same physical component when applied to different systems.

When analyzing, for example, a mechanical system, the modeler may be interested in considering masses of massless elements (springs, shafts, etc.), the shape of the gear tooth, the static deflections of flexible parts, the physical working conditions ... and the dynamic behavior of the system.

Taking into consideration each of these unique viewpoints will lead to the development of a variety of different models. In contrast, we will limit ourselves in this book to models that describe the behavior of dynamic systems, including the above-mentioned viewpoints, or at least some of them.

The essence of the art of modeling is to choose, from among the numerous accessible qualities, only those that are required and adequate to accurately characterize the system and achieve the goals set by the modeler.

We can understand the fundamentals of a frequently unpredictable system because the models have been simplified. Therefore, modeling is extracting the fundamental problem from the real system and expressing it in an abstract language, such as a mathematical model. Modeling, an integral part of various methodologies applied, makes it possible to grasp the dynamic system, even if only in part. As a result, modeling is an essential stage of all areas of research and technology, including but not limited to mechanical engineering, electrical engineering, fluid engineering, thermal engineering, chemical engineering, civil engineering, and physics.

A system can be considered a collection of elements that define a particular area and serve as the subject of the study. It is a subjective entity with boundaries (assumptions) that include things considered most significant to our objectives and excludes those deemed less valuable. A model is a way in which we want to describe the essential aspects of the system that is being investigated. It needs to have some representation of the objects that are part of the system and should accurately reflect the actions in which these entities participate. Therefore, a model mirrors the modeler's comprehension of reality, its constituents, and their interrelationships.

During the creation of models, one of the most critical decisions must be to select the system boundary. This boundary defines which aspects of reality, in addition to the system itself, will be considered. If the chosen boundary is too broad, it will become tricky. Such a model may become impossible to examine, and many unnecessary details may hide crucial results. If the boundary that has been chosen is narrow enough, then the model will only account for some of the essential characteristics of the system. This will not produce satisfactory outcomes and will not fulfill the requirements that have been outlined. Therefore, a detailed description of the underlying physical rules is required. So, a model with less precision can be utilized, and consequently, the accuracy requirements differ depending on how the models will be used. In this book, we will discuss some techniques that can assist in deriving appropriate models.

1.1.1 MODEL BUILDING APPROACHES

Getting our hands on a model can be done in several different ways. It is essential to develop a model based on prior knowledge of the system and the behavior of dynamical systems. This can be accomplished by combining these two sets of information. This knowledge is priceless. The previous knowledge of the system might be based on the physical laws that explain the mutual relations between the variables of an idealized system with idealized physical components. Alternatively, the preliminary information can be based on the system itself. For example, if a model contains a body mass, it is assumed to be a point mass without any physical dimensions; a mass flow rate in the fluid system will be assumed as laminar; concentrations will be considered homogeneous throughout the tank in fluid and thermal systems, and so on.

Equations, either algebraic or differential, could describe the physical laws governing the system. Differential equations are utilized when evidence of dynamic behavior describes a dynamic model. For example, the laws of conservation of energy, mass, momentum, and various other quantities serve as the basis for these equations; the law of conservation of mass plays an important role. The structure of the model can be better understood with prior knowledge about the system, which can also provide extra information. It becomes much easier to see the model's structure and the input-output interactions between the variables.

It is possible to explain dynamical systems using various models; one is the mathematical model in which differential and algebraic equations can characterize the system's linear or nonlinear behavior. This book does not present any models other than mathematical ones using the modeling approaches it discusses. These models can be further broken down into static and dynamic models. The dynamic models can be broken down into continuous or discrete models depending on the variable's state events. When a current i suddenly becomes zero, this is an example of a state event. Another example would be when the displacement x of a moving element reaches its boundary. In the continuous model, a state event may trigger a switch, while in the discrete model, a state event may activate some timing mechanism. Timing systems under the discrete event model are responsible for generating time events. They can trigger a switch or additional actions inside the continuous model. In this book, we will limit ourselves to modeling only continuous dynamic systems and discuss these variances further.

1.1.2 DYNAMIC CONTINUOUS MODELS

When it comes to modeling continuous dynamical systems, one has a wide variety of mathematical tools, some of which are the following:

- Lumped-parameter models (ordinary differential equations)
- Distributed-parameter models (partial differential equations)

1.1.2.1 Lumped-parameter models

Ordinary differential equations, known as ODEs, represent lumped-parameter models in which explicit differential equations are used to formulate the ODEs. Each derivative is defined on an equation's right-hand side as follows:

$$\dot{x}_i(t) = f_n(x_i(t), u(t)) \tag{1.5}$$

where $x_i(t)$ is the system variable and $u(t)$ is the input. In this case, the value of $\dot{x}_i(t)$ can be calculated explicitly for each value of t, as indicated in equation (1.4).

1.1.2.2 Distributed-parameter models

Because physical systems are continuous, they can typically be effectively characterized using either lumped or distributed parameter models. For example, calculating the temperature gradient for different materials that make up a wall is how one may describe the heat transfer process that occurs through the wall. This system can be modeled using distributed parameters and described by partial differential equations. In this case, the temperature T_i and the time t are considered independent variables in the model.

$$\frac{\partial q}{\partial t} = a_i \frac{\partial q}{\partial T_i} \tag{1.6}$$

where a_i and q are the system parameters and the amount of heat transferred through walls, respectively, while T_i are the temperatures of the materials of the walls.

1.2 SIMULATION

To investigate and understand the dynamic behavior of dynamic models, simulation is a method that may be used to acquire their responses. It is possible to determine the time response using various methods. One such method is the inverse Laplace transformation, which converts the model into a format from which the time responses can be calculated straightforwardly.

One of the most widespread applications of simulation technology is the generation of simulations through computer software. Imagine that a complicated model must find an analytical solution to the problem. It is possible to get an idea of how a model will behave when it is put through its tests in a situation representative of the actual application if one solves the numerical equations of the model using simulation software and then runs the model to get the response data. The model's behavior will become apparent to us due to these data.

Simulation software is a computer program that simulates the behavior of an analog computer on a digital platform. Therefore, all of the analog computer's functions, including gains, integrators, function generators, adders, subtractors, and multipliers, are accessible. Furthermore, when issuing the Run command, the solution to the problem with the output variable is computed and shown immediately.

Simulation software is an attractive option for simulation tools because of its appealing user interfaces and the numerous simulation features it provides. The most fundamental constraint they have is that it is impossible to add any new blocks, and the simulation model also cannot be expanded by adding statements in a higher level of programming language. Each building block that makes up a program has just one output. This limitation prevents the use of matrices and vectors; dealing with this restriction while dealing with increasingly complex models can be highly unpleasant. However, modern simulation programs come with additional capabilities, such as blocks that may be programmed by the user or the capability of working with vectors and matrices.

1.3   RESPONSE (SYSTEM ANALYSIS)

The parametric identification is provided here as the system analysis (response); this analysis deals with identifying parametric models. System analysis and control design may all benefit from parametric models. Obtaining a model must never serve as a goal in and of itself, and it is always done for a particular reason. System analysis and control design will frequently be the goals of these endeavors. System analysis aims to predict how a system will behave under given conditions. Simulation is used to carry out this analysis. In this case, the model has to be able to predict how the system will behave accurately.

A demonstration of the steps for modeling, simulation, and response discussed in this book can be found in Figure 1.2.

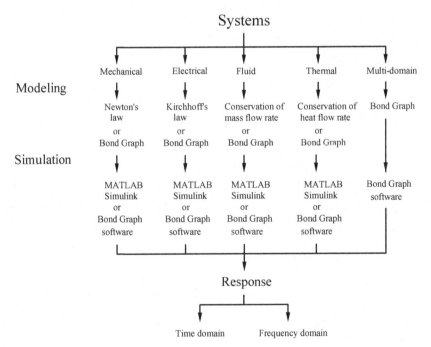

Figure 1.2 Iterative phases needed for modeling, simulation, and response of dynamic systems

2 Mechanical Systems

2.1 INTRODUCTION

Every mechanical system involves an infinite number of elements. So, it is necessary to have a fundamental knowledge of these elements and their overall behavior to acquire competence in modeling their dynamic motions. While working with practical uses, we regularly come across systems of a particular type, such as mechanical or electrical, and situations where multiple systems coexist in a single system. An electrical motor, for example, can be powered by current from an electrical circuit while simultaneously producing torque for mechanical systems. A survey of such devices will be discussed later.

Returning to the subject of this chapter, we will explore the fundamental elements utilized to describe components of systems that involve solid body motion (mechanical systems). This covers many potentially life-threatening conditions, ranging from a watch mechanism's accuracy to a rapid aircraft's riding qualities. To represent the characteristics of such systems, three passive (non-energy-producing) elements are required: the inertia element (mass), the storage element (spring), and the dissipative element (damper). All these elements are available in both translational and rotational mechanical systems, depending on the type of motion. Besides these three elements, we will consider the driving inputs of mechanical systems: forces, torques, and translational motion, or angular motion, that cause the system elements to respond. The analysis of both translational and rotational systems will be considered in this chapter.

2.2 MASS (INERTIA) ELEMENT

The actual dynamics of processes in natural mechanical systems are associated with the property of inertia, and this property must be reflected in one way or another in the design scheme. Note that when dealing with translational systems, the inertia element is represented by the mass m, whereas when dealing with rotating systems, it is represented by the moment of inertia J. In modeling dynamic system elements, they are generally made of simplified schemes. In such design schemes, some system elements are assumed to be massless and represented as deformable, massless constraints. The elements of the system for which the property of inertia is retained in the design scheme are then considered to be concentrated masses or rigid bodies. When we introduce a component into a system to enhance inertia, the inertia element frequently produces an unwanted, inescapable effect because all elements in mechanical systems have inertia properties. Furthermore, the mass itself has several functional purposes. Newton's second law is the physical law that governs the force-motion behavior of mass elements, which is given by

$$\sum \text{Force} = \text{mass} * \text{acceleration} \tag{2.1}$$

DOI: 10.1201/9781032685656-2

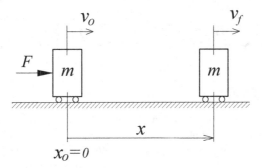

Figure 2.1   Translational mass element

The concept of the inertia element is introduced to apply this law directly to practical situations. It is a term that refers to an idealized point mass that occupies an infinite volume. Every point in the mass element moves in the same direction with the same velocity throughout the purely translational or rotational motion. As a result, when equation (2.1) is applied to the mass shown in Figure 2.1, the following result is obtained:

$$F = ma = m\ddot{x} \tag{2.2}$$

where m is the body mass and $a = \ddot{x}$ is its acceleration. Similarly, when Newton's second law applies to a body moving in a pure rotational motion along a fixed axis, as shown in Figure 2.2, the following results are obtained:

$$T = J\ddot{\theta} \tag{2.3}$$

where J is the moment of inertia of the rotating body and $\ddot{\theta}$ is its angular acceleration. Regarding the definition of the moment of inertia, the rotation is considered fully rigid at all times. We can compute the moment of inertia J with relative simplicity for simple geometrical and homogeneous masses. Table 2.1 provides some relevant results.

When a force is applied to a mass element, we will find that it accelerates the mass and gives it energy. This energy, called kinetic energy $K.E.$, has been stored

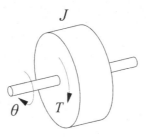

Figure 2.2   Torsional mass element

Table 2.1

The moment of inertia of rigid bodies of common shapes

Mass element	Moment of inertia
Solid desk: 	$t \ll R$ $J_x = J_y = \frac{1}{4}mR^2$ $J_z = J_x + J_y = \frac{1}{2}mR^2$
Solid cylinder: 	$J_x = J_y = \frac{1}{12}m(3R^2 + l^2)$ $J_z = \frac{1}{2}mR^2$
Rod: 	$J_{xx} = \frac{1}{12}ml^2$ $J_{rr} = \frac{1}{12}m(l\sin\theta)^2$
Mass element connected to a massless rod: 	$J_x = J_y = ml^2$

in the mass. When the force is removed, the mass decelerates under its inertia force, which always acts in the opposite direction of its movement, and thus its velocity approaches zero. As a result, the mass moves a distance x, as shown in Figure 2.1, and the amount of energy transmitted to it is given by

$$K.E. = (m\ddot{x})x \qquad (2.4)$$

The acceleration of the mass is obtained from Newton's first law as follows:

$$\ddot{x} = \frac{v_f^2 - v_o^2}{2x} \tag{2.5}$$

Substituting equation (2.5) in equation (2.4) yields

$$K.E. = m(\frac{v_f^2 - v_o^2}{2x})x \tag{2.6}$$

or,

$$K.E. = \frac{1}{2}mv_f^2 - \frac{1}{2}mv_o^2 \tag{2.7}$$

Suppose the mass moves from rest ($v_o = 0$), equation (2.7) becomes

$$K.E. = \frac{1}{2}mv_f^2 \tag{2.8}$$

Similarly, for torsional mass, shown in Figure 2.2, the amount of kinetic energy transmitted to the rotating mass is given by

$$K.E_t. = \frac{1}{2}J\dot{\theta}^2 \tag{2.9}$$

We will consider this energy for translational and rotational mechanical system elements.

2.3 EQUIVALENT MASS

Different inertia elements, such as pulleys, levers, bars, gears, etc., and springs and dampers, are frequently used in mechanical systems. Despite the significant differences in their physical appearance, all mechanical systems perform the same fundamental function: they transform the motion of an input component into the motion of an output component that is kinematically related to and depends on the type of the input component. However, while analyzing systems incorporating such motion transformers does not necessitate new components or techniques, reducing the existing system to an imaginary but dynamically equivalent one may be simpler in many situations. Just for simplicity, in the following sections, we will be dealing with the equivalence of the system by considering the equivalent of their elements separately (masses, springs, and dampers) for the time being. We will deal with the equivalency of the overall system later on.

We will begin by looking at the equivalent mass. Consider two masses arranged and fixed at specific positions on a massless bar, as shown in Figure 2.3. It is required to determine an expression representing their equivalent mass, m_{eq}. Assume the bar is deflecting an angle θ. Figure 2.3a shows that the mass m_1 moves a displacement x_1, whereas the mass m_2 moves a displacement x_2. We assume that the equivalent mass m_{eq} is placed instead of the mass m_1, as shown in Figure 2.3b. Applying the energy

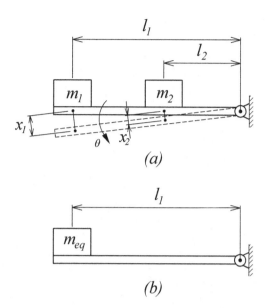

Figure 2.3 Equivalent translational mass

equivalency to the motion of both masses, which should be equal to the energy of the equivalent one, yields

$$\frac{1}{2}m_{eq}\dot{x}_1^2 = \frac{1}{2}m_1\dot{x}_1^2 + \frac{1}{2}m_2\dot{x}_2^2 \tag{2.10}$$

The displacements x_1 and x_2 are given by

$$x_1 = l_1 \sin\theta \text{ and } x_2 = l_2 \sin\theta \tag{2.11}$$

Assuming the angle θ is small ($\sin\theta \approx \theta$), yields

$$x_1 = l_1\theta \text{ and } x_2 = l_2\theta \tag{2.12}$$

Taking the derivative of this equation yields

$$\dot{x}_1 = l_1\dot{\theta} \quad \text{and} \quad \dot{x}_2 = l_2\dot{\theta} \tag{2.13}$$

Substituting equation (2.13) in equation (2.10) yields

$$\frac{1}{2}m_{eq}l_1^2\dot{\theta}^2 = \frac{1}{2}m_1l_1^2\dot{\theta}^2 + \frac{1}{2}m_2l_2^2\dot{\theta}^2 \tag{2.14}$$

Rearranging this equation, the equivalent mass m_{eq} of the system related to the angular deflection of the bar, shown in Figure 2.3b, is given by

$$m_{eq} = \left(m_1 + \left(\frac{l_2}{l_1}\right)^2 m_2\right) \tag{2.15}$$

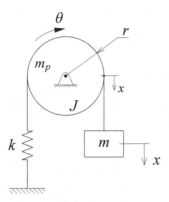

Figure 2.4 Equivalent translational-rotational mass

Let us look at the equivalent of the translational and rotational masses in the system shown in Figure 2.4. We will ignore the spring; it is not our subject now; we will only be concerned with the masses. At this point, we are providing the choice of two locations for the equivalent mass, which may be placed instead of the mass m or placed at the center of the pulley. Let us focus on the equivalent mass instead of the mass m.

Assuming that the mass m moves a distance x and that there is no slip between the cored and the pulley. The pulley rotates at an angle θ, as shown in Figure 2.4. Applying the energy equivalency to motions of both masses, m and m_p, which should be equal to the energy of the corresponding equivalent one, yields the following result:

$$\frac{1}{2}m_{eq}\dot{x}^2 = \frac{1}{2}m\dot{x}^2 + \frac{1}{2}J\dot{\theta}^2 \tag{2.16}$$

Assuming small angle, θ, yields

$$x = r\theta \quad \text{or} \quad \theta = \frac{x}{r} \tag{2.17}$$

Taking the derivative of this equation gives

$$\dot{\theta} = \frac{\dot{x}}{r} \tag{2.18}$$

Substituting equation (2.18) in equation (2.16) yields

$$m_{eq}\dot{x}^2 = m\dot{x}^2 + J\frac{1}{r^2}\dot{x}^2 \tag{2.19}$$

Note that the moment of inertia of the pulley about its center is equal to $J = \frac{1}{2}m_p r^2$, see Table 2.1, then

$$m_{eq}\dot{x}^2 = m\dot{x}^2 + \frac{1}{2}m_p r^2\frac{1}{r^2}\dot{x}^2 \tag{2.20}$$

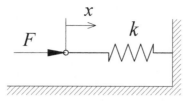

Figure 2.5 Translational spring element

Then, the equivalent mass m_{eq} is given by

$$m_{eq} = \left(m + \frac{1}{2} m_p \right)$$ (2.21)

Now, let us determine the equivalent mass at the center of the pulley. Again, applying the energy equivalency to the motion of the system gives

$$\frac{1}{2} J_{eq} \dot{\theta}^2 = \frac{1}{2} m \dot{x}^2 + \frac{1}{2} J \dot{\theta}^2$$ (2.22)

Substituting equation (2.18) in equation (2.22) yields

$$J_{eq} \dot{\theta}^2 = m r^2 \dot{\theta}^2 + J \dot{\theta}^2$$ (2.23)

Rearranging this equation gives

$$J_{eq} = (m r^2 + J)$$ (2.24)

Set the moment of inertia of the mass $J = \frac{1}{2} m_p r^2$ in equation (2.24), the equivalent inertia J_{eq} of the system is given by

$$J_{eq} = \left(m r^2 + \frac{1}{2} m_p r^2 \right)$$ (2.25)

2.4 SPRING ELEMENT

Suppose we intend to perform the spring function alone; we will find that the spring has some inertia and damping that are unnecessary to the system's function but still exist. Functionally, if the spring is a nonlinear spring, it presents mathematical difficulties one must be willing to deal with. On the other hand, the linear spring has a strictly linear input-output relationship that defines its characteristics. Consider a linear translational spring with force applied to its endpoint as a function of time, as shown in Figure 2.5. If k is the stiffness of the spring, the displacement x of its endpoint is defined by the ordinary static formula

$$x = \frac{F}{k}$$ (2.26)

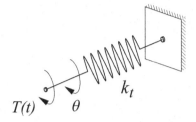

Figure 2.6   Torsional spring element

For a linear torsional spring shown in Figure 2.6, the angular displacement θ of its endpoint is defined by

$$\theta = \frac{T}{k_t} \tag{2.27}$$

Note that equations (2.26) and (2.27) are not dynamic, even though the displacement obtained is not constant but reflects the function of time. The spring could also be described in terms of its capacitance, C (compliance), which is defined as the reciprocal of k ($C = \frac{1}{k}$) or ($C_t = \frac{1}{k_t}$). In terms of compliance, equations (2.26) and (2.27) become

$$x = CF \quad \text{or} \quad F = \frac{1}{C}x \tag{2.28}$$

$$\theta = C_t T \quad \text{or} \quad T = \frac{1}{C_t}\theta \tag{2.29}$$

When the force, or torque, is gradually applied to a spring element and then held constant, we will find that the force has done work in deflecting the spring, and this energy has been stored in the spring, waiting to be retrieved when the spring is relaxed, indicating that the spring is a storage element. Since power is defined as the amount of energy extracted from the force and transmitted to the spring, which is called potential energy $P.E.$ (conservative energy) the definition of power gives the instantaneous power transmitted to the spring as,

- for translational motion

$$P.E. = Fv = F\frac{dx}{dt} = (kx)\frac{dx}{dt} \tag{2.30}$$

or,

$$P.E.\,dt = kx\,dx \tag{2.31}$$

The total energy transmitted to the translational spring is given by

$$\int P.E.\,dt = \frac{1}{C}\int x\,dx \tag{2.32}$$

or,

$$\int P.E.\,dt = k\int x\,dx \tag{2.33}$$

Carrying out the integration of equations (2.32) and (2.33) gives the total stored potential energy $P.E.$,

$$P.E. = \frac{1}{2}\frac{1}{C}x^2 \tag{2.34}$$

$$P.E. = \frac{1}{2}kx^2 \tag{2.35}$$

- for rotational motion

Similarly, for rotational motion, the total potential energy $P.E_t.$ transmitted to the torsional spring is given by

$$\int P.E_t.\,dt = \frac{1}{C_t}\int \theta\,dx \tag{2.36}$$

or,

$$\int P.E_t.\,dt = k_t \int \theta\,dx \tag{2.37}$$

Carrying out the integration of equations (2.36) and (2.37) gives the total stored potential energy $P.E_t.$, as

$$P.E_t. = \frac{1}{2}\frac{1}{C_t}\theta^2 \tag{2.38}$$

or,

$$P.E_t. = \frac{1}{2}k_t\theta^2 \tag{2.39}$$

2.4.1   EQUIVALENT SPRING

The combination of several springs is often utilized in practical applications. These springs can be joined together to form a single equivalent spring. In mechanical systems, there are two fundamental arrangements of spring elements: parallel and series.

2.4.1.1   Springs in parallel

To derive an expression for the equivalent spring stiffness of springs connected in parallel, consider the two springs shown in Figure 2.7a. The system undergoes a static deflection δ when a force is applied, as shown in Figure 2.7b. From the free body diagram shown in Figure 2.7c, the equilibrium equation is given by

$$F = k_1\delta + k_2\delta \tag{2.40}$$

Note that when the force is applied to the equivalent spring, it deflects the same distance δ as both springs deflected, as shown in Figure (2.7d). So,

$$F = k_{eq}\delta \tag{2.41}$$

Substituting equation (2.41) in (2.40) gives

$$k_{eq}\delta = k_1\delta + k_2\delta \tag{2.42}$$

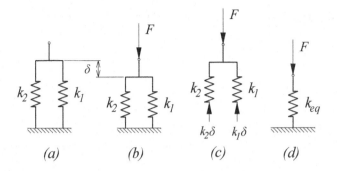

Figure 2.7 Spring elements connected in parallel

So,

$$k_{eq} = (k_1 + k_2) \tag{2.43}$$

In general, if we have i spring elements connected in parallel with spring stiffnesses $k_1, k_2, \ldots, k_i$, then the equivalent spring stiffness k_{eq} is given by

$$k_{eq} = k_1 + k_2 + \cdots + k_i = \sum k_i \tag{2.44}$$

This equation indicated that the equivalent spring stiffness of springs connected in parallel equals the sum of their stiffnesses. Similar to this, the equivalent torsional spring stiffness k_{teq} of parallel torsional springs is

$$k_{teq} = k_{t1} + k_{t2} + \cdots + k_{ti} = \sum k_{ti} \tag{2.45}$$

2.4.1.2 Springs in series

To derive an expression for the equivalent spring stiffness of springs connected in series, consider the two springs shown in Figure 2.8a. As a result of a force applied to the two springs, both are deflected at different rates, δ_1 and δ_2, as shown in Figure 2.8b. The total deflection is given by

$$\delta_{eq} = \delta_1 + \delta_2 \tag{2.46}$$

Since both springs are subjected to the same force F as shown in Figure 2.8, we have

$$F = k_1 \delta_1 \text{ and } F = k_2 \delta_2 \tag{2.47}$$

Since the equivalent spring is subjected to the same force, we have

$$F = k_{eq} \delta_{eq} \tag{2.48}$$

Combining equations (2.47) and (2.48) gives

$$k_{eq} \delta_{eq} = k_1 \delta_1 = k_2 \delta_2 \tag{2.49}$$

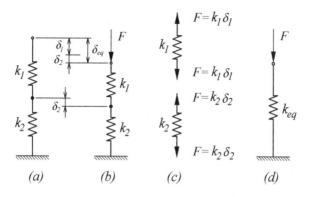

Figure 2.8 Spring elements connected in series

or,

$$\delta_1 = \frac{k_{eq}\delta_{eq}}{k_1} \text{ and } \delta_2 = \frac{k_{eq}\delta_{eq}}{k_2} \tag{2.50}$$

Substituting equation (2.50) in equation (2.46) yields

$$\delta_{eq} = \frac{k_{eq}\delta_{eq}}{k_1} + \frac{k_{eq}\delta_{eq}}{k_2} \tag{2.51}$$

or,

$$\delta_{eq} = k_{eq}\left(\frac{\delta_{eq}}{k_1} + \frac{\delta_{eq}}{k_2}\right) \tag{2.52}$$

So,

$$\frac{\delta_{eq}}{k_{eq}} = \frac{\delta_{eq}}{k_1} + \frac{\delta_{eq}}{k_2} \tag{2.53}$$

This is

$$\frac{1}{k_{eq}} = \frac{1}{k_1} + \frac{1}{k_2} \tag{2.54}$$

Then, the equivalent spring stiffness of the two springs connected in series shown in Figure 2.8d is given by

$$k_{eq} = \left(\frac{k_1 k_2}{k_1 + k_2}\right) \tag{2.55}$$

In general, if we have i spring elements connected in series with spring stiffnesses $k_1, k_2, \ldots, k_i$, then the equivalent spring stiffness k_{eq} is given by

$$\frac{1}{k_{eq}} = \frac{1}{k_1} + \frac{1}{k_2} + \cdots + \frac{1}{k_i} = \sum \frac{1}{k_i} \tag{2.56}$$

Similarly, when torsional springs connect in series, their equivalent torsional spring stiffness is given by

$$\frac{1}{k_{teq}} = \frac{1}{k_{t1}} + \frac{1}{k_{t2}} + \cdots + \frac{1}{k_{ti}} = \sum \frac{1}{k_{ti}} \tag{2.57}$$

Springs are probably connected to solid components in many applications. The equivalent spring stiffness is determined in this situation, as indicated in the following two examples.

Example (2.1):
A rigid hinged bar of length l_1 is connected by two springs of stiffnesses k_1 and k_2, as shown in Figure 2.9. Assuming that the angular deflection of the bar θ is small, determine the equivalent spring stiffness of the system that relates to the angular deflection of the bar at three different locations:

1. Instead of the spring of stiffness, k_1.
2. Instead of the spring of stiffness, k_2.
3. At the midpoint between both springs.

Solution:
In general, the equivalent spring stiffness k_{eq} can be determined in two ways:

1. by considering the energy equivalency between the stored energy in the system caused by the spring's forces and the spring force corresponding to the equivalent.
2. by taking the moment, the balance of the spring forces, including the equivalent, about a fixed point.

For small angular deflection θ, the points of attachment of the two springs, points A and B shown in Figure 2.9, undergo a horizontal displacement, x_1 and x_2, respectively. These displacements are given by

$$x_1 = l_1 \sin \theta \text{ and } x_2 = l_2 \sin \theta \tag{2.58}$$

For small θ ($\sin \theta \approx \theta$), these relations become

$$x_1 = l_1 \theta \text{ and } x_2 = l_2 \theta \tag{2.59}$$

- equivalent spring instead of the spring of stiffness k_1
- by considering energy equivalency
Applying energy equivalency on the motion of the system and placing the equivalent spring at point A shown in Figure 2.10a yields

$$\frac{1}{2} k_{eq} x_1^2 = \frac{1}{2} k_1 x_1^2 + \frac{1}{2} k_2 x_2^2 \tag{2.60}$$

Substituting equation (2.59) in equation (2.60) gives

$$\frac{1}{2} k_{eq} (l_1 \theta)^2 = \frac{1}{2} k_1 (l_1 \theta)^2 + \frac{1}{2} k_2 (l_2 \theta)^2 \tag{2.61}$$

$$k_{eq} l_1^2 = k_1 l_1^2 + k_2 l_2^2 \tag{2.62}$$

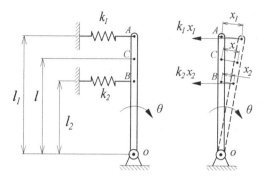

Figure 2.9 Springs connected to the rigid bar

Rearranging this equation yields

$$k_{eq} = \left(k_1 + (\frac{l_2}{l_1})^2 k_2 \right) \tag{2.63}$$

This equation represents the equivalent stiffness k_{eq} of the equivalent spring of the system concerning the location at point A given by

- by considering moment balance:
The equivalent spring stiffness k_{eq} of the system can also be determined by considering the moment the balance of spring forces about the pivot point o. Here, we will determine the equivalent spring stiffness concerning point A shown in Figure 2.10a. Taking the moment balance about the pivot point o and assuming small angle θ yields

$$(k_{eq}x_1)l_1 = (k_1x_1)l_1 + (k_2x_2)l_2 \tag{2.64}$$

Substituting equation (2.59) in equation (2.64) yields

$$k_{eq}(l_1\theta)l_1 = k_1(l_1\theta)l_1 + k_2(l_2\theta)l_2 \tag{2.65}$$

$$k_{eq}l_1^2 = k_1l_1^2 + k_2l_2^2 \tag{2.66}$$

Rearranging this equation gives

$$k_{eq} = \left(k_1 + \left(\frac{l_2}{l_1}\right)^2 k_2 \right) \tag{2.67}$$

By comparing equations (2.67) and (2.63), it is found that the equivalent spring stiffness resulted in the same results for both cases.

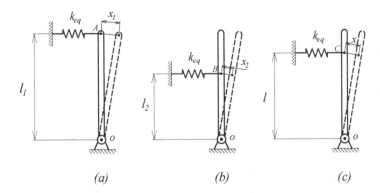

Figure 2.10 The equivalent spring is positioned at different locations

- equivalent spring instead of the spring of stiffness k_2
- **by considering energy equivalency**

Applying energy equivalency on the motion of the system and placing the equivalent spring at point B shown in Figure 2.10b yields

$$\frac{1}{2}k_{eq}x_2^2 = \frac{1}{2}k_1x_1^2 + \frac{1}{2}k_2x_2^2 \tag{2.68}$$

Substituting equation (2.59) in equation (2.68) gives

$$\frac{1}{2}k_{eq}(l_2\theta)^2 = \frac{1}{2}k_1(l_1\theta)^2 + \frac{1}{2}k_2(l_2\theta)^2 \tag{2.69}$$

Rearranging this equation yield

$$k_{eq} = \left(\left(\frac{l_1}{l_2}\right)^2 k_1 + k_2\right) \tag{2.70}$$

This equation represents the equivalent stiffness k_{eq} of the equivalent spring of the system concerning the location at point A given by

- **by considering moment balance**

Here, we will determine the equivalent spring stiffness concerning point B shown in Figure 2.10b. So, taking the moment balance about the pivot point O and assuming small angle θ yields

$$(k_{eq}x_2)l_2 = (k_1x_1)l_1 + (k_2x_2)l_2 \tag{2.71}$$

Substituting equation (2.59) in equation (2.71) yields

$$k_{eq}(l_2\theta)l_2 = k_1(l_1\theta)l_1 + k_2(l_2\theta)l_2 \tag{2.72}$$

$$k_{eq}l_2^2 = k_1l_1^2 + k_2l_2^2 \tag{2.73}$$

Rearranging this equation yields

$$k_{eq} = \left(\left(\frac{l_1}{l_2}\right)^2 k_1 + k_2\right) \tag{2.74}$$

By comparing equations (2.74) and (2.70), it is found that the equivalent spring stiffness resulted in the same results for both cases.

- equivalent spring at the midpoint between the two springs
- by considering energy equivalency
We will assume that the midpoint between the two springs is placed at point C, as shown in Figure 2.10c, which is a distance l away from the bar's pivot point o. As the location is positioned at the midpoint between the two springs, the distance l is given by

$$l = \frac{l_1 + l_2}{2} \tag{2.75}$$

Applying energy equivalency to the system's motion and placing the equivalent spring at point C results in

$$\frac{1}{2}k_{eq}x^2 = \frac{1}{2}k_1 x_1^2 + \frac{1}{2}k_2 x_2^2 \tag{2.76}$$

Assuming small angle θ gives

$$x_1 = l_1\theta, \ \ x_2 = l_2\theta \ \text{ and } x = l\theta \tag{2.77}$$

Substituting equation (2.77) in equation (2.76) gives

$$k_{eq}l^2\theta^2 = k_1 l_1^2 \theta^2 + k_2 l_2^2 \theta^2 \tag{2.78}$$

So,

$$k_{eq}l^2 = k_1 l_1^2 + k_2 l_2^2 \tag{2.79}$$

Rearranging this equation yields

$$k_{eq} = \frac{1}{l^2}(k_1 l_1^2 + k_2 l_2^2) \tag{2.80}$$

Substituting equation (2.75) in equation (2.80) and rearranging it gives

$$k_{eq} = \left(\frac{4(k_1 l_1^2 + k_2 l_2^2)}{(l_1 + l_2)^2} \right) \tag{2.81}$$

- by considering moment balance
We will determine the equivalent spring stiffness concerning the midpoint between both springs at point C shown in Figure 2.10c. Now, taking the moment balance about the pivot point o and assuming small angle θ yields

$$(k_{eq}x)l = (k_1 x_1)l_1 + (k_2 x_2)l_2 \tag{2.82}$$

Substituting equations (2.77) in equation (2.82) gives

$$k_{eq}(l\theta)l = k_1(l_1\theta)l_1 + k_2(l_2\theta)l_2 \tag{2.83}$$

Rearranging this equation gives

$$k_{eq}l^2 = k_1 l_1^2 + k_2 l_2^2 \qquad (2.84)$$

So,

$$k_{eq} = \left(\frac{1}{l^2}(k_1 l_1^2 + k_2 l_2^2) \right) \qquad (2.85)$$

Substituting equation (2.75) in equation (2.85) and rearranging the results gives

$$k_{eq} = \left(\frac{4(k_1 l_1^2 + k_2 l_2^2)}{(l_1 + l_2)^2} \right) \qquad (2.86)$$

By comparing equations (2.86) and (2.81), it is found that the equivalent spring stiffness resulted in the same results for both cases.

Example (2.2):
Two pulleys of radius r_1 and r_2 are rotated about their center point by an angle θ. Two springs are connected with the pulleys and arranged as shown in Figure 2.11. Assuming that the rotating angle θ is small, find the equivalent spring stiffness of the system that relates to the angular rotation of the pulleys at three different locations:

1. Instead the spring of stiffness k_1 (use the energy equivalency method).
2. Instead the spring of stiffness k_2 (use the energy equivalency method).
3. At the center point of the pulleys (use both energy equivalency and moment balance methods).

Solution:
- equivalent spring instead of the spring of stiffness k_1
Applying energy equivalency to the system's motion and placing the equivalent spring instead of the spring of stiffness k_1, at point A shown in Figure 2.12a gives

$$\frac{1}{2}k_{eq}x_1^2 = \frac{1}{2}k_1 x_1^2 + \frac{1}{2}k_2 x_2^2 \qquad (2.87)$$

Assuming small θ, gives

$$x_1 = r_1 \theta \text{ and } x_2 = r_2 \theta \qquad (2.88)$$

Substituting equation (2.88) in equation (2.87) yields

$$\frac{1}{2}k_{eq}r_1^2\theta^2 = \frac{1}{2}k_1 r_1^2 \theta^2 + \frac{1}{2}k_2 r_2^2 \theta^2 \qquad (2.89)$$

So,

$$k_{eq}r_1^2 = r_1^2 k_1 + r_2^2 k_2 \qquad (2.90)$$

Rearranging this equation yields

$$k_{eq} = \left(k_1 + \left(\frac{r_2}{r_1} \right)^2 k_2 \right) \qquad (2.91)$$

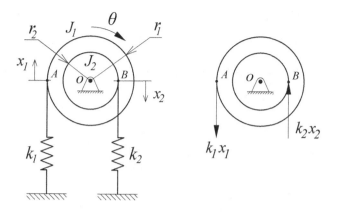

Figure 2.11 Rotating pulleys connected with translational springs

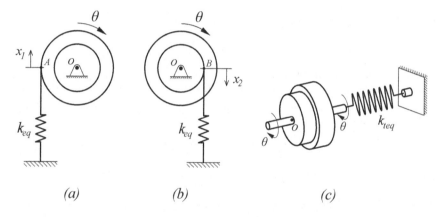

Figure 2.12 System involved translational and torsional springs

- equivalent spring instead of the spring of stiffness k_2

Similarly, applying the energy equivalency to the motion of the system and placing the equivalent spring instated of the spring of stiffness k_2, at point B shown in Figure 2.12b yields

$$\frac{1}{2}k_{eq}x_2^2 = \frac{1}{2}k_1x_1^2 + \frac{1}{2}k_2x_2^2 \tag{2.92}$$

Substituting equation (2.88) in equation (2.92) yields

$$\frac{1}{2}k_{eq}r_2^2\theta^2 = \frac{1}{2}k_1r_1^2\theta^2 + \frac{1}{2}k_2r_2^2\theta^2 \tag{2.93}$$

So,

$$k_{eq}r_2^2 = r_1^2k_1 + r_2^2k_2 \tag{2.94}$$

Rearranging this equation yields

$$k_{eq} = \left(\left(\frac{r_1}{r_2}\right)^2 k_1 + k_2\right) \tag{2.95}$$

- equivalent spring at the center point of the pulleys
- by considering energy equivalency
When the equivalent spring is placed at the pulley's center point, it will act as a torsional spring, as shown in Figure 2.12c. Applying energy equivalency on the rotating motion of the pulleys about their center point o yields

$$\frac{1}{2}k_{teq}\theta^2 = \frac{1}{2}k_1x_1^2 + \frac{1}{2}k_2x_2^2 \tag{2.96}$$

Substituting equation (2.88) in equation (2.96) yields

$$k_{teq}\theta^2 = k_1(r_1\theta)r_1 + k_2(r_2\theta)r_2 \tag{2.97}$$

So,

$$k_{teq}\theta = k_1r_1^2\theta + k_2r_2^2\theta \tag{2.98}$$

Rearranging this equation yields

$$k_{teq} = (k_1r_1^2 + k_2r_2^2) \tag{2.99}$$

- by considering moment balance
We will assume that the equivalent spring is placed in the center of the pulley, acting as a torsional spring, as shown in Figure 2.12c. Taking a moment balance about the pulley center point o yields

$$k_{teq}\theta = (k_1x_1)r_1 + (k_2x_2)r_2 \tag{2.100}$$

Substituting equation (2.88) in equation (2.100) gives

$$k_{teq}\theta = (k_1r_1\theta)r_1 + (k_2r_2\theta)r_2 \tag{2.101}$$

Rearranging this equation yields

$$k_{teq} = (k_1r_1^2 + k_2r_2^2) \tag{2.102}$$

By comparing equations (2.102) and (2.99), it is found that the equivalent spring stiffness resulted in the same results for both cases.

Referring to Examples (2.1) and (2.2), using the same procedures, we will attempt to rearrange the position of one of the two springs while keeping their acting location points, as shown in Figure 2.13, and compare the equivalent spring stiffness findings with the results obtained in the previous two examples for each system individually. The equivalent spring stiffnesses of both systems will turn out to be the same as the corresponding one discussed before, implying that as long as their acting locations are considered, the corresponding equivalent spring stiffness can be obtained regardless of how the springs are set up in the system (exercise problem).

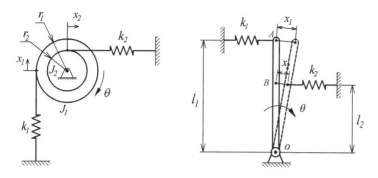

Figure 2.13   Different positions of spring connections

2.5 DAMPER ELEMENT

Because energy cannot be destroyed, energy dissipation involves converting mechanical energy to thermal energy (heat), which flows away and is no longer available for practical work. Functionally, nonlinear dampers present mathematical difficulties that one must be willing to deal with. On the other hand, the linear damper has a strictly linear input-output relationship that defines its characteristics. When a force is applied to the endpoint of a damper element, as shown in Figure 2.14, the output-input relationship provides the specification of the translational damper element given by

$$F_d = b\frac{dx}{dt} = b\dot{x} = bv \qquad (2.103)$$

where b is the damping factor. This equation indicates that the damping force F_d is directly proportional to the velocity of its endpoint. Similarly, for the torsional damper shown in Figure 2.15, the damping torque is given by

$$T_{td} = b_t\frac{d\theta}{dt} = b_t\dot{\theta} = b_t\omega \qquad (2.104)$$

where b_t is the torsional damping factor and ω is the angular velocity of the torsional damper. Since instantaneous power equals the product of instantaneous force and velocity, a damper element entirely dissipates the mechanical energy supplied to it. For translational damper shown in Figure 2.14, it is given by

$$P_d = \text{Power dissipation} = F_d\frac{dx}{dt} = b\frac{dx}{dt}\frac{dx}{dt} = b\left(\frac{dx}{dt}\right)^2 = b\dot{x}^2 = bv^2 \qquad (2.105)$$

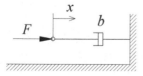

Figure 2.14   Translational damper element

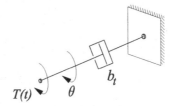

Figure 2.15 Torsional damper element

For the torsional damper shown in Figure 2.15, it is given by

$$P_{td} = \text{Power dissipation} = F_{dt}\frac{d\theta}{dt} = b_t\frac{d\theta}{dt}\frac{d\theta}{dt} = b_t\left(\frac{d\theta}{dt}\right)^2 = b_t\,\dot{\theta}^2 = b_t\,\omega^2 \quad (2.106)$$

The time integral of the power P_d or P_{td} represents the total energy dissipated (non-conservative energy) over any time. Note that any force or torque applied to the damper generates a velocity in the same direction, indicating that it absorbs all the energy supplied. Therefore, the source of force must also give power to the damper, as when the force on an element and its velocity have the same sign, the power input to the element should be positive. When using a damper, the applied force or torque and the resultant velocity cannot have the opposite sign. So, a damper element dissipates all of the energy provided to it. As a result, the damper is never used to provide power to another element.

2.5.1 EQUIVALENT DAMPER

Whether the dampers appear in combination connections (parallel or series), we can use procedures that are similar to those used to determine the equivalent stiffness of several springs to get the equivalent damping factor of a single equivalent damper (see Section 2.4.1), no need to repeat the previous analysis.

2.5.1.1 Dampers in parallel

When two translational dampers with damping factors b_1 and b_2 appear in combinations in parallel connection, as shown in Figure 2.16, the equivalent damping factor b_{eq} is given by

$$b_{eq} = (b_1 + b_2) \quad (2.107)$$

In general form, it is given by

$$b_{eq} = b_1 + b_2 + \cdots + b_i = \sum b_i \quad (2.108)$$

2.5.1.2 Dampers in series

For series connections shown in Figure 2.17, the equivalent damping factor b_{eq} is given by

$$\frac{1}{b_{eq}} = \frac{1}{b_1} + \frac{1}{b_2} \quad (2.109)$$

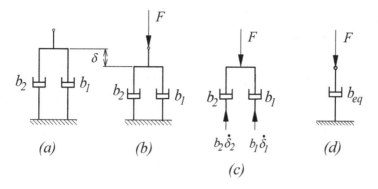

Figure 2.16 Damper elements connected in parallel

In general form, it is given by

$$\frac{1}{b_{eq}} = \frac{1}{b_1} + \frac{1}{b_2} + \cdots + \frac{1}{b_i} = \sum \frac{1}{b_i} \qquad (2.110)$$

Similarly, for torsional dampers, the general form of the equivalent damping factor b_{teq} for dampers connected in parallel is given by

$$b_{teq} = b_{t1} + b_{t2} + \cdots + b_{ti} = \sum b_{ti} \qquad (2.111)$$

For torsional dampers, the general form of the equivalent damping factor b_{teq} for dampers connected in series is given by

$$\frac{1}{b_{teq}} = \frac{1}{b_{t1}} + \frac{1}{b_{t2}} + \cdots + \frac{1}{b_{ti}} = \sum \frac{1}{b_{ti}} \qquad (2.112)$$

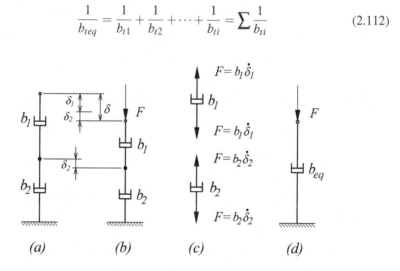

Figure 2.17 Damper elements connected in series

Example (2.3):
In this example, we consider the same system as in Example (2.1), but we replace the springs with two dampers positioned at the same locations as the springs, as shown in Figure 2.18. The bar is connected by two dampers of damping factors b_1 and b_2. Assuming that the angular deflection of the bar θ is small, determine the equivalent damping factor of the system related to the angular deflection of the bar at the following two different locations:

1. Instead the damper element of damping factor b_1.
2. Instead the damper element of damping factor b_2.

Solution
For a small angular deflection of the bar θ, the points of attachment of the dampers, points A and B shown in Figure 2.18, undergo a horizontal displacement, x_1 and x_2, respectively. Since θ is small, the horizontal displacement of points A and B can be approximated as

$$x_1 = l_1\theta \text{ and } x_2 = l_2\theta \tag{2.113}$$

Taking the derivative of this equation gives

$$\dot{x}_1 = l_1\dot{\theta} \text{ and } \dot{x}_2 = l_2\dot{\theta} \tag{2.114}$$

The system's equivalent damping factor can be determined as follows. Since damper elements are dissipating energy elements, we cannot use the energy equivalency to determine the equivalent damping factor for such elements. In this situation, the equivalent damping factor is determined by taking the moment of damping forces about the bar pivot point o.

- equivalent damper instead of the damper element of damping factor b_1
Taking a moment, the balance of damping forces around the pivot point o yields

$$(b_{eq}\dot{x}_1)l_1 - (b_1\dot{x}_1)l_1 + (b_2\dot{x}_2)l_2 \tag{2.115}$$

Substituting equation (2.114) in equation (2.115) gives

$$b_{eq}(l_1\dot{\theta})l_1 = b_1(l_1\dot{\theta})l_1 + b_2(l_2\dot{\theta})l_2 \tag{2.116}$$

or,

$$b_{eq}l_1^2\dot{\theta} = b_1l_1^2\dot{\theta} + b_2l_2^2\dot{\theta} \tag{2.117}$$

Then, the equivalent damping factor of the equivalent damper concerning the location at point A is given by

$$b_{eq} = \left(b_1 + \left(\frac{l_2}{l_1}\right)^2 b_2\right) \tag{2.118}$$

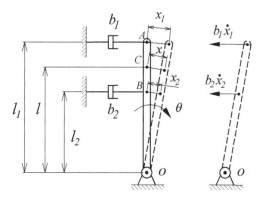

Figure 2.18 Dampers connected to the rigid bar

- equivalent damper instead of the damper element of damping factor b_2
Similarly, considering the moment of the damping forces about pivot point o, yields

$$(b_{eq}\dot{x}_2)l_2 = (b_1\dot{x}_1)l_1 + (b_2\dot{x}_2)l_2 \qquad (2.119)$$

Substituting equations (2.114) in equation (2.119) gives

$$b_{eq}(l_2\dot{\theta})l_2 = b_1(l_1\dot{\theta})l_1 + b_2(l_2\dot{\theta})l_2 \qquad (2.120)$$

or

$$b_{eq}l_2^2\dot{\theta} = b_1l_1^2\dot{\theta} + b_2l_2^2\dot{\theta} \qquad (2.121)$$

Then, the equivalent damping factor of the equivalent damper concerning the location at point B is given by

$$b_{eq} = \left(\left(\frac{l_1}{l_2}\right)^2 b_1 + b_2\right) \qquad (2.122)$$

Example (2.4):
Two pulleys of radius r_1 and r_2 are rotated about their center point by an angle θ. Two dampers are connected with the pulleys and arranged as shown in Figure 2.19. Assuming that the rotating angle θ is small, determine the equivalent damping factor of the system that relates to the angular rotation of the pulleys at three different locations:

1. Instead of the damper element of the damping factor b_1.
2. Instead of the damper element of the damping factor b_2.
3. At the center point of the pulleys.

Solution
- equivalent damper instead of the damper of damping factor b_1
Taking the moment of the damping forces about pulleys center point o yields

$$(b_{eq}\dot{x}_1)r_1 = (b_1\dot{x}_1)r_1 + (b_2\dot{x}_2)r_2 \qquad (2.123)$$

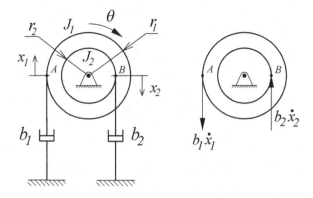

Figure 2.19 Rotating pulleys connected with translational dampers

For small θ, we have

$$\dot{x}_1 = r_1\dot{\theta} \text{ and } \dot{x}_2 = r_2\dot{\theta} \tag{2.124}$$

Substituting equation (2.124) in equation (2.123) gives

$$b_{eq}r_1^2\dot{\theta} = b_1r_1^2\dot{\theta} + b_2r_2^2\dot{\theta} \tag{2.125}$$

Then, the equivalent damping factor of the equivalent damper of the system concerning the location at point A is given by

$$b_{eq} = \left(b_1 + \left(\frac{r_2}{r_1}\right)^2 b_2\right) \tag{2.126}$$

- equivalent damper instead of the damper of damping factor b_2
Similarly, taking the moment the balance of the damping forces about the center point of the pulleys yields

$$(b_{eq}\dot{x}_2)r_2 = (b_1\dot{x}_1)r_1 + (b_2\dot{x}_2)r_2 \tag{2.127}$$

Substituting equations (2.124) in equation (2.127) yields

$$b_{eq}(r_2\dot{\theta})r_2 = b_1(r_1\dot{\theta})r_1 + b_2(r_2\dot{\theta})r_2 \tag{2.128}$$

or,

$$b_{eq}r_2^2\dot{\theta} = b_1r_1^2\dot{\theta} + b_2r_2^2\dot{\theta} \tag{2.129}$$

Then, the equivalent damping factor of the equivalent damper of the system concerning the location at point B is given by

$$b_{eq} = \left(\left(\frac{r_1}{r_2}\right)^2 b_1 + b_2\right) \tag{2.130}$$

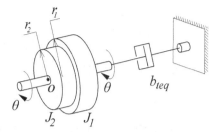

Figure 2.20 The equivalent torsional damper positioned at the center of the pulley

- equivalent damper at the center point of the pulleys

When the equivalent damper is placed at the pulleys' center point, it will act as a torsional damper, as shown in Figure 2.20. Taking the moment of the damping forces about the center point of the pulleys o yields

$$b_{teq}\dot{\theta} = (b_1\dot{x}_1)r_1 + (b_2\dot{x}_2)r_2 \tag{2.131}$$

Substituting equation (2.124) in equation (2.131) yields

$$b_{teq}\dot{\theta} = b_1(r_1\dot{\theta})r_1 + b_2(r_2\dot{\theta})r_2 \tag{2.132}$$

or,

$$b_{teq}\dot{\theta} = b_1 r_1^2\dot{\theta} + b_2 r_2^2\dot{\theta} \tag{2.133}$$

Then, the equivalent damping factor of the equivalent damper of the system concerning the location at the center point o of the pulleys is given by

$$b_{teq} = (r_1^2 b_1 + r_2^2 b_2) \tag{2.134}$$

2.6 FORCE AND MOTION INPUTS

The various elements of mechanical systems driven by input forces or motions are demonstrated, but how these inputs occur in actual situations still needs to be specified. Before going further, it is essential to understand that the primary driving force in every mechanical system is always a force, not a motion. Following Newton's second law, the force causes the acceleration, but the acceleration does not cause the force. Motion input may be preferable to force input in some systems, but we should not mislead ourselves into thinking that motion occurs without force first. The question is: what do we know about the input of the system, force, or motion? We should not prevent postulating an issue with a motion input if, for example, the motion is known and produced by an unknown force.

Two types of forces are available to drive the equations of motion of mechanical systems: forces associated with physical contact between two bodies and forces associated with arbitrary action at a distance, namely gravitational, magnetic, and electrostatic forces. In other words, when Newton's second law is used, the terms entered into the force summation should come from physical contact or gravitational, magnetic, or electrostatic interaction; there are no other forces. All these input forces are time-dependent, and the standard input forces are classified into three types: impulse, step, and ramp. In addition, we may generate our forcing input functions, meaning time-creating functions (for example, sinusoidal input). Later on, we will provide detailed information on such types of input forces; see Chapters 8 and 9.

2.7 DYNAMIC MODELING OF MECHANICAL SYSTEMS

In mechanical systems, the positions of their elements vary with time, so the independent coordinates, which uniquely determine the positions of these elements, are functions of time. The fundamental analysis of the dynamics of mechanical systems is to find these functions, i.e., to determine the system's motion. Therefore, it is easy to find the internal forces of the system components. They are developed in that their constraints are widely different and in the part they play in the dynamic process. According to the specifications of mechanical systems, these forces are classified as storage and dissipative forces and, of course, input forces.

2.7.1 STORAGE FORCES

These forces occur when the system is displaced from its equilibrium position and restored to that position. This may mean that the dynamic characteristics of mechanical systems result from the action of these varying forces. As previously established, the mass and spring are storage elements, implying that all their forces are stored. Consider that the mass and the spring are suspended vertically, as shown in Figure 2.21a. In this position, the mass is acted on by its weight, mg and the reaction of the spring is equal to $k\delta_{st}$, where δ_{st} is the static extension of the spring corresponding to the mass weight, mg. Then, the condition of equilibrium of the mass at rest ($x = 0$) is

$$mg - k\delta_{st} = 0 \qquad (2.135)$$

or

$$mg = k\delta_{st} \qquad (2.136)$$

If this equilibrium position is disturbed and the mass begins to move under the effect of an input force $F(t)$, as shown in Figure 2.21b, the weight of the mass will be neglected. This means that if the mass-spring system is suspended vertically and begins to move away from its equilibrium position, the effect of mass weight mg will be ignored. Of course, there are many systems in which the weight of the mass should be considered in their motions in different ways, as will be demonstrated later; see Example (2.6). If both the mass and the spring are connected horizontally, as shown in Figure 2.21c, then the mass weight has no effect on the system's motion. Applying Newton's second law on the systems shown in Figures 2.21b and 2.21c yields

$$m\ddot{x} + F_s = F(t) \qquad (2.137)$$

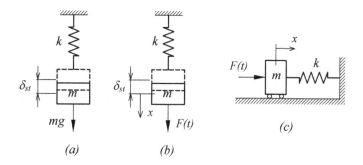

Figure 2.21 Translational mass-spring system

where $F_s = -kx$ is the projection of the reaction of the spring along x axis, and $\ddot{x}$ is the acceleration of the mass m. Thus, we obtain

$$m\ddot{x} + kx = F(t) \tag{2.138}$$

Unlike equation (2.26), which serves to calculate x directly, equation (2.138) represents a differential equation in a function of the displacement $x(t)$. To find the form of this function, it is necessary to integrate equation (2.138). After solving this equation, the function $x(t)$ is used to find the internal forces of the system. It may be said that, in the above system shown in Figure 2.21, the displacement $x(t)$ alone defines the dynamic motion of both the mass and the spring at any instant of time. Similarly, for the torsional mass-spring system shown in Figure 2.22, the differential equation representing the equation of the motion of this system is given by

$$J\ddot{\theta} + F_{ts} = T(t) \tag{2.139}$$

where $F_{ts} = -k_t\theta$ is the projection of the reaction of the torsional spring along the center of the pulley axis of rotation θ and $\ddot{\theta}$ is the acceleration of the rotating mass J. Thus, we obtain

$$J\ddot{\theta} + k_t\theta = T(t) \tag{2.140}$$

After solving this equation, the function $\theta(t)$ is used to find the internal forces of the system. This function defines the dynamic motion of the rotating mass and the torsional spring at any instant. It is worth noting that when the torsional mass-spring

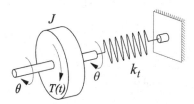

Figure 2.22 Torsional mass-spring system

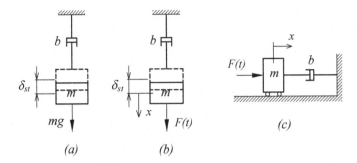

Figure 2.23 Translational mass-damper system

is in an equilibrium position ($\theta = 0$), the weight of the rotating mass is applied on its axis of rotation, implying that the system motion is not affected by the rotating mass weight.

2.7.2 DISSIPATIVE FORCES

As stated previously, the dissipation elements do irrecoverable work, resulting in mechanical energy dissipation; such forces are called dissipative forces. Among these damping elements is the friction force of the simplest friction element, the viscous dashpot. Similarly, as we demonstrated in the mass-spring system shown in Figure 2.23, when both the mass and the damper are suspended vertically, as shown in Figures 2.23a and 2.23b, the reaction of the damper is equal to $b\delta_{st}$, where δ_{st} is the static extension of the damper corresponding to the mass weight mg. This means that if the mass-damper system is suspended vertically and begins to move away from its equilibrium position ($x = 0$), the effect of mass weight mg is to be ignored. Also, no mass weight affects the system motion if the mass and the damper are connected horizontally, as shown in Figure 2.23c. Now, applying Newton's second law to both systems shown in Figures 2.23b and 2.23c yields

$$F(t) + F_d = m\ddot{x} \tag{2.141}$$

where $F_d = -b\dot{x}$ is the projection of the reaction of the damper along x axis and $\dot{x}$ is the velocity of the mass m. Thus, we obtain

$$m\ddot{x} + b\dot{x} = F(t) \tag{2.142}$$

where b is the damping factor. Whatever the nature of friction, the direction of dissipative forces at any instant during the motion is opposite to the velocity, and the magnitude of this force is related in one way or another to the magnitude of the velocity $\dot{x}$. Similarly, for the torsional damper shown in Figure 2.24, applying Newton's second law yields

$$T(t) + F_{td} = J\ddot{\theta} \tag{2.143}$$

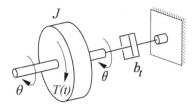

Figure 2.24   Torsional mass-damper system

where $F_{td} = -b_t \dot{\theta}$ is the projection of the reaction of the torsional damper along the center of the pulley axis of rotation θ and $\dot{\theta}$ is the angular velocity of the rotating mass J. Thus, we obtain

$$J\ddot{\theta} + b_t \dot{\theta} = T(t) \tag{2.144}$$

where b_t is the damping factor of the torsional damper. When the torsional mass-damper system is in an equilibrium position ($\theta = 0$), the weight of the rotating mass is applied on its axis of rotation, implying that the system's motion is not affected by the mass weight.

Finally, the characteristics of storage and dissipative forces are determined exclusively by the properties of the system itself, and the corresponding forces not only affect the motion but are themselves governed by this motion since they depend on the displacement $x(t)$ or $\theta(t)$ and the velocity $\dot{x}(t)$ or $\dot{\theta}(t)$.

Now, connect the three passive elements of the translational mechanical systems (mass, spring, and damper) as shown in Figure 2.25a. To drive the equation of motion describing the dynamic motion of a mechanical system, the following two general steps should be followed:

1. Draw the free-body diagram of the system.
2. Deduce the equation of motion by applying the physical law (Newton's second law) and considering justified simplifying assumptions (the spring and damper are linear elements).

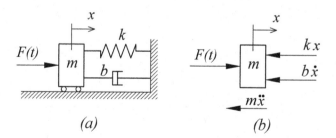

Figure 2.25   Transnational mass-spring-damper system

Considering the free body diagram shown in Figure 2.25b and applying Newton's second law yields

$$\sum F = m\ddot{x} \qquad (2.145)$$

So,

$$F(t) - b\dot{x} - kx = m\ddot{x} \qquad (2.146)$$

Rearranging this equation yields

$$m\ddot{x} + b\dot{x} + kx = F(t) \qquad (2.147)$$

Note that the inertia force $m\ddot{x}$ always acts in the opposite direction of the mass movement, as shown in Figure 2.25b. Equation (2.147) is an ordinary second-order differential equation, known as the governing equation, which completely characterizes the dynamic motion of the given system shown in Figure 2.25a. This equation indicates that only one independent coordinate, the displacement x, implies that the system has one degree of freedom. Note that the displacement variable $x(t)$ captures the potential energy stored in the spring. In contrast, the velocity variable $\dot{x}(t)$ captures the kinetic energy stored by the mass, while the damper only dissipates energy. Similarly, for the torsional system shown in Figure 2.25a, applying Newton's second law yields

$$\sum T = J\ddot{\theta} \qquad (2.148)$$

$$T(t) - b_t\dot{\theta} - k_t\theta = J\ddot{\theta} \qquad (2.149)$$

As stated previously, the rotating inertia force $J\ddot{\theta}$ always acts in the opposite direction of the rotating mass, as shown in Figure 2.25b. Rearranging equation (2.149)

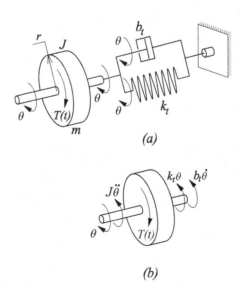

(a)

(b)

Figure 2.26 Torsional mass-spring-damper system

yields

$$J\ddot{\theta} + b_t\dot{\theta} + k_t\theta = T(t) \tag{2.150}$$

Since the mass m rotates around its center, referring to Table 2.1, the moment of inertia J is given by

$$J = \frac{1}{2}mr^2 \tag{2.151}$$

Substituting this equation in equation (2.150) gives

$$\frac{1}{2}mr^2\ddot{\theta} + b_t\dot{\theta} + k_t\theta = T(t) \tag{2.152}$$

Equation (2.152) is an ordinary second-order differential equation, which completely characterizes the dynamic motion of the mechanical torsional system shown in Figure 2.26a. Similarly, as in the translational system, the angular displacement variable $\theta(t)$ captures the potential energy stored in the torsional spring, while the angular velocity variable $\dot{\theta}(t)$ captures the kinetic energy stored by the rotating mass.

Example (2.5):
Consider a pendulum of chord length l with a bob of mass m shown in Figure 2.27. The chord is connected to a spring at a distance l_1 away from the pivot point o. Considering the angular displacement θ of the pendulum is of small value,

1. Draw the free-body diagram.
2. Drive the equation of motion of the system.

- Solution:

- Free-body diagram:
When the pendulum undergoes an angular displacement θ, the mass m moves by a distance x along the horizontal direction, while at the same time, the spring is deflected by a distance x_1 in the same direction, as shown in Figure 2.28. Here, we will consider the weight mg of the bob's mass m. Notice that when the pendulum undergoes an angular displacement θ, it is subjected to a restoring moment created

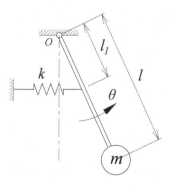

Figure 2.27   Pendulum system

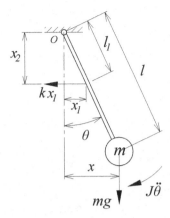

Figure 2.28   The free-body diagram of the pendulum system

by the weight of the mass mg about the pivot point o. The analysis of forces that are applied to the pendulum system for angular displacement θ is shown in Figure 2.28.

- Equations of motion:
Newton's second law for angular motion is given by

$$\sum M = J\ddot{\theta} \qquad (2.153)$$

So, taking a moment about the pivot point o yields

$$-(mg)x - (kx_1)x_2 = J\ddot{\theta} \qquad (2.154)$$

From Figure 2.28 we get

$$x = l \sin\theta \quad \text{and} \quad x_1 = l_1 \sin\theta \qquad (2.155)$$

Assuming small angular displacement, θ, gives

$$x = l\theta \quad \text{and} \quad x_1 = l_1\theta \qquad (2.156)$$

Substituting equation (2.156) in equation (2.154) and rearranging the results yields

$$J\ddot{\theta} + (kl_1^2 + mgl)\theta = 0 \qquad (2.157)$$

Since the mass m rotates about the pivot point o, the moment of inertia J is given by, see Table 2.1,

$$J = ml^2 \qquad (2.158)$$

Substituting equation (2.158) in equation (2.157) gives

$$ml^2\ddot{\theta} + (kl_1^2 + mgl)\theta = 0 \qquad (2.159)$$

This equation is the equation of motion of the given system shown in Figure 2.27. It indicates only one coordinate θ, implying that the system has one degree of freedom.

Example (2.6):
Considering the translational mechanical system shown in Figure 2.29,

1. Draw the free-body diagram.
2. Drive the equations of motion.
3. Put the results in matrix form.

Solution
- Free body diagram:
As shown in Figure 2.29, the spring of stiffness k_1 is connected between two masses, implying that its endpoints are both moving, and the spring force will be determined based on the relative movement of the masses $(x_1 - x_2)$. Note that both masses are subjected to friction damping by damping factors b_1 and b_2 as shown in Figure 2.29. The analysis of forces (free body diagram) that are applied to both masses is shown in Figure 2.30.

- Equations of motion:
Applying Newton's second law for both masses m_1 and m_2, yields
- for mass m_1:

$$\sum F = m_1 \ddot{x}_1 \tag{2.160}$$

So,

$$F(t) - b(\dot{x}_1 - \dot{x}_2) - b_1\dot{x}_1 - k(x_1 - x_2) - k_1 x_1 = m_1 \ddot{x}_1 \tag{2.161}$$

Rearranging this equation yields

$$m_1 \ddot{x}_1 + b(\dot{x}_1 - \dot{x}_2) + b_1\dot{x}_1 + k(x_1 - x_2) + k_1 x_1 = F(t) \tag{2.162}$$

- for mass m_2:

$$\sum F = m_2 \ddot{x}_2 \tag{2.163}$$

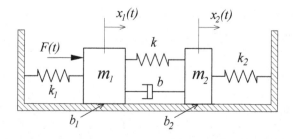

Figure 2.29 Translational system of two degrees of freedom

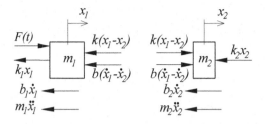

Figure 2.30 The free-body diagram of the system

So,

$$k(x_1 - x_2) + b(\dot{x}_1 - \dot{x}_2) - b_2\dot{x}_2 - k_2x_2 = m_2\ddot{x}_2 \qquad (2.164)$$

Rearranging this equation yields

$$m_2\ddot{x}_2 + b(\dot{x}_2 - \dot{x}_1) + b_2\dot{x}_2 + k(x_2 - x_1) + k_2x_2 = 0 \qquad (2.165)$$

Equations (2.162) and (2.165) are equations of motion describing the dynamic motion of the given system shown in Figure 2.29. These two equations indicated two independent coordinates, x_1 and x_2, implying that the system has two degrees of freedom.

- in matrix form

Rearranging equations (2.162) and (2.156) yields

$$m_1\ddot{x}_1 + (b + b_1)\dot{x}_1 - b\dot{x}_2 + (k + k_1)x_1 - kx_2 = F(t) \qquad (2.166)$$

$$m_2\ddot{x}_2 + (b + b_2)\dot{x}_2 - b\dot{x}_1 + (k + k_2)x_2 - kx_1 = 0 \qquad (2.167)$$

So, the matrix form of equations (2.166) and (2.167) is given by

$$\begin{bmatrix} m_1 & 0 \\ 0 & m_2 \end{bmatrix} \begin{bmatrix} \ddot{x}_1 \\ \ddot{x}_2 \end{bmatrix} + \begin{bmatrix} (b+b_1) & -b \\ -b & (b+b_2) \end{bmatrix} \begin{bmatrix} \dot{x}_1 \\ \dot{x}_2 \end{bmatrix}$$
$$+ \begin{bmatrix} (k+k_1) & -k \\ -k & (k+k_2) \end{bmatrix} \begin{bmatrix} x_1 \\ x_2 \end{bmatrix} = \begin{bmatrix} F(t) \\ 0 \end{bmatrix}$$

Example (2.7):

Considering the mechanical rotational system shown in Figure 2.31,

1. Draw the free-body diagram.
2. Drive the equations of motion.
3. Put the results in matrix form.

Solution:

- Free body diagram:

Figure 2.31 shows that the first spring of stiffness k_{t1} is connected between two rotating masses. This implies that its endpoints are both moving, and the spring force

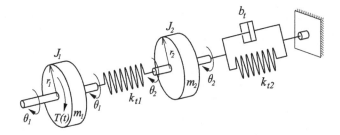

Figure 2.31 Torsional system of two degrees of freedom

will be determined based on the relative angular displacement of the masses $(\theta_1 - \theta_2)$ as shown in Figure 2.32. On the other hand, the left endpoints of both the spring and damper rotate in the direction of the rotation of mass m_2.

- Equations of motion:
Applying Newton's second law for both rotating masses J_1 and J_2 yields
- for mass inertia J_1:

$$\sum T = J_1 \ddot{\theta}_1 \tag{2.168}$$

So,

$$T(t) - k_{t1}(\theta_1 - \theta_2) = J_1 \ddot{\theta}_1 \tag{2.169}$$

Rearranging this equation yields

$$J_1 \ddot{\theta}_1 + k_{t1}(\theta_1 - \theta_2) = T(t) \tag{2.170}$$

Since the mass m_1 rotates around its center, referring to Table 2.1, the moment of inertia J_1 is given by

$$J_1 = \frac{1}{2} m_1 r_1^2 \tag{2.171}$$

Substituting this equation in equation (2.170) gives

$$\frac{1}{2} m_1 r_1^2 \ddot{\theta}_1 + k_{t1}(\theta_1 - \theta_2) = T(t) \tag{2.172}$$

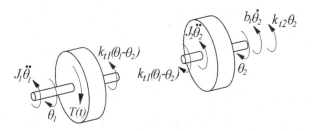

Figure 2.32 The free-body diagram of the given system

- for mass inertia J_2:

$$\sum T = J_2\ddot{\theta}_2 \tag{2.173}$$

So,

$$k_{t1}(\theta_1 - \theta_2) - b_t\dot{\theta}_2 - k_{t2}\theta_2 = J_2\ddot{\theta}_2 \tag{2.174}$$

Rearranging this equation yields

$$J_2\ddot{\theta}_2 + b_t\dot{\theta}_2 + k_{t2}\theta_2 + k_{t1}(\theta_2 - \theta_1) = 0 \tag{2.175}$$

The moment of inertia J_2 of the mass m_2 is given by (see Table 2.1)

$$J_2 = \frac{1}{2}m_2 r_2^2 \tag{2.176}$$

Substituting this equation in equation (2.175) gives

$$\frac{1}{2}m_2 r_2^2\ddot{\theta}_2 + b_t\dot{\theta}_2 + k_{t2}\theta_2 + k_{t1}(\theta_2 - \theta_1) = 0 \tag{2.177}$$

Equations (2.170) and (2.175) are the equations of motion of the given system. These two equations indicate that two independent coordinates, θ_1 and θ_2, imply that the system has two degrees of freedom.

- in matrix form

Rearranging equations (2.170) and (2.175) yields

$$J_1\ddot{\theta}_1 + k_{t1}\theta_1 - k_{t1}\theta_2 = T(t) \tag{2.178}$$

$$J_2\ddot{\theta}_2 + b_t\dot{\theta}_2 + (k_{t1} + k_{t2})\theta_2 - k_{t1}\theta_1 = 0 \tag{2.179}$$

So, the matrix form of equations (2.178) and (2.179) is given by

$$\begin{bmatrix} J_1 & 0 \\ 0 & J_2 \end{bmatrix}\begin{bmatrix} \ddot{\theta}_1 \\ \ddot{\theta}_2 \end{bmatrix} + \begin{bmatrix} 0 & 0 \\ 0 & b_t \end{bmatrix}\begin{bmatrix} \dot{\theta}_1 \\ \dot{\theta}_2 \end{bmatrix} + \begin{bmatrix} k_{t1} & -k_{t1} \\ -k_{t1} & (k_{t1}+k_{t2}) \end{bmatrix}\begin{bmatrix} \theta_1 \\ \theta_2 \end{bmatrix} = \begin{bmatrix} T(t) \\ 0 \end{bmatrix}$$

Example (2.8):

A pulley is attached to torsional and translational springs, as shown in Figure 2.33. The pulley is subjected to a torque $T(t)$ causing it to lift the mass m a distance x. Considering small rotating angle θ of the pulley,

1. Draw the free-body diagram.
2. Drive the equation of motion.
3. Put the results in matrix form.

Solution:

- Free-body diagram:

Figure 2.33 shows that the spring of stiffness k is connected between the pulley and the mass m. This implies that its endpoints are both moving, and its force will be

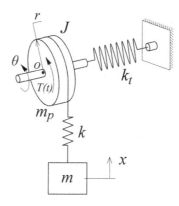

Figure 2.33 Translational-torsional system

determined based on the relative movement of the pulley and the mass $(x_1 - x)$, as shown in Figure 2.34. Note that the displacement x_1 is a function of the rotating angle θ. Assuming small θ, it is given by

$$x_1 = r\theta \tag{2.180}$$

- Equations of motion:
Assuming the system moves from equilibrium, the static spring force cancels out with the weight mg. Applying Newton's second law on the system motion yields
- for translational mass m:

$$\sum F = m\ddot{x} \tag{2.181}$$

So,

$$k(x_1 - x) = m\ddot{x} \tag{2.182}$$

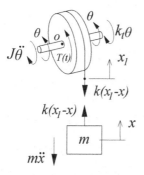

Figure 2.34 The free-body diagram of the system

Rearranging this equation gives

$$m\ddot{x} + k(x - x_1) = 0 \qquad (2.183)$$

Substituting equation (2.180) in equation (2.183) gives

$$m\ddot{x} + k(x - r\theta) = 0 \qquad (2.184)$$

This equation is the equation of motion of the mass m.
- **for rotating mass m_p:**

$$\sum T = J\ddot{\theta} \qquad (2.185)$$

So,

$$T(t) - k_t\theta - k(x_1 - x) = J\ddot{\theta} \qquad (2.186)$$

Rearranging this equation gives

$$J\ddot{\theta} + k_t\theta + k(x_1 - x) = T(t) \qquad (2.187)$$

Substituting equation (2.180) in equation (2.187) gives

$$J\ddot{\theta} + k_t\theta + k(r\theta - x) = T(t) \qquad (2.188)$$

Since the mass m_p rotates around its center, the moment of inertia J is given by (see Table 2.1),

$$J = \frac{1}{2}m_p r^2 \qquad (2.189)$$

Substituting equation (2.189) in equation (2.188) gives

$$\frac{1}{2}m_p r^2\ddot{\theta} + k_t\theta + k(r\theta - x) = T(t) \qquad (2.190)$$

Equations (2.184) and (2.190) are the equations of motion of the given system shown in Figure 2.33. These two equations indicated two independent coordinates, θ and x, implying that the system has two degrees of freedom.
- **in matrix form:**
Rearranging equations (2.184) and (2.190) yields

$$m\ddot{x} + kx - rk\theta = 0 \qquad (2.191)$$

$$\frac{1}{2}m_p r^2\ddot{\theta} + (k_t + rk)\theta - kx = T(t) \qquad (2.192)$$

So, the matrix form of equations (2.191) and (2.192) is given by

$$\begin{bmatrix} m & 0 \\ 0 & \frac{1}{2}m_p r^2 \end{bmatrix}\begin{bmatrix} \ddot{x} \\ \ddot{\theta} \end{bmatrix} + \begin{bmatrix} k & -rk \\ -k & (k_t + rk) \end{bmatrix}\begin{bmatrix} x \\ \theta \end{bmatrix} = \begin{bmatrix} 0 \\ T(t) \end{bmatrix}$$

Example (2.9):
For the system shown in Figure 2.35,

1 Draw the free body diagram.
2 Derive the equation(s) of motions.
3 Put the results in matrix form.

Solution:
- Free body diagram:
Figure 2.35 shows that the mass m_1 and the upper end of a massless lever rod are connected by the spring of stiffness k, indicating that both endpoints are moving. So, the force of the spring will be defined by the relative movement of the mass and the rod's upper end $(x_1 - x_3)$, as shown in Figure 2.36. The mass m_1 is connected to a damper of a damping factor b_1, as in the same figure. On the other side, the endpoint of the lever rod that is lowered, the mass m_2, and the right endpoint of the damper of damping factor b_2 are moving a displacement x_2. It is important to note that the motion of the mass m_1 is transmitted to the mass m_2 via the moment the balance of the lever rod around the pivot point o. The analysis of the forces (free body diagram) that apply to the lever and both masses is shown in Figure 2.36.
 - Equations of motion:
Since the motion of the mass m_1 is transmitted to the mass m_2 through the lever rod, the relation between the displacements of its upper point x_3 and lower point x_2 is given by

$$\tan\theta = \frac{x_3}{b} = \frac{x_2}{a} \quad \text{so} \quad x_3 = \frac{b}{a}x_2 \qquad (2.193)$$

As shown in Figure 2.36, applying Newton's second law for both masses yields
- For mass m_1:

$$\sum F = m_1\ddot{x}_1 \qquad (2.194)$$

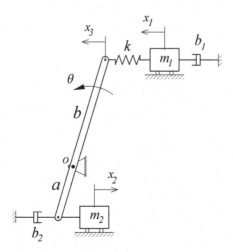

Figure 2.35 Mechanical system

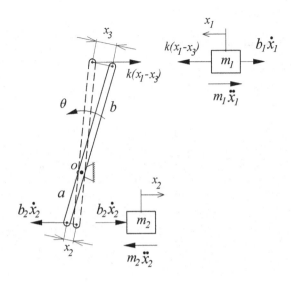

Figure 2.36 The free-body diagram of the given system

So,

$$-k(x_1 - x_3) - b_1\dot{x}_1 = m\ddot{x}_1 \tag{2.195}$$

Substituting equation (2.193) in equation (2.195) and rearranging the results gives

$$m_1\ddot{x}_1 + b_1\dot{x}_1 + k\left(x_1 - \frac{b}{a}x_2\right) = 0 \tag{2.196}$$

This equation represents the equation of motion of the mass m_1.

- For mass m_2:
Taking a moment, the balance of the lever rod around the pivot point o gives

$$[k(x_1 - x_3)]b = (m_2\ddot{x}_2 + b_2\dot{x}_2)a \tag{2.197}$$

So,

$$m_2\ddot{x}_2 + b_2\dot{x}_2 = \frac{b}{a}[k(x_1 - x_3)] \tag{2.198}$$

Substituting equation (2.193) in equation (2.198) and rearranging the results yields

$$m_2\ddot{x}_2 + b_2\dot{x}_2 + k\frac{b}{a}\left(\frac{b}{a}x_2 - x_1\right) = 0 \tag{2.199}$$

This equation represents the equation of motion of the mass m_2. Equations (2.196) and (2.199) indicated two independent coordinates, x_1 and x_2, implying that the system has two degrees of freedom.

Example (2.10):
For the system shown in Figure 2.37, derive the equation(s) of motion.

Solution:
- Free body diagram:
The analysis of the forces (free body diagram) that apply to the system components is shown in Figure 2.38. Since the right end of the spring of stiffness, k_1, and the left end of the damper are linked, they would move the same displacement, y.

- Equations of motion:
Applying Newton's second law to the motion of the system gives
- for snubber plate:
Since the snubber plate is a massless element, applying Newton's second law gives

$$\sum F = 0 \qquad (2.200)$$

So,

$$F(t) - k_2 x - k_1 y = 0 \qquad (2.201)$$

Rearranging this equation yields

$$k_2 x + k_1 y = F(t) \qquad (2.202)$$

- for mass m:

$$\sum F = m\ddot{x} \qquad (2.203)$$

So,

$$-b(\dot{x} - \dot{y}) - k_2 x = m\ddot{x} \qquad (2.204)$$

Rearranging this equation yields

$$m\ddot{x} + b(\dot{x} - \dot{y}) + k_2 x = 0 \qquad (2.205)$$

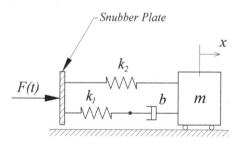

Figure 2.37 Mechanical system

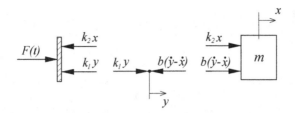

Figure 2.38 The free-body diagram of the system

When we look at these equations, we observe a few issues, which we will go through in the following outlines.

- Since the snubber plate is a massless element, no inertia force is considered, and the forces generated by both springs are equal to simply the external force $F(t)$, as indicated in equation (2.202).
- Because the displacement y of the linked point of the spring with stiffness k_1 and the damper is an interactive variable, it is not an independent variable. It should be eliminated from the system equations of motion. In Bond Graph modeling technique, this situation is called derivative causality, which will be discussed in detail in Chapter 3.

Now, to get the equation of motion of the given system, let us eliminate y and $\dot{y}$ from equations (2.202) and (2.205). From equation (2.202) we get

$$y = \frac{F(t) - k_2 x}{k_1} \tag{2.206}$$

Taking the derivative of this equation gives

$$\dot{y} = \frac{\dot{F}(t) - k_2 \dot{x}}{k_1} \tag{2.207}$$

Substituting equation (2.207) in equation (2.205) gives

$$m\ddot{x} + b\left(\dot{x} - \left(\frac{\dot{F}(t) - k_2 \dot{x}}{k_1}\right)\right) + k_2 x = 0 \tag{2.208}$$

Rearranging this equation yields

$$m\ddot{x} + b\left(1 + \frac{k_2}{k_1}\right)\dot{x} + k_2 x = \frac{b}{k_1}\dot{F}(t) \tag{2.209}$$

This equation is the equation of motion of the given system shown in Figure 2.37.

2.8 PROBLEMS

Problem (2.1):
For the system shown in Figure 2.39, assuming the lever rod is massless, and its angular displacement θ is small, find the equivalent mass, m_{eq} considering the following two positions:

a. Instead of the mass m_1.
b. At the midpoint between m_2 and m_3.

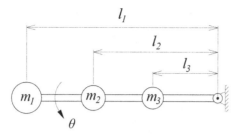

Figure 2.39 Mechanical system

Problem (2.2):
The stepped pulley, shown in Figure 2.40, is made of two homogeneous masses m_1 and m_2 fastened together. Assuming its rotating angle θ is small, find the equivalent mass m_{eq} considering the following two positions:

a. At the point of application of the force mg (instead of mass m).
b. An equivalent torsional mass at the center of the pulley.

Problem (2.3):
The stepped pulley, shown in Figure 2.41, is made of three homogeneous masses, m_1, m_2, and m_3, fastened together. Assuming the rotating angle θ is small, find the equivalent spring stiffness k_{eq} considering the following two positions.

a. Instead of the spring k_2.
b. An equivalent torsional spring at the center of the pulley.

Problem (2.4):
For the system shown in Figure 2.42, assuming the lever rod is massless, and its angular displacement θ is small, find the equivalent spring stiffness k_{eq} considering the following two positions:

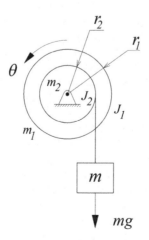

Figure 2.40 Mechanical system

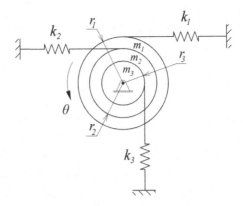

Figure 2.41 Mechanical system

a. At the point of application of the spring k_1.
b. At the midpoint between springs k_1 and k_2.

Problem (2.5):
For the same system shown in Figure 2.41, replace all springs with dampers at the same positions. Find the equivalent damper factor, b_{eq}, considering the following two positions:

a. Instead of the damper b_2.
b. An equivalent torsional damper at the center of the pulley.

Problem (2.6):
For the same system shown in Figure 2.42, replace all springs with dampers at the same positions. Considering the following two positions, find the equivalent damper factor, b_{eq}.

a. Instead of the damper b_1.
b. At the midpoint between dampers b_1 and b_2.

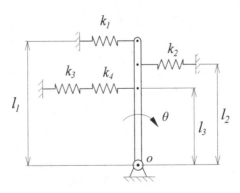

Figure 2.42 Mechanical system

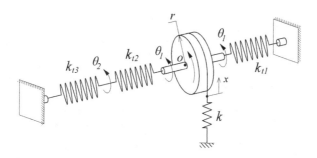

Figure 2.43 Mechanical system

Problem (2.7):
For the system shown in Figure 2.43, assuming the rotating angles θ_1 and θ_2 are small, obtain the equivalent torsional spring and derive the equations of motion.

Problem (2.8):
For each system shown in Figure 2.44, draw the free-body diagram and derive the equation(s) of motion.

Problem (2.9):
For the system shown in Figure 2.45, draw the free-body diagram and derive the equation(s) of motion.

Problem (2.10):
For the system shown in Figure 2.46,

a. Draw the free-body diagram.
b. Derive the equation(s) of motion.
c. Put the results in matrix form.

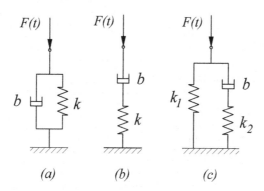

Figure 2.44 Mechanical system

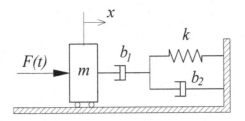

Figure 2.45 Mechanical system

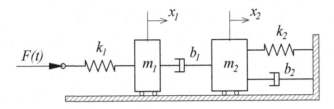

Figure 2.46 Mechanical system

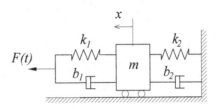

Figure 2.47 Mechanical system

Problem (2.11):
For the system shown in Figure 2.47, draw the free-body diagram and derive the equation(s) of motion.

Problem (2.12):
For the system shown in Figure 2.48, draw the free-body diagram and derive the equation(s) of motion.

Problem (2.13):
For the system shown in Figure 2.49,

a. Draw the free-body diagram.
b. Derive the equation(s) of motion.
c. Put the results in matrix form.

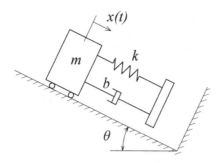

Figure 2.48   Mechanical system

Problem (2.14):
For the system shown in Figure 2.50,

a. Draw the free-body diagram.
b. Derive the equation(s) of motion.
c. Put the results in matrix form.

Problem (2.15):
For the system shown in Figure 2.51,

a. Draw the free-body diagram.
b. Derive the equation(s) of motion.
b. Put the results in matrix form.

Problem (2.16):
For the geared system shown in Figure 2.52, draw the free-body diagram and derive
the equations of motion. Assume that the angular displacement θ_1 and θ_2 are small.

Problem (2.17):
For the system shown in Figure 2.53, assuming the angular displacement θ is small,
draw the free-body diagram and derive the equation(s) of motion. Put the results in
matrix form.

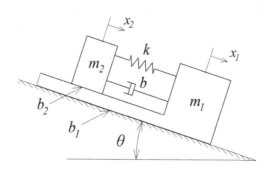

Figure 2.49   Mechanical system

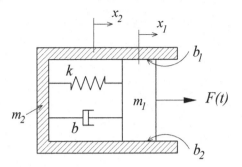

Figure 2.50   Mechanical system

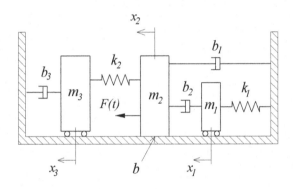

Figure 2.51   Mechanical system

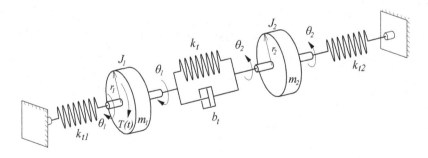

Figure 2.52   Mechanical system

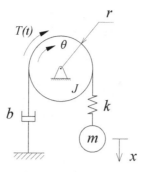

Figure 2.53   Mechanical system

Problem (2.18):
Draw the free-body diagrams and derive the equation(s) of motion for the rack and pinion system shown in Figure 2.54. Assume the rotating angle of the pinion θ is small.

Problem (2.19):
The stepped pulley, shown in Figure 2.55, is made of two homogeneous masses J_1 and J_2 fastened together. Assuming the rotating angle θ is small,

a. Draw the free-body diagram.
b. Derive the equation(s) of motion.
c. Put the results in matrix form.

Problem (2.20):
For the system shown in Figure 2.56, assuming no slip between the pulley and the cord, the lever rod is massless, and its angular displacement θ is small, derive the equation(s) of motion.

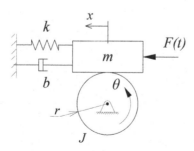

Figure 2.54   Mechanical system

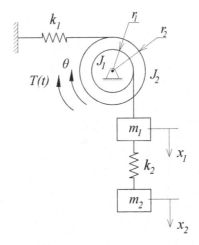

Figure 2.55 Mechanical system

Problem (2.21):
For the system shown in Figure 2.57, assuming the angular displacement θ is small,

a. Draw the free-body diagram.
b. Derive the equation(s) of motion.
c. Put the results in matrix form.

Problem (2.22):
For the system shown in Figure 2.58,

a. Draw the free-body diagram.
b. Derive the equation(s) of motion.
c. Put the results in matrix form.

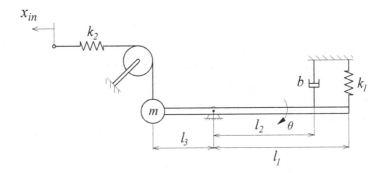

Figure 2.56 Mechanical system

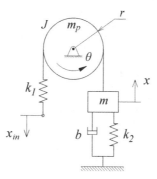

Figure 2.57 Mechanical system

Problem (2.23):
For the system shown in Figure 2.59, assuming the lever rod is massless, and its angular displacement θ is small, draw the free-body diagram and derive the equation(s) of motion.

Problem (2.24):
For the system shown in Figure 2.60, assuming that the rod is massless, derive the equation(s) of motion. Assume the angular displacement θ is small.

Problem (2.25):
For the system shown in Figure 2.61, draw the free-body diagram and derive the equation(s) of motion. The pendulum's angular displacement θ is assumed to be small. Put the results in matrix form.

Problem (2.26):
For the system shown in Figure 2.62, draw the free-body diagram and derive the equation(s) of motion. Assume the angular displacement θ is small.

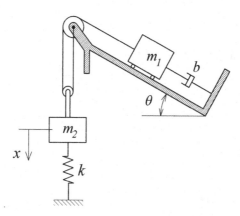

Figure 2.58 Mechanical system

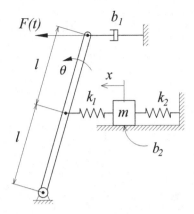

Figure 2.59 Mechanical system

Problem (2.27):
For the system shown in Figure 2.63, draw the free-body diagram and derive the equation(s) of motion. The angular displacement θ of the inverted pendulum is assumed to be small.

Problem (2.28): For the system shown in Figure 2.64, assuming the rod is massless, and its angular displacement θ is small,

a. Draw the free-body diagram.
b. Derive the equation(s) of motion.
c. Put the results in matrix form.

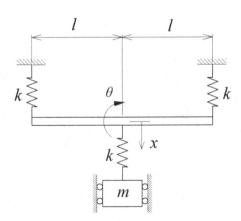

Figure 2.60 Mechanical system

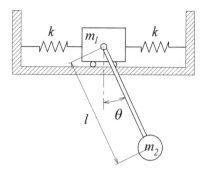

Figure 2.61 Mechanical system

Problem (2.29):
For the system shown in Figure 2.65, assume the angular displacements θ_1 and θ_1 are small. Draw the free-body diagram and derive the equation(s) of motion. Put the results in matrix form.

Problem (2.30):
For the system shown in Figure 2.66, assuming the rod is massless and its angular displacement θ is small,

a. Draw the free-body diagram.
b. Derive the equation(s) of motion.
c. Put the results in matrix form.

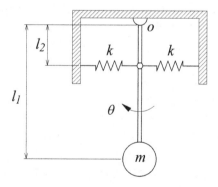

Figure 2.62 Mechanical system

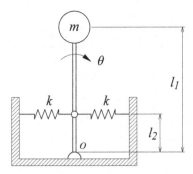

Figure 2.63 Mechanical system

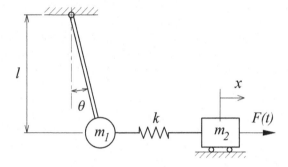

Figure 2.64 Mechanical system

Problem (2.31):
For the inverted pendulum system shown in Figure 2.67, assuming the pendulum rod is massless, and its angular displacement θ is small,

a. Draw the free-body diagram.
b. Derive the equation(s) of motion.
c. Put the results in matrix form.

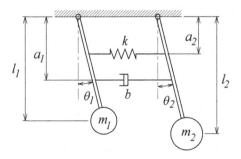

Figure 2.65 Mechanical system

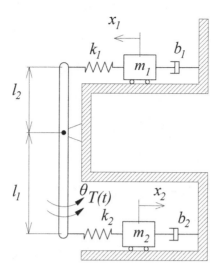

Figure 2.66 Mechanical system

Problem (2.32):

For the system shown in Figure 2.68, assuming the lever rod is massless, and its angular displacement θ is small, draw the free-body diagram and derive the equation(s) of motion.

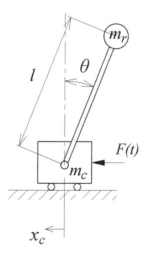

Figure 2.67 Mechanical system

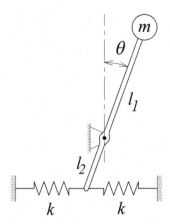

Figure 2.68 Mechanical system

3 Bond Graph Modeling Technique

3.1 INTRODUCTION

Modeling is an important activity in many research areas in scientific engineering fields. As a result, a substantial amount of literature on modeling dynamic systems is accessible. We seek a modeling technique that considers modeling a general-purpose tool rather than a specific use alone. We may have deduced equations of motion using Newton's laws for a purely mechanical system (translational or rotational), previously explained in Chapter 2. Kirchoff's voltage and current laws are often used to derive equations representing the model of electrical systems, which will be presented in Chapter 4. We may have to use Bernoulli's equation and law of the conservation of mass flow rate (continuity equation) to derive a dynamic model for a fluid field system, as explained in Chapter 5. We may use the heat transfer laws to derive differential equations representing thermal systems, as briefly described in Chapter 5.

Our formulation of equations of motion for modeling dynamic systems was based on these traditional techniques. A modeling technique known as Bond Graphs can efficiently represent the dynamics of all the systems mentioned above in the same way. In this specific circumstance, we do not need to use the traditional techniques; instead, we will use Bond Graph. The construction of Bond Graphs is an efficient approach for developing dynamic models for multi-domain structures in a very organized manner.

First, we must ask and answer the following question: Why learn Bond Graphs when the traditional modeling methods work just fine? In other words, what are the advantages of Bond Graphs? Bond Graphs are particularly well suited to dealing with multiple discipline systems, implying that it is a universal technique for modeling different systems (mechanical, electrical, fluid, and thermal, or a combination of them) in the same way, which is one of the great benefits of using Bond Graph as a modeling technique. The following are some of the other benefits of using Bond Graphs.

1. Bond Graphs can be processed in a standard manner to generate equations of motion of the dynamic systems (a very algorithmic approach).
2. Bond Graphs have an analogy built-in between systems from various domains, allowing physical insight to be transferred.
3. Bond Graphs are very precise model statements. Precise schematic diagrams for specific dynamic systems are complicated to generate using traditional diagrams.

DOI: 10.1201/9781032685656-3

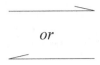

Figure 3.1 The bond

4. Apart from making it easier to derive system equations, representing the system
   by a Bond Graph reveals hidden interactions between elements and gives the de-
   signer a better physical understanding of the dynamics of systems.

Finally, there are two approaches for simulating the Bond Graph model. The first is
solving the equations of motion generated from the Bond Graph model, followed by
the traditional methods. The second is carried out using the software directly from
the structure of the Bond Graph model. In this way, it is not necessary to generate
the system equations. Note that this is also another significant benefit of using Bond
Graph as a modeling technique. Aside from the benefits of utilizing Bond Graphs,
gaining access to Bond Graphs requires learning a new modeling language and associated simulation methodologies.

Now, we will start describing the structural components of the Bond Graph technique.

3.2 BOND GRAPH STRUCTURES

3.2.1 BONDS

Bond Graphs are made up of interconnected components that transmit power. These
connections are called Bonds. The bond is represented by a line, and each bond
represents a transition of power as shown in Figure 3.1.

3.2.2 POWER FLOW DIRECTION

The direction of the positive power flow is denoted by a half arrow located on one
side of the bond, as shown in Figure 3.1. It indicates that it points in the direction that
the power of the bond moves positively. It is characterized by how the components
of the system interact with one another physically.

3.2.3 VARIABLES

The Bond Graph involved two generalized power variables, effort e and flow f. The
power is the result of multiplying these two variables. Bonds represent the mutual
interaction between the dynamics of the system variables, which is why they are used
in system dynamics. Consequently, these variables will interact. It simply means that
the effort and flow leaving one port and entering another are the same. Bond Graphs
are made up of five generalized elements: SE (source of effort) and SF (source of
flow), I (inertia), C (capacitor), and R (resistance). These five elements, as well as

the two power variables e (effort) and f (flow), can be used for various dynamical systems. In addition to the two generalized power variables, effort e and flow f, two other generalized energy variables, q (displacement) and p (momentum), can be estimated. The relations between the power variables e, f and the energy variables p, q are given by

$$p = \int e\,dt \text{ and } q = \int f\,dt \tag{3.1}$$

or

$$e = \dot{p} \text{ and } f = \dot{q} \tag{3.2}$$

In the following section, the components of Bond Graphs will be described in detail. We will first define the Bond Graph power and energy variables as follows.

3.2.3.1 Power variables

Power variables e and f are so-called because

$$P = e f \tag{3.3}$$

This equation indicates that power is the product of effort e and flow f.

3.2.3.2 Energy variables

Energy variables p and q are so-called because

$$E(p) = \int f(p)\,dp \tag{3.4}$$

and

$$E(q) = \int e(q)\,dq \tag{3.5}$$

These two equations indicated that the dynamics of a system can be completely described in terms of e, f, p, and q, as illustrated in the following analysis.
- **For mass:**
Since the mass is a storage energy element, the amount of the stored energy by the mass m is given by

$$P_m = m v \text{ and } f = v \tag{3.6}$$

Then, the energy transmitted to the mass element (conservative energy) is given by

$$E(p) = \int m f\,dv = \int m v\,dv \tag{3.7}$$

Carrying out this integration, the total amount of energy (Kinetic Energy KE) that is transmitted to the mass is given by

$$K.E. \text{ (conservative energy)} = \frac{1}{2} m f^2 = \frac{1}{2} m v^2 \tag{3.8}$$

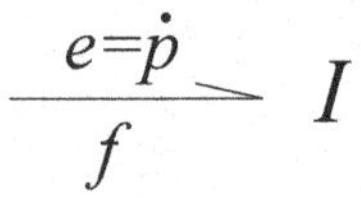

$$\frac{e=\dot{p}}{f} \searrow\; I$$

Figure 3.2 Inertia element

- For spring:
Since the spring is a storage energy element, the amount of the stored energy by the spring of stiffness k is given by

$$P_s = e = kx \text{ and } q = x \tag{3.9}$$

Then, the energy transmitted to the spring element (Conservative Energy) is given by

$$E(q) = \int kq\,dx = \int kx\,dx \tag{3.10}$$

Carrying out this integration, the total amount of energy (Potential Energy) that is transmitted to the spring is given by

$$P.E.\,(\text{conservative energy}) = \frac{1}{2}kq^2 = \frac{1}{2}kx^2 \tag{3.11}$$

3.2.4 PORTS

In the Bond Graph technique, ports are where the power can flow between Bond Graph elements. They are classified into 1-port, 2-ports, and multi-ports. This section will describe the ports of Bond Graphs in detail as follows:

- 1-port I – Inertia
As stated, the 1-port inertia, shown in Figure 3.2, is a storage element. Through this element, the energy is stored and retrieved from the system. It is used to model mass elements. Its general relations are given by

$$p = If \text{ or } f = \frac{1}{I}p \tag{3.12}$$

and

$$p = mv \tag{3.13}$$

where I is the Bond Graph inertia element representing the mass, while p and v are the momentum and velocity of the mass, respectively.

- 1-port C – Capacitance
The 1-port capacitor, shown in Figure 3.3, is a storage element. Through this element, the energy is stored and retrieved from the system. It is used to model spring elements. Its general relations are given by:

$$q = Ce \text{ or } e = \frac{1}{C}q \tag{3.14}$$

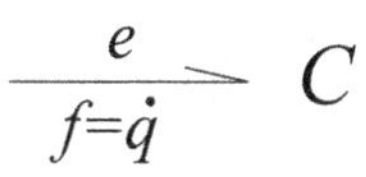

$$\frac{e}{f=\dot{q}} \rightharpoonup C$$

Figure 3.3 Capacitor element

and

$$F_s = \frac{1}{C}q = kx \quad \text{and} \quad x = q \tag{3.15}$$

From this equation, we get:

$$F_s = Cx = Cq \quad \text{and} \quad C = \frac{1}{k} \tag{3.16}$$

where C is the capacitance of the Bond Graph capacitor element representing the spring and q is the displacement of the spring. Equation (3.16) states that the stiffness of a spring is equal to the reciprocal of its capacitance, or that the spring's capacitance is equal to the reciprocal of its stiffness.

- 1-port R - Resistance
Resistance is not a storage element; it is a dissipation of energy. Because total energy is conserved, this energy is not destroyed but converted into a form from which it cannot be recovered. So, the 1-port resistance, shown in Figure 3.4, is a dissipated element. It is used to model damping elements (dampers). Through this element, the energy is dissipated and withdrawn from the system. Its general relations are given by

$$e = Rf \quad \text{or} \quad f = \frac{1}{R}e \tag{3.17}$$

and

$$F_d = bv \tag{3.18}$$

where R is the Bond Graph resistance element representing the damper and b is the damping factor.

- Tetrahedron of state
The state of a system is described by two generalized coordinates, p (momentum) and q (displacement). These are simply momentum and displacement in the mechanical domain; however, they are different in other domains. Figure 3.5 shows these states and the relationship between e, f, q, and p. This is known as the Tetrahedron of State.

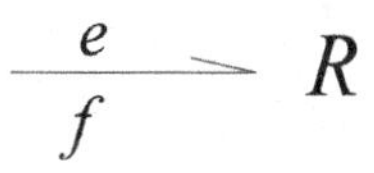

$$\frac{e}{f} \rightharpoonup R$$

Figure 3.4 Resistance element

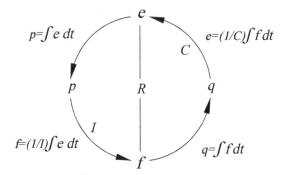

Figure 3.5 Tetrahedron of state of 1-port elements

- 1-port *SE* - Source of Effort
The source of effort $e(t)$, shown in Figure 3.6, is a time-dependent function, while f is arbitrary. It provides a predetermined amount of effort without requiring a certain amount of flow. It means that an unlimited amount of power can be transmitted.
- 1-port *SF* - Source of flow
Source of flow $f(t)$, shown in Figure 3.7, is a time-dependent function, while e is arbitrary. It generates the prescribed flow regardless of the effort required.

In both cases, *SE* and *SF*, the half-arrow points away from the source, indicating that it flows from the source to the system, as shown in Figures 3.6 and 3.7.

- 2-ports *TF* – Transformer
The transformer element *TF* has two ports, as shown in Figure 3.8. It transmits power from one of its ports to the other. As a result, the product of effort e and flow f of both ports are equal. The size of the effort and flow, rather than the power flow, is influenced by the variable m. Note that *TF* does not store or absorb energy. Its constitutive relations are given by

$$e_1 = m e_2 \tag{3.19}$$

$$SE \ \frac{e=F(t)}{f} \quad\diagdown$$

Figure 3.6 Source of effort

$$SF \ \frac{e}{f=v(t)} \quad\diagdown$$

Figure 3.7 Source of flow

$$m f_1 = f_2 \tag{3.20}$$

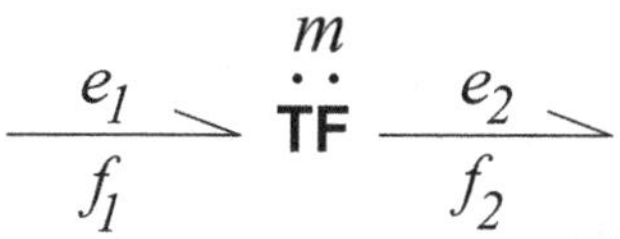

$$\dfrac{e_1}{f_1} \searrow \overset{\overset{\displaystyle m}{\cdot\,\cdot}}{\text{TF}} \dfrac{e_2}{f_2} \searrow$$

Figure 3.8 Two-ports transformer element

- 2-ports GY – gyrator

The gyrator element GY has two ports, as shown in Figure 3.9. It transmits power from one of its ports to the other. As a result, the product of effort e and flow f of both ports are equal. The size of the effort and flow, rather than the power flow, is influenced by the variable n. Note that the gyrator element GY does not store or absorb energy. Its constitutive relations are given by

$$e_1 = n f_2 \tag{3.21}$$

$$n f_1 = e_2 \tag{3.22}$$

- Multi-ports 1-junction – effort junction

The multi-port junction has more than two ports. The serial connection of elements is represented by a 1-junction, as shown in Figure 3.10. In this junction, the sum of efforts is zero because all flows are equal. So, the 1-junction is a common flow (velocity) junction. Its constitutive relations are given by

$$f_1 = f_2 = f_3 \tag{3.23}$$

$$e_1 + e_2 + e_3 = 0 \tag{3.24}$$

- Multi-ports 0-junction – flow junction

A parallel connection of elements is represented by a 0-junction, as shown in Figure 3.11. With a 0-junction, all efforts are equal, and the overall flow amount is zero. Therefore, the 0-junction is a common effort (force) junction. Its constitutive relations are given by

$$e_1 = e_2 = e_3 \tag{3.25}$$

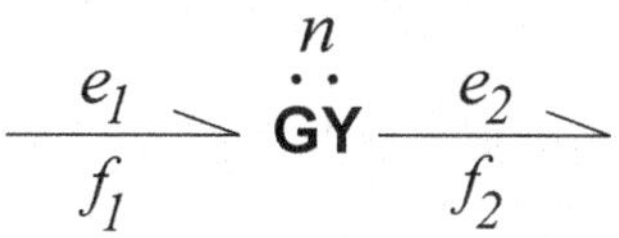

$$\dfrac{e_1}{f_1} \searrow \overset{\overset{\displaystyle n}{\cdot\,\cdot}}{\text{GY}} \dfrac{e_2}{f_2} \searrow$$

Figure 3.9 Two-ports gyrator element

Figure 3.10 Multi-ports 1-junction

$$f_1 + f_2 + f_3 = 0 \tag{3.26}$$

3.2.5 JUNCTIONS REPRESENTATIONS IN MECHANICAL SYSTEMS

Considering the description of the Bond Graph junctions above, the simplest way to determine if the junction is a 0-junction or a 1-junction is to check if all efforts at the junction's elements are identical (0-junction) or if all flows have the same value (1-junction). In other words, if some elements have the same velocity (flow), they should be connected to a 1-junction, while if they have the same force (effort), they should be connected to a 0-junction. These situations are explained as follows.

- When a force is applied to a mass, 1-junction will connect a source of effort *SE* element, which represents the applied force $F(t)$, and inertia *I* element, which represents the mass, as shown in Figure 3.12a, satisfying the physical relations indicated in equations (3.12) and (3.13).
- When a force is applied to a spring, which is a massless element, the components connected to its two ends have the same force and, consequently, effort; otherwise, there would be a resultant force, generating an infinite acceleration of the massless spring, which is unrealistic when using a linear spring. The spring itself, as well as the elements at its two ends, share the same effort. Furthermore, the perceived spring displacement equals the relative displacement of the two ends. As a result, in the Bond Graph, the spring will be represented by a capacitor element *C* connected by a 0-junction, which is a common effort junction, as shown in Figure 3.12b, satisfying the physical relations indicated in equations (3.15) and (3.16).
- Similarly, the resistance element *R* shown in Figure 3.12c, which represents the damper element, would utilize the same junction structure (0-junction) as the

Figure 3.11 Multi-ports 0-junction

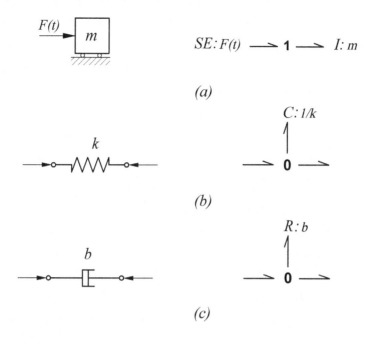

(a)

(b)

(c)

Figure 3.12 Mechanical elements and the corresponding Bond Graph junctions

spring element and satisfy the physical relations indicated in equations (3.17) and (3.18). Note that the velocity (flow) variables in the resistance element will be positive or negative based on only the power direction half arrows, which will be demonstrated later.

Now, let's look at some specific examples of junctions representing basic linkages between mechanical components (mass, spring, and damper).

- When the spring is linked to a mass, as shown in Figure 3.13a, its capacitor element C has to be connected to a 0-junction that is connected to the 1-junction created for the mass.
- Similarly, as the spring, when the damper is linked to a mass, as shown in Figure 3.13b, its resistance element R has to be connected to a 0-junction that is connected to the 1-junction created for the mass.
- When the spring is connected between two masses, as shown in Figure 3.14a, a capacitor C is connected to the 0-junction that must be positioned between two 1-junctions created for the two inertia elements representing both masses.
- Similarly, when the damper is connected between two masses, its resistance R is connected to the 0-junction that must be positioned between two 1-junctions created for the two inertia elements representing both masses, as shown in Figure 3.14b.

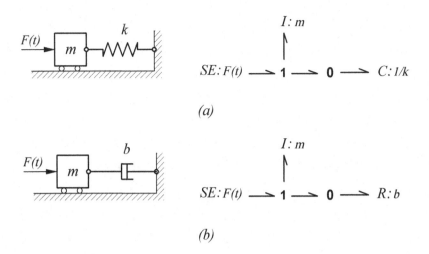

Figure 3.13 Mechanical elements and the corresponding Bond Graph junctions

3.3 CAUSALITY

Causality is not a natural process that can be predicted using the physical charac-teristics of dynamic system components. Bonds are connections between elements that affect one another, whereas causality is a choice made artificially and then incor-porated into the model computations. Mathematical expressions are frequently used to express a relationship between cause and effect, or input and output, concerning specific phenomena. It can be represented by a causal stroke at one of the two ends of the bond, as shown in Figure 3.15. It is worth noting that the causal stroke has no relation with the bond's half arrow, which shows the positive flow of power.

Bonds connecting to an element in a Bond Graph include causal strokes that show whether the effort is placed on or imposed by the element. If the causal stroke is located opposite to the element side of the bond, as shown in Figure 3.15a, it imposes an effort on the system, responding to that effort with a flow. If the causal stroke is placed on the element side of the bond, as shown in Figure 3.15b, the effort is being imposed on it, responding with a flow.

3.4 CAUSALITY OF 1-PORT ELEMENT

In general, sources have a necessary causality specified by the kind of source. In contrast, junctions, transformers, and gyrators have rules controlling which possi-ble causal stroke combinations are allowed, and other components have the desired causality dictated by the element type. The criteria for each element, as well as the preferred causal strokes, are explained in detail in the following sections.

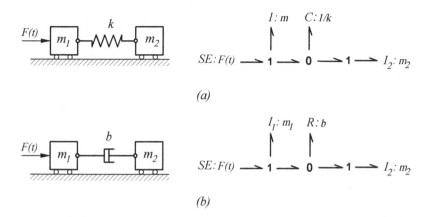

Figure 3.14 Mechanical elements and the corresponding Bond Graph junctions

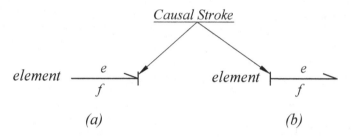

(a)   *(b)*

Figure 3.15 Causal stroke locations

3.4.1 SOURCE OF EFFORT CAUSALITY

Causality is always described throughout the sources. The effort is imposed on the system by the source of effort SE. In this case, the causal stroke should be located opposite to the element side of the bond, where the effort e points, as shown in Figure 3.16.

3.4.2 SOURCE OF FLOW CAUSALITY

The flow would be imposed on the system by the source of flow SF. Since the flow exits the source and the effort enters the source, the causal stroke must be placed on the SF element side of the bond, as shown in Figure 3.17.

$$SE \quad \frac{e=F(t)}{f}$$

Figure 3.16 Source of effort causality

$$SF \vdash \frac{e}{f=v(t)}$$

Figure 3.17 Source of flow causality

$$\frac{e}{f} \rightarrowtail I$$

Figure 3.18 Inertia element causality

3.4.3 INERTIA CAUSALITY

Since the inertia element I can be described using an integral and a differential equation, the solution of this element requires an integral equation. A differentiate, on the other hand, cannot be realized using simulation tools. By this definition, the effort will be the input of the inertia I. As a result, the causal stroke showing the causality of the inertia should be imposed on the element side, as shown in Figures 3.18. This means that the flow f is the cause, and the effort e is the effect. When looking for the causal stroke, keep in mind that the power flow half arrow indicating the direction of e points towards the causal stroke, as shown in Figure 3.18.

3.4.4 CAPACITANCE CAUSALITY

A capacitor element C can be described using both an integral equation and a differential equation; the solution of this element requires only the integral equation, as stated above. By this definition, the flow will be the input of the capacitor C. As a result, the causal stroke showing causality for the capacitor element, shown in Figure 3.19, should be located opposite to the element side of its bond.

3.4.5 RESISTANCE CAUSALITY

Figure 3.20 shows the causality of the resistance R element. Based on the causality requirements, the causality of the resistance R element is often chosen in practice. The concepts of integral and derivative causality do not apply to resistance elements, and any of the causalities shown in Figure 3.20 are equally valid. Since the causal strokes of the resistance element affect the equation structure, there is an algebraic loop concept, but it is beyond the scope of this course. A brief explanation of the resistance element algebraic loop will be presented at the end of this chapter.

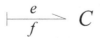

Figure 3.19 Capacitor element causality

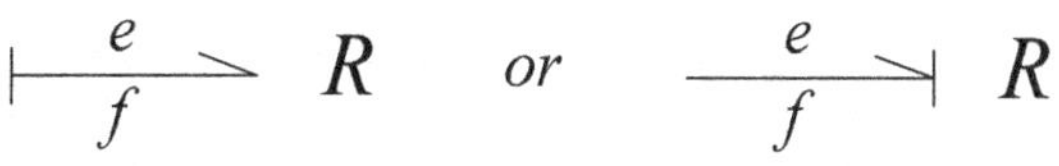

Figure 3.20 Resistance element causality

3.5 CAUSALITY OF 2-PORTS ELEMENT

3.5.1 TRANSFORMER CAUSALITY

The causality of a 2-ports transformer TF can be realized in two ways.

1. If e_1 is the cause of a transformer, then both e_2 ($e_2 = me_1$) and f_1 will be the effect, and subsequently f_2 will be the cause as well.
2. If e_1 is chosen to be an effect of the cause e_2, f_1 will be the cause and f_2 will be the effect.

These analyses will be understandable with the help of the solved examples in this chapter. Figures 3.21a and 3.21b show the two possible realizations of the transformer's causality.

3.5.2 GYRATOR CAUSALITY

Also, the causality of the two-port Gyrator GY can be realized in two ways.

1. If e_1 is the cause of a Gyrator, both f_1 and $f_2(f_2 = e_1/n)$ will be the effects, and e_2 will be the cause as well.
2. If e_1 is chosen as an effect of f_1 ($e_2 = nf_1$), so e_2 is chosen as well, and both f_1 and f_2 are causes.

Figure 3.22a and 3.22b depict the two possible realizations of Gyrator causality.

3.6 CAUSALITY OF MULTI-PORTS ELEMENT

3.6.1 1-JUNCTION CAUSALITY

the 1-junction represents serial connections with the same value for all flows. This means that there will be only one input flow, as shown in Figure 3.23a. The efforts

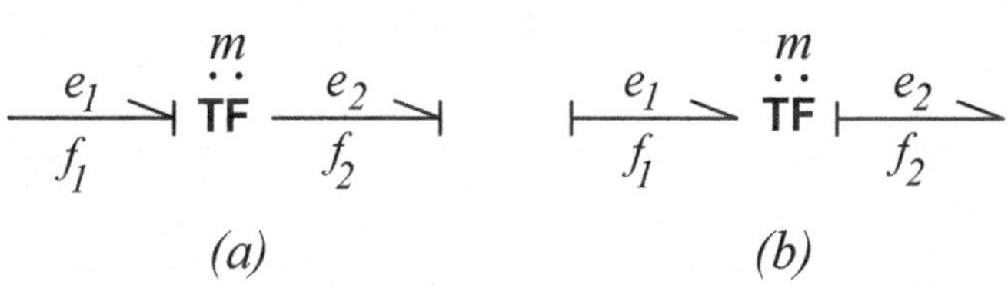

(a)   *(b)*

Figure 3.21 Transformer element causality

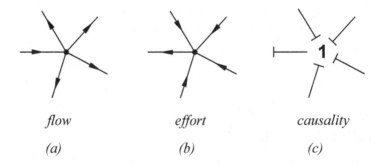

$$\xrightarrow[f_1]{e_1} \overset{\overset{\cdot\cdot}{n}}{GY} \xrightarrow[f_2]{e_2}$$

(a)

$$\xrightarrow[f_1]{e_1} \overset{\overset{\cdot\cdot}{n}}{GY} \xrightarrow[f_2]{e_2}$$

(b)

Figure 3.22 Gyrator element causality

placed forth in the other bonds in the junction would be used as inputs, as shown in Figure 3.23b. In this situation, all bonds would have causal strokes at the side of the 1-junction, except for only one bond, as shown in Figure 3.23c. Figure 3.24 shows all realizations of the causality of the 1-junction.

3.6.2   0-JUNCTION CAUSALITY

The 0-junction represents parallel connections with the same value for all efforts. In this case, the summation of flows is equal to zero, meaning that there will only one output flow, as shown in Figure 3.25a. The efforts placed forth in the other bonds in the junction would be used as output, as shown in Figure 3.25b. In this situation, all bonds would have causal strokes at the opposing sides of the 0-junction side, except for only one bond, as shown in Figure 3.23c. Figure 3.26 depicts all realizations of the causality of the 0-junction.

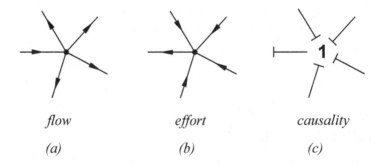

flow   effort   causality

(a)   (b)   (c)

Figure 3.23 Effort and flow directions of the 1-junction

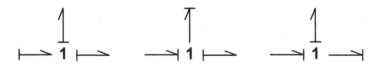

Figure 3.24 Different possible locations of the causality on 1-junction

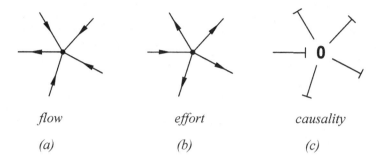

flow

(a)

effort

(b)

causality

(c)

Figure 3.25 Effort and flow directions of the 0-junction

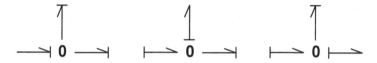

Figure 3.26 Different possible locations of the causality on 0-junction

3.7 BOND GRAPH CONSTRUCTION FOR MECHANICAL SYSTEM

In general, to create the Bond Graph model for a mechanical system, either translational or rotational, should consider the following general steps:

1. For each absolute velocity, create a 1-junction. Connect inertia elements to their absolute velocities at 1-junctions. For springs and dampers, relative velocities are essential; thus, building 1-junctions for relative velocities could be helpful.
2. Unless you have already constructed a relative velocity 1-junction for each in step (1), connect capacitor element C and resistance element R between 1-junction pairs using 0-junctions.
3. Connect sources of forces and velocities as required.
4. To impose velocity restrictions between 1-junctions, transformers or 0-junctions are used. Check to see that the force relations that have resulted are accurate.
5. Half-arrows must be added to bonds to show the direction of positive power flow. Note that the energy storage elements (masses and springs) and energy dissipative elements (dampers) receive power simultaneously while sources supply power. The relationships between relative velocities determine the other power bonds.
6. Wherever possible, simplify the Bond Graph. Deleting all of the connections that hold ground velocity. Simple bonds can also be used to exchange 2-port variants of 1-junction and 0-junction with in-out bonds (through power flow directions), as shown in Figure 3.27. This figure shows all possible power-through flow directions.
7. Finally, to finish, you should add causality strokes and bond numbers. In this case, the final Bond Graph, referred to as the Augmented Bond Graph, is ready to derive the equations of motion for the dynamic system.

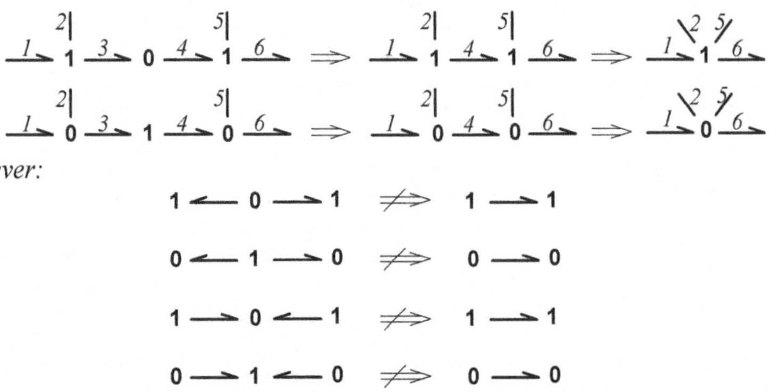

Figure 3.27 Different possible locations of the causality on 0-junction and 1-junction

3.8 HOW TO ASSIGN CAUSALITY FOR A BOND GRAPH MODEL

This section will discuss in detail the causality choices for different Bond Graph elements and how they will be assigned to a Bond Graph model. As previously stated, when analyzing causality, it is essential to keep the points of the following outline in mind:

1. Sources have defined causality.
2. 0-junction, 1-junction, transformer TF and gyrator GY have restricted causality possibilities.
3. Inertia I and capacitor C elements should have integral causality.
4. Resistance R element has arbitrary causality.

To assign causality to the Bond Graph model, we have to follow the following steps:

1. In the first step, we will consider sources, selecting the source SE and/or SF accountable for the system's excitation and assigning the necessary causality. The causal stork of the source of effort SE is located on the bond's junction side (opposed to the SE element side), whereas it is located on the bond's source side for the flow source. Extend the scope of causal implications by incorporating all relevant 0-junction, 1-junction, transformer TF, and gyrator GY limitations.
2. Repeat step (1) until all causalities of sources have been assigned.
3. Every I and C element should be assigned integral causality. Extend the causal implications by incorporating all previously applied 0-junction, 1-junction, transformer TF, and gyrator GY restrictions.
4. Repeat step (3) until all inertia I and capacitor C elements have causality assigned.
5. Select any unassigned R element and assign it arbitrary causality. Extend the causal implications using all 0-junction, 1-junction, transformer TF and gyrator GY restrictions that applied.

6. Repeat step (5) until all resistance R elements have been assigned.

After completing the causality assignment and numbering all bonds, we now have an Augmented Bond Graph of the system. Once this Bond Graph has been created, we must answer the question: What do we know about the system represented by the Augmented Bond Graph?. The answer are

1. Equations of motion will be a series of first-order differential equations. Since we are dealing with linear elements, the equations of motion are ready for matrix form (state space form).
2. State variables will be energy variables (p and q), which means that each inertia element I will have a momentum p and each capacitance element C will have a displacement q. Keep in mind that all inertia I and capacitor C elements must have integral causality.
3. The constitutive relations define the flow and effort variables of the storage elements with integral causalities.
4. Number of energy storage elements, with integral causality, will define the order of the system.
5. The system's inputs would come from sources of effort and/or flow, with the total number of inputs equals to the total number of sources, implying that the system could have a single or multiple inputs.

3.9 STATE EQUATIONS FROM BOND GRAPHS MODEL

We are now showing how we can use causality as a guide to state equation derivations. State equations will take the form:

$$\dot{x}_i = f_i(x_1, x_2, \ldots, x_i) \qquad (3.27)$$

where x_i variables are either p or q variables of integrally causal energy storage elements. They should find only states, inputs, and system parameters on the right-hand side of these equations. Consequently, these equations will be ordered according to their bond numbers. The final results may be expressed in matrix form as

$$\dot{x}_i = Ax_i + Bu \qquad (3.28)$$

where x_i is a column vector of state variables ordered according to the incoming bond numbers of the element that have integral causality, and u is the input vector.

3.10 FORMULATING DIFFERENTIAL EQUATIONS FROM BOND GRAPHS MODEL

As the Augmented Bond Graph model is implemented, the system's equations of motion can be derived very schematically. Note that the fundamental differential equation of motion comes from each storage element's integral causality (inertia and capacitor elements). The first-order differential equation for the inertia element I is given by

$$\dot{p} = e \tag{3.29}$$

where $\dot{p}$ is the rate of change of the momentum of the inertia element, and e is the summation of all effort variables associated with linked junctions. For the capacitor element C, the first-order differential equation is given by

$$\dot{q} = f \tag{3.30}$$

where $\dot{q}$ is the velocity of the capacitor element, and f is the summation of all flow variables associated with linked junctions.

The mass momentum p and the spring displacement q are often used as generalized state variables to generate the system's first-order differential equations. As a result, we can estimate that the number of equations equals the number of storage elements with integral causalities. We cannot consider the elements with differential causality for some reasons, which will be briefly discussed at the end of this chapter. Therefore, the storage elements with integral causalities serve the state variables p and q. To formulate the equations of motion from the Bond Graph model, we have to follow the following two steps:

3.10.1   KEY VARIABLES

To derive the system's equation of motion, we must first define the key variables: inputs, state variables, and co-energy variables, as follows:
- Inputs: SE and/or SF
- State variables: p and q
- Co-energy (Power) variables: e and f
and then the constitutive relations, as follows:

3.10.2   CONSTITUTIVE RELATIONS

Write equations describing the relationship between power variables and state variables for each energy storage element, as follows:
- For inertia element I :

$$f = \frac{1}{I}p \tag{3.31}$$

- For capacitor element C:

$$e = \frac{1}{C}q \tag{3.32}$$

Equations (3.31) and (3.32) are called constitutive or physical relations. During the derivation of the system equations, power variables would gradually be replaced by energy variables, significantly reducing the need for constitutive relations.

Table 3.1
Bond Graph elements

Symbol	Name	Function
Sources: $SE \dfrac{e}{f}$ $SF \dfrac{e}{f}$	Source of effort Source of flow	$e(t)$ $f(t)$
One-ports: $\dfrac{e}{f}\ I$ $\dfrac{e}{f}\ C$ $\dfrac{e}{f}\ R$	Inertia Capacitor Resistance	$f = \frac{1}{I}\int e\,dt$ $e = \frac{1}{C}\int f\,dt$ $e = Rf$
Two-ports: $\dfrac{e_1}{f_1}\ \overset{m}{TF}\ \dfrac{e_2}{f_2}$ $\dfrac{e_1}{f_1}\ \overset{n}{GY}\ \dfrac{e_2}{f_2}$	Transformer Gyrator	$e_2 = me_1$ and $f_1 = mf_2$ $e_2 = nf_1$ and $e_1 = nf_2$
Multi-ports: $\dfrac{f_2\|e_2}{}$ $\dfrac{e_1}{f_1}\ 1\ \dfrac{e_3}{f_3}$ $\dfrac{f_2\|e_2}{}$ $\dfrac{e_1}{f_1}\ 0\ \dfrac{e_3}{f_3}$	1-Junction (series) 0-Junction (parallel)	$f_1 = f_2 = f_3$ and $e_1 + e_2 + e_3 = 0$ $e_1 = e_2 = e_3$ and $f_1 + f_2 + f_3 = 0$

Generally, when the system model has two first-order differential equations, the second-order system has two degrees of freedom. This depends on the number of elements that have integral causality in the Bond Graph model. This means the system's degree of freedom depends on the number of elements with integral causality.

Table 3.1 lists all the Bond Graph elements, along with their names and functions, making them easy to identify.

Example (3.1):
For the system shown in Figure 3.28,

- Create the preliminary Bond Graph model and simplify it.
- Create the Augmented Bond Graph model.

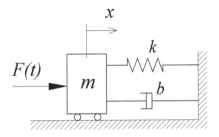

Figure 3.28 Mass-spring-damper system

- Write the key variables (State and Co-energy variables) and their constitutive re-
  lations.
- Derive the equation(s) of motion.
- Check if Newton's law is embedded in the Augmented Bond Graph of the system.

Solution:
A mass-spring-damper system, shown in Figure 3.28, demonstrates how to use the
Bond Graph technique in modeling dynamic systems. The goal is to create a Bond
Graph model that defines the velocity $v = \dot{x}$ and displacement x of the mass m. Note
that the external force $F(t)$ and the forces coming from the spring and damper act
on the mass. We must follow the steps outlined in Section 3.7 to build a Bond Graph
model for this system.

1- Create a 1-junction for each absolute velocity
The system has two absolute velocities, one for mass v and another for ground v_o,
that we will consider for the time being. The ground can generally be represented
in the preliminary Bond Graph model as either a source of effort SE or a source of
flow SF. If we consider a source of effort SE, it implies that we will consider ground
displacement! If we consider the ground as a source of flow SF, then it follows that
we should similarly consider the ground's velocity! Both will be removed later for
the simplicity of the preliminary Bond Graph model of the system. Consequently,
the preliminary Bond Graph model considers two common flow 1-junctions with
velocities v and v_o, as shown in Figure 3.29.

2- Connect sources of forces and velocities as required
It is important to note that both the spring and the damper are linked between the
mass and the ground, indicating that their ends are subjected to two forces: one from
the external force $F(t)$ applied to the mass, the other from the ground. As a result,

$$SE\text{:}\,F(t) \; \underline{\quad\quad} \; \mathbf{1} \qquad\qquad \mathbf{1} \; \underline{\quad\quad} \; SE\text{:}\,ground$$
$$\qquad\qquad\quad v \qquad\qquad\qquad\qquad\quad v_o = 0$$

Figure 3.29 Preliminary Bond Graph model: step 1 and step 2

Figure 3.30 Preliminary Bond Graph model of the system: step 3

we have two sources of effort: one is connected to a 1-junction associated with the common velocity v, and the other is connected to a 1-junction associated with the common velocity v_o, as shown in Figure 3.29.

3- Connect inertia elements to their absolute velocities 1-junctions
Because the mass is moving at a constant velocity v, it will be represented by an inertia element I connected to the 1-junction of common velocity v, as shown in Figure 3.30.

4- Connect capacitor and resistance elements to their relative velocities 0-junctions
The spring and the damper elements are connected to the mass and the ground, implying that one end of each element moves while the other is stationary. In this case, a capacitor element C should be used to represent the spring, and a resistance element R should be used to represent the damper. They should be linked between the two 1-junctions that have already been made, as shown in Figure 3.31. Now, we have the preliminary Bond Graph model representing the given system.

5- Wherever possible, simplify the Bond Graph
Firstly, all the connections that hold a ground velocity must be removed. So, the source of effort SE associated with the ground velocity v_o and its bonds are removed,

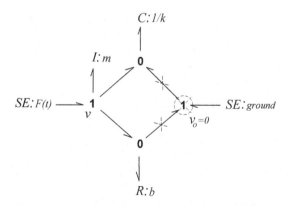

Figure 3.31 Preliminary Bond Graph model: step 4

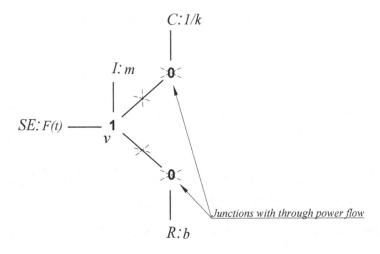

Figure 3.32 Preliminary Bond Graph model: step 5

as shown in Figure 3.31. Now, we can use an in-out bond to swap the 2-ports variants of 0-junctions connected with the spring and connected with the damper, as shown in Figure 3.32. As a result, we will remove the two 0-junctions from the preliminary Bond Graph, implying that both spring and damper are connected directly to the 1-junction of a common velocity v, as shown in Figures 3.32 and 3.33.

6- Add half-arrows to bonds to show the positive direction of power flow
To indicate the direction of positive power flow, half-arrows must be added to all bonds. It is important to note that the energy storage components (mass and spring) and the dissipated energy element (damper) receive power, while the source of effort supplies power. As a result, the half-arrow power directions of the three elements, I, C, and R should be added on the element side of their bonds, while they will be located at the junction side for the source of effort, SE, as shown in Figure 3.33.

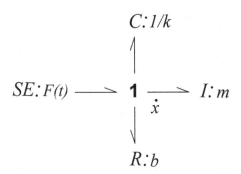

Figure 3.33 Bond Graph model of the system: step 6

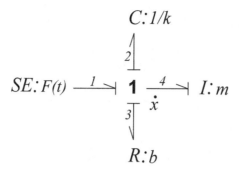

Figure 3.34 Augmented Bond Graph model of the system

7- Add causality stroke marks and bond numbers

To assign causality to the Bond Graph model, we have to follow the following steps, see Figure 3.34:

1. Sources have to define causality. We have only one source of effort SE representing the external force $F(t)$. So, the causal stroke mark should be located on the opposite side of the SE side on the bond (junction side). Here, we must apply the causality restrictions explained in Section 3.3.
2. Inertia and capacitor elements should have integral causality. So, the causalities of the inertia I element and the capacitor C element should be located on their bonds.
3. Resistance element R has arbitrary causality. Applying the restriction causality of the 1-junction, which represented serial connections of bonds with the same flow, the causal stroke mark of the element R should be located at the opposite side of the R bond (junction side).

To finish numbering all bonds, we get the Augmented Bond Graph model shown in Figure 3.34. To derive the equation of motion representing the dynamics of the given system, we are forced to estimate the effort and flow variables of the system elements, as we stated previously. As a result, the effort variable can be selected as $F(t)$, and the flow variable can be chosen as the velocity of the mass, v. Now, we must determine the relationships between four elements: the external force, mass, spring, and damper; all should be a function of force (effort) and velocity (flow). The source of effort SE can represent the external force $F(t)$. From the Augmented Bond Graph model shown in Figure 3.34, we have two storage elements of integral causality represented by bonds 2 and 4.

To derive the system's equation of motion from the Bond Graph model, we must first define the key variables: Inputs, State, and Co-energy (power) variables and their constitutive relations.

- Key variables:
- Inputs: $e_{(t)} = SE = F(t)$.
- State variables: q_2 and p_4 (integral causality).
- Co-energy (Power) variables: e_2 and f_4 (integral causality).

- **Constitutive relations:**
Write the constitutive relations that describe the relationship between co-energy variables and state variables for each energy storage element with integral causality as follows:

$$e_2 = \frac{1}{C}q_2 \text{ and } f_4 = \frac{1}{I}p_4 \qquad (3.33)$$

where p_4 and q_2 are the momentum and displacement of the mass m, respectively.

Now that we know the connections between flow and effort variables at the model junctions let's determine the equations that reflect them.

The effort variables of the 1-junction of common velocity v are given by

$$f_1 = f_2 = f_3 = f_4 \qquad (3.34)$$

$$e_4 = \dot{p}_4 = e_1(t) - e_2 - e_3 = F(t) - e_2 - e_3 \qquad (3.35)$$

and

$$e_2 = \frac{1}{C}q_2 \qquad (3.36)$$

$$e_3 = Rf_3 = R\frac{1}{I}p_4 = \frac{R}{I}p_4 \qquad (3.37)$$

Substituting equations (3.36) and (3.37) in equation (3.35), yields

$$\dot{p}_4 = F(t) - \frac{1}{C}q_2 - \frac{R}{I}p_4 \qquad (3.38)$$

or,

$$\dot{p}_4 + \frac{1}{C}q_2 + \frac{R}{I}p_4 = F(t) \qquad (3.39)$$

where $\dot{p}_4$ is the mass momentum change rate. Equation (3.39) is the equation of motion created from the Bond Graph model, representing the given system's dynamic motion.

Now, we will check if the system's equation of motion is embedded in the Augmented Bond Graph model shown in Figure 3.34. In other words, we will check if Newton's law of motion applied to the mass satisfies the Bond Graph model. The following analyses are what we get when using the constitutive relations indicated in equation (3.33).

- the mass momentum is given by

$$p_4 = If_4 = Iv = I\dot{x} \qquad (3.40)$$

Taking the derivative of this equation gives

$$\dot{p}_4 = I\dot{f}_4 = I\dot{v} = m\ddot{x} \qquad (3.41)$$

- the displacement of the mass m can be determined as follows

$$\dot{q}_2 = f_4 = \dot{x} \tag{3.42}$$

or

$$q_2 = \int \dot{x}\,dt = x \tag{3.43}$$

Now, substituting equations (3.40), (3.41), and (3.43) in equation (3.39) gives

$$m\ddot{x} + \frac{1}{C}x + \frac{b}{I}I\dot{x} = SE \tag{3.44}$$

Rearranging this equation and inserting $SE = F(t)$, $R = b$ and $1/C = k$ gives

$$m\ddot{x} + b\dot{x} + kx = F(t) \tag{3.45}$$

By comparing equation (3.45) with equation (2.147), it is found that Newton's law of motion applied to the system is satisfied in its Augmented Bond Graph model. This example shows that the system's equation of motion can be derived from the Bond Graph model without the requirement to create the system's free body diagram, which may be another significant benefit of using the Bond Graph as a modeling technique of dynamic systems.

Example (3.2):
For the torsional mechanical system shown in Figure 3.35, create the Augmented Bond Graph model and derive the equation of motion.

Solution:
The system is a single degree of freedom excited by an external torque of $T(t)$. It contains only a single flow variable: the angular velocity $\omega = \dot{\theta}$. The torque is represented by a source of effort (SE). The rotating mass m is represented by an angular moment of inertia J. The torsional spring is provided by a capacitor whose capacitance is $C_t = (1/k_t)$. A resistance of $R_t = b_t$ represents the torsional damper. All four components, SE, J, C_t, and R_t elements should be connected through a 1-junction. In this case, the sum of all efforts is zero at this junction, and all flow (angular velocities) is equal. Create a Bond Graph model by connecting these four elements to this 1-junction, as shown in Figure 3.36. Similar to example (3.1), indicate the half-arrows representing the positive directions of the power flow and assign the causality stork marks to all bonds following the same steps.

To derive the system's equation of motion, we must define the Key variables: Inputs, State, and Co-energy variables and their constitutive relations.

- Key Variables:
- Inputs: $e_1 = SE = T(t)$.
- State variables: q_2 and p_4.
- Co-energy (Power) variables: e_2 and f_4.

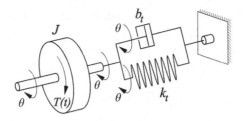

Figure 3.35 Torsional mass-spring-damper system

- Constitutive relations:

Write the constitutive relations that describe the relationship between co-energy variables and state variables for each energy storage element with integral causality as follows:

$$e_2 = \frac{1}{C_t}q_2 \text{ and } f_4 = \frac{1}{J}p_4 \tag{3.46}$$

where p_4 and q_2 are the angular momentum and the angular displacement of the mass, respectively. Now that we know the connections between flow and effort variables at the model junctions let's determine the equations that reflect them.

The flow and effort variables of the 1-junction of a common angular flow $\dot{\theta}$ are given by

$$f_1 = f_2 = f_3 = f_4 \tag{3.47}$$

$$e_1 = e_2 + e_3 + e_4 \tag{3.48}$$

As shown in Figure 3.36, the effort variables of the 1-junction are given by

$$e_4 = \dot{p}_4 = e_1(t) - e_2 - e_3 = T(t) - e_2 - e_3 \tag{3.49}$$

$$e_2 = \frac{1}{C_t}q_2 \tag{3.50}$$

$$e_3 = R_t f_3 = R_t f_4 \tag{3.51}$$

$$C_t : 1/k_t$$

$$SE: T(t) \xrightarrow{\ 1\ } \mathbf{1} \xrightarrow{\ 4\ } I:J$$

$$R_t : b_t$$

Figure 3.36 Augmented Bond Graph model

where $\dot{p}_4$ is the rate of change of the angular momentum of the mass. Note that the flow $\dot{q}_2$ of the capacitor C equals f_4 given by the constitutive relation indicated in equation (3.46). So, the state variable q_2 equals $\int f_4 dt$. Considering these, equations (3.50) and (3.51) become

$$e_2 = \frac{1}{C_t} \int \frac{1}{J} p_4 \, dt \tag{3.52}$$

$$e_3 = R_t f_3 = R_t f_4 = R_t \frac{1}{J} p_4 \tag{3.53}$$

Substituting equations (3.52) and (3.53) in equation (3.49) gives

$$\dot{p}_4 = SE - R_t \frac{1}{J} p_4 - \frac{1}{C_t} \int \frac{1}{J} p_4 \, dt \tag{3.54}$$

Rearranging this equation gives

$$\dot{p}_4 + R_t \frac{1}{J} p_4 + \frac{1}{C_t} \int \frac{1}{J} p_4 \, dt = SE \tag{3.55}$$

This equation is the equation of motion of the given system.

Let us check if Newton's law is embedded in the Augmented Bond Graph model shown in Figure (3.36). The following analyses are what we get when using the constitutive relations indicated in equation (3.46).

The momentum of the inertia element is given by

$$p_4 = J f_4 = J \dot{\theta} \tag{3.56}$$

Taking the derivative of this equation gives

$$\dot{p}_4 = J \dot{f}_4 = J \ddot{\theta} \tag{3.57}$$

Substituting equations (3.56) and (3.57) in equation (3.55) gives

$$J \ddot{\theta} + R_t \frac{1}{J} J \dot{\theta} + \frac{1}{C_t} \int \frac{1}{J} J \dot{\theta} \, dt = SE \tag{3.58}$$

Rearranging this equation, carrying out the integration term, and inserting $SE = T(t)$, $R_t = b_t$ and $1/C_t = k_t$ yields

$$J \ddot{\theta} + b_t \dot{\theta} + k_t \theta = T(t) \tag{3.59}$$

Comparing equation (3.59) with equation (2.150) indicates Newton's law of motion applied to the torsional system shown in Figure 3.35 is satisfied in the Augmented Bond Graph model shown in Figure 3.36.

Example (3.3):
For the system shown in Figure 3.37,

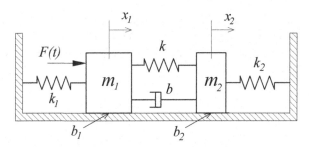

Figure 3.37 Schematic diagram of a Two-degree of freedom translational mechanical system

a. Create the preliminary Bond Graph model and impose it on the schematic diagram.
b. Simplify it and create the Augmented Bond Graph model.
c. Write the State and Co-energy variables and their constitutive relations, and then derive the equations of motion. Put the results in matrix form.
d. Check if Newton's law is embedded in the Augmented Bond Graph of the system.

Solution:

- Physical description of the system:
Figure 3.37 illustrates that the mass m_1 moves in the same direction as the external force, $F(t)$. This mass is linked to the left ends of the spring of stiffness k and the damper of a damping factor b, as well as the right end of the spring of stiffness k_1, implying that they should move in the same direction as the external force $F(t)$ and share the same flow v_1. Note that the external force $F(t)$, the forces coming from the spring of stiffness k and the damper at their left ends, and the force coming from the spring of stiffness k_1 at its right end all act on the mass m_1. On the other hand, the mass m_2 is linked to the right ends of the spring of stiffness k, and the damper as well as the left end of the spring of stiffness k_2 implies that they share the same flow $v_2 = \dot{x}_2$. Both masses are subjected to frictional contact generated between them and the ground, represented by damping forces with damping factors of b_1 and b_2, as shown in Figure 3.38. Note that the left end of the spring of stiffness k_1 and the right end of the spring of stiffness k_2 are connected to the ground. For now, we will assume that the ground generates absolute velocity $v = v_o$! The goal is to create a Bond Graph model that determines the velocity v_1 and displacement x_1 of the mass m_1, as well as the velocity v_2 and displacement x_2 of the mass m_2.
- Preliminary Bond Graph model:
In this example, we will illustrate how to generate the Bond Graph model by superimposing it on the schematic diagram of the given system. Because there are two masses in the system, two common flows 1-junctions have been imposed on both masses, each connected to the inertia elements I_1 and I_2, representing the two masses as shown in Figure 3.38. The inertia element I_1 represents the mass m_1 is connected

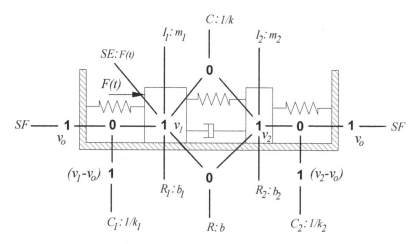

Figure 3.38 Preliminary Bond Graph model of the system imposed on its schematic diagram

to the 1-junction of a common velocity v_1, while the inertia element I_2 is connected to the 1-junction of a common velocity v_2. The external force $F(t)$ applied to the mass m_1 is represented by a source of effort SE connected to the 1-junction of a common velocity v_1. As previously stated, the ground can generally be represented in the preliminary Bond Graph model as either a source of effort SE or a source of flow SF. If we consider a source of effort SE, it implies that we will consider ground displacement! If we consider the ground as a source of flow SF, then it follows that we should similarly consider the ground's velocity! Because there are two connections with the ground on both the diagram's right and left sides, we will consider two flow sources SF, as shown in Figure 3.38. Because the spring of stiffness k_1 on the left side of the diagram is linked between the mass m_1 and the ground, its capacitor element C_1 should be connected to a 1-junction with a common velocity of $(v_1 - v_o)$, via a 0-junction that is positioned between the two 1-junctions that are already created, as illustrated in Figure 3.38. Similarly, the same argument applies to the capacitor C_2 of the spring of stiffness k_2 located on the diagram's right side. Since the left and right ends of the spring of stiffness k share the same effort, its capacitor element C is connected through a 0-junction of a common effort that is connected between the two 1-junctions of common velocities v_1 and v_2 which are already created, as shown in Figure 3.38. The same argument applies to the resistance R element representing the damper. Note that two resistance elements represent the friction (damping) forces generated between both masses, and the ground is considered in the model represented by R_1 and R_2, as shown in Figure 3.38. The resistance R_1 is located at the 1-junction of the common flow v_1 for the mass m_1, while the resistance R_2 is located at the 1-junction of the common flow v_2 for the mass m_2.

At this point, the preliminary Bond Graph model of the system is created and superimposed on the system schematic diagram, as shown in Figure 3.38. To

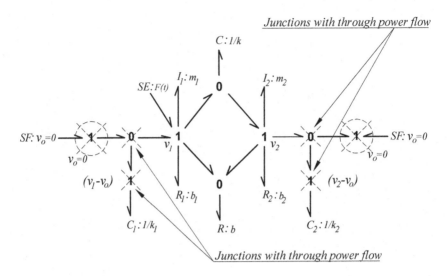

Figure 3.39   Preliminary Bond Graph model

indicate the direction of positive power flow, a half-arrow must be added to all bonds, as shown in Figure 3.39. It depends on the movement of the system's components and how they interact with one another, as described before. Remember that when we select the power direction, the sources supply power, while the inertia I, capacitors C, and resistances R receive power.

To simplify this Bond Graph model, all connections that hold ground velocity v_o with their associated bonds must be removed, and then any 2-port 1-junctions or 0-junctions with through power flow should be swapped with single bonds. As a result, sources of flow SF located on both sides of the preliminary Bond Graph model associated with the ground velocity v_o and their bonds should be removed, as shown in Figure 3.39. Now we can use an in-out bond to swap 2-port and 3-port variants of 0-junctions and 1-junctions connected with the capacitors and sources of flow, C_1, C_2 and SF on both sides of the Bond Graph model, as shown in Figure 3.40. In addition to that, the left and right ends of the spring of stiffness k and the damper of damping factor b share the same effort, implying that capacitor C and resistance R elements are connected to a 0-junction of a common effort as shown in Figure 3.40. Since both ends of this combination move, we are implying that they share the same flow and are connected to the 1-junction of common flow $(v_1 - v_2)$. The resulting simplified Bond Graph model of the given system is shown in Figure 3.40.

- Augmented Bond Graph model:
Now, let us assign the causality of the Bond Graph model. Since there is only one source of effort, SE, the causal stroke mark should be out of its bond, located on the junction side, as shown in Figure 3.40. Considering the causality restrictions in Section 3.3, apply the causal stork marks to all elements and junctions. It is essential to

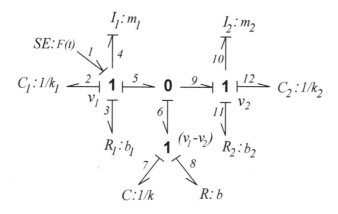

Figure 3.40   Augmented Bond Graph model

remember that as we work through assigning causality, we must consider the integral causality that applies to all associated elements and bonds. We are finally numbing all bonds to get the Augmented Bond Graph model of the system, as shown in Figure 3.40. This figure shows five storage elements of integral causality of bonds 2, 4, 7, 10, and 12.

- Equations of motion:
To derive the system's equations of motion from the Augmented Bond Graph model, we must first define the model's key variables: input, State, and Co-energy (power) variables and their constitutive relations.

- Key variables:
- Inputs: $e_1(t) = SE = F(t)$.
- State variables: q_2, p_4, q_7, p_{10} and q_{12}.
- Co-energy (power) variables: e_2, f_4, e_7, f_{10} and e_{12}.

- Constitutive relations:
Write the constitutive relations that describe the relationship between co-energy variables and state variables for each energy storage element with integral causality as follows:

$$e_2 = \frac{1}{C_1}q_2, \quad f_4 = \frac{1}{I_1}p_4, \quad e_7 = \frac{1}{C}q_7, \quad f_{10} = \frac{1}{I_2}p_{10}, \quad \text{and} \quad e_{12} = \frac{1}{C_2}q_{12} \qquad (3.60)$$

where p_4 and p_{10} are the momentum of the mass m_1 and m_2 respectively, while q_2, q_7 and q_{12} are the displacements of the capacitances C_1, C and C_2, respectively.
It is essential to keep in mind that resistances are elements that dissipate energy rather than storage elements. That being the case, there are no state variables, and the consequent constitutive relations have been considered. Now that we have the connections between flow and effort variables at the model junctions let us determine

the equations that reflect them.

- As shown in Figure 3.40, the flow variables of the 1-junction of a common velocity v_1 are given by

$$f_1 = f_2 = f_3 = f_4 = f_5 \qquad (3.61)$$

and the effort variables are given by

$$e_4 = \dot{p}_4 = e_1(t) - e_2 - e_3 - e_5 = SE - e_2 - e_3 - e_5 \qquad (3.62)$$

where $\dot{p}_4$ is the rate of change of the momentum of the mass m_1.
- The effort variables of the 0-junction are given:

$$e_5 = e_6 = e_9 \qquad (3.63)$$

and the flow variables are given by

$$f_5 = f_6 + f_9 \qquad (3.64)$$

- The flow variables of the 1-junction of a common velocity $(v_1 - v_2)$ is given by

$$f_6 = f_7 = f_8 \qquad (3.65)$$

and the effort variables are given by

$$e_6 = e_5 = e_9 = e_7 + e_8 \qquad (3.66)$$

- The flow variables of the 1-junction of a common velocity v_2 are given by

$$f_9 = f_{10} = f_{11} = f_{12} \qquad (3.67)$$

and the effort variables are given

$$e_{10} = \dot{p}_{10} = e_9 - e_{11} - e_{12} \qquad (3.68)$$

where $\dot{p}_{10}$ is the rate of change of the momentum of the mass m_2.

Now, let us derive the equations of motion of the given system.

- For mass m_1:
Substituting equation (3.66) in equation (3.62) gives

$$\dot{p}_4 = SE - e_2 - e_3 - (e_7 + e_8) \qquad (3.69)$$

This equation indicates that to determine the momentum change $\dot{p}_4$ of the mass m_1, we must first determine the efforts, e_2, e_3, e_7, and e_8 as follows.
The efforts e_2 and e_7 of the capacitances C_1 and C are given by the constitutive relation indicated in equation (3.60) which are

$$e_2 = \frac{1}{C_1} q_2 \text{ and } e_7 = \frac{1}{C} q_7 \qquad (3.70)$$

The effort e_3 of the friction damping of resistance R_1 that is generated between the mass m_1 and the ground is given by

$$e_3 = R_1 f_3 = R_1 f_4 \tag{3.71}$$

Considering the constitutive relation indicated in equation (3.60), equation (3.71) becomes

$$e_3 == R_1 f_4 = \frac{R_1}{I_1} p_4 \tag{3.72}$$

The effort e_8 of the damper of resistance R is given by

$$e_8 = R f_8 = R f_6 = R(f_5 - f_9) = R(f_4 - f_{10}) \tag{3.73}$$

Considering the constitutive relations indicated in equation (3.60), equation (3.73) becomes

$$e_8 = R\left(\frac{1}{I_1} p_4 - \frac{1}{I_2} p_{10}\right) \tag{3.74}$$

Considering the constitutive relations indicated in equation (3.60) and substituting equations (3.70), (3.72), and (3.74) in equation (3.69) gives

$$\dot{p}_4 = SE - \frac{1}{C_1} q_2 - \frac{R_1}{I_1} p_4 - \frac{1}{C} q_7 - R\left(\frac{1}{I_1} p_4 - \frac{1}{I_2} p_{10}\right) \tag{3.75}$$

Rearranging this equation gives

$$\dot{p}_4 + R\left(\frac{1}{I_1} p_4 - \frac{1}{I_2} p_{10}\right) + \frac{R_1}{I_1} p_4 + \frac{1}{C_1} q_2 + \frac{1}{C} q_7 = SE \tag{3.76}$$

Note that the displacement q_7 of the capacitance C is equal to the difference between $(q_2 - q_{12})$, which are the displacements of the capacitances C_1 and C_2, respectively. Considering this, equation (3.76) becomes

$$\dot{p}_4 + R\left(\frac{1}{I_1} p_4 - \frac{1}{I_2} p_{10}\right) + \frac{R_1}{I_1} p_4 + \frac{1}{C_1} q_2 + \frac{1}{C}(q_2 - q_{12}) = SE \tag{3.77}$$

This equation represents the equation of motion of the mass m_1.

- **For mass m_2:**
Substituting equation (3.66) in equation (3.68) gives

$$\dot{p}_{10} = (e_7 + e_8) - e_{11} - e_{12} \tag{3.78}$$

This equation indicates that to determine the momentum change $\dot{p}_{10}$ of the mass m_2, we must first determine the efforts, e_7, e_8, e_{11} and e_{12}. The effort e_7 is given by the constitutive relations indicated in equation equations (3.70), while e_8 is given by equation (3.74). The effort e_{11} of the damping friction of resistance R_2 that is generated between the mass m_2 and the ground is given by

$$e_{11} = R_2 f_{11} \tag{3.79}$$

Equation (3.67) indicated that the flow f_{11} is equal to f_{10}. Considering this and using the constitutive relations indicated in equation (3.60), equation (3.79) becomes

$$e_{11} = R_2 f_{10} = \frac{R_2}{I_2} p_{10} \tag{3.80}$$

Considering the constitutive relations, the effort e_{12} of the capacitance C_2 is given by

$$e_{12} = \frac{1}{C_2} q_{12} \tag{3.81}$$

Substituting equations (3.70), (3.74), (3.80), and (3.81) in equation (3.78) gives

$$\dot{p}_{10} = \frac{1}{C} q_7 + R(\frac{1}{I_2} p_4 - \frac{1}{I_1} p_{10}) - \frac{R_2}{I_2} p_{10} + \frac{1}{C_2} q_{12} \tag{3.82}$$

Rearranging this equation yields

$$\dot{p}_{10} + R(\frac{1}{I_2} p_{10} - \frac{1}{I_1} p_4) + \frac{R_2}{I_2} p_{10} - \frac{1}{C} q_7 + \frac{1}{C_2} q_{12} = 0 \tag{3.83}$$

As mentioned previously, the displacement q_7 of the capacitance C is equal to the difference between $(q_2 - q_{12})$, which are the displacements of the capacitances C_1 and C_2, respectively. Considering this, equation (3.83) becomes

$$\dot{p}_{10} + R(\frac{1}{I_1} p_4 - \frac{1}{I_2} p_{10}) + \frac{R_2}{I_2} p_{10} + \frac{1}{C_2} q_{12} + \frac{1}{C}(q_{12} - q_2) = 0 \tag{3.84}$$

This equation represents the equation of motion of the mass m_2.

- In state matrix form:
Rearranging equations (3.77) and (3.84) yields

$$\dot{p}_4 = SE - (\frac{R}{I_1} + \frac{R_1}{I_1}) p_4 + \frac{R}{I_2} p_{10} - (\frac{1}{C} + \frac{1}{C_1}) q_2 + \frac{1}{C} q_{12} \tag{3.85}$$

$$\dot{p}_{10} = -(\frac{R}{I_2} + \frac{R_2}{I_2}) p_{10} + \frac{R}{I_1} p_4 - (\frac{1}{C} + \frac{1}{C_2}) q_{12} + \frac{1}{C} q_2 \tag{3.86}$$

So, the state matrix form of equations (3.85) and (3.86) is given by

$$\begin{bmatrix} \dot{p}_4 \\ \dot{p}_{10} \end{bmatrix} = \begin{bmatrix} -(\frac{R}{I_1} + \frac{R_1}{I_1}) & \frac{R}{I_2} \\ \frac{R}{I_1} & -(\frac{R}{I_2} + \frac{R_2}{I_2}) \end{bmatrix} \begin{bmatrix} q_2 \\ q_{12} \end{bmatrix} +$$

$$\begin{bmatrix} -(\frac{1}{C} + \frac{1}{C_1}) & \frac{1}{C} \\ \frac{1}{C} & -(\frac{1}{C} + \frac{1}{C_2}) \end{bmatrix} \begin{bmatrix} q_2 \\ q_{12} \end{bmatrix} + \begin{bmatrix} SE \\ 0 \end{bmatrix}$$

- Check if Newton's law is embedded in the Augmented Bond Graph model:
Now, we will check if Newton's law of motion is embedded in the Augmented Bond
Graph model shown in Figure 3.40. The following analyses are what we get when
using the constitutive relations indicated in equation (3.60).

Note that the flows f_2, f_7, and f_{12} of the capacitances C_1, C, and C_2 are equal to
their velocities $\dot{q}_2$, $\dot{q}_7$, and $\dot{q}_{12}$, respectively.

- The momentum of the mass m_1 is given by

$$p_4 = I_1 f_2 = I_1 v_1 = I_1 \dot{x}_1 \tag{3.87}$$

Taking the derivative of this equation gives

$$\dot{p}_4 = I_1 \dot{v}_1 = m_1 \ddot{x}_1 \tag{3.88}$$

- The momentum of the mass m_2 is given by

$$p_{10} = I_2 f_{10} = I_2 v_2 = I_2 \dot{x}_2 \tag{3.89}$$

Taking the derivative of this equation gives

$$\dot{p}_{10} = I_2 \dot{v}_2 = m_2 \ddot{x}_2 \tag{3.90}$$

- Since only the right end of the spring of stiffness k_1 is moving, its flow (velocity)
$\dot{q}_2$ is given by

$$\dot{q}_2 = f_2 = f_4 \tag{3.91}$$

Considering the constitutive relations, equation (3.91) becomes

$$\dot{q}_2 = \frac{1}{I_2} p_4 \tag{3.92}$$

To determine the displacement q_2 of this spring, take the integration of equation
(3.92) as follows.

$$q_2 = \int \frac{1}{I_2} p_4 dt \tag{3.93}$$

- Since both ends of the spring of stiffness k are moving, the flow (relative velocity)
f_7, which is equal to f_6 as indicated in equation (3.64), is given by

$$\dot{q}_7 = f_7 = f_6 = (f_5 - f_9) \tag{3.94}$$

Equations (3.61) and (3.67) indicate that the flows f_5 and f_9 are equal to f_4 and
f_{10}, respectively. Considering these, equation (3.94) becomes

$$\dot{q}_7 = (f_4 - f_{10}) \tag{3.95}$$

Considering the constitutive relations indicated in equation (3.60), equation (3.95)
becomes

$$\dot{q}_7 = (\frac{1}{I_1} p_4 - \frac{1}{I_2} p_{10}) \tag{3.96}$$

To determine the displacement q_7 of the spring of stiffness k, take the integration of equation (3.96) as follows.

$$q_7 = \int (\frac{1}{I_1} p_4 - \frac{1}{I_2} p_{10}) dt \tag{3.97}$$

- Since only the left end of the spring of stiffness k_2 is moving, its flow (velocity) $\dot{q}_{12}$ is given by

$$\dot{q}_{12} = f_{10} \tag{3.98}$$

Considering the constitutive relation indicated in equation (3.60), equation (3.98) becomes

$$\dot{q}_{12} = \frac{1}{I_2} p_{10} \tag{3.99}$$

To determine this spring's displacement q_{12}, take the integration of equation (3.99) as follows.

$$q_{12} = \int \frac{1}{I_2} p_{10} dt \tag{3.100}$$

Substituting equations (3.93), (3.97), and (3.100) in equations (3.77) and (3.84) yields

$$\dot{p}_4 + R(\frac{1}{I_1} p_4 - \frac{1}{I_2} p_{10}) + \frac{R_1}{I_1} p_4 + \frac{1}{C} \int (\frac{1}{I_1} p_4 - \frac{1}{I_2} p_{10}) dt + \frac{1}{C_1} \int \frac{1}{I_1} p_4 dt = SE \tag{3.101}$$

and

$$\dot{p}_{10} + R(\frac{1}{I_1} p_{10} - \frac{1}{I_2} p_4) + \frac{R_2}{I_2} p_{10} + \frac{1}{C} \int (\frac{1}{I_1} p_{10} - \frac{1}{I_2} p_4) dt + \frac{1}{C_2} \int \frac{1}{I_2} p_{10} dt = 0 \tag{3.102}$$

Now, substituting equations (3.87) to (3.90) in equations (3.101) and (3.102) yields

$$m_1 \ddot{x}_1 + R(\frac{1}{I_1} I_1 \dot{x}_1 - \frac{1}{I_2} I_2 \dot{x}_2) + R_1 \frac{1}{I_1} I_1 \dot{x}_1 + \frac{1}{C} \int \frac{1}{I_1} I_1 \dot{x}_1 dt +$$
$$\frac{1}{C} \int (\frac{1}{I_1} I_1 \dot{x}_1 - \frac{1}{I_2} I_2 \dot{x}_2) dt = SE \tag{3.103}$$

and

$$m_2 \ddot{x}_2 + R(\frac{1}{I_2} I_2 \dot{x}_2 - \frac{1}{I_1} I_1 \dot{x}_1) + R_2 \frac{1}{I_2} I_2 \dot{x}_2 + \frac{1}{C} \int (\frac{1}{I_2} I_2 \dot{x}_2 - \frac{1}{I_1} I_1 \dot{x}_1) dt +$$
$$\frac{1}{C_2} \int (\frac{1}{I_2} I_2 \dot{x}_2 - \frac{1}{I_1} I_1 \dot{x}_1) dt = 0 \tag{3.104}$$

Rearrange equations (3.103) and (3.104) and inserting $SE = F(t)$, $R = b$, $R_1 = b_1$, $R_2 = b_2$, $1/C = k$, $1/C_1 = k_1$ and $1/C_2 = k_2$ yields

$$m_1 \ddot{x}_1 + b(\dot{x}_1 - \dot{x}_2) + b_1 \dot{x}_1 + k_1 \int \dot{x}_1 dt + k \int (\dot{x}_1 - \dot{x}_2) dt = F(t) \tag{3.105}$$

and

$$m_2\ddot{x}_2 + b(\dot{x}_2 - \dot{x}_1) + b_2\dot{x}_2 + k\int(\dot{x}_2 - \dot{x}_1)\,dt + k_2\int\dot{x}_2\,dt = 0 \qquad (3.106)$$

Now, carrying out the integration terms that are involved in equations (3,105) and (3.106) gives

$$m_1\ddot{x}_1 + b(\dot{x}_1 - \dot{x}_2) + b_1\dot{x}_1 + k(x_1 - x_2) + k_1x_1 = F(t) \qquad (3.107)$$

and

$$m_2\ddot{x}_2 + b(\dot{x}_2 - \dot{x}_1) + b_2\dot{x}_2 + k(x_2 - x_1) + k_2x_2 = 0 \qquad (3.108)$$

Comparing equations (3.107) and (3.108) with equations (2.162) and (2.165) indicated that Newton's law of motion applied to the system is satisfied in the Augmented Bond Graph model of the system shown in Figure 3.40. This example illustrates that the system's equations of motion derivation can be obtained from the Bond Graph model without creating the system's free-body diagram.

Example (3.4):
For the system shown in Figure 3.41, considering the bar is massless,

1 Create the preliminary Bond Graph model and impose it on the schematic diagram.
2 Simplify it and create the Augmented Bond Graph model.
3 Write the State and Co-energy variables and their constitutive relations.
4 Derive the equation(s) of motion.
5. Check if Newton's law is embedded in the Augmented Bond Graph model.

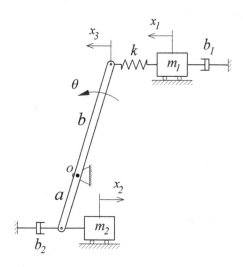

Figure 3.41 Mechanical system

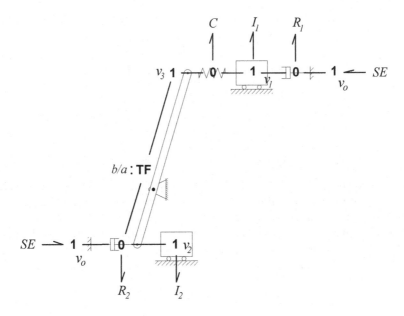

Figure 3.42 Preliminary Bond Graph model

SOLUTION:

Let us jump directly into the process of creating the Bond Graph model and superimposing it on the schematic diagram of the given system shown in Figure 3.41.
Because the system has two masses, two common flow 1-junctions have been imposed on both masses. The inertia element I_1 representing the mass m_1 is connected
to the 1-junction of a common flow v_1, while the inertia element I_2 representing the
mass m_2 is connected to the 1-junction of a common flow v_2 as shown in Figure 3.42.
However, the system is not subjected to external forces but is excited by moving the
lever to a small angle θ, as shown in Figure 3.42. Hence, there are no sources in the
system's Bond Graph model. Because both the right-upper and left-lower corners
of the system are connected to the ground, two sources of effort SE are considered.
Both are connected to two 1-junctions; each is of a common flow v_o as shown in Figure 3.42. Because the damper of the damping factor b_1 in the right-upper corner of
the diagram is linked between the mass m_1 and the ground, its resistance R_1 should
be connected to a 0-junction of a common effort that is positioned between the two
1-junctions that are already created, as shown in Figure 3.42. The same argument
applies to the damper of the damping factor b_2, located at the left lower corner of
the diagram, represented by the resistance R_2. Note that the lever is stationary at a
certain point, so the motion at its upper end is transferred to its lower end using a
transformer element TF of a transformation factor equal to (b/a), as shown in Figure 3.42. Because the left end of the spring of stiffness k is connected to the upper
end of the lever, its capacitor C is connected to a 0-junction of a common effort po-

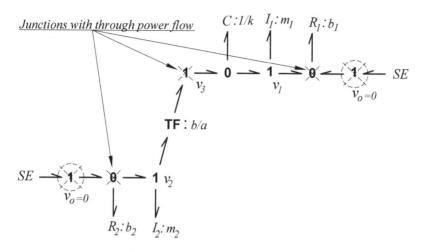

Figure 3.43 Simplification of the Bond Graph model

sitioned between the 1-junction of a common flow v_1 that has been already created and a 1-junction that should be created for the common velocity v_3 of the upper end of the lever, as shown in Figure 3.42.

At this point, the preliminary Bond Graph model of the system is created and superimposed on the system schematic diagram, as shown in Figure 3.43. In order to indicate the direction of positive power flow, a half-arrow must be added to all bonds, as shown in the same Figure 3.43. As previously stated, it depends on the movement of the system's components and how they interact with one another, as described before. Remember that when we select the power direction, the sources supply power, while the inertia I, capacitors C, and resistances R receive power. In order to simplify this Bond Graph model, all connections that hold ground velocity v_o with their associated bonds must be removed, and then any 2-port 1-junctions or 0-junctions with through power half arrows should be swapped with single bonds. As a result, we should remove both sources of effort SE and their bonds located on the right and left corners of the Bond Graph model associated with the ground velocity v_o and their bonds, as shown in Figure 3.43. Since the velocity of the upper end of the lever is transmitted to the lower end of the lever, the 1-junction of the common velocity v_3 is removed. We can now utilize an in-out bond to swap 2-port variants, 1-junctions linked to ground, and the 0-junctions associated with them. Considering the causality restrictions stated in Section 3.3, apply the causal stork marks to all elements and junctions. It is essential to remember that as we work through the process of assigning causality, we need to consider the integral causality that applies to all junctions and associated elements. We are finally numbing all bonds to get the Augmented Bond Graph model of the system, as shown in Figure 3.44. This figure shows three storage elements of integral causality of bonds 1, 4, and 8.

To derive the equations of motion of the system from the Augmented Bond Graph model, we must first define the key variables of the model; Inputs, State variables, and Co-energy (power) variables and their constitutive relations.

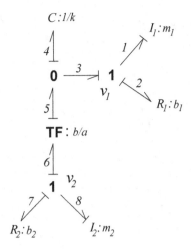

Figure 3.44 Augmented Bond Graph model

- Key variables
- Inputs: none
- State variables: p_1, q_4, p_8.
- Co-energy (power) variables: f_1, e_4, f_8.

- Constitutive relations:
Write the constitutive relations that describe the relationship between co-energy variables and state variables for each energy storage element with integral causality as follows.

$$f_1 = \frac{1}{I_1} p_1, \quad e_4 = \frac{1}{C} q_4 \quad \text{and} \quad f_8 = \frac{1}{I_2} p_8 \tag{3.109}$$

where p_1 and p_8 are the momentum of the mass m_1 and m_2 respectively, while q_4 is the displacement of the spring of stiffness k. It is essential to keep in mind that resistances are elements that dissipate energy rather than storage elements. Therefore, no state variables exist, and consequent constitutive relations have been considered for resistance elements.

Now that we have the connections of the flow and effort variables of the model junctions let's determine the equations that reflect them.

- As shown in Figure 3.44, the flow variables of the 1-junction of a common velocity v_1 are given by

$$f_1 = f_2 = f_3 \tag{3.110}$$

and the effort variables are given by

$$e_3 = e_1 + e_2 \quad \text{or} \quad e_1 = \dot{p}_1 = e_3 - e_2 \tag{3.111}$$

where $\dot{p}_1$ is the rate of change of the momentum of the mass m_1.

- The effort variables of the 0-junction are given,

$$e_3 = e_4 = e_5 \tag{3.112}$$

and the flow variables are given by

$$f_5 = f_3 + f_4 \tag{3.113}$$

- As stated above, the motion of the upper end of the lever is transmitted to its lower end through a transformer element TF of a transformation factor of (b/a). Since the effort exerted on one side of the transformer element, TF relates to the effort on the other side, and vice versa, the flow and effort variables of the transformer TF are given by

$$f_5 = \frac{b}{a} f_6 \quad \text{and} \quad e_6 = \frac{b}{a} e_5 \tag{3.114}$$

Equation (3.114) indicates that the power balance $(e_5 f_5 = e_6 f_6)$ at both sides of the transformer is satisfied.

- The flow variables of the 1-junction of a common velocity v_2 are given by

$$f_6 = f_7 = f_8 \tag{3.115}$$

and the effort variables are given

$$e_6 + e_7 + e_8 = 0 \quad \text{or} \quad e_8 = \dot{p}_8 = -e_6 - e_7 \tag{3.116}$$

where $\dot{p}_8$ is the rate of change of the momentum of the mass m_2.

Now, let us derive the equations of motion of the given system.

- For mass m_1:

Considering equation (3.111), to determine the momentum change $\dot{p}_1$ of the mass m_1, we must first determine the efforts, e_2 and e_3 as follows. The effort e_2 of the damper of resistance R_1 is given by

$$e_2 = R_1 f_2 \tag{3.117}$$

Equation (3.110) indicates that the flow f_2 equals f_1. Considering this and using the constitutive relations indicated in equation (3.109), equation (3.117) becomes

$$e_2 = R_1 f_1 = \frac{R_1}{I_1} p_1 \tag{3.118}$$

Equation (3.112) indicates that the effort e_3 equals e_4. Considering this, using the constitutive relations indicated in equation (3.109), the effort e_3 is given by

$$e_3 = e_4 = \frac{1}{C} q_4 \tag{3.119}$$

Substituting equations (3.118) and (3.119) in equation (3.111) gives

$$\dot{p}_1 = \frac{1}{C}q_4 - \frac{R_1}{I_1}p_1 \qquad (3.120)$$

Rearranging this equation yields

$$\dot{p}_1 + \frac{R_1}{I_1}p_1 - \frac{1}{C}q_4 = 0 \qquad (3.121)$$

This equation represents the equation of motion of the mass m_1.

- **For mass m_2:**
Considering equation (3.116), to determine the momentum change $\dot{p}_8$ of the mass m_2, we must first determine the efforts, e_6 and e_7. Considering the constitutive relations and using transformer relations indicated in equation (3.114), the effort e_6 of the transformer is given by

$$e_6 = \frac{b}{a}e_5 = \frac{b}{a}e_4 = \frac{b}{a}\frac{1}{C}q_4 \qquad (3.122)$$

As indicated in equation (3.113), the flow $f_4 = \dot{q}_4$ of the capacitance C is given by

$$\dot{q}_4 = f_4 = (f_5 - f_3) \qquad (3.123)$$

The effort e_7 of the damper of resistance R_2 is given by

$$e_7 = R_2 f_7 \qquad (3.124)$$

Equation (3.115) indicates that the flow f_7 equals f_8. Considering this and using the constitutive relations indicated in equation (3.109), equation (3.124) becomes

$$e_7 = R_2 f_8 = \frac{R_2}{I_2}p_8 \qquad (3.125)$$

Substituting equations (3.122) and (3.125) in equation (3.116) yields

$$\dot{p}_8 = \frac{-b}{a}q_4 - \frac{R_2}{I_2}p_8 \qquad (3.126)$$

Rearranging this equation yields

$$\dot{p}_8 + \frac{R_2}{I_2}p_8 + \frac{b}{a}\frac{1}{C}q_4 = 0 \qquad (3.127)$$

This equation represents the equation of motion of the mass m_2. Now, we will check if Newton's law of motion is embedded in the Augmented Bond Graph model shown in Figure 3.44. The following analyses are what we get when using the constitutive relations indicated in equation (3.109).

- The momentum of the mass of m_1 is given by

$$p_1 = I_1 f_1 = I_1 v_1 = I_1 \dot{x}_1 \tag{3.128}$$

Taking the derivative of this equation gives

$$\dot{p}_1 = I_1 \dot{v}_1 = m_1 \ddot{x}_1 \tag{3.129}$$

- The momentum of the mass of m_2 is given by

$$p_8 = I_2 f_8 = I_2 v_2 = I_2 \dot{x}_2 \tag{3.130}$$

Taking the derivative of this equation gives

$$\dot{p}_8 = I_2 \dot{v}_2 = m_2 \ddot{x}_2 \tag{3.131}$$

- Since both ends of the spring of stiffness k are moving, its velocity is given by equation (3.123). Equation (3.110) indicates that the flow f_3 equals f_1. Considering this, equation (3.123) becomes

$$\dot{q}_4 = (f_5 - f_1) \tag{3.132}$$

Substituting equation (3.114) in equation (3.132) gives

$$\dot{q}_4 = (\frac{b}{a} f_6 - f_1) \tag{3.133}$$

Equation (3.115) indicated that f_6 equals f_8. Considering this and using the constitutive relations indicated in equation (3.109), equation (3.133) becomes

$$\dot{q}_4 = (\frac{b}{a} f_8 - f_1) = (\frac{b}{a} \frac{1}{I_2} p_8 - \frac{1}{I_1} p_1) \tag{3.134}$$

To determine the displacement q_4 of the spring, take the integration of equation (3.134) as follows.

$$q_4 = \int (\frac{b}{a} \frac{1}{I_2} p_8 - \frac{1}{I_1} p_1) dt \tag{3.135}$$

Now, substituting equations (3.28) to (3.131) and (3.135) in equations (3.121) and (3.127) gives

$$m_1 \ddot{x}_1 + R_1 \frac{1}{I_1} I_1 \dot{x}_1 + \frac{1}{C} \int (\frac{1}{I_1} I_1 \dot{x}_1 - \frac{b}{a} \frac{1}{I_2} I_2 \dot{x}_1) dt = 0 \tag{3.136}$$

and

$$m_2 \ddot{x}_2 + R_2 \frac{1}{I_2} I_2 \dot{x}_2 + \frac{b}{a} \frac{1}{C} \int (\frac{b}{a} \frac{1}{I_2} I_2 \dot{x}_2 - \frac{1}{I_1} I_1 \dot{x}_1) dt = 0 \tag{3.137}$$

Rearranging equations (3.136) and (3.137) and inserting $R_1 = b_1$, $R_2 = b_2$, and $1/C = k$ yields

$$m_1 \ddot{x}_1 + b_1 \dot{x}_1 + k \int (\dot{x}_1 - \frac{b}{a} \dot{x}_2) dt = 0 \tag{3.138}$$

and

$$m_2\ddot{x}_2 + b_2\dot{x}_2 + \frac{b}{a}k \int \left(\frac{b}{a}\dot{x}_2 - \dot{x}_1\right) dt = 0 \tag{3.139}$$

Carrying out the integration terms that are involved in equations (3.138) and (3.139) yields

$$m_1\ddot{x}_1 + b_1\dot{x}_1 + k\left(x_1 - \frac{b}{a}x_2\right) = 0 \tag{3.140}$$

and

$$m_2\ddot{x}_2 + b_2\dot{x}_2 + k\frac{b}{a}\left(\frac{b}{a}x_2 - x_1\right) = 0 \tag{3.141}$$

Comparing equations (3.140) and (3.141) with equations (2.196) and (2.199) indicated that they are identical, implying that Newton's law of motion applied to the system is satisfied in the Augmented Bond Graph model of the system shown in Figure 3.44. As stated previously, this example shows that the system's equations of motion can be derived from the Bond Graph model without creating the system's free-body diagram.

Example (3.5):
For the system shown in Figure (3.45), create the Augmented Bond Graph model and derive the equations of motion considering the following two movement possibilities of the mass m_2.

1 When m_2 moves in the same direction as the mass m_1.
2 When m_2 moves in the opposite direction of mass m_1.

Solution
This example will demonstrate the implication of the direction of half-arrow positive power flow selections. In other words, we will study the effect of changing the direction of the positive power flow, pointing to the equations of motion for the system.
The system contains two masses, m_1 and m_2, linked by a spring-damper combination. The external force $F(t)$ is applied to the mass m_1, implying that both are always moving in the same direction.

1. When m_2 moves in the same direction as the mass m_1
As shown in Figure 3.45, the mass m_1 moves in the same direction as the external force $F(t)$ and the left ends of the spring-damper combination, whereas the mass m_2

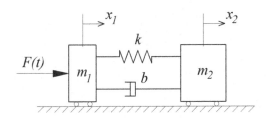

Figure 3.45 Mechanical system

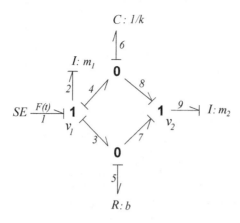

Figure 3.46 First Bond Graph model of the system

move with the right ends of the spring-damper combination in the same direction. This means that the mass m_1 with the left ends of the spring-damper combination shares the same flow v_1, while the mass m_2 with the right ends of the spring-damper combination shares the same flow v_2. Since both masses share different flows, each inertia element should be connected to a 1-junction, as shown in Figure 3.46. The external force $F(t)$ is represented by a source of effort SE located at the 1-junction of common velocity v_1. Since the left and right ends of the spring share the same effort, a capacitor element C, representing the spring, is connected through a 0-junction of a common effort. The same argument applies to the R element. The half-arrow positive power flow directions are chosen, as shown in Figure 3.46. Since both masses move in the same direction, the half-arrow power directions of bonds 7 and 8 should be towards the 1-junction of a common velocity v_2. At the same time, the direction of the half-arrow of bonds 3 and 4 towards both 0-junctions. Considering the causality restrictions, apply the causal stork marks to all elements and junctions. Since there is only one source of effort, SE, the power flow and the causal stroke causality mark should be out of their bond, located on the 1-junction side of a common velocity v_1. Remember that during the assignment process of causality, we should consider the integral causality of all elements and associated junctions. We are finally numbing all bonds to get the Augmented Bond Graph of the given system shown in Figure 3.46. This figure shows three storage elements of integral causality of bonds 2, 6, and 9.

To derive the equations of motion from the Augmented Bond Graph model, we must first define the key variables: Inputs, State variables, and Co-energy variables and their constitutive relations.

- Key variables
- Inputs: $e_1 = SE = F(t)$.
- State variables: p_2, q_6, and p_9.
- Co-energy (power) variables: f_2, e_6, and f_9.

- **Constitutive relations:**
Write the constitutive relations that describe the relationship between co-energy variables and state variables for each energy storage element with integral causality.

$$f_2 = \frac{1}{I_1}p_2, \ e_6 = \frac{1}{C}q_6 \ \text{and} \ f_9 = \frac{1}{I_2}p_9 \qquad (3.142)$$

where p_2 and p_9 are the momentum of the mass m_1 and m_2 respectively, while q_6 is the displacement of the spring of stiffness k. As stated previously, resistances dissipate energy rather than storage elements. Therefore, no state variables exist, and consequent constitutive relations have been considered for resistance elements.

Now that we have the connections of the model junctions' flow and effort variables let us determine the equations that reflect them.

- As shown in Figure 3.46, the flow variables of the 1-junction of a common flow v_1 are given by

$$f_1 = f_2 = f_3 = f_4 \qquad (3.143)$$

and the effort variables are given by

$$e_2 = \dot{p}_2 = e_1 - e_3 - e_4 = SE - e_3 - e_4 \qquad (3.144)$$

where $\dot{p}_2$ is the rate of change of the momentum of the mass m_1.
- The effort variables of the 0-junction of a common effort that is connected to the damper are given by

$$e_3 = e_5 = e_7 \qquad (3.145)$$

and the flow variables are given by

$$f_3 = f_5 + f_7 \qquad (3.146)$$

- The effort of the 0-junction of a common effort that is connected to the spring is given by

$$e_4 = e_6 = e_8 \qquad (3.147)$$

and the flow variables are given by

$$f_4 = f_6 + f_8 \qquad (3.148)$$

- The flow variables of the 1-junction of a common flow v_2 are given by

$$f_7 = f_8 = f_9 \qquad (3.149)$$

and the effort variables are given by

$$e_9 = \dot{p}_9 = e_7 + e_8 \qquad (3.150)$$

where $\dot{p}_9$ is the rate of change of the momentum of the second mass m_2.

Now, let us derive the equations of motion of the given system.

- For mass m_1:
Considering equation (3.144), to determine the momentum change $\dot{p}_2$ of the mass m_1, we must first determine the efforts e_3 and e_4 as follows. Equations (3.145) and (3.147) indicated that the efforts e_3 and e_4 equal the efforts e_5 and e_6, respectively. Considering these, equation (3.144) becomes

$$\dot{p}_2 = SE - e_5 - e_6 \tag{3.151}$$

The effort e_5 of the damper of resistance R is given by

$$e_5 = Rf_5 \tag{3.152}$$

Equation (3.146) indicates that the flow f_5 equals $(f_3 - f_7)$. Considering this, equation (3.152) becomes

$$e_5 = Rf_5 = R(f_3 - f_7) \tag{3.153}$$

Equations (3.143) and (3.149) indicate that the flows f_3 and f_7 equal the flows f_2 and f_9, respectively. Considering these, the effort e_5 is given by

$$e_5 = R(f_2 - f_9) \tag{3.154}$$

Using the constitutive relations indicated in equation (3.142), equation (3.154) becomes

$$e_5 = R(\frac{1}{I_1} p_2 - \frac{1}{I_2} p_9) \tag{3.155}$$

The effort e_6 of the capacitance C is given by the constitutive relations indicated in equation (3.142), which is

$$e_6 = \frac{1}{C} q_6 \tag{3.156}$$

Since both sides of the spring are moving, implying that they share different flows, which are f_4 and f_8, as indicated in equation (3.148). So, its relative flow $\dot{q}_6$ is given by

$$\dot{q}_6 = f_6 = (f_4 - f_8) \tag{3.157}$$

Equations (3.143) and (3.149) indicated that the flows f_4 and f_8 equal the flows f_2 and f_9, respectively. Considering this, equation (3.157) becomes

$$\dot{q}_6 = (f_2 - f_9) \tag{3.158}$$

To determine the displacement q_6 of the capacitance C, take the integration of equation (3.158). Note that the flows f_2 and f_9 are given by the constitutive relations indicated in equation (3.142). Now, substituting equations (3.155) and (3.156) in equation (3.151) gives

$$\dot{p}_2 = SE - R(\frac{1}{I_1} p_2 - \frac{1}{I_2} p_9) - \frac{1}{C} q_6 \tag{3.159}$$

Rearranging this equation yields

$$\dot{p}_2 + R(\frac{1}{I_1} p_2 - \frac{1}{I_2} p_9) + \frac{1}{C} q_6 = SE \qquad (3.160)$$

This equation represents the equation of motion of the mass m_1 generated from the first Bond Graph model shown in Figure 3.46.

- For mass m_2:
Considering equation (3.150), to determine the momentum change $\dot{p}_9$ of the mass m_2, we must first determine the efforts e_7 and e_8 as follows. Equations (3.145) and (3.147) indicate that the effort e_7 and e_8 equal the effort e_5 and e_6, respectively. Considering these, equation (3.150) becomes

$$\dot{p}_9 = e_7 + e_8 = e_5 + e_6 \qquad (3.161)$$

Substituting equations (3.155) and (3.156) in equation (3.161) gives

$$\dot{p}_9 = R(\frac{1}{I_1} p_2 - \frac{1}{I_2} p_9) + \frac{1}{C} q_6 \qquad (3.162)$$

Rearrange this equation gives

$$\dot{p}_9 + R(\frac{1}{I_2} p_9 - \frac{1}{I_1} p_2) - \frac{1}{C} q_6 = 0 \qquad (3.163)$$

This equation represents the equation of motion of the mass m_2 generated from the first Bond Graph model shown in Figure 3.46.

2. When m_2 moves in the opposite direction of mass m_1
Let us assume that the mass m_2 moves in the opposite direction as the mass m_1. So, the half-arrow directions are chosen, as shown in Figure 3.47. Since the mass m_2 moves in the opposite direction as m_1, the positive power flow direction of bonds 7 and 8 will be towards the two 0-junctions, implying that the direction of bonds 3 and 4 must be towards the 1-junction of a common velocity v_1. As a result, the directions of the positive power flow of bonds 3, 4, 7, and 8 in the second Bond Graph are opposite to those in the first Bond Graph, as shown clearly in Figures 3.46 and 3.47.

- For mass m_1:
As shown in Figure 3.47, the effort variables of the common velocity v_1 are given by

$$\dot{p}_2 = e_2 = e_1 + e_4 + e_3 \qquad (3.164)$$

Using equations (3.145) and (3.147), equation (3.164) becomes

$$\dot{p}_2 = SE + e_5 + e_6 \qquad (3.165)$$

Using equations (3.143) and (3.149), the effort e_5 of the damper of resistance R is given by

$$e_5 = R f_5 = R(f_7 - f_3) = R(f_9 - f_2) \qquad (3.166)$$

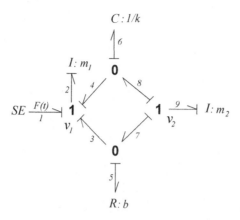

Figure 3.47   Second Bond Graph model of the system

Using the constitutive relations indicated in equation (3.142), equation (3.166) becomes

$$e_5 = R(\frac{1}{I_2} p_9 - \frac{1}{I_1} p_2) \tag{3.167}$$

Since both sides of the spring are moving, implying that they share different flows. So, its relative velocity $\dot{q}_6$ is given by

$$\dot{q}_6 = (f_8 - f_4) \tag{3.168}$$

Using equations (3.143) and (3.149), equation (3.168) becomes

$$\dot{q}_6 = (f_9 - f_2) \tag{3.169}$$

To determine the displacement, take the integration of this equation. Comparing equations (3.169) and (3.158), we noted that the direction of the power flow causes a change in the sign of the shared flows f_2 and f_9. This means that when assuming that the mass m_2 moves in the opposite direction of the mass m_1, changing the signs of both flows f_7 and f_9. Using the constitutive relations indicated in equation (3.142) and substituting equation (3.167) in equation (3.165) gives

$$\dot{p}_2 = SE + R(\frac{1}{I_2} p_9 - \frac{1}{I_1} p_2) + \frac{1}{C} q_6 \tag{3.170}$$

Rearranging this equation yields

$$\dot{p}_2 + R(\frac{1}{I_1} p_2 - \frac{1}{I_2} p_9) - \frac{1}{C} q_6 = SE \tag{3.171}$$

This equation represents the equation of motion of the mass m_1 generated from the second Bond Graph model shown in Figure 3.47.

- For mass m_2:
As shown in Figure 3.47, the effort variables of the common velocity v_2 are given by

$$\dot{p}_9 = e_9 = -e_7 - e_8 \tag{3.172}$$

Using equations (3.145) and (3.147), equation (3.172) becomes

$$\dot{p}_9 = -e_5 - e_6 \tag{3.173}$$

The efforts e_5 and e_6 are given by equations (3.167) and (3.142), respectively. Substituting these equations in equation (3.173) gives

$$\dot{p}_9 = -\frac{1}{C}q_6 - R(\frac{1}{I_2}p_9 - \frac{1}{I_1}p_2) \tag{3.174}$$

Rearranging this equation yields

$$\dot{p}_9 + R(\frac{1}{I_2}p_9 - \frac{1}{I_1}p_2) + \frac{1}{C}q_6 = 0 \tag{3.175}$$

This equation represents the equation of motion of the mass m_2 generated from the second Bond Graph model shown in Figure 3.47.

From this example, it is to be noted that the order of term $\dot{q}_6$ indicated in equations (3.158) and (3.169) varies depending on half-arrow direction choice. As this changes, sign on $\frac{1}{C}q_6$ term changes appropriately. When the system equations are derived using Newton's law, this situation will show up when the free-body diagrams of both masses are considered.

Finally, when comparing equation (3.160) with equation (3.171) and comparing equation (3.163) with equation (3.175), the equations that the two different Bond Graph models generated provide identical results. Therefore, altering the direction in which the mass m_2 moves does not affect the equations of motion that govern the system.

Example (3.6):
To understand more about how to derive the equations of motion from a Bond Graph model, we will consider the Bond Graph model of a specific system shown in Figure 3.48. Assuming all elements are linear,

1. Create the Augmented Bond Graph model (add the half-arrow positive power flow directions, assign causality, and number all bonds).
2. Derive the equations of motion and put the results in state matrix form.

Solution
Figure 3.48 shows that the system includes two inputs, SE_1 and SE_2, and two energy storage elements, I_1 and I_2. The system also includes a dissipation element R. Considering the causality restrictions, apply the causal stroke marks to all elements and

Figure 3.48 Bond Graph model

junctions. By numbering all bonds, we get the Augmented Bond Graph model shown in Figure 3.49. To derive the system's equations of motion from the Augmented Bond Graph model, we must first define the key variables and their constitutive relations.

- Key variables
- Inputs: $e_1 = SE_1$ and $e_6 = SE_2$.
- State variables: p_2 and p_7.
- Co-energy (power) variables: f_2 and f_7.

- Constitutive relations
Write the constitutive relations that describe the relationship between co-energy variables and state variables for each energy storage element with integral causality

$$f_2 = \frac{1}{I_1}p_2 \text{ and } f_7 = \frac{1}{I_2}p_7 \tag{3.176}$$

where p_2 and p_7 are the momentum of inertia elements I_1 and I_2, respectively. As stated previously, resistances dissipate energy rather than storage elements. Therefore, no state variables exist, and consequent constitutive relations have been considered for resistance elements.

Now that we have the connections between the flow and effort variables of the model junctions, let us determine the equations that reflect them.

- As shown in Figure 3.49, the flow variables of the 1-junction of a common flow f_2 are given by

$$f_1 = f_2 = f_3 \tag{3.177}$$

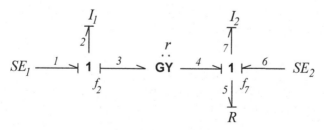

Figure 3.49 Augmented Bond Graph model of the system

and the effort variables are given by

$$e_1 = e_2 + e_3 \quad \text{or} \quad e_2 = \dot{p}_2 = e_1 - e_3 = SE_1 - e_3 \tag{3.178}$$

where $\dot{p}_2$ is the momentum change of the inertia I_1.
- Since the effort exerted on one side of the gyrator element GY relates to the flow on the other side and vice versa, and a sense of direction is not required, the effort variables of the Gyrator GY of a gyration factor r are given by

$$e_4 = rf_3 \quad \text{and} \quad e_3 = rf_4 \tag{3.179}$$

Equation (3.179) indicates that the power balance $(e_3 f_3 = e_4 f_4)$ at both sides of the Gyrator is satisfied.
- The flow variables of the 1-junction of common flow f_7 are given by

$$f_4 = f_5 = f_6 = f_7 \tag{3.180}$$

and the effort variables are given by

$$e_4 + e_6 = e_5 + e_7 \quad \text{or} \quad e_7 = \dot{p}_7 = e_4 - e_5 + e_6 = SE_2 + e_4 - e_5 \tag{3.181}$$

Now, let us derive the equations of motion of the given system.

- For inertia element I_1:
Considering equation (3.178), to determine the momentum change $\dot{p}_2$ of the inertia I_1, we must determine the efforts e_3. Considering the gyration relations indicated in equation (3.179), the effort e_3 is given by

$$e_3 = rf_4 \tag{3.182}$$

Equation (3.180) indicates that the flow f_4 equals the flow f_7. Considering this, equation (3.182) becomes

$$e_3 = rf_4 = rf_7 \tag{3.183}$$

Using the constitutive relations indicated in equation (3.176), equation (3.183) becomes

$$e_3 = \frac{r}{I_2} p_7 \tag{3.184}$$

Substituting this equation in equation (3.178) and rearranging it gives

$$\dot{p}_2 + \frac{r}{I_2} p_7 = SE_1 \tag{3.185}$$

This equation represents the equation of motion of the inertia element I_1.

- For inertia element I_2:
Considering equation (3.181), to determine the momentum change $\dot{p}_7$ of the inertia element I_2, we must determine the efforts e_4 and e_5. Considering the gyration relations indicated in equation (3.179), the effort e_4 is given by

$$e_4 = rf_3 \tag{3.186}$$

Equation (3.177) indicates that the flow f_3 equals the f_2. Considering this, equation (3.186) becomes

$$e_4 = rf_3 = rf_2 \tag{3.187}$$

Considering the constitutive relations indicated in equation (3.176), equation (3.187) becomes

$$e_4 = \frac{r}{I_1}p_2 \tag{3.188}$$

The effort e_5 of the resistance R is given by

$$e_5 = Rf_5 \tag{3.189}$$

Equation (3.180) indicated that f_5 equals f_7. Considering this and using the gyration relations indicated in equation (3.179), equation (3.189) becomes

$$e_5 = \frac{R}{I_2}p_7 \tag{3.190}$$

Substituting equations (3.188) and (3.190) in equation (3.181) and rearranging them gives

$$\dot{p}_7 + \frac{R}{I_2}p_7 - \frac{r}{I_1}p_2 = SE_2 \tag{3.191}$$

This equation represents the equation of motion of the inertia element I_2.

- In the state matrix form:
Rearranging equations (3.185) and (3.191) yields

$$\dot{p}_2 = SE_1 - \frac{r}{I_2}p_7 \tag{3.192}$$

$$\dot{p}_7 = SE_2 + \frac{r}{I_1}p_2 - \frac{R}{I_2}p_7 \tag{3.193}$$

So, the matrix form of equations (3.192) and (3.193) is given by

$$\begin{bmatrix} \dot{p}_2 \\ \dot{p}_7 \end{bmatrix} = \begin{bmatrix} 0 & \frac{-r}{I_2} \\ \frac{r}{I_1} & \frac{-R}{I_2} \end{bmatrix} \begin{bmatrix} p_2 \\ p_7 \end{bmatrix} + \begin{bmatrix} 1 & 0 \\ 0 & 1 \end{bmatrix} \begin{bmatrix} SE_1 \\ SE_2 \end{bmatrix}$$

3.11 DERIVATIVE CAUSALITY

As stated previously, the integral relationship should give the actual sequence of movements for energy storage elements. However, this might not be possible in some situations, and we might be forced to reverse the position of the integral causal stroke mark and apply it to the other end. The energy storage elements' reversed causality refers to *Derivative Causality* or *Differential Causality*.

Let us now investigate the conditions that must be met for derivative causality. Consider the mass-spring system shown in Figure 3.50a. It is applied to displacement input x_{in}. In this situation, we have respected the causality of the flow source

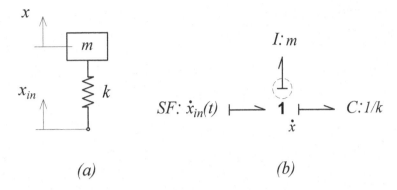

Figure 3.50 Mass-spring system

and the 1-junction, and it must be located at the source element side of the bond, as shown in the Bond Graph model Figure 3.50b. As a result, considering the causality restrictions, the inertia element I indicates derivative causality, as shown inside a dotted circle in Figure 3.50b. The question is: Where did we go wrong? We know that the spring element does not allow its displacement to change instantly, implying that it will continuously fluctuate and turn in the mass m. This means that if the input source dictated an instantaneous change, the induced force on the spring would be infinite. This taught us that a displacement source should not be connected in series with only the spring element under any circumstances.

3.12 EQUATIONS FORMULATION WHEN DERIVATIVE CAUSALITY OCCURS

If the system's Bond Graph model includes derivative causality, the generated equations of motion take the following form.

$$\dot{x}_i = f_{ni}(x_i, \dot{x}_d, u) \tag{3.194}$$

and

$$x_d = f_{nd}(x_i, u) \tag{3.195}$$

where, x_i are sets of the independent state variables (integral causality), x_d are sets of non-independent state variables (derivative causality), and u is the input. Taking the derivative of equation (3.195) gives

$$\dot{x}_d = f_{ndd}(x_i, \dot{x}_i, u, \dot{u}) \tag{3.196}$$

Substituting this equation in equation (3.194) gives

$$\dot{x}_i = f_{nii}(x_i, \dot{x}_i, u, \dot{u}) \tag{3.197}$$

This equation shows the implicit term $\dot{x}_i$ on the right-hand side. We have to move this term to the left side of this equation to get an explicit expression for $\dot{x}_i$ (this may not be possible for nonlinear equations).

When the Augmented Bond Graph exhibits derivative causality, the equations of motion can be derived by following the outlined steps.

1. Identify x_i and x_d variables:

    - x_i energy variables from the capacitor C and inertia I elements having integral causality.

    - x_d energy variables from the capacitor C and the inertia I elements having derivative causality.

2. Write the equations in the initial form as indicated in equations (3.194) and (3.195).
3. From equation (3.195), compute $\dot{x}_d$ as indicated in equation (3.196) and substituting it in equation (3.194) gives equation (3.197).
4. If possible, reduce equation (3.197) for $\dot{x}_i$ into explicit form as follows.

$$\dot{x}_i = f_n(x_i, u, \dot{u}) \tag{3.198}$$

In this case, the derived equations of motion of the system should not include the variable x_d of the energy storage element with a derivative causality, as indicated in equation (3.198).

Example (3.7) (Derivative Causality):
Consider the same system presented in Example (2.10), as shown in Figure 3.51,

1. Create the Augmented Bond Graph model, then determine which storage element has a derivative causality.
2. Derive the equation(s) of motion.
3. Create the Augmented Bond Graph model considering the following two cases and describe what happens in both cases to establish derivative causality:
    - the input force $F(t)$ is replaced by a velocity input $v(t)$.
    - when the snubber plate has a small mass.

Solution:
- Augmented Bond Graph:
This example illustrates systems that do not include independent energy storage elements that handle derivative causality. As shown in Figure 3.51, because the snubber plate and the left ends of both springs move in the direction that the external force $F(t)$ is acting, they share the same flow. Therefore, a 1-junction is created and connected to the source of effort SE representing the force $F(t)$ and two bonds representing the left sides of both springs, as shown in the Bond Graph model of the system shown in Figure 3.52. Hence, since the snubber plate does not have mass, no inertia element is considered in this 1-junction, as shown in the same figure. On the

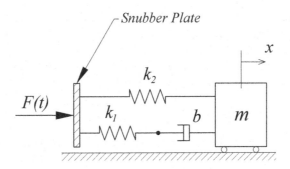

Figure 3.51   Mechanical system

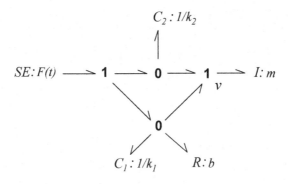

Figure 3.52   Bond Graph model of the system.

other hand, the mass m is linked to the right ends of the spring of stiffness k_2 and the damper of a damping factor b, implying that they share the same flow. Therefore, a 1-junction of a common flow $v = \dot{x}$ is created and connected to two bonds representing the right ends of the spring and the damper, as shown in Figure 3.52. Because the spring of stiffness k_2 is connected between the snubber plate and the mass m, its capacitor C_2 is connected to a 0-junction of a common effort positioned between the two 1-junctions that have already been created. Since the right end of the spring of stiffness, k_1, and the left end of the damper are linked, they should share the same effort, implying that both ends should be represented by a 0-junction of a common effort that is positioned between the two 1-junctions. As stated previously, the direction of the positive power flow depends on the internal interaction between the system elements. As a result, the half-arrow positive power flow directions are selected, as shown in Figure 3.52. Considering the causality restrictions, apply the causal stork marks to all elements and junctions. It is essential to remember that as we work through assigning causality, we must consider the integral causality that applies to all junctions and associated elements. As a result, it is found that the capacitor element C_1 has a derivative causality shown inside a dotted circle, as shown in Figure 3.53. We are finally numbing all bonds to get the Augmented Bond Graph model of the system shown in Figure 3.53.

- Equations of motion:
To derive the equations of motion for the system, considering that a Bond Graph model has derivative causality, we must specify the state variables for the energy storage elements of integral causality and derivative variables of the energy storage elements of derivative causality separately. Note that the energy storage element with derivative causality should be eliminated from the system's equations of motion because it is not a variable that can be considered independent of other energy storage variables.

- Key variables (integral causality):
- Inputs: $e_1 = SE = F(t)$
- State variables: q_6, p_9
- Power variables: e_6, f_9

- Constitutive relations:

$$e_6 = \frac{1}{C_2}q_6, \quad \text{and} \quad f_9 = \frac{1}{I}p_9 \tag{3.199}$$

- Key variables (derivative causality):
- Derivative variables: q_4
- Power variables: e_4

- Constitutive relations:

$$e_4 = \frac{1}{C_1}q_4 \tag{3.200}$$

As shown in Figure 3.53, the flow and effort variables of the 1-junction of the massless snubber plate element are given by

$$f_1 = f_2 = f_3 \tag{3.201}$$

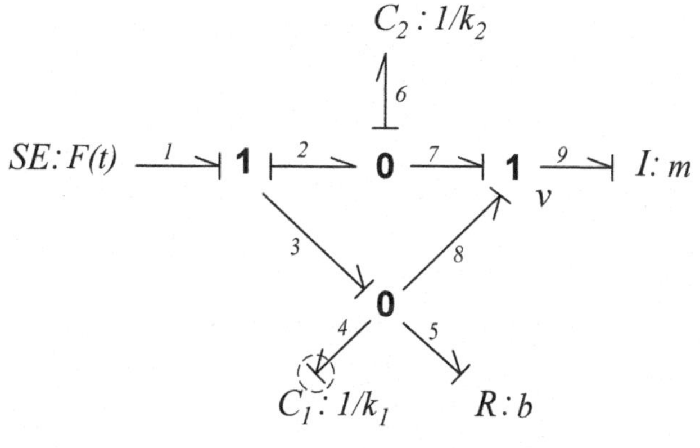

Figure 3.53 Augmented Bond Graph model of the system

and

$$e_1 = e_2 + e_3 \quad \text{or} \quad e_3 = e_1 - e_2 = SE - e_2 \tag{3.202}$$

The flow and effort variables of the 0-junction of a common effort that is connected to the capacitance C_2 are given by

$$f_2 = f_6 + f_7 \quad \text{or} \quad f_6 = f_2 - f_7 \tag{3.203}$$

and

$$e_2 = e_6 = e_7 \tag{3.204}$$

Using equation (3.204), equation (3.202) becomes

$$e_3 = SE - e_6 \tag{3.205}$$

Using equation (3.203), the flow f_6 of the capacitance C_2 is given by

$$f_6 = \dot{q}_6 = f_2 - f_7 \tag{3.206}$$

Using equation (3.201), equation (3.206) becomes

$$\dot{q}_6 = f_3 - f_7 \tag{3.207}$$

where $\dot{q}_6$ is the flow (velocity) of both mass m and capacitance C_2. The flow and effort variables of the 0-junction of a common effort that is connected to the capacitance C_1 and the resistance R are given by

$$f_3 = f_4 + f_5 + f_9 \tag{3.208}$$

and

$$e_3 = e_4 = e_5 = e_8 \tag{3.209}$$

The flow and effort variables of the 1-junction of a common flow v are given by

$$f_7 = f_8 = f_9 \tag{3.210}$$

and

$$e_9 = \dot{p}_9 = e_7 + e_8 = e_6 + e_3 = e_6 + (e_1 - e_6) = e_1 = SE \tag{3.211}$$

where $\dot{p}_9$ is the momentum change of the mass m. Using equations (3.208) and (3.210), equation (3.207) becomes

$$\dot{q}_6 = (f_4 + f_5 + f_9) - f_9 = f_4 + f_5 \tag{3.212}$$

The flow f_4 of the capacitance C_1, which has derivative causality, is given by

$$f_4 = \dot{q}_4 \tag{3.213}$$

To determine the flow $\dot{q}_4$, considering equation (3.209), we have

$$e_4 = e_3 \tag{3.214}$$

Substituting equations (3.200) and (3.205) in equation (3.114) gives

$$\frac{1}{C_1}q_4 = (SE - \frac{1}{C_2}q_6) \tag{3.215}$$

Rearranging this equation gives

$$q_4 = C_1 SE - \frac{C_1}{C_2}q_6 \tag{3.216}$$

Taking the derivative of this equation yields

$$\dot{q}_4 = C_1 \dot{SE} - \frac{C_1}{C_2}\dot{q}_6 \tag{3.217}$$

The flow f_5 of the damper of resistance R is given by

$$f_5 = \frac{1}{R}e_5 \tag{3.218}$$

Using equation (3.209), equation (3.218) becomes

$$f_5 = \frac{1}{R}e_3 \tag{3.219}$$

Substituting equation (3.205) in equation (3.219) gives

$$f_5 = \frac{1}{R}(SE - e_6) \tag{3.220}$$

Using the constitutive relation indicated in equation (3.199), equation (3.220) becomes

$$f_5 = \frac{1}{R}(SE - \frac{1}{C_2}q_6) \tag{3.221}$$

To determine the flow $\dot{q}_6$ of the capacitance C_2, substituting equations (3.213) and (3.221) in equation (3.212) gives

$$\dot{q}_6 = f_4 + f_5 = \dot{q}_4 + \frac{1}{R}(SE - \frac{1}{C_2}q_6) \tag{3.222}$$

Now, substituting equation (3.217) in equation (3.222) gives

$$\dot{q}_6 = C_1 \dot{SE} - \frac{C_1}{C_2}\dot{q}_6 + \frac{1}{R}(SE - \frac{1}{C_2}q_6) \tag{3.223}$$

Rearranging this equation yields

$$\dot{q}_6 = \frac{1}{(1 + \frac{C_1}{C_2})}[\frac{1}{R}(SE - \frac{1}{C_2}q_6) + C_1 \dot{SE}] \tag{3.224}$$

Note that the momentum changes $\dot{p}_9$ of the mass equal the source of effort SE as indicated in equation (3.211). Considering this, equation (3.224) becomes

$$\dot{q}_6 = \frac{1}{(1+\frac{C_1}{C_2})}[\frac{1}{R}(\dot{p}_9 - \frac{1}{C_2}q_6) + C_1 \dot{SE}] \qquad (3.225)$$

This equation represented the equation of motion for the given system when the derivative causality occurred. It indicates that the flow $\dot{q}_4$ of the capacitance C_1, which has a derivative causality, is not considered in the system's equation of motion that satisfies the condition when it has a derivative causality.

Now, we have to ask the following two questions:

1. What happens if a velocity input is applied to the massless snubber plate rather than the input force?
2. What happens if we assume that the snubber plate has some significant mass? How does this affect the Bond Graph model and the equations of motion? Is this a more realistic assumption?

- The input force $F(t)$ is replaced by a velocity input $v(t)$:
Let us answer the first question. Consider the source of flow SF representing the velocity input instead of the source of effort SE as shown in Figure 3.54. Once the causal stroke marks are applied to all elements and junctions of the Bond Graph model, it will no longer have derivative causality, as shown in Figure 3.55. Since we have established that the model does not exhibit derivative causality, we can now start deriving the equations of motion for the system (exercise problem).

- The snubber plate has a small mass:
Regarding the second question, considering that the snubber plate possesses some significant mass, an inertia element I_s representing the snubber mass m_s is connected to the 1-junction, which is connected to the source of effort SE, as shown in Figure

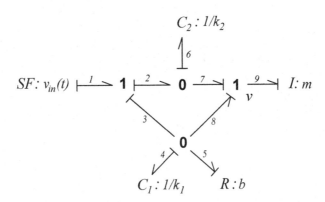

Figure 3.54 The Bond Graph model of the system considers a source of flow instead of the source of effort

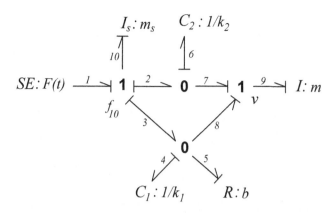

Figure 3.55 The Bond Graph model of the system considers snubber plate mass inertia

3.55. The derivative causality is no longer present when the causal stroke marks have been applied to all elements and junctions, as shown in Figure 3.55. Since the model does not exhibit derivative causality, we can derive the system's equations of motion (homework problem). Since adding the snubber plate inertia, I_s, to the model is done only to avoid derivative causality, it is essential to remember that this mass must be lightweight to avoid changing the model of the given system. In this situation, the assumption that the snubber plate has a lightweight mass is realistic, implying that it does not significantly affect the system model.

From this example, we can conclude that the derivative causality results from decisions taken at various points during the modeling process. In some situations, altering the system model may be beneficial for removing derivative causality and achieving possible advantages by changing the system model.

Example (3.8) (Derivative Causality):
Now, let us illustrate how we can change the system's model itself to eliminate the derivative causality without affecting its equations of motion. Consider, for example, the system shown in Figure 3.56a. In this system, a mass m_1, a spring of stiffness k, and a damper with damping factor b are linked to the left end of a lever rod, implying that they share the same flow. As a result, a 1-junction is used to connect the inertia element I_1 representing the mass m_1, the capacitor C representing the spring, and the resistance R representing the damper to a transformer element TF of a transformation factor (l_2/l_1) representing the lever rod movement, as shown in Figure 3.56b. On the other hand, a mass m_2 is linked directly to the right end of the lever rod, meaning that they share the same flow. As a result, a 1-junction connects an inertia element I_2 representing the mass m_2 to the transformer element TF, as shown in Figure 3.56b. Note that the external force $F(t)$ is represented by a source of effort SE. Considering the causality restrictions, apply the causality stroke marks to all elements and junctions, as shown in Figure 3.56b. From this figure, we noted that if we assign the inertia element I_1 in an integrally causal manner, the

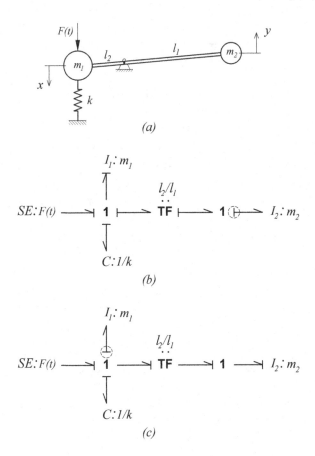

Figure 3.56 System model without considering the lever rod's flexibility.

input to the transformer element TF will come from its left side bond. In this case, the inertia element I_2 will be derivative causally, as shown inside a dotted circle in Figure 3.56b. Now, it is up to the user to decide whether he assigns integral causality to the inertia element I_1 before assigning derivative causality to the inertia element I_2. If he decides to assign integral causality to the inertia element I_2, the inertia element I_1 will be derivative causally, as shown in Figure 3.56c. In this case, the input to the transformer element TF will come from its right-side bond, as shown in the same figure. In both cases, the two masses are not moving independently, meaning they cannot choose how the forces acting on them will affect their flows. As a result, two different suggestions of causality are established, as shown in Figures 3.56b and 3.56c.

Two different techniques may be used to get rid of this difficulty.

1. We may reasonably assume that the lever rod has some flexibility, and assuming that it is of linear stiffness, it is similar to having a linear spring with stiffness k_l

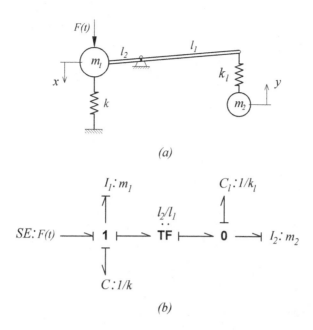

(a)

(b)

Figure 3.57 System model considers the lever rod's flexibility.

linked to it at its right end that separates the mass m_2 from the lever rod, as shown in Figure 3.57a. In this case, the down end of this spring and the mass m_2 move in the same direction and share the same effort. As a result, a 0-junction is used to connect the inertia I_2 and the capacitor C_1, representing the spring stiffness k_l of the lever rod to the transformer element TF, as shown in Figure 3.57b. The model no longer has derivative causality once the causal stroke marks are applied to all elements and junctions. This indicates that the derivative causality is eliminated by considering the lever rod's stiffness in the model. Consequently, it is now possible to consider the actual movement of the mass m_2 as an independent energy storage variable. Now, we are in a position to derive the equations of motion (exercise problem). Since adding the stiffness k_l of the lever rod to the model only eliminates the derivative causality, it is essential to note that this spring must have a very low stiffness to prevent altering the given system model.

2. If we consider the equivalent mass m_{eq} of both masses as a replacement for the mass m_1 (neglecting the mass of the lever rod). In this case, we may use the equivalent inertia I_{eq} to represent the equivalent mass m_{eq} as shown in Figure 3.58. Let us now determine the equivalent mass m_{eq} that represents the linear relationship between the masses m_1 and m_2. Applying the energy equivalence to the motion of both masses, which should be equal to the energy of the equivalent one, yields the following result:

$$\frac{1}{2}m_{eq}\dot{x}^2 = \frac{1}{2}m_1\dot{x}^2 + \frac{1}{2}m_2\dot{y}^2 \qquad (3.226)$$

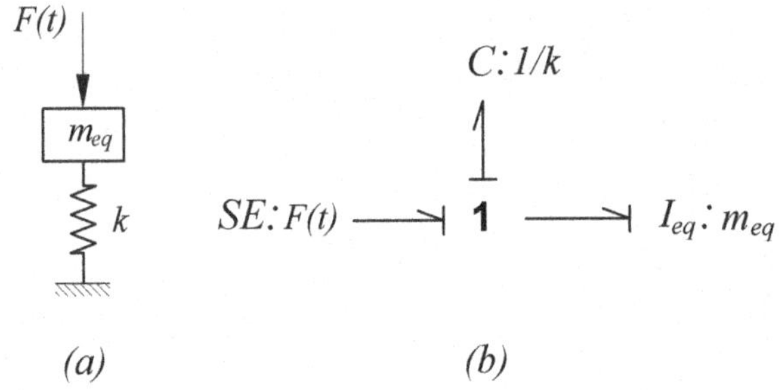

Figure 3.58 Equivalent system model

As shown in Figure 3.57, the displacement y of the mass m_2 as a function of the displacement x of the mass m_1 is given by

$$l_2 x = l_1 y \quad \text{or} \quad y = \frac{l_2}{l_1} x \tag{3.227}$$

Taking the derivative of this equation yields

$$\dot{y} = \frac{l_2}{l_1} \dot{x} \tag{3.228}$$

Substituting this equation in equation (3.226) gives

$$\frac{1}{2} m_{eq} \dot{x}^2 = \frac{1}{2} m_1 \dot{x}^2 + \frac{1}{2} m_2 \left(\frac{l_2}{l_1} \right)^2 \dot{x}^2 \tag{3.229}$$

Rearranging this equation yields

$$m_{eq} = \left(m_1 + \left(\frac{l_2}{l_1} \right)^2 m_2 \right) \tag{3.230}$$

This equation indicates that it is possible to determine the actual movement of the mass m_2 on its own by utilizing a linear relationship with the motion of the mass m_1, which allows us to do so independently. As equation (3.230) indicates, both masses merge to form the equivalent mass m_{eq}. This means that we do not need to use the transformer element TF that is shown in the Bond Graph model of the system in Figure 3.57, as shown in Figure 3.58.

As stated previously, one can conclude that the source of the derivative causality is the decision made at a particular stage throughout the modeling process. Changing the system's model can efficiently eliminate derivative causality and access various potential benefits depending on the situation.

3.13 ALGEBRAIC LOOPS

An issue with the causal structure may arise when attempting to assign causalities to particular systems. Consider, for instance, the Bond Graph model shown in Figure 3.59a. After the assignments of the integral causality of the sources and storage elements have been made, we must continue arbitrarily assigning one of the resistances, R_1 or R_2, and then the causality to the other to complete the causal structure of the Bond Graph model. This allows for the causal determination of the Bond Graph's remaining bonds. On the other hand, if we had assigned the opposite causality to one of both resistances, the causalities of the rest of the bonds would have also been reversed, leading to an entirely different causal structure. If we had given the resistance incorrect causality, this would have been the case for the *algebraic loop*. Figures 3.59a and 3.59b show the two alternative approaches. Note that these two causal structures can both be considered valid representations of the system, which leads to an issue that is pretty typical of the kind that arises when one is attempting to derive the equations that govern the given system. Let us now demonstrate how this works by deriving the equations of motion for this system. Let us begin with the Bond Graph model shown in Figure 3.59a. To derive the equations of motion, we first defined the key variables and their constitutive relations.

- **Key variables:**
- Inputs: $e_1 = SE$
- State variables: p_4 and q_7
- Power variables: f_4 and e_7

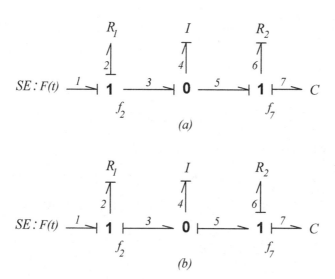

Figure 3.59 Bond Graph model with algebraic loop

- Constitutive relations:

$$f_4 = \frac{1}{I}p_4 \quad \text{and} \quad e_7 = \frac{1}{C}q_7 \tag{3.231}$$

As shown in Figure 3.59a, the flow and effort variables of the 1-junction of a common flow f_2 are given by

$$f_1 = f_2 = f_3 \tag{3.232}$$

and

$$e_1 = e_2 + e_3 \quad \text{or} \quad e_3 = e_1 - e_2 = SE - e_2 \tag{3.233}$$

The effort e_2 of the resistance R_1 is given by

$$e_2 = R_1 f_2 \tag{3.234}$$

Substituting this equation in equation (3.233), gives

$$e_3 = SE - R_1 f_2 \tag{3.235}$$

The flow and effort variables of the 0-junction of a common effort are given by

$$f_3 = f_4 + f_5 \quad \text{or} \quad f_5 = f_3 - f_4 \tag{3.236}$$

and

$$e_3 = e_4 = e_5 \tag{3.237}$$

The flow and effort variables of the 1-junction of a common flow f_7 are given by

$$f_5 = f_6 = f_7 \tag{3.238}$$

and

$$e_5 = e_6 + e_7 \tag{3.239}$$

Using equations (3.237), equation (3.239) becomes

$$e_5 = e_3 = e_6 + e_7 \tag{3.240}$$

Substituting equation (3.240) in equation (3.235) gives

$$SE - R_1 f_2 = e_6 + e_7 \tag{3.241}$$

Using the constitutive relations in equation (3.231), equation (3.241) becomes

$$e_6 + \frac{1}{C}q_7 = SE - R_1 f_2 \tag{3.242}$$

The effort e_6 of the resistance R is given by

$$e_6 = R_2 f_6 \tag{3.243}$$

Using equation (3.238), equation (3.243) becomes

$$e_6 = R_2 f_5 \tag{3.244}$$

Using equation (3.236), equation (3.244) becomes

$$e_6 = R_2(f_3 - f_4) \tag{3.245}$$

Using equation (3.232), equation (3.245) becomes

$$e_6 = R_2(f_2 - f_4) \tag{3.246}$$

Using the constitutive relations indicated in equation (3.231), equation (3.246) becomes

$$e_6 = R_2(f_2 - \frac{1}{I}p_4) \tag{3.247}$$

When we look at equations (3.242) and (3.247), we find that they are two simultaneous equations of two variables, f_2 and e_6, which indicate that they are algebraically related to each other. As stated before, this is an *algebraic loop*. However, this is not an effective barrier. We can easily obtain these two variables f_2 and e_6 by solving these two equations since we have two equations representing the two unknowns. Now let us solve equations (3.242) and (3.247) for f_2 and e_6. From equation (3.242), we get

$$f_2 = \frac{1}{R_1}(SE - \frac{1}{C}q_7 - e_6) \tag{3.248}$$

Substituting equation (3.247) in equation (3.248) and solving for f_2 we get

$$f_2 = \frac{1}{R_1 + R_2}(SE - \frac{1}{C}q_7 + R_2\frac{1}{I}p_4) \tag{3.249}$$

Substituting equation (3.249) in equation (3.247), we get

$$e_6 = \frac{R_2}{R_1 + R_2}(SE - \frac{1}{C}q_7 - R_1\frac{1}{I}p_4) \tag{3.250}$$

When three or more resistances are coupled in such an algebraic loop, solving the coupled algebraic equations produced by applying the matrix inversion method is much easier. It is essential to note that the same outcomes can be anticipated when utilizing the Bond Graph model of the same system, as shown in Figure 3.59b (exercise problem).

3.14 PROBLEMS

Problem (3.1):
For each of the systems shown in Figure 3.60,

1. Create the preliminary Bond Graph model and simplify it.
2. Create the Augmented Bond Graph (add the half-arrow positive power flow directions, assign causalities, and number all bonds). Check the derivative causality.
3. Derive the equations of motion (write the key variables and their constitutive relations).
4. Check if Newton's law is embedded in the Augmented Bond Graph model.

Problem (3.2):
For the system shown in Figure 3.61,

1 Create the preliminary Bond Graph model. Impose it on the system diagram and simplify it.
2 Create the Augmented Bond Graph model (add the half-arrow positive power flow directions, assign causalities, and number all bonds). Check the derivative causality.
3 Derive the equations of motion (write the key variables and their constitutive relations). Put the results in matrix form.
4 Check if Newton's law is embedded in the Augmented Bond Graph.

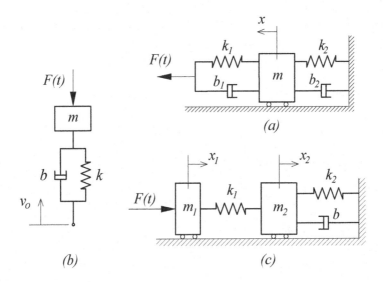

Figure 3.60 Mechanical system

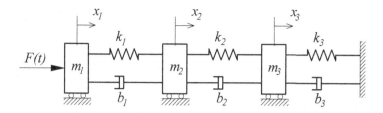

Figure 3.61 Mechanical system

Problem (3.3):
For the system shown in Figure 3.62,

1 Create the preliminary Bond Graph model. Impose it on the system diagram and simplify it.
2 Create the Augmented Bond Graph model (add the half-arrow positive power flow directions, assign causalities, and number all bonds). Check the derivative causality.
3 Derive the equations of motion (write the key variables and their constitutive relations). Put the results in matrix form.
4 Check if Newton's law is embedded in the Augmented Bond Graph.

Problem (3.4):
For the system shown in Figure 3.63, assuming the angular displacements θ_1 and θ_2 are small,

1 Create the preliminary Bond Graph model. Impose it on the system diagram and simplify it.
2 Create the Augmented Bond Graph model (add the half-arrow positive power flow directions, assign causalities, and number all bonds). Check the derivative causality.
3 Derive the equations of motion (write the key variables and their constitutive relations). Put the results in matrix form.
4 Check if Newton's law is embedded in the Augmented Bond Graph.

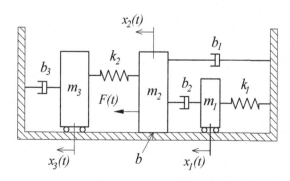

Figure 3.62 Mechanical system

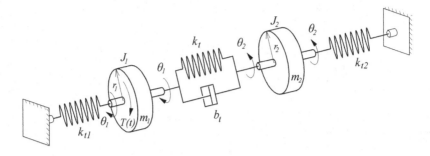

Figure 3.63 Mechanical system

Problem (3.5):

For the system shown in Figure 3.64,

1 Create the preliminary Bond Graph model and simplify it.
2 Create the Augmented Bond Graph model (add the half-arrow positive power flow directions, assign causalities, and number all bonds). Check the derivative causality.
3 Derive the equations of motion (write the key variables and their constitutive relations). Put the results in matrix form.
4 Check if Newton's law is embedded in the Augmented Bond Graph.

Problem (3.6):

For the rack and pinion system shown in Figure 3.65, assuming the angular displacement θ is small,

1 Create the preliminary Bond Graph model and simplify it.

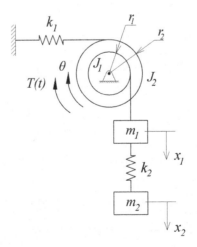

Figure 3.64 Mechanical system

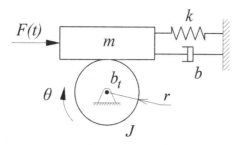

Figure 3.65 Mechanical system

2 Create the Augmented Bond Graph model (add the half-arrow positive power flow directions, assign the causalities, and number all bonds). Check the derivative causality.
3 Derive the equations of motion (write the key variables and their constitutive relations).
4 Check if Newton's law is embedded in the Augmented Bond Graph.

Problem (3.7):
For the system shown in Figure 3.66, assuming the angular displacement θ is small,

1 Create the preliminary Bond Graph model and simplify it.
2 Create the Augmented Bond Graph model (add the half-arrow positive power flow directions, assign causalities, and number all bonds). Check the derivative causality.
3 Derive the equations of motion (write the key variables and their constitutive relations).
4 Check if Newton's law is embedded in the Augmented Bond Graph.

Problem (3.8): :
For the system shown in Figure 3.67, assuming no slip between the cord and the pulley, the rod is massless, and its angular displacement θ is small

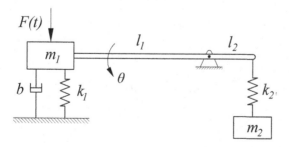

Figure 3.66 Mechanical system

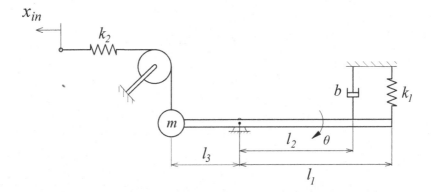

Figure 3.67 Mechanical system

1 Create the preliminary Bond Graph model and simplify it.
2 Create the Augmented Bond Graph model (add the half-arrow positive power flow directions, assign causalities, and number all bonds). Check the derivative causality.
3 Derive the equations of motion (write the key variables and their constitutive relations).
4 Check if Newton's law is embedded in the Augmented Bond Graph.

Problem (3.9):
For the system shown in Figure 3.68, assuming the angular displacement θ is small,

1 Create the preliminary Bond Graph model and simplify it.
2 Create the Augmented Bond Graph model (add the half-arrow positive power flow directions, assign causalities, and number all bonds). Check the derivative causality.
3 Derive the equations of motion (write the key variables and their constitutive relations).
4 Check if Newton's law is embedded in the Augmented Bond Graph.

Problem (3.10):
For the system shown in Figure 3.69,

1 Create the preliminary Bond Graph model and simplify it.
2 Create the Augmented Bond Graph model (add the half-arrow positive power flow directions, assign causalities, and number all bonds). Check the derivative causality.
3 Derive the equations of motion (write the key variables and their constitutive relations).
4 Check if Newton's law is embedded in the Augmented Bond Graph.

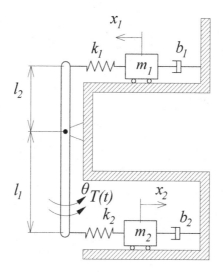

Figure 3.68 Mechanical system

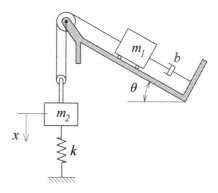

Figure 3.69 Mechanical system

4 Electrical Systems

4.1 INTRODUCTION

Electrical systems are more familiar than mechanical systems and are frequently presented in introductory physics courses. This chapter presents material covered in more depth in most first-year classes. Our point of view is that students are majoring in mechanical engineering, and to deal with the electrical and electronic components of mixed systems, they need to have a fundamental understanding of the behavior of the electrical circuits in addition to some simple analysis tools. We will quickly overview Kirchhoff's laws, which can be used to find differential equations that describe the dynamics of electrical systems. A brief overview of the major parts of an electrical system may be necessary before beginning the dynamic modeling of these systems.

4.2 ELECTRICAL SYSTEM ELEMENTS

The main components of an electrical circuit are comprised of three passive elements (inductance, capacitance, and resistance) and two energy sources (voltage and current).

4.2.1 INDUCTANCE ELEMENT

It is possible to divide inductive effects into two categories: self-inductance and mutual inductance. Mutual inductance is the property of a circuit that allows a changing current in one circuit to generate a voltage in another. The self-inductance element, shown in Figure 4.1a, is a characteristic of a single coil caused by the magnetic field created by the coil current. The induced inductance voltage is instantaneously related to the derivative of its current in the mutual inductance element, although the relationship might be nonlinear. In the self-inductance element, the voltage e_L is directly proportional to the di/dt, meaning it is linear and free from the capacitance and resistance. Therefore, the behavior of the self-inductance element is given by

$$e_L = L\frac{di}{dt} \tag{4.1}$$

where L is the inductance factor. Since a voltage-current rather than a voltage-charge relation is often more desirable, we can write the definition of current i as

$$i = \frac{dq}{dt} = \dot{q} \tag{4.2}$$

Considering this equation, equation (4.1) in electrical charge form is given by,

$$e_L = L\frac{d^2q}{dt^2} = L\ddot{q} \tag{4.3}$$

DOI: 10.1201/9781032685656-4

136

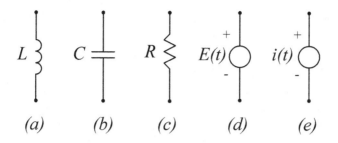

L 〈 *C* 〉 *R* 〈 *E(t)* *i(t)*

(a)   (b)   (c)   (d)   (e)

Figure 4.1 Electrical circuit elements

This equation indicates that the electrical conductor element is a storage element equivalent to the mechanical system's inertia (mass) element.

4.2.2 CAPACITANCE ELEMENT

The ideal capacitor element, shown in Figure 4.1b, is constructed from two parallel, separate conductor plates. When a current passes through it, the amount of electricity produced is directly proportional to the potential difference between the two plates, as calculated by the formula.

$$q = Ce \qquad (4.4)$$

or

$$e = \frac{1}{C}q \qquad (4.5)$$

where q is the amount of charging electricity of the capacitor (quantity of electricity), and C is the capacitance factor. Recalling the definition of the current indicated in equation (4.2) and differentiating equation (4.5), yields

$$\frac{de}{dt} = \frac{1}{C}\frac{dq}{dt} = \frac{1}{C}i \qquad (4.6)$$

So,

$$de = \frac{1}{C}i\,dt \qquad (4.7)$$

Integrating this equation gives

$$\int_{e_o}^{e_C} de = \frac{1}{C}\int_0^t i\,dt \qquad (4.8)$$

So,

$$(e_C - e_o) = \frac{1}{C}\int_0^t i\,dt \qquad (4.9)$$

where e_o is the voltage drop across the capacitor due to the initial charge, which existed at a time equal to zero. If e_o is zero (the capacitor initially uncharged), we would have

$$e_C = \frac{1}{C}\int i\,dt \qquad (4.10)$$

This equation indicates that the capacitor element is a storage element equivalent to the mechanical system's spring element.

4.2.3 RESISTANCE ELEMENT

The fundamental characteristics of this element are the strict linearity between the voltage and the current and the fact that all of the electrical energy provided is lost as heat. This indicates that the electrical resistance R is a dissipation element (like a damping element in the mechanical system). The ideal resistance element, shown in Figure 4.1c, follows Ohm's law, which provides the current-voltage relationship as

$$i = \frac{e_R}{R} \tag{4.11}$$

or

$$e_R = Ri \tag{4.12}$$

where e_R is the voltage drop across the resistance, and R is the resistance factor. Recalling the definition of the current $i = dq/dt$, equation (4.12) in electrical charge q form is given by,

$$e_R = Ri = R\frac{dq}{dt} = R\dot{q} \tag{4.13}$$

Equation (4.13) indicates that the resistance element is a dissipation element and is equivalent to the damper element in the mechanical system, as mentioned above.

4.2.4 CURRENT AND VOLTAGE SOURCES

Voltage source $E(t)$ and current source $i(t)$, shown in Figures 4.1d and 4.1e, respectively, are the main energy sources that drive the electrical system. An ideal voltage source will supply the proper voltage regardless of how much current the circuit requires (and thus the power); conversely, for a current source. In the same way that the force source is fundamental in mechanical systems because of the cause-and-effect relationship established by Newton's law (a force causes a motion - not the reverse), we could say that voltage sources are fundamental in electrical systems because it is the electromotive force (voltage) that causes a current to flow. However, using a current source may be valid in some systems since real devices that behave similarly to current sources are readily available.

4.3 DYNAMIC MODELING OF ELECTRICAL SYSTEMS

The basic physical laws governing electrical circuit dynamics are Kirchoff's voltage loop law and Kirchoff's current node law. Kirchoff's voltage loop law states that the summation of the voltage drops around a closed loop must be zero at every instant. In contrast, Kirchoff's current node law is based on the fact that at any node in a circuit, there can be no accumulation of electrical charge. Since the current i is defined as the flow of charge $\dot{q}$, we may say that at every instant of time, the summation of currents into a node must equal the summation of currents out.

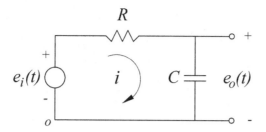

Figure 4.2 Electrical circuit

- Sign Conventions

Regarding mechanical systems, we need sign conventions for forces and motions (free-body diagram). When it comes to electrical systems, we need them for voltages and currents. Let us consider the electrical circuit shown in Figure 4.2.

1. At the beginning, the positive direction of a current considered to be flowing can, in most cases, be chosen arbitrarily. Suppose the response to the unknown current turns out to be a positive value at any given instant. In this case, we know that the current, i, flows in the same direction indicated by the arrow used for sign conventions, as shown in Figure 4.2. When it is a negative value, the direction of the current will be counter to that of its arrow. Because the meaning of positive and negative currents is not defined, performing an orderly analysis at the beginning of a problem is impossible if the assumed positive direction of a current has not been specified. Furthermore, it cannot appropriately interpret any results obtained because of this lack of definition.
2. Regarding voltages, the sign conventions consist of a plus sign and a minus sign placed at the terminals where the voltage is present, as shown in Figure 4.2. Once more, selecting which terminal should be given the plus sign at the beginning of an analysis is completely arbitrary. If the voltage at some time is a positive number when the solution is reached, then the actual polarity is the same as that given by the sign-convention marks. If the output voltage is negative, the real polarity will be the opposite of what is being selected by the sign convention marks.

Whenever the sign conventions for all of the currents and voltages have been chosen, combining the rules of Kirchoff's law with the known voltage/current relations that characterize the circuit elements brings us immediately to the system differential equations that describe the system's dynamics.

Now that we have everything established, let us begin our analysis of an electrical circuit. Consider the circuit shown in Figure 4.2,

1. We initially select the positive sense of e_i, as shown in this figure. Although we could have, of course, chosen the opposite sense, we went with the former. For e_o, we also have a free choice; however, after the positive direction of e_i has been fixed, there may be some motivation to choose e_o such that a positive e_i produces

a positive e_o. So, we have a free choice for e_o. This is not required, although doing so would be beneficial; that is what has been done in Figure 4.2. Because we assume that the output terminals are not experiencing any current flow (no load circuit), there is only one current in the system, and we will refer to it as i. It is possible to choose to go in a positive direction. However, the desire to have a positive e_i, to create a positive e_i, and to pick i positive by moving in the direction indicated once more influences.

2. Because a positive value for i leads to a positive value for e_R, it was chosen to give the voltage across the resistance R a positive value. This choice was made such that a negative value for i would generate a negative value for e_R.
3. Since the capacitor voltage, e_C is also e_o; there is no need to define it again.
4. If the circuit contains an inductance element L, it can be thought of using the same concept as the capacitor element, which implies that a positive value for the current results in a positive value for the inductance voltage.

The question is how to identify which Kirchoff's law (voltage or current) is most appropriate for the particular circuit. Since there is only one current, as shown in Figure 4.2, Kirchoff's voltage loop law looks more relevant for this circuit than the current node law. Once a loop has been defined, we can select the starting point and the direction to traverse it in any way we see fit. When putting this rule into practice, one must choose where in the circuit to begin, the path around the loop that will take, and the direction in which to travel around the loop. The loop in concern is evident in our preset circuit, as shown in Figure 4.2. The starting point is completely arbitrary; however, we can choose how to cross the circuit to make writing the equations easier. Kirchoff's voltage loop law, which is the summation of voltage drop at any instance, is determined by moving clockwise around the loop, starting at point o (which could be chosen anywhere on the loop) in Figure 4.2 and moving in the direction of positive current. So, applying Kirchoff's voltage loop in our preset circuit, shown in Figure 4.2, gives

$$-e_i + e_R + e_C = 0 \qquad (4.14)$$

Note that because we add drops, the input voltage e_i enters this equation with a negative sign, following the sign convention set for that particular variable. Recalling equations (4.10) and (4.12) and substituting them in equation (4.14) yields

$$-e_i + Ri + \frac{1}{C}\int i\,dt = 0 \qquad (4.15)$$

As shown in Figure 4.2, the circuit voltage e_o is equal to the voltage of the capacitor element e_C, and it is given by

$$e_o = \frac{1}{C}\int i\,dt \qquad (4.16)$$

Equation (4.15) is the differential equation that describes the dynamics of the given circuit shown in Figure 4.2.

The question establishes which of Kirchoff's laws, voltage or current, is most applicable to the particular circuit. Examples (4.1) and (4.2) are provided below to

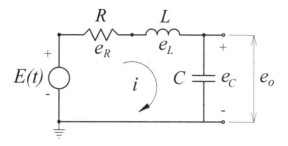

Figure 4.3 Electrical circuit

demonstrate how to accomplish this. In these two examples, we will examine RLC connections in both parallel and series.

Example (4.1): (Kirchhoff's voltage law)
Derive the differential equation(s) that describe the dynamics of the electrical circuit shown in Figure 4.3.

Solution:
As shown in Figure 4.3, all passive elements are in series connections and driven by a voltage source $E(t)$. Due to the constraint of the series connection, only one current i flows through the circuit, as shown in this figure. We assume no current flows at the output terminals (no load); only one current flows in the circuit i. As a result, Kirchhoff's voltage law is more appropriate than the current node law. Recalling this law, we write the summation of the voltage drop around the loop as

$$\sum e = 0 \tag{4.17}$$

$$E(t) - e_R - e_L - e_C = 0 \tag{4.18}$$

where

$$e_R = Ri = R\dot{q}, \quad e_C = \frac{1}{C}\int i\,dt = \frac{1}{C}q \text{ and } e_L = L\frac{di}{dt} = L\ddot{q} \tag{4.19}$$

where q is the electrical charge. Substituting equation (4.19) in equation (4.18) gives

$$L\frac{di}{dt} + iR + \frac{1}{C}\int i\,dt = E(t) \tag{4.20}$$

In electrical charge form, equation (4.20) becomes

$$L\ddot{q} + R\dot{q} + \frac{1}{C}q = E(t) \tag{4.21}$$

Equation (4.20) or (4.21) is the differential equation that describes the dynamics of the given circuit shown in Figure 4.3. How does the dynamic equation change if the input voltage $E(t)$ is replaced by the input current $i(t)$? (exercise problem).

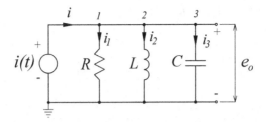

Figure 4.4 Electrical circuit

Example (4.2): (Kirchhoff's current law)
Derive the differential equation(s) that describe the dynamics of the electrical circuit
shown in Figure 4.4.

Solution:
As shown in Figure 4.4, all passive elements are in parallel connection and driven by
the current source $i(t)$. In this case, Kirchhoff's current node law is often the most
direct approach to analysis. Applying it at nodes 1, 2, and 3 yields

$$\sum i = 0 \tag{4.22}$$

$$i(t) - i_1 - i_2 - i_3 = 0 \tag{4.23}$$

The voltage across each element is given by

$$e_R = i_1 e_R, \quad e_L = L\frac{di_2}{dt} \quad \text{and} \quad e_C = \frac{1}{C}\int i\,dt \tag{4.24}$$

Rearranging this equation gives

$$i_1 = \frac{1}{R}e_R, \quad i_2 = \frac{1}{L}\int e_L\,dt \quad \text{and} \quad i_3 = C\frac{de_C}{dt} \tag{4.25}$$

Due to the constraints of the parallel connection, all elements provide the same voltage drop e_o. Considering this, $e_R = e_L = e_C = e_o$. Substituting equation (4.24) in
equation (4.23) and rearranging the results gives

$$C\frac{de_o}{dt} + \frac{1}{R}e_o + \frac{1}{L}\int e_o\,dt = i(t) \tag{4.26}$$

This equation is the differential equation that describes the dynamics of the given
circuit shown in Figure 4.4. How does the dynamic equation change if the input
current $i(t)$ is replaced by the input voltage $E(t)$? (exercise problem).

4.3.1 EQUIVALENT RESISTANCE

Ohm's law states that the current in the circuit is proportional to the electromotive
force (*emf*) acting in the circuit and inversely proportional to the circuit's total resistance. It is given by:

$$i = \frac{e}{R} \tag{4.27}$$

where i (amp.) is the current, e (volt) is the *emf*, and R (ohms) is the resistance.

4.3.1.1 Resistances in parallel

When resistances are connected in parallel, as shown in Figure 4.5a, the total voltage across the resistances is the same; however, the current that passes through each resistance is different. Applying Kirchhoff's current law, the total current i is given by

$$i = i_1 + i_2 + i_3 \tag{4.28}$$

where

$$i_1 = \frac{e}{R_1}, \quad i_2 = \frac{e}{R_2} \quad \text{and} \quad i_3 = \frac{e}{R_3} \tag{4.29}$$

Substituting this equation in equation (4.29) and inserting $i = (e/R_p)$ gives

$$i = \frac{e}{R_p} = \frac{e}{R_1} + \frac{e}{R_2} + \frac{e}{R_3} \tag{4.30}$$

where R_p is the equivalent resistance for resistances connected in parallel. Rearranging equation (4.30) gives

$$\frac{1}{R_p} = \frac{1}{R_1} + \frac{1}{R_2} + \frac{1}{R_3} \tag{4.31}$$

So,

$$R_p = \frac{1}{\frac{1}{R_1} + \frac{1}{R_2} + \frac{1}{R_3}} = \frac{R_1 R_2 R_3}{R_1 R_2 + R_2 R_3 + R_1 R_3} \tag{4.32}$$

In its general form, it is given by

$$R_{pi} = \sum_{i=2}^{n} \left(\frac{1}{\frac{1}{R_1} + \frac{1}{R_2} + \cdots + \frac{1}{R_n}} \right) \tag{4.33}$$

4.3.1.2 Resistances in series

When resistances are connected in series, as shown in Figure 4.5b, the total current across the resistances is the same; however, the voltage that passes through each resistance is different. Applying Kirchhoff's voltage law, the total voltage e is given by

$$e = e_1 + e_2 + e_3 \tag{4.34}$$

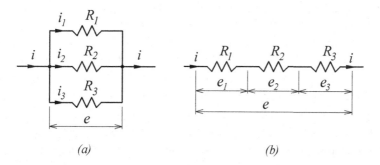

(a)   *(b)*

Figure 4.5 Resistance elements: (a) parallel connection; (b) series connection

where

$$e_1 = iR_1, \quad e_2 = iR_2 \quad \text{and} \quad e_3 = iR_3 \qquad (4.35)$$

Substituting this equation (4.35) in equation (4.34) and inserting $e = (R_s i)$ gives

$$R_s i = R_1 i + R_2 i + R_3 i \qquad (4.36)$$

where R_s is the equivalent resistance of resistances connected in series. So,

$$R_s = R_1 + R_2 + R_3 \qquad (4.37)$$

In its general form, it is given by

$$R_{si} = \sum_{i=2}^{n} (R_1 + R_2 + \cdots + R_n) \qquad (4.38)$$

4.3.2 EQUIVALENT CAPACITOR

One of the fundamental elements used in electronic circuits is the capacitor, and sophisticated configurations of capacitors are frequently found in practical electrical circuits. These configurations can be in series, parallel, or both.

4.3.2.1 Capacitors in parallel

As shown in Figure 4.6a, the voltage difference e remains the same when capacitors are connected in parallel. However, the total charge, q, being held in capacitors must be distributed equally to keep the voltage across all capacitors the same. Since the capacitors C_1, C_2, and C_3 have different capacitances, the charges q_1, q_2, and q_3 may be also different. Consider the relationship of the voltage-charge as follows

$$e = \frac{1}{C}q \quad \text{or} \quad C = \frac{q}{e} \quad \text{or} \quad \frac{1}{C} = \frac{e}{q} \qquad (4.39)$$

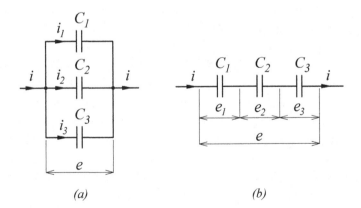

(a)   *(b)*

Figure 4.6 Capacitors elements: (a) parallel connection; (b) series connection

Considering this equation, the equivalent capacitance C_p of capacitors connected in parallel is given by

$$C_p = \frac{q}{e} = \frac{q_1}{e} + \frac{q_2}{e} + \frac{q_3}{e} = C_1 + C_2 + C_3 \tag{4.40}$$

So,

$$C_p = C_1 + C_2 + C_3 \tag{4.41}$$

In its general form, it is given by

$$C_{pi} = \sum_{i=2}^{n} (C_1 + C_2 + \cdots + C_n) \tag{4.42}$$

4.3.2.2 Capacitors in series

When the capacitors are connected in series, as shown in Figure 4.6b, it is important to realize that the charge q stored in the capacitors is the same. However, the voltages across the capacitors are different, and their summation equals the total voltage drop e across the input and the output. Considering equation (4.39), the equivalent capacitance C_s of capacitors connected in series is given by

$$\frac{1}{C_s} = \frac{e}{q} = \frac{e_1}{q} + \frac{e_2}{q} + \frac{e_3}{q} = \frac{1}{C_1} + \frac{1}{C_2} + \frac{1}{C_3} \tag{4.43}$$

So,

$$\frac{1}{C_s} = \left(\frac{1}{C_1} + \frac{1}{C_2} + \cdots + \frac{1}{C_n} \right) \tag{4.44}$$

In its general form, it is given by

$$\frac{1}{C_{si}} = \sum_{i=2}^{n} \left(\frac{1}{C_1} + \frac{1}{C_2} + \cdots + \frac{1}{C_n} \right) \tag{4.45}$$

4.3.3 EQUIVALENT INDUCTOR

The interconnections of inductors produce a more complex circuit whose overall inductance is a combination of the individual inductors. However, there are specific rules for connecting inductors in parallel or series because we assumed that the inductors were self-induced and that no magnetic coupling existed between the individual inductors.

4.3.3.1 Inductors in parallel

Figure 4.7b shows that the voltage drop across each inductor will be the same when inductors are connected in parallel. Applying Kirchhoff's current law, the total current i is thus provided as follows:

$$i = i_1 + i_2 + i_3 \tag{4.46}$$

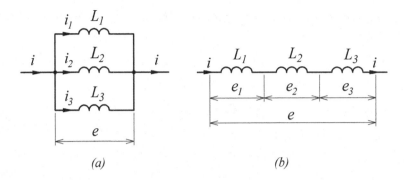

Figure 4.7 Inductance elements: (a) parallel connection; (b) series connection

Since we assumed that the inductances are self-induced, the voltage across an inductor is given by

$$e = L\frac{di}{dt} \quad \text{or} \quad \frac{di}{dt} = \frac{e}{L} \tag{4.47}$$

Substituting equation (4.47) in equation (4.46), the total voltage across the inductors is given by

$$e = L_p\frac{di}{dt} = L_p\frac{d}{dt}(i_1 + i_2 + i_3) = L_p\left(\frac{di_1}{dt} + \frac{di_2}{dt} + \frac{di_3}{dt}\right) \tag{4.48}$$

Considering equation (4.47), equation (4.48) becomes

$$\frac{e}{L_p} = \left(\frac{e}{L_1} + \frac{e}{L_2} + \frac{e}{L_3}\right) \tag{4.49}$$

Rearranging this equation gives

$$\frac{1}{L_p} - \left(\frac{1}{L_1} + \frac{1}{L_2} + \frac{1}{L_3}\right) \tag{4.50}$$

where L_p is the equivalent capacitance for capacitors connected in parallel. In its general form, it is given by

$$\frac{1}{L_{pi}} = \sum_{i=2}^{n}\left(\frac{1}{L_1} + \frac{1}{L_2} + \cdots + \frac{1}{L_n}\right) \tag{4.51}$$

4.3.3.2 Inductors in series

Applying Kirchhoff's voltage law to the circuit shown in Figure 4.7b gives

$$e = e_1 + e_2 + e_3 \tag{4.52}$$

Since the same current flows in each conductor, considering equation (4.47), equation (4.52) becomes

$$L_s\frac{di}{dt} = L_1\frac{di}{dt} + L_2\frac{di}{dt} + L_3\frac{di}{dt} \tag{4.53}$$

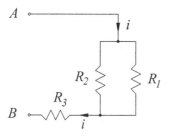

Figure 4.8 Resistance elements

So,

$$L_s = L_1 + L_2 + L_3 \tag{4.54}$$

where L_s is the equivalent inductance of inductors connected in series. In its general form, it is given by

$$L_{si} = \sum_{i=2}^{n}(L_1 + L_2 + \cdots + L_n) \tag{4.55}$$

Example (4.3):
For resistances connected as shown in Figure 4.8, obtain their equivalent resistance.
Solution:
Figure 4.8 shows that R_1 and R_2 are in a parallel connection. Considering equation (4.32), their equivalent resistance R_{12} is given by

$$R_{12} = \frac{1}{\frac{1}{R_1} + \frac{1}{R_2}} = \frac{R_1 R_2}{R_1 + R_2} \tag{4.56}$$

So, the resistance R_{12} becomes in series with the resistance R_3, as shown in Figure 4.9. As a result, considering equation (4.37), the equivalent resistance R_{123} of the system is given by

$$R_{123} = R_3 + \frac{R_1 R_2}{R_1 + R_2} \tag{4.57}$$

Similarly, determine the equivalent capacitance and inductance of the circuits shown in Figure 4.10. The connections of the elements in both circuits are chosen to be similar to the resistances shown in Figure 4.8. Compare the results obtained in the three cases (exercise problem).

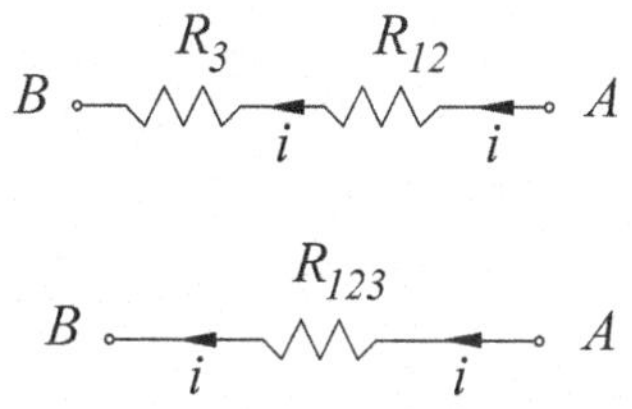

Figure 4.9 Equivalent resistance

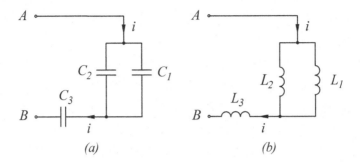

Figure 4.10 Electrical elements: (a) capacitors; (b) inductors

Example (4.4):
Considering the circuit shown in Figure 4.11, assume the switch S is open at $t < 0$, derive the differential equation(s) that describe the dynamics of the circuit when the switch is closed at $t \geq 0$, and put the results in matrix form.

Solution:
Applying Kirchhoff's voltage law to the first loop of the current i_1 gives

$$L\frac{di_1}{dt} + R_a i_1 + \frac{1}{C}\int (i_1 - i_2)\, dt = E(t) \tag{4.58}$$

When the switch S is closed, the second loop will be closed, and the current i_2 will pass through the resistance R_b. So, applying Kirchhoff's voltage law to the second loop gives

$$\frac{1}{C}\int (i_2 - i_1)\, dt + R_b i_2 = 0 \tag{4.59}$$

Equations (4.58) and (4.59) are the differential equations representing the dynamics of the given circuit. In terms of the electrical charge q, inserting $i = dq/dt$ in these equations and carrying out the integration terms gives

$$L\ddot{q}_1 + R_a\dot{q}_1 + \frac{1}{C}(q_1 - q_2) = E(t) \tag{4.60}$$

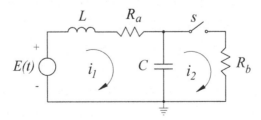

Figure 4.11 Electrical circuit

and

$$R_b\dot{q}_2 + \frac{1}{C}(q_2 - q_1) = 0 \tag{4.61}$$

Rearranging these equations yields

$$L\ddot{q}_1 + R_a\dot{q}_1 + \frac{1}{C}q_1 - \frac{1}{C}q_2 = E(t) \tag{4.62}$$

and

$$R_b\dot{q}_2 + \frac{1}{C}q_2 - \frac{1}{C}q_1 = 0 \tag{4.63}$$

In matrix form,

$$\begin{bmatrix} L & 0 \\ 0 & 0 \end{bmatrix} \begin{bmatrix} \ddot{q}_1 \\ \ddot{q}_2 \end{bmatrix} + \begin{bmatrix} R_a & 0 \\ 0 & R_b \end{bmatrix} \begin{bmatrix} \dot{q}_1 \\ \dot{q}_2 \end{bmatrix} + \begin{bmatrix} \frac{1}{C} & -\frac{1}{C} \\ -\frac{1}{C} & \frac{1}{C} \end{bmatrix} \begin{bmatrix} q_1 \\ q_2 \end{bmatrix} = \begin{bmatrix} E(t) \\ 0 \end{bmatrix} \tag{4.64}$$

Example (4.5):
Derive the differential equation(s) that describe the dynamics of the electrical circuit shown in Figure 4.12.

Solution:
Applying Kirchhoff's voltage law to the circuit shown in Figure 4.12 gives

$$V_s = R_1 i + V_o \tag{4.65}$$

and

$$V_o = \frac{1}{C}\int i\, dt + R_2 i \tag{4.66}$$

Taking the derivative of equation (4.66) gives

$$\frac{dV_o}{dt} + \frac{1}{C}i + R_2\frac{di}{dt} \tag{4.67}$$

From equation (4.65), we get

$$i = \frac{V_s - V_o}{R_1} \tag{4.68}$$

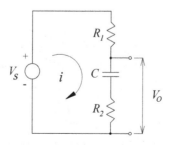

Figure 4.12 Electrical circuit

Taking the derivative of this equation gives

$$\frac{di}{dt} = \frac{1}{R_1}(\frac{dV_s}{dt} - \frac{dV_o}{dt})$$

(4.69)

Substituting equations (4.68) and (4.69) in equation (4.67) gives

$$\frac{dV_o}{dt} = \frac{1}{R_1 C}(V_s - V_o) + \frac{R_2}{R_1}(\frac{dV_s}{dt} - \frac{V_o}{dt})$$

(4.70)

Rearranging this equation gives

$$C\frac{dV_o}{dt} + \frac{1}{(R_1 + R_2)}V_o = \frac{R_2 C}{(R_1 + R_2)}\frac{dV_s}{dt} + \frac{1}{(R_1 + R_2)}V_s$$

(4.71)

This equation represents the dynamics of the given electrical circuit shown in Figure 4.12.

Example (4.6):
Derive the differential equation(s) that describe the dynamics of the electrical circuit shown in Figure 4.13 and put the results in matrix form.

Solution:
Applying Kirchhoff's voltage law to the loops of the given circuit shown in Figure 4.13 gives

$$R_1(i_1 - i_2) + \frac{1}{C_2}\int (i_1 - i_3)\,dt = E(t)$$

(4.72)

$$\frac{1}{C_1}\int i_2\,dt + R_1(i_2 - i_1) = 0$$

(4.73)

and

$$\frac{1}{C_2}\int (i_3 - i_1)\,dt + R_2 i_3 = 0$$

(4.74)

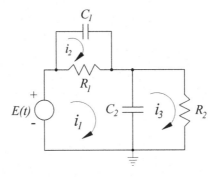

Figure 4.13 Electrical circuit.

Equations (4.72), (4.73), and (4.74) are the differential equations representing the dynamics of the given circuit. In terms of the electrical charge q, inserting $i = dq/dt$ in these equations and carrying out the integration terms gives

$$R_1(\dot{q}_1 - \dot{q}_2) + \frac{1}{C_2}(q_1 - q_3) = 0 \tag{4.75}$$

$$\frac{1}{C_1}q_2 + R_1(\dot{q}_2 - \dot{q}_1) = 0 \tag{4.76}$$

and

$$\frac{1}{C_2}(q_3 - q_1) + R_2\dot{q}_3 = 0 \tag{4.77}$$

Rearranging these equations yields

$$R_1\dot{q}_1 - R_1\dot{q}_2 + \frac{1}{C_2}q_1 - \frac{1}{C_2}q_3 = 0 \tag{4.78}$$

$$\frac{1}{C_1}q_2 + R_1\dot{q}_2 - R_1\dot{q}_1 = 0 \tag{4.79}$$

and

$$\frac{1}{C_2}q_3 - \frac{1}{C_2}q_1 + R_2\dot{q}_3 = 0 \tag{4.80}$$

In matrix form:

$$\begin{bmatrix} R_1 & -R_1 & 0 \\ -R_1 & R_1 & 0 \\ 0 & 0 & R_2 \end{bmatrix} \begin{bmatrix} \dot{q}_1 \\ \dot{q}_2 \\ \dot{q}_3 \end{bmatrix} + \begin{bmatrix} \frac{1}{C_2} & 0 & -\frac{1}{C_2} \\ 0 & \frac{1}{C_1} & 0 \\ -\frac{1}{C_2} & 0 & \frac{1}{C_2} \end{bmatrix} \begin{bmatrix} q_1 \\ q_2 \\ q_3 \end{bmatrix} = \begin{bmatrix} E(t) \\ 0 \\ 0 \end{bmatrix} \tag{4.81}$$

Example (4.7):
Derive the differential equation(s) that describe the dynamics of the electrical circuit shown in Figure 4.14 and put the results in matrix form.

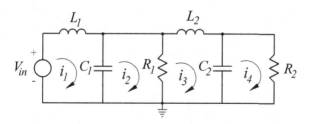

Figure 4.14 Electrical circuit

Solution:
Applying Kirchhoff's voltage law to the loops of the given circuit shown in Figure 4.14 gives

$$L_1\frac{di_1}{dt} + \frac{1}{C_1}\int (i_1 - i_2)\,dt = V_{in} \tag{4.82}$$

$$\frac{1}{C_1}\int (i_2 - i_1)\,dt + R_1(i_2 - i_3) = 0 \tag{4.83}$$

$$L_2\frac{di_3}{dt} + \frac{1}{C_2}\int (i_3 - i_4)\,dt + R_1(i_3 - i_2) = 0 \tag{4.84}$$

and

$$\frac{1}{C_2}\int (i_4 - i_3)\,dt + R_2 i_4 = 0 \tag{4.85}$$

Equations (4.82) to (4.85) are the differential equations representing the dynamics of the given circuit. In terms of the electrical charge q, inserting $i = dq/dt$ in these equations and carrying out the integration terms gives

$$L_1\ddot{q}_1 + \frac{1}{C_1}q_1 - \frac{1}{C_1}q_2 = V_{in} \tag{4.86}$$

$$R_1\dot{q}_2 - R_1\dot{q}_3 + \frac{1}{C_1}q_2 - \frac{1}{C_1}q_1 = 0 \tag{4.87}$$

$$L_2\ddot{q}_3 + \frac{1}{C_2}q_3 - \frac{1}{C_2}q_4 + R_1\dot{q}_3 - R_1\dot{q}_2 = 0 \tag{4.88}$$

and

$$\frac{1}{C_2}q_4 - \frac{1}{C_2}q_3 + R_2\dot{q}_4 = 0 \tag{4.89}$$

In matrix form:

$$\begin{bmatrix} L_1 & 0 & 0 & 0 \\ 0 & 0 & 0 & 0 \\ 0 & 0 & L_2 & 0 \\ 0 & 0 & 0 & 0 \end{bmatrix}\begin{bmatrix} \ddot{q}_1 \\ \ddot{q}_2 \\ \ddot{q}_3 \\ \ddot{q}_4 \end{bmatrix} + \begin{bmatrix} 0 & 0 & 0 & 0 \\ 0 & R_1 & -R_1 & 0 \\ 0 & -R_1 & R_1 & 0 \\ 0 & 0 & 0 & R_2 \end{bmatrix}\begin{bmatrix} \dot{q}_1 \\ \dot{q}_2 \\ \dot{q}_2 \\ \dot{q}_3 \end{bmatrix} + $$

$$\begin{bmatrix} \frac{1}{C_1} & -\frac{1}{C_1} & 0 & 0 \\ -\frac{1}{C_1} & \frac{1}{C_1} & 0 & 0 \\ 0 & 0 & \frac{1}{C_2} & -\frac{1}{C_2} \\ 0 & 0 & -\frac{1}{C_2} & \frac{1}{C_2} \end{bmatrix}\begin{bmatrix} q_1 \\ q_2 \\ q_3 \\ q_4 \end{bmatrix} = \begin{bmatrix} V_{in} \\ 0 \\ 0 \\ 0 \end{bmatrix} \tag{4.90}$$

Example (4.8):
Derive the differential equation(s) that describe the dynamics of the electrical circuit shown in Figure 4.15 and put the results in matrix form.

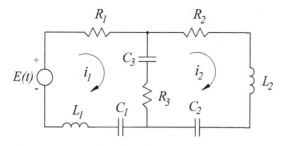

Figure 4.15 Electrical circuit

Solution:
Applying Kirchhoff's voltage law to the current loops of the given circuit shown in Figure 4.15 gives

$$L_1 \frac{di_1}{dt} + R_1 i_1 + R_3(i_1 - i_2) + \frac{1}{C_3} \int (i_1 - i_2)\, dt + \frac{1}{C_1} \int i_1\, dt = E(t) \qquad (4.91)$$

and

$$L_2 \frac{di_2}{dt} + R_2 i_2 + R_3(i_2 - i_1) + \frac{1}{C_3} \int (i_2 - i_1)\, dt + \frac{1}{C_2} \int i_2\, dt = 0 \qquad (4.92)$$

Equations (4.91) and (4.92) are the differential equations representing the dynamics of the given circuit. In terms of the electrical charge q, inserting $i = dq/dt$ in these equations and carrying out the integration terms gives

$$L_1 \ddot{q}_1 + R_1 \dot{q}_1 + R_3(\dot{q}_1 - \dot{q}_2) + \frac{1}{C_3}(q_1 - q_2) + \frac{1}{C_1} q_1 = E(t) \qquad (4.93)$$

and

$$L_2 \ddot{q}_2 + R_2 \dot{q}_2 + R_3(\dot{q}_2 - \dot{q}_1) + \frac{1}{C_3}(q_2 - q_1) + \frac{1}{C_1} q_2 = 0 \qquad (4.94)$$

Rearranging these equations gives

$$L_1 \ddot{q}_1 + (R_1 + R_3)\dot{q}_1 - R_3 \dot{q}_2 + (\frac{1}{C_1} + \frac{1}{C_3})q_1 - \frac{1}{C_3} q_2 = E(t) \qquad (4.95)$$

and

$$L_2 \ddot{q}_2 + (R_2 + R_3)\dot{q}_2 - R_3 \dot{q}_1 + (\frac{1}{C_2} + \frac{1}{C_3})q_2 - \frac{1}{C_3} q_1 = E(t) \qquad (4.96)$$

In the matrix:

$$\begin{bmatrix} L_1 & 0 \\ 0 & L_2 \end{bmatrix} \begin{bmatrix} \ddot{q}_1 \\ \ddot{q}_2 \end{bmatrix} + \begin{bmatrix} (R_1 + R_3) & -R_3 \\ -R_3 & (R_2 + R_3) \end{bmatrix} \begin{bmatrix} \dot{q}_1 \\ \dot{q}_2 \end{bmatrix} + $$
$$\begin{bmatrix} (\frac{1}{C_1} + \frac{1}{C_3}) & -\frac{1}{C_3} \\ -\frac{1}{C_3} & (\frac{1}{C_2} + \frac{1}{C_3}) \end{bmatrix} \begin{bmatrix} q_1 \\ q_2 \end{bmatrix} = \begin{bmatrix} E(t) \\ 0 \end{bmatrix} \qquad (4.97)$$

4.3.4 ELECTRICAL TRANSFORMER MODEL

The transformer is not an actual element in the usual sense, but it is made up of some passive elements so that it can be thought of as such. Because of its enormous utility, we have included it in this chapter. It is utilized in power distribution systems such as cell phones, laptops, power tools, and small appliances. All have a transformer incorporated into their plug-in unit that transforms the voltage from ± 120 or ± 240 volts AC into whatever voltage the item requires.

A transformer comprising two separated inductors transmits the voltage from one value to another, as shown in Figure 4.16. The coils of the inductor are referred to as the primary and secondary. During typical operation, the input voltage source $E(t)$ is connected to the primary, and the secondary generates the transformed output voltage. In the case of a simple transformer, the value of the output voltage e_s is nearly entirely determined by the value of the input voltage e_p, as well as the ratio between the numbers of turns n_p and n_s in the primary and secondary coils, as shown in Figure 4.16. Faraday's law of induction for the secondary coil gives its induced voltage e_s (assuming the coil resistance is negligible) to be

$$e_s = -n_s \frac{\Delta\lambda}{\Delta t} \tag{4.98}$$

where n_s is the number of turns of the secondary coil, and $\Delta\lambda/\Delta t$ is the rate of change of magnetic flux. The input primary voltage e_p is also related to changing the magnetic flux by

$$e_p = -n_p \frac{\Delta\lambda}{\Delta t} \tag{4.99}$$

Kirchhoff's law states that the induced voltage e_p exactly equals the input voltage $E(t)$ (neglecting the coil resistance). Because the two inductors are similar except for the turns on their coils, as is the magnetic field strength, $\Delta\lambda/\Delta t$ is the same on both sides. So, taking the ratio of equations (4.98) and (4.99) yields

$$\frac{e_s}{e_p} = \frac{n_s}{n_p} \tag{4.100}$$

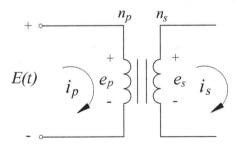

Figure 4.16 Electrical circuit

As indicated in this equation, the ratio of the secondary voltage e_s to the primary voltage e_p equals the ratio of the number of turns in the secondary and primary coils n_1/n_2. Depending on this ratio, the output voltage of a transformer may be lower, equal, or greater than the input voltage $E(t)$. Now, let us determine the ratio of the currents in both the primary and secondary circuits. The electrical power output of a transformer is equal to its input if the coil resistance is negligible. When the power input and output are equalized, we get

$$i_p e_p = i_s e_s \tag{4.101}$$

Rearranging this equation gives

$$\frac{e_s}{e_p} = \frac{i_p}{i_s} \tag{4.102}$$

Substituting this equation in equation (4.100), we get

$$\frac{i_s}{i_p} = \frac{n_p}{n_s} \tag{4.103}$$

This is the relationship between a transformer's output and input currents. According to equation (4.102), if the voltage increases, the current will decrease, and vice versa.

Example (4.9):
Find the relationship between the output voltage e_o and the input voltage $E(t)$ of the transformer circuit shown in Figure 4.17. Discuss how to change the transformation ratio n_1/n_2 to reduce the transformer output voltage e_o.

Solution:
As shown in Figure 4.17, the circuit consists of two separate serial circuits, primary and secondary, joined together by a transformer with two inductance coils with a turns ratio of n_1/n_2. Applying Kirchhoff's voltage law gives

- Primary circuit:

$$L\frac{di_1}{dt} + e_p = E(t) \tag{4.104}$$

where e_p is the primary inductance voltage.

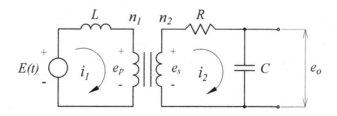

Figure 4.17 Transformer circuit

- Secondary circuit:

$$e_s = Ri_2 + \frac{1}{C} \int i_2 \, dt \qquad (4.105)$$

where e_s is the secondary inductance voltage.

Recalling equations (4.100) and (4.103) as follows

$$e_p = \frac{n_1}{n_2} e_s \qquad (4.106)$$

$$i_2 = \frac{n_1}{n_2} i_1 \qquad (4.107)$$

Substituting equation (4.106) in equation (4.104) gives

$$L\frac{di_1}{dt} + \frac{n_1}{n_2} e_s = E(t) \qquad (4.108)$$

Solving this equation for e_s gives

$$e_s = \frac{n_2}{n_1}[E(t) - L\frac{di_1}{dt}] \qquad (4.109)$$

Substituting equations (4.107) and (4.109) in equation (4.105) gives

$$\frac{n_2}{n_1}[E(t) - L\frac{di_1}{dt}] = \frac{n_1}{n_2} Ri_1 + \frac{1}{C} \int i_2 \, dt = 0 \qquad (4.110)$$

Note that the capacitance voltage equals the transformer output voltage e_o. So,

$$e_o = \frac{1}{C} \int i_2 \, dt \qquad (4.111)$$

Substituting this equation in equation (4.110) and rearranging the results gives

$$\frac{n_1}{n_2} Ri_1 + e_o = \frac{n_2}{n_1}[E(t) - L\frac{di_1}{dt}] \qquad (4.112)$$

Solving this equation for e_o, gives

$$e_o = \frac{n_2}{n_1}[E(t) - L\frac{di_1}{dt}] - \frac{n_1}{n_2} Ri_1 \qquad (4.113)$$

This equation represents the relationship between the transformer output voltage e_o and the input voltage $E(t)$. Based on this equation, if we want to lower the transformer's output voltage, we need to make the secondary coil's turns n_2 lower than that of the primary coil n_1.

4.3.5 OPERATIONAL AMPLIFIER MODEL

The operational amplifier is not an actual element in the usual sense but comprises some passive elements, often called op-amp. It is likely the most popular component in any integrated circuit. We can construct instrumentation amplifiers, filters, controllers, analog signal processing devices, and more with the help of the operational amplifier.

As shown in Figure 4.18, the input consists of two terminals: e_1 with a minus sign (for inverting) and e_2 with a plus sign (for non-inverting). One of these terminals is often connected to the ground. The output voltage e_o is given by

$$e_o = A(e_2 - e_1) \quad \text{or} \quad e_o = -A(e_1 - e_2) \tag{4.114}$$

The voltages e_1 and e_2 can be either DC or AC signals, and A is an adjustable gain known as differential gain or voltage gain, and its value is established by what we choose. As equation (4.114) indicates, the op-amp amplifies the difference between e_1 and e_2 by a gain A. Since the op-amp amplifies the input signal, the gain A should be of a very high value, implying that it will be inherently unstable. In this case, it is necessary to have feedback from the output to the input to stabilize it. Since the feedback signal should be in a negative sign, it is connected from the output to the inverted input, the e_1 terminal, as shown in Figure 4.18. In general, for ideal operation of the op-amp, no current flows into the input terminals, and the load connected to the output terminal does not affect the output voltage. In the following analysis, we will assume that the operation of the op-amp is ideal.

Example (4.10):
For the operational amplifier circuit shown in Figure 4.19, obtain the output voltage e_o.

Solution:
As shown in Figure 4.19, the input voltage $e(t)$ is supplied to the amplifier input terminals. We aim to establish a relationship between e_o and $e(t)$. Applying Kirchhoff's current node law to the summing junction of voltage e_m, we get

$$i_1 = i_2 + i_3 \tag{4.115}$$

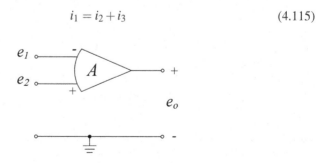

Figure 4.18 Operational amplifier

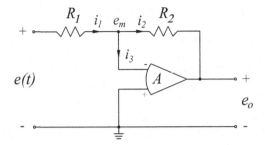

Figure 4.19 Inverting op-amp

It is assumed that no current flows into the amplifier input terminals, meaning that
$i_3 = 0$. Considering this, equation (4.115) becomes

$$i_1 = i_2 \tag{4.116}$$

and i_1 and i_2 are given by

$$i_1 = \frac{(e(t) - e_m)}{R_1} \quad \text{and} \quad i_2 = \frac{(e_m - e_o)}{R_2} \tag{4.117}$$

Note that the voltage e_m is related to e_o by $e_o = -Ae_m$, thus

$$e_m = \frac{-e_o}{A} \tag{4.118}$$

Considering this equation and Substituting equation (4.117) in equation (4.116) gives

$$\frac{(e(t) - (-e_o/A))}{R_1} = \frac{((-e_o/A) - e_o)}{R_2} \tag{4.119}$$

Solving this equation for e_o gives

$$e_o = -\frac{R_2}{R_1} e(t) \tag{4.120}$$

This equation indicated that the voltage at the output of the op-amp, shown in Figure
4.19, is the negative of the voltage at its input. As a result, this circuit is known as an
inverting operational amplifier.

4.3.6 MULTI-DOMAIN SYSTEM MODELING

When modeling systems, it is essential to distinguish between single-domain and
multi-domain systems. Mechanical, electrical, hydraulic, thermal, or other systems
are typical single-domain systems. Their description languages are available, which
can be used to model and analyze each system separately. A system is called a multi-
domain system when it has multiple single-domain systems that differ from the other

types (such as electro-mechanical, electro-hydraulic, or mechatronic systems). It is essential to understand how they physically interact. Establishing mathematical relationships between domains is difficult since doing so requires figuring out the variables in each domain, which makes the task challenging. Controlling the rotational speed of a machine, such as a motor shaft, is required in a wide variety of commercial applications. A DC motor is commonly utilized for this purpose. It transforms electrical power into mechanical rotation, implying that it operates on electrical and mechanical power, making it a typical multi-domain system. In such a system, we may need to keep the speed of the motor shaft constant regardless of the disturbing torques, or we need to change the speed in response to a control accurately. In general, we can control the speed of a DC motor by changing the field current i_f, in which case the motor is called field-controlled, or the armature current i_a, in which case the field current is assumed constant, and the motor is called armature-controlled, as shown in Figure 4.20. The armature-controlled DC motor with a constant field current will be the only focus of this analysis. The dynamics of a DC motor will be described as a multi-domain system, and the corresponding dynamics equations will be derived in the following example.

Example (4.11):
For the armature-controlled DC motor shown in Figure 4.20, deduce the differential equations describing its dynamic behavior.

Solution:
Figure 4.20 shows that a DC motor transforms electrical power into a mechanical rotating motion. The electrical circuit and the mechanical part can both be considered separate subsystems when looking at this figure. The torque T_m exerted by the electrical side on the mechanical side and the back emf voltage e_b exerted by the mechanical side on the electrical side are how these two subsystems interact. The back emf and the torque T_m are given the status of external forces on the appropriate subsystems since the electrical and mechanical subsystems are considered separately. A mathematical model can represent the electrical domain as a simple serial circuit,

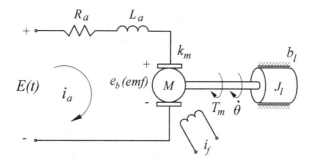

Figure 4.20   Electro-mechanical DC motor system

as shown in Figure 4.20. Since the DC motor is armature-controlled, the field current i_f is assumed to be constant and has a negligible effect. The voltage source $E(t)$, the resistance in the armature winding R_a, and the inductance in the armature winding L_a make up the electrical elements of this circuit. The constant flow of current i_a in the circuit is because the circuit is a serial connection; nevertheless, the voltage of each electrical element is different. The mechanical domain covers concepts like the rotational inertia J_l of a rotating mass, the motor's electro-mechanical conversion of a motor-torque constant k_m and a bearing of damping factor b_l as shown in Figure 4.20. Now, let us derive the equations of motion for the given DC motor.

- Electrical domain:
Applying Kirchhoff's voltage law, the dynamic equation of the electrical circuit shown in Figure 4.20 is given by

$$L_a \frac{di_a}{dt} + R_a i_a + e_b = E(t) \tag{4.121}$$

The back *emf* voltage e_b is directly proportional to the angular speed $\dot{\theta}$ of the motor, which is given by
$$e_b = k_b \dot{\theta} \tag{4.122}$$

where k_b is the back *emf* constant. Considering this and substituting equation (4.122) in equation (4.121) gives

$$L_a \frac{di_a}{dt} + R_a i_a + k_b \dot{\theta} = E(t) \tag{4.123}$$

This is the dynamic equation of the electric circuit (electrical domain).

- Mechanical domain:
Since the torque T_m developed to the motor, applying Newton's law to the mechanical component shown in Figure 4.20 gives

$$J_l \ddot{\theta} + b_l \dot{\theta} = T_m \tag{4.124}$$

Considering the field current i_f is constant, only the armature current i_a produces the torque T_m that is applied to the inertia element J_l, which is directly proportional to the armature current i_a, so
$$T_m = k_m i_a \tag{4.125}$$

where k_m is the motor-torque constant. Substituting equation (4.125) in equation (4.124) gives

$$J_l \ddot{\theta} + b_l \frac{d\theta}{dt} = k_m i_a \tag{4.126}$$

This equation is the differential equation of motion for the rotating mass J_l (mechanical domain). During the steady-state operation of the motor, when the rotor resistance is ignored, the bake emf k_b constant will equal the motor-torque constant

k_m. The power balance during the steady-state operation reveals this equality. That is, the power input to the rotor is $e_b i_a = (k_b \dot{\theta}) i_a$, and the power delivered to the motor shaft is $(k_b \dot{\theta}) i_a = T_m \dot{\theta}$; since $T_m = k_m i_a$, we deduce that $k_b = k_m$.

Equations (4.123) and (4.126) are the differential equations describing the dynamics of the multi-domain DC motor system shown in Figure 4.20. Now, the question is: How do these equations change if the motor is driven with a current source $i(t)$? (Homework exercise).

4.4 ANALOGOUS SYSTEMS

While analogous systems can be represented using the same mathematical model, their physical properties differ. This is how an analogous system is defined. As a result, differential equations can represent a wide variety of systems. The advantage of analogous systems is that the results of equations representing one system can be directly transferred to another analogous system, even though the two systems are physically different. This is true regardless of the subject area in which the system is being investigated. This section will examine the potential analogies between mechanical and electrical systems and their opposites, from electrical to mechanical.

4.4.1 MECHANICAL-ELECTRICAL ANALOGIES

It is essential to understand mechanical systems by studying their electrical analogs, which, in many cases, can be constructed with less effort than models of the corresponding mechanical systems. The force-voltage analogy and the force-current analogy are the two types of electrical analogies that can be used to describe mechanical systems.

It is essential to remember that analogies between two systems become discredited when the regions of operation are extended beyond their natural boundaries. So, because the mathematical models on which the analogies are based are only approximations of the dynamic behavior of physical systems, the analogy may struggle if the operating zone of the system is vast. Despite this, if the operating zone of the given mechanical system is large, it can be divided into two or more subregions, and equivalent electrical systems can be constructed for each subregion. The concept of analogy, of course, is not restricted to comparisons of mechanical and electrical systems; instead, it includes analogies of any system, physical or otherwise. Some advantages of creating analogous systems are outlined below

1. Analogous systems have a transfer that is identical to one another or has mathematical models identical.
2. Analogous systems will always display the same outcome with the same input. Applying well-established findings from one region to another can benefit from using analogy as a conceptual framework.
3. It is beneficial when a particular physical system (mechanical, fluid, thermal, etc.) is complex and would benefit from being analyzed using the lens of an analog electrical circuit.

Table 4.1

Force-current and force-voltage analogy

Mechanical Systems	Force-Current Analogy	Force-Voltage Analogy
Force $F(t)$ or Torque $T(t)$	Current i	Voltage $E(t)$
Mass m or moment of inertia J	Capacitance C	Inductance L
Viscous-friction coefficient b	Reciprocal of resistance $1/R$	Resistance R
Spring Constance k	Reciprocal of inductance $1/L$	Reciprocal of capacitance $1/C$
Displacement x (angular displacement θ)	Charge q	Electrical Systems
Velocity $\dot{x}$ (angular velocity $\dot{\theta}$)	Current i	

Finally, it is worth emphasizing that all the concepts mentioned above are considered whenever the need arises to create a mechanical analog system for electrical systems. As stated above, two methods can generate an analog electrical system: Force-voltage and Force-current. The terms known as analogous variables refer to the terms that occupy corresponding positions in the differential equations. A list of analog variables based on Kirchhoff's voltage law may be found in Table 4.1.

- Force-voltage analogy:

Considering the mechanical system shown in Figure 4.21, derive the equation of motion and create the corresponding analog electrical circuit using the Force-voltage analogy. As shown in the same figure, the system consists of a mass m connected to a combination of a spring with stiffness k and a damper with a damping factor b positioned in a parallel connection. When the external force $F(t)$ is applied to the system, the mass, damper, and spring move a displacement x. Applying Newton's

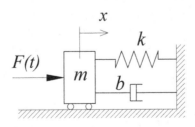

Figure 4.21 Mechanical system

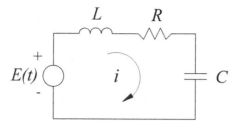

Figure 4.22 Electrical circuit

law, the equation of motion of the system is given by

$$m\frac{d^2x}{dt^2} + b\frac{dx}{dt} + kx = F(t) \tag{4.127}$$

Let us create the corresponding analogous electrical circuit using the Force-voltage analogy. Applying the Force-voltage analogy presented in Table 4.1 and using the formulas indicated in equation (4.128), the external force $F(t)$ corresponds to the voltage source $E(t)$ in the electrical system, while the mass m, the damper b, and the spring k correspond, respectively, to the conductor L, the resistance R, and the capacitor C, as clearly indicated in the following equation.

$$|m_i| = |L_i|, \ |k_i| = |\frac{1}{C_i}|, \ |b_i| = |R_i|, \ |F_i| = |E_i|, \ \text{and} \ |x_i| = |q_i| \tag{4.128}$$

Since the mass, damper, and spring all move the same distance x, the corresponding analog electrical circuit must consist of a single loop carrying a single current i, as shown in Figure 4.22. Applying Kirchhoff's voltage law, the dynamic equation for the electrical system is given by

$$L\frac{di}{dt} + Ri + \frac{1}{C}\int i\,dt = E(t) \tag{4.129}$$

In electrical charge, q this equation becomes

$$L\frac{d^2q}{dt^2} + R\frac{dq}{dt} + \frac{1}{C}q = E(t) \tag{4.130}$$

Equations (4.127) and (4.130) can be compared to show that the differential equations for both systems have the same form. Hence, these two systems are analogs of one another. The analogy here is called Force-voltage or mass-inductance.

- Force-current analogy:

Consider the same mechanical system shown in Figure 4.21. Here, we will apply the Force-Current analogy. It is clear that the external force $F(t)$ in a mechanical system is represented by the current source $i(t)$ in the electrical system. The capacitor C,

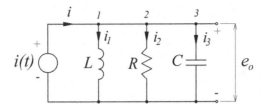

Figure 4.23 Electrical circuit

the resistance R, and the conductor L correspond to the mass m, damper b, and the spring k, respectively, as presented in Table 4.1 and shown in Figure 4.23, given by the following formulas.

$$|m_i| = |C_i|, \; |k_i| = |\frac{1}{L_i}|, \; |b_i| = |\frac{1}{R_i}|, |F_i| = |E_i| \text{ and } |x_i| = |q_i| \tag{4.131}$$

Note that the equation of motion of the mechanical system shown in Figure 4.21 is indicated in equation (4.127). Applying Kirchhoff's current law, the dynamic equation for the electrical system shown in Figure 4.23 is given by

$$i_1 + i_2 + i_3 = i(t) \tag{4.132}$$

where

$$i_1 = \frac{1}{L} \int e_o \, dt, \quad i_2 = \frac{e_o}{R} \quad \text{and} \quad i_3 = C\frac{de_o}{dt} \tag{4.133}$$

Substituting equation (4.133) in equation (4.132) gives

$$\frac{1}{L} \int e_o \, dt + \frac{e_o}{R} + C\frac{de_o}{dt} = i(t) \tag{4.134}$$

Since the magnetic flux linkage λ is related to the voltage e_o by equation

$$\dot{\lambda} = e_o \tag{4.135}$$

Considering this equation in terms of λ, equation (4.134) becomes

$$C\frac{d^2\lambda}{dt^2} + \frac{1}{R}\frac{d\lambda}{dt} + \frac{1}{L}\lambda = i(t) \tag{4.136}$$

Comparing equations (4.136) and (4.127), we found that both systems are analogous. The analogous variables are listed in Table 4.1. The analogy here is called the force-current analogy or mass-capacitance analogy.

4.4.2 ELECTRICAL-MECHANICAL ANALOGIES

In the same way, we can turn an electrical circuit into a mechanical system using Table 4.1.

Example (4.12):
Using the force-voltage analogy, obtain a mechanical analog of the electrical system shown in Figure 4.24.

Solution:
Applying Kirchhoff's voltage law to the given circuit, shown in Figure 4.24 gives

$$L_1\frac{di_1}{dt} + R_1 i_1 + R_3(i_1 - i_2) + \frac{1}{C_3}\int (i_1 - i_2)dt + \frac{1}{C_1}\int i_1 dt = E(t) \qquad (4.137)$$

$$L_2\frac{di_2}{dt} + R_2 i_2 + R_3(i_2 - i_1) + \frac{1}{C_3}\int (i_2 - i_1)dt + \frac{1}{C_2}\int i_2 dt = 0) \qquad (4.138)$$

Recalling the definition of the current, $i = dq/dt$, equations (4.137) and (4.138) become

$$L_1\ddot{q}_1 + R_1\dot{q}_1 + R_3(\dot{q}_1 - \dot{q}_2) + \frac{1}{C_3}(q_1 - q_2) + \frac{1}{C_1}q_1 = E(t) \qquad (4.139)$$

$$L_2\ddot{q}_2 + R_2\dot{q}_2 + R_3(\dot{q}_2 - \dot{q}_1) + \frac{1}{C_3}(q_2 - q_1) + \frac{1}{C_1}q_2 = 0 \qquad (4.140)$$

By converting equations (4.139) and (4.140) to mechanical system differential equations using the Force-voltage analogy, we get

$$m_1\ddot{x}_1 + b_3(\dot{x}_1 - \dot{x}_2) + k_3(x_1 - x_2) + k_1 x_1 + b_1\dot{x}_1 = F \qquad (4.141)$$

$$m_2\ddot{x}_2 + b_2\dot{x}_2 + b_3(\dot{x}_2 - \dot{x}_1) + k_3(x_2 - x_1) + k_2 x_2 = 0 \qquad (4.142)$$

By converting these two equations into a mechanical system, we get the mechanical system shown in Figure 4.25. This figure represents a mechanical analogy of the electrical system shown in Figure 4.24.

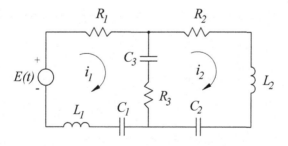

Figure 4.24 Electrical circuit

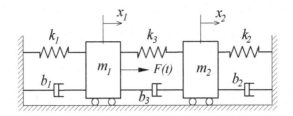

Figure 4.25 Mechanical system

4.5 BOND GRAPH CONSTRUCTIONS OF ELECTRICAL SYSTEMS

Constructing Bond Graphs for electrical circuits is relatively simple; we must understand how the circuit elements correspond to the various Bond Graph elements.

- The passive electrical elements, inductance L, capacitance C, and resistance R, are represented by inertia I, capacitance C, and resistance R Bond Graph elements, respectively. As stated previously, inductance integrates the effort (voltage), which indicates that the effort determines the flow (current). Its voltage-current relationship is given by (see Figure 4.26),

$$e_L = L\frac{di_L}{dt} = L\frac{d^2 q_L}{dt^2} \tag{4.143}$$

while its effort-flow relationship is given by

$$e_l = I f_l = I \lambda \tag{4.144}$$

where q_L is the electrical charge, i_L is the current passing through the inductance, λ is the magnetic flux linkage of the inductance, e_l is the inductance effort, and f_l is the inductance flow. Note that the magnetic flux linkage λ in the electrical system is equivalent to the momentum p of the mass in the mechanical system.

The effect of the capacitor is represented by the time integral of the flow, implying that the effort is calculated based on the flow. Its voltage-current is given by (see Figure 4.27)

$$e_C = \frac{1}{C} \int i_C \, dt = \frac{1}{C} \int \dot{q}_C \, dt = \frac{1}{C} q_C \tag{4.145}$$

while its effort-flow relationship is given by

$$e_c = \frac{1}{C} \int f_c \, dt = \frac{1}{C} \int \dot{q}_c \, dt = \frac{1}{C} q_c \tag{4.146}$$

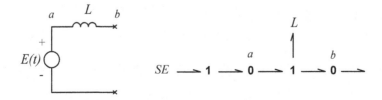

Figure 4.26 Electrical inductance element and the corresponding Bond Graph junctions

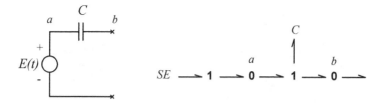

Figure 4.27 Electrical capacitance element and the corresponding Bond Graph junctions

where q_C is the electrical charge, i_C is the current passing through the capacitor C, q_c is the capacitance state variable, e_c is the capacitance effort, and f_c is the capacitance flow.

Since the resistance is a dissipation element, its voltage-current relationship is given by (see Figure 4.28)

$$e_R = Ri_R = R\dot{q}_R \tag{4.147}$$

while its effort-flow relationship is given by

$$e_r = Rf_r = R\dot{q}_r \tag{4.148}$$

where q_R is the electrical charge, i_R is the current passing through the resistance R, q_r is the Bond Graph resistance variable, e_r is the resistance effort, and f_r is the resistance flow.

- The voltage source $E(t)$ is represented by a source of effort SE, while the current source $i(t)$ is represented by a source of flow SF, see Figures 4.26 to 4.28.
- A wire without branches is referred to as a 1-junction, whereas a node in which wires are connected refers to a 0-junction.
- The transformer TF and the Gyrator GY are used for their appropriate functions. Considering the transformation factor n_1/n_2 for the transformer, ensure the power balance across the bonds is satisfied. Its voltage-current relationship is given by (see Figure 4.29)

$$e_{T_1} = \frac{n_1}{n_2}e_{T_2} \text{ and } i_1 = \frac{n_1}{n_2}i_2 \tag{4.149}$$

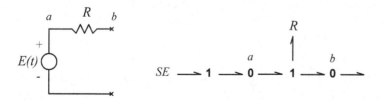

Figure 4.28 Electrical resistance element and the corresponding Bond Graph junctions

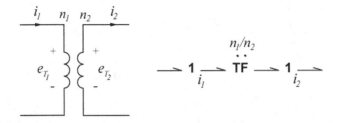

Figure 4.29 Electrical transformer element and the corresponding junctions

while its effort-flow relationship is given by

$$e_{t_1} = \frac{n_1}{n_2} e_{t_2} \text{ and } f_{t_1} = \frac{n_1}{n_2} f_{t_2} \tag{4.150}$$

For the gyrator, considering the gyration factor k_m, check the resulting electromotive force relations by checking them out. Its voltage-current relationship is given by (see Figure 4.30)

$$e_{T_1} = k_m i_2 \text{ and } i_1 = k_m e_{T_2} \tag{4.151}$$

while its effort-flow relationship is given by

$$e_{t_1} = k_m f_{t_2} \text{ and } f_{t_1} = k_m e_{t_2} \tag{4.152}$$

4.5.1 JUNCTIONS REPRESENTATIONS IN ELECTRICAL SYSTEMS

As previously stated, considering the description of the Bond Graph junctions, the simplest way to determine if the junction is a 0-junction or a 1-junction is to check if all voltages (efforts) at the junction's elements are identical (0-junction) or if all currents (flow) have the same value (1-junction). In other words, if some elements have the same current, they should be connected to a 1-junction, while if they have the same voltage, they should be connected to a 0-junction. Considering these situations, to create a Bond Graph model for an electrical circuit, the following steps should be considered.

1. Create a 0-junction for each unique voltage.
2. Connect each 1-port element L, C, R, SE, or SF to a 1-junction before bonding that 1-junction to two 0-junctions, as shown in Figures 4.26 to 4.28.

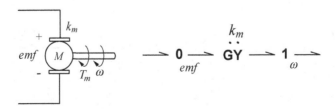

Figure 4.30 Electrical gyrator element and the corresponding Bond Graph junctions

3. Insert Transformer element TF to set the transformation required as shown in Figure 4.29.
4. Insert Gyrator element GY to set the inherent limitations between 0-junction and 1-junction, as shown in Figure 4.30.
5. Assign a power flow half arrow using the through sign convention for the 1-junction and two 0-junction parts created in points (1) to (4), as shown in Figures 4.26 to 4.30. Remember that sources SE and SF supply power while L, C, and R elements receive it.
6. The ground is allocated to a specific node, and all bonds related to it must be removed to simplify the model. To simplify the Bond Graph further, any 2-port 1-junctions or 0-junctions with through power half arrows should be swapped by single bonds, as shown in Figure 3.27.
7. Finally, to finish, we should add causality strokes and bond numbers. As a result, the Augmented Bond Graph is ready to derive the differential equations describing the dynamics of electrical systems. It is worth noting that the causality restrictions applied to the Bond Graph model of electrical systems will be the same as those applied to the mechanical system (Section 3.3).

4.5.2 CAUSALITY FOR BOND GRAPH MODEL OF ELECTRICAL SYSTEM

As stated above, the processes and restrictions limiting causality assignment in electrical systems are identical to those we followed when studying mechanical systems. Remember that:

- Source elements SE and SF have defined causality. The causal stock mark is placed at the junction side for the voltage source $E(t)$ and at the element side of the bond for the current source $i(t)$.
- 1-port elements 0-junction, 1-junction, TF, and GY have restricted causality possibilities.
- The inductance L and capacitance C elements must have integral causality.
- Resistances have arbitrary causality.

To assign the causality of the Bond Graph model of an electrical system, the following steps must be taken into consideration:

1. Choose any SE or SF and assign the required causality. Extend causal implications using all 0-junction, 1-junction, TF, and GY restrictions that apply.
2. Repeat step (1) until all sources have causality assigned.
3. Choose any C or L and assign integral causality. Again, extend causal implications using the 0-junction, 1-junction, TF, and GY restrictions that apply.
4. Repeat step (3) until all C or L elements have causality assigned.
5. Choose any resistance element R that is unassigned and give it arbitrary causality. Extend the causal implications.
6. Repeat step (5) until all R elements have been assigned.

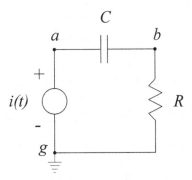

Figure 4.31 Electrical elements and the corresponding Bond Graph junctions

To understand how to create a Bond Graph model for an electrical circuit, let us look at applying the above procedures to a simple circuit shown in Figure 4.31, as follows:

- The circuit has three distinct potentials, including an explicit ground node a, b, and g, as shown in Figure 4.31.
- These nodes are represented by three 0-junctions, as shown in Figure 4.32.
- The three 1-port elements, SF, C, and R have been inserted using three 1-junctions positioned between the three 0-junctions, which are already created as shown in Figure 4.32. Note that powers have been directed through each 1-junction. Figure 4.32 shows the preliminary Bond Graph model of the given circuit.

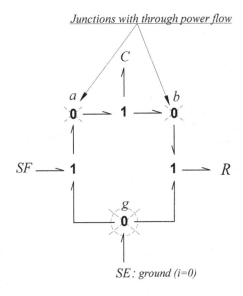

Figure 4.32 Preliminary Bond Graph model of the given circuit

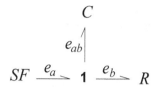

Figure 4.33 Simplified Bond Graph model of the given circuit

- To simplify the Bond Graph model shown in Figure 4.32, first, the ground 0-junction g and its bonds should be eliminated, and second, condensing the 2-port 0-junctions and 1-junction to simple bonds, meaning that any 1-junction and 0-junction that has a through power flow direction should be removed. As a result, the simplified Bond Graph model of the given circuit is just what we expect: a model of a simple series connection, as shown in Figure 4.33.

The RLC circuits shown in Figures 4.34 and 4.38 are the ones we will use to demonstrate how to implement the Bond Graph technique for modeling the dynamics of electrical circuits.

Example (4.13):
For each electrical circuit shown in Figures 4.34 and 4.38,

- Create the preliminary Bond Graph model and simplify it.
- Create the Augmented Bond Graph model (choose the half arrows positive power directions, assign the causality marks, and number the bonds).
- Write the key variables (Input, State, and Co-energy variables) and their constitutive relations.
- Derive the dynamic equation and check if Kirchhoff's law is embedded in the Augmented Bond Graph model.

Solution:
- RLC circuit in series connection:

Let us begin with the RLC circuit with a series connection shown in Figure 4.34. To create the circuit's Bond Graph model, we must follow the processes outlined in Section 4.4.1. As shown in Figure 4.34, each element has its own way of interacting with the other elements. That is, the voltage source $E(t)$ is the element that provides power, and the capacitor C and the inductance L serve as temporary stores of energy, while the resistance R is the element that consumes power. Due to the constraint imposed by the series connection, only one current i flows through the circuit, which indicates that all elements share the same flow i, as shown in Figure 4.34. In contrast, each element provides a different voltage drop. So, three 0-junctions are created for the voltage drop across the three elements, as shown in Figure 4.35.
Since all elements share the same flow i, three 1-junctions were created, as shown in the same figure. Each element is connected with a 1-junction positioned between

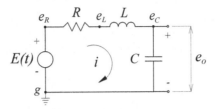

Figure 4.34 RLC circuit in series connection

two 0-junctions. This provides for the voltage drop across each element. The source
of effort *SE* is connected to a 1-junction of a common flow *i*. Since the capacitor, *C*
has established a link to the ground, represented by a source of effort, *SE* : *ground*,
and is connected through a 0-junction that is positioned between two 1-junctions, as
shown in Figure 4.35. A half-arrow must be added to all bonds to indicate the direction of positive power flow. It is important to note that the energy storage elements
(inductance and capacitance) and the dissipating energy element (resistance) receive
power, while the source of effort, *SE*, supplies power. As a result, the half-arrow positive power directions of all bonds are selected, as shown in Figure 4.36. This figure
shows the preliminary Bond Graph model of the given circuit. To simplify this Bond
Graph model, all connections that hold ground with their associated bonds must be
removed, and then any 2-port 1-junctions or 0-junctions with through power half arrows should be swapped with single bonds, as shown in Figure 4.36. The resulting
simplified Bond Graph model of the given circuit is shown in Figure 4.37.

Now, let us assign the model causality marks. As mentioned previously, the
causality restrictions applied to the Bond Graph model of electrical circuits will be
the same as those applied to mechanical systems. So, to assign causality to the Bond
Graph model shown in Figure 4.37, we have to follow the following steps:

1. Sources have to define causality. We have only one source of effort, *SE* representing the input voltage, $E(t)$. So, the causal stroke mark should be located at the
   junction side of its bond.
2. inductance and capacitor elements should have integral causality. So, the causality
   of the inductance *L* element and the capacitor *C* element should be located on their
   bonds.

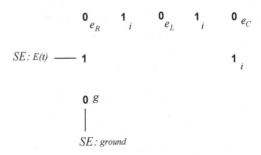

Figure 4.35 Junction structures of Bond Graph model of RLC circuit in parallel connection

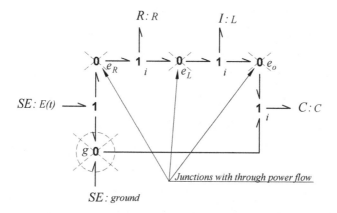

Figure 4.36 Preliminary Bond Graph model of RLC circuit in series connection

3. Resistance element R has arbitrary causality. Applying the restriction causality of the 1-junction, which represented serial connections of bonds with the same flow i, the causal stroke mark of the element R should be located at the junction side of the R bond.

To finish numbering all bonds, we get the Augmented Bond Graph model shown in Figure 4.37. To derive the differential equation representing the dynamics of the given circuit, we are forced to estimate the effort and flow variables of the circuit elements, as we stated previously. As a result, the flow variable can be selected as i, and the effort variable can be chosen as the voltage drop across each element.

Now, we must determine the relationships between four elements: the source of effort SE, inductance L, capacitance C, and resistance R; all should be functions of effort and flow. From the Augmented Bond Graph model shown in Figure 4.37, we have two storage elements of integral causalities, inductance L and capacitance C of bonds 2 and 4, respectively, and one dissipating element R of bond 3.

To derive the equations representing the circuit dynamics from the Augmented Bond Graph model, we must first define the key variables: Inputs, State variables, and Co-energy (power) variables and their constitutive relations.

$$C : C$$
$$\overset{\displaystyle 2}{\big\uparrow}$$
$$SE : E(t) \xrightarrow{\;\;1\;\;} \mathbf{1} \xrightarrow{\;\;4\;\;} I : L$$
$$\underset{\displaystyle 3}{\big\downarrow} \; i$$
$$R : R$$

Figure 4.37 Augmented Bond Graph model of the RLC circuit in series connection

- Key variables:
- Inputs: $e_1 = E(t) = SE$.
- State variables: q_2 and p_4 (integral causality).
- Co-energy (Power) variables: e_2 and f_4.

Constitutive relations:
Write equations describing the relationship between co-energy variables and state variables for each energy storage element, as follows:

$$e_2 = \frac{1}{C}q_2 \text{ and } f_4 = \frac{1}{I}\lambda_4 \tag{4.153}$$

Now that we have the connections between flow and effort variables at the model junctions, let us derive the equations that reflect them.

As shown in Figure 4.37, the flow variables of the 1-junction of a common flow i are given by

$$f_1 = f_2 = f_3 = f_4 \tag{4.154}$$

The effort variables of the 1-junction of a common flow i are given by

$$e_4 = \dot{\lambda}_4 = e_1 - e_2 - e_3 = SE - e_2 - e_3 \tag{4.155}$$

where $\dot{\lambda}_4$ is the rate of change of the inductance flux linkage. The effort e_3 of the resistance R is given by

$$e_3 = Rf_3 \tag{4.156}$$

Equation (4.154) indicates that flow f_3 equals flow f_4. Considering this, equation (4.156) becomes

$$e_3 = Rf_4 \tag{4.157}$$

Using the constitutive relations indicated in equation (4.153) and substituting equations (4.157) in equation (4.155) yields

$$\dot{\lambda}_4 = SE - \frac{R}{I}\lambda_4 - \frac{1}{C}q_2 \tag{4.158}$$

Rearranging this equation gives

$$\dot{\lambda}_4 + \frac{R}{I}\lambda_4 + \frac{1}{C}q_2 = SE \tag{4.159}$$

This equation is the differential equation created from the Augmented Bond Graph model shown in Figure 4.37 that represents the dynamics of the given circuit shown in Figure 4.34.

Now, we will check if Kirchhoff's law is embedded in the Augmented Bond Graph model. The following analyses are what we get when using the constitutive relations indicated in equation (4.153).

- The inductance flux linkage is calculated as follows

$$\lambda_4 = I f_4 = I i \tag{4.160}$$

Considering the definition of the current $i = dq/dt$, equation (4.160) becomes

$$\lambda_4 = I \dot{q} \tag{4.161}$$

Taking the derivative of this equation gives

$$\dot{\lambda}_4 = I \dot{f}_4 = L \frac{di}{dt} = L \ddot{q} \tag{4.162}$$

- The state variable q_2 of the capacitor C is calculated as follows

$$\dot{q}_2 = f_4 = i \tag{4.163}$$

Integrating this equation gives

$$q_2 = \int i \, dt \tag{4.164}$$

Substituting equations (4.161), (4.162), and (4.164) in equation (4.159) and inserting $SE = E(t)$ yields

$$L \frac{di}{dt} + \frac{R}{I} I i + \frac{1}{C} \int i \, dt = E(t) \tag{4.165}$$

Rearranging this equation yields

$$L \frac{di}{dt} + R i + \frac{1}{C} \int i \, dt = E(t) \tag{4.166}$$

As a function of the electrical charge q, equation (4.166) becomes

$$L \ddot{q} + R \dot{q} + \frac{1}{C} q = E(t) \tag{4.167}$$

By comparing equations (4.166) and (4.167) with equations (4.20) and (4.21), respectively, it is found that they are identical, implying that Kirchhoff's voltage law applied to the given circuit is satisfied in the Augmented Bond Graph model.

- RLC elements in parallel connection:
Let us consider the parallel connection of the RLC element shown in Figure 4.38. Due to the constraints of the parallel connections, each element's voltage (effort) will be the same. The given circuit comprises three currents, i_1, i_2, and i_3, and three voltage nodes, 1, 2, and 3, as shown in Figure 4.38. It is important to note that the sum of the currents flowing through each element equals the current flowing from the current source, $i(t)$. The source of flow SF, which represents the current source $i(t)$, is connected to a 1-junction of a common flow i. Each element should be connected

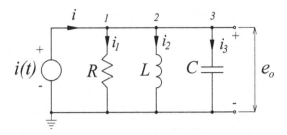

Figure 4.38 Electrical circuit

to a 0-junction positioned between two 1-junctions. As a result, three 0-junctions of common efforts, e_R, e_L, and e_C, are inserted, as shown in Figure 4.39. All elements have established a link to the ground, represented by a source of effort *SE* : *ground*, and are connected through a 0-junction of a common effort V_o. Now, connect all junctions by bonds, as shown in the same figure. A half-arrow must be added to all bonds to indicate the direction of positive power flow, as shown in Figure 4.40. Note that the source of flow *SF* supplies power, while the resistance *R*, the inductance *L*, and the capacitor *C* receive power. This figure shows the preliminary Bond Graph model of the given circuit. This Bond Graph model might be simplified by removing the ground-connecting connections shown in a dashed circle in Figure 4.40, and then any 2-port 1-junctions or 0-junctions with through power half arrows should be swapped with single bonds. As a result, the connection at the 1-junction of the common flow i_1, the connection at the 1-junction of the common flow i_2, and the connection at the 1-junction of the common flow i_3 all must be removed, as shown in Figure 4.40. The resulting simplified Bond Graph model of the given circuit is shown in Figure 4.41.

Now, let us assign the causality of the Bond Graph model. Since there is only one source of flow, *SF*, the causal stroke marks should be located on its bond at the element side, as shown in Figure 4.41. It is essential to remember that as we work through assigning causality, we must consider the integral causality that applies to all

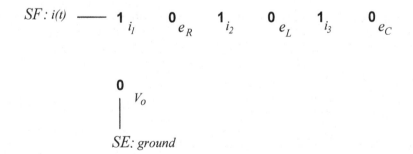

Figure 4.39 Junction structures of Bond Graph model of RLC circuit in parallel connection

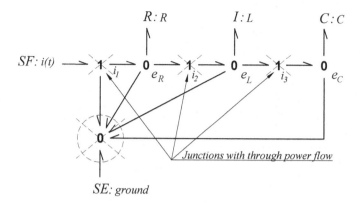

Figure 4.40 Preliminary Bond Graph model of the RLC circuit in parallel connection

associated elements and bonds. Considering the causality restrictions in Section 3.3, apply the causal stork marks to all elements and junctions. As a result, the integral causality of the inductance L and capacitance C should be located as shown in Figure 4.41. Considering the restriction causality of the 0-junction, which represented a parallel connections of bonds with the common effort e_o, the causal stroke mark of the element R should be located at the element side, as shown in the same figure. We finally number all bonds to get the Augmented Bond Graph model shown in Figure 4.41. This figure shows two integrally causal storage elements, L, and C represented by bonds 2 and 4, and one dissipating element, R represented by bond 3.

To derive the dynamic equations of the given circuit, we must first define the key variables of the model: Input, State variables, and Co-energy (power) variables and their constitutive relations.

Key variables:
- Inputs: $f_1 = SF = i(t)$.
- State variables: q_2 and p_4 (integral causality).
- Co-energy (Power) variables: e_2 and f_4.

$$C:C$$
$$SF:i(t) \longmapsto^{1} \mathbf{0} \xrightarrow{4} I:L$$
$$R:R$$

Figure 4.41 Augmented Bond Graph model of the RLC circuit in parallel connection

Constitutive relations:
Write equations describing the relationship between co-energy variables and state variables for each energy storage element, as follows:

$$e_2 = \frac{1}{C}q_2 \text{ and } f_4 = \frac{1}{I}\lambda_4 \tag{4.168}$$

Now that we have the connections of flow and effort variables of the model junctions, let us derive the equations that reflect them.

As shown in Figure 4.41, the effort variables of the 0-junction of common effort e_o are given by

$$e_1 = e_2 = e_3 = e_4 \tag{4.169}$$

and the flow variables are given by

$$f_1 = SF = f_2 + f_3 + f_4 = \dot{q}_2 + f_3 + f_4 \tag{4.170}$$

Using the constitutive relations indicated in equation (4.168), the state variable q_2 of the capacitor C can be calculated as follows

$$q_2 = Ce_2 \tag{4.171}$$

Taking the derivative of this equation gives

$$\dot{q}_2 = f_2 = C\frac{de_2}{dt} \tag{4.172}$$

The flow f_3 of the resistance R is given by

$$f_3 = \frac{1}{R}e_3 \tag{4.173}$$

Using the constitutive relations indicated in equation (4.168) and substituting equations (4.172) and (4.173) in equation (4.170) gives

$$C\frac{de_2}{dt} + \frac{1}{R}e_3 + \frac{1}{I}\lambda_4 = SF \tag{4.174}$$

Remember that the rate of change of the inductance flux linkage $\dot{\lambda}_4$ is equal to the voltage drop across it. So,

$$\dot{\lambda}_4 = \frac{d\lambda}{dt} = e_4 \text{ or } d\lambda_4 = e_4\, dt \tag{4.175}$$

Taking the integration of this equation gives

$$\lambda_4 = \int e_4\, dt \tag{4.176}$$

Substituting this equation in equation (4.174) gives

$$C\frac{de_2}{dt} + \frac{1}{R}e_3 + \frac{1}{I}\int e_4\, dt = i(t) \tag{4.177}$$

Equation (4.177) is the equation derived from the Bond Graph model that represents the dynamics of the given circuit shown in Figure 4.41. Considering the parallel connection of RLC elements, it is essential to be aware of the fact that the efforts e_2, e_3, and e_4 are equivalent to the circuit voltage e_o and that the electrical resistance R and capacitor C are equivalent to the Bond Graph element resistance R and capacitor C respectively. Considering these and inserting $I = L$, equation (4.177) becomes

$$C\frac{de_o}{dt} + \frac{1}{R}e_o + \frac{1}{L}\int e_o \, dt = i(t) \tag{4.178}$$

Comparing equation (4.178) with equation (4.26), it is found that Kirchhoff's current node law applied to the given circuit is satisfied in its Augmented Bond Graph model. Now the question is: what happens if the input current $i(t)$ is replaced by the input voltage $E(t)$? (exercise problem)

Example (4.14):
For the electrical circuit shown in Figure 4.42,

a. Create the preliminary Bond Graph model and simplify it.
b. Create the Augmented Bond Graph model (choose the half arrows positive power directions, assign the causality marks, and number the bonds).
c. Write the State and Co-energy variables and their constitutive relations, then derive the dynamic equations.

Solution:
The given circuit comprises three current loops i_1, i_2, and i_3, and three voltage nodes V_1, V_2, and V_3, as shown in Figure 4.42. Note that the resistance R_1, the capacitance C, and the resistance R_2 have established a link with the ground. We assume that the output terminals are not experiencing any current (no load circuit); only one current i_3 passes through the resistance R_2, implying that it has the same voltage drop as the capacitance C. The following procedures must be followed to create the preliminary Bond Graph model of the given circuit concerning Section 4.4.1; we will start on the left side of the circuit and continue.

1. The three current loops are represented by three 1-junctions of common flows i_1, i_2 and i_3, while the three 0-junctions of common efforts V_1, V_2, and V_3 are created for the voltage nodes, as shown in Figure 4.43.

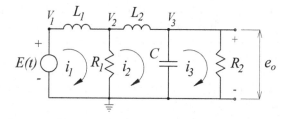

Figure 4.42 Electrical circuit

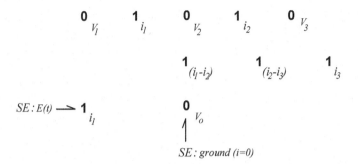

Figure 4.43 Junction structures of the Bond Graph model of the given circuit

2. Since the input voltage $E(t)$ supplies the power, it is represented by a source of effort *SE* connected to the 1-junction of a common flow i_1, while the ground is represented by *SE : ground* of a common effort V_o, as shown in Figure 4.43.
3. The source of effort *SE* and the inductance L_1 share the same flow i_1; nevertheless, each provides a different effort (different voltage drop). As a result, the inductance L_1 is connected to the 1-junction of common flow i_1 that is positioned between the two 0-junctions of common efforts V_1 and V_2, as shown in Figure 4.43.
4. Because the resistance R_1 is located between the first and second current loops and linked to the ground, the current that flowed through it was equal to the difference between i_1 and i_2 ($i_1 - i_2$). As a result, it is connected to a 1-junction of common flow ($i_1 - i_2$), positioned between the two 0-junctions of common efforts V_2 and V_o, as shown in Figure 4.43.
5. The inductance, L_2, is connected to the 1-junction positioned between the two 0-junctions of effort, V_2 and V_3, as shown in Figure 4.43.
6. Because the capacitance C is located between the second and third current loops, the current that flowed through it was equal to the difference between i_2 and i_3 ($i_2 - i_3$). As a result, it is connected to a 1-junction of common flow ($i_2 - i_3$) that is positioned between the two 0-junctions of common efforts V_3 and V_o, as shown in Figure 4.43.
7. Since the resistance R_2 shares the same effort as the capacitance C, it is connected to the 1-junction of common flow i_3 that is positioned between the two 0-junctions of efforts V_3 and V_o, as shown in Figure 4.43.

At this point, the preliminary Bond Graph model of the circuit is constructed, as shown in Figure 4.44. The half-arrow positive power flow directions are selected, as shown in the same figure. To simplify this Bond Graph model, all connections that hold ground with their associated bonds must be removed, as shown in Figure 4.45, and then any 2-port 1-junctions or 0-junctions with through power half arrows should be swapped with single bonds, as shown in Figure 4.45. As a result, the connection at the 0-junction of the common effort V_1, the connection at the 1-junction that is connected to the resistance R_1, the connection at the 1-junction that is connected to the capacitance C, and the connection at the 1-junction that is connected to the

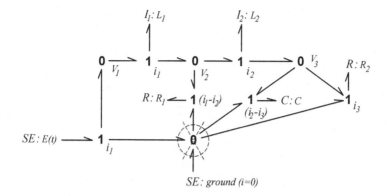

Figure 4.44 Preliminary Bond Graph model of the given circuit

resistance R_2 all must be removed, as shown in Figure 4.45. The resulting simplified Bond Graph model of the given circuit is shown in Figure 4.46.

Now, let us assign the causality of the Bond Graph model. Since there is only one source of effort, SE, the causal stroke marks should be out of their bond, located on their 1-junction side, as shown in Figure 4.46. Considering the causality restrictions in Section 4.5, apply the causal stork marks to all elements and junctions. It is essential to remember that as we work through the process of assigning causality, we must consider the integral causality that applies to all associated elements and bonds. We finally number all bonds to get the Augmented Bond Graph model of the circuit shown in Figure 4.46. This figure shows three integral causalities storage elements, I_1, I_2 and C represented by bonds 2, 6 and 8 and two dissipating elements, R_1 and R_2, represented by bonds 4 and 9.

To derive the dynamic equations of the given circuit, we must first define the Key variables of the model: Input, State variables, and Co-energy (power) variables and their constitutive relations.

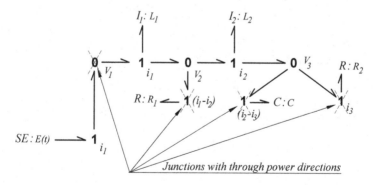

Figure 4.45 Simplifications of the preliminary Bond Graph model

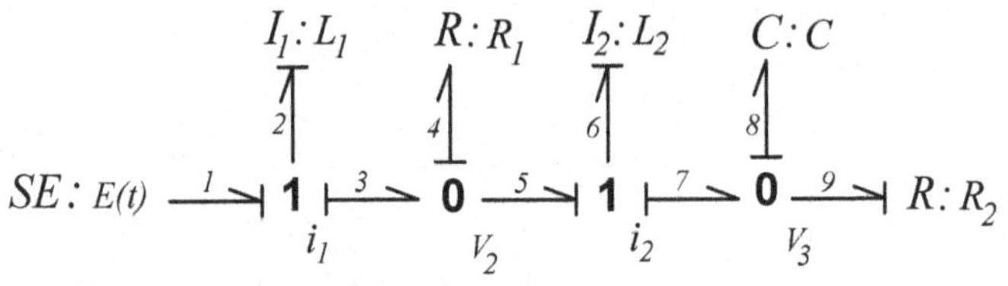

$$I_1 : L_1 \qquad R : R_1 \qquad I_2 : L_2 \qquad C : C$$

Figure 4.46 Augmented Bond Graph model

Key variables:
- Inputs: $e_1 = SE = E(t)$
- State variables: λ_2, λ_6 and q_8
- Co-energy (power) variables: f_2, f_6 and e_8

Constitutive relations:
Write the equations describing the relationship between the co-energy variables and the state variables for each energy storage element as follows:

$$f_2 = \frac{1}{I_1}\lambda_2, \quad f_6 = \frac{1}{I_2}\lambda_6 \quad \text{and} \quad e_8 = \frac{1}{C}q_8 \qquad (4.179)$$

Now that we have the connections of flow and effort variables of the model junctions let us derive the equations that reflect them.

- As shown in Figure 4.46, the flow variables of the 1-junction of common flow i_1 are given by

$$f_1 = f_2 = f_3 \qquad (4.180)$$

and the effort variables are given by

$$e_1 = SE = e_2 + e_3 \quad \text{or} \quad e_2 = \dot{\lambda}_2 = SE - e_3 \qquad (4.181)$$

where $\dot{\lambda}_2$ is the flux linkage change of the inductance L_1.
- The effort variables of the 0-junction of common effort V_2 are given by

$$e_3 = e_4 = e_5 \qquad (4.182)$$

and the flow variables are given by

$$f_3 = f_4 + f_5 \qquad (4.183)$$

- The flow variables of the 1-junction of common flow i_2 are given by

$$f_5 = f_6 = f_7 \qquad (4.184)$$

and the effort variables are given by

$$e_5 = e_6 + e_7 \quad \text{or} \quad e_6 = \dot{\lambda}_6 = e_5 - e_7 \qquad (4.185)$$

where $\dot{\lambda}_6$ is the flux linkage change of the inductance L_2.

- The effort variables of the 0-junction of common effort V_3 are given by

$$e_7 = e_8 = e_9 \tag{4.186}$$

and the flow variables are given by

$$f_7 = f_8 + f_9 \tag{4.187}$$

Now, let us derive the equations describing the dynamics of the given circuit.

- For the first loop of common flow i_1:
Considering equation (4.181), to determine the flux linkage change $\dot{\lambda}_2$ of the inductance L_1, we must determine the effort e_3 as follows. Equation (4.182) indicated that the effort e_3 equals e_4 of the resistance R_1 which is given by

$$e_3 = e_4 = R_1 f_4 \tag{4.188}$$

Equation (4.183) indicates that the flow f_4 equals $(f_3 - f_5)$. Considering this, equation (4.188) becomes

$$e_4 = R_1(f_3 - f_5) \tag{4.189}$$

Equation (4.180) indicates that the flow f_3 equals the flow f_2, while equation (4.184) indicates that the flow f_5 equals f_6, both f_2 and f_6 are given by the constitutive relations indicated in equation (4.179). Considering these and using the constitutive relations, equation (4.184) becomes

$$e_3 = e_4 = R_1(f_2 - f_6) = R_1\left(\frac{1}{I_1}\lambda_2 - \frac{1}{I_2}\lambda_6\right) \tag{4.190}$$

Substituting equation (4.190) in equation (4.181) gives

$$\dot{\lambda}_2 = SE - R_1\left(\frac{1}{I_1}\lambda_2 - \frac{1}{I_2}\lambda_6\right) \tag{4.191}$$

Rearranging this equation gives

$$\dot{\lambda}_2 + R_1\left(\frac{1}{I_1}\lambda_2 - \frac{1}{I_2}\lambda_6\right) = SE \tag{4.192}$$

This equation is the dynamic equation of the first loop of a common flow i_1.

- For the second loop of common flow i_2:
Considering equation (4.185), to determine the rate of change of the inductance flux linkage $\dot{\lambda}_6$ of the inductance L_2, we must determine the efforts e_5 and e_7 as follows. Equation (4.182) indicates that the effort e_5 equals the effort e_4 given by the constitutive relations in equation (4.179), whereas equation (4.186) indicates

that the effort e_7 equals the effort e_8 of the capacitance C given by the constitutive relations. Considering these, equation (4.185) becomes

$$\dot{\lambda}_6 = e_4 - e_8 = R_1 \left(\frac{1}{I_1} \lambda_2 - \frac{1}{I_2} \lambda_6 \right) - \frac{1}{C} q_8 \tag{4.193}$$

Rearranging this equation gives

$$\dot{\lambda}_6 + R_1 \left(\frac{1}{I_2} \lambda_6 - \frac{1}{I_1} \lambda_2 \right) + \frac{1}{C} q_8 = 0 \tag{4.194}$$

This is the dynamic equation of the second loop of a common flow i_2. Equations (4.192) and (4.194) are the dynamic equations of the given circuit shown in Figure 4.42, derived from the Augmented Bond Graph model shown in Figure 4.46.

Example (4.15):
For the electrical circuit shown in Figure 4.47,

a. Create the preliminary Bond Graph model and impose it on the circuit diagram.
b. Simplify the preliminary Bond Graph model and create the Augmented Bond Graph model (choose the half arrows positive power directions, assign the causality marks, and number the bonds).
c. Write the state and power variables and their constitutive relations, and then derive the dynamic equations.

Solution:
This example demonstrates building the Bond Graph model by superimposing it on the circuit diagram. We will begin on the left side of the circuit and continue as follows. The input voltage $E(t)$ is represented by a source of effort SE, as shown in Figure 4.47. The inductance L and the resistance R_a share the same flow i_1, but each element provides a different voltage drop. As a result, each element is connected to a 1-junction positioned between two 0-junctions, as shown in Figure 4.48. Since the capacitance C and resistance R_b share the same effort, both are connected to a 1-junction of common flow i_2 through a 0-junction of a common effort e_o. The source of effort SE, the capacitance C, and the resistance R_b have established a connection

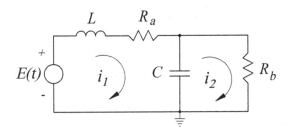

Figure 4.47 Electrical circuit

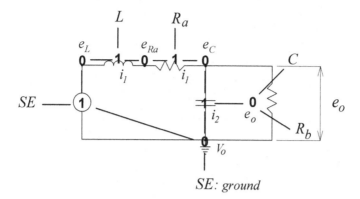

Figure 4.48 Preliminary Bond Graph model imposed on the circuit diagram

to the ground, represented by a source of effort $SE : ground$, which is connected to a
0-junction of common effort V_o, as shown in Figure 4.48.

At this point, the preliminary Bond Graph model of the circuit is constructed and
superimposed on the circuit diagram. The half-arrow positive power flow directions
are selected, as shown in Figure 4.49. It depends on the positive directions of the
currents through the circuit elements. To simplify this Bond Graph model, all connections that hold ground and are associated with bonds need to be removed, and
then any 2-port 1-junctions or 0-junctions with through power half arrows should be
swapped with single bonds, as shown in Figure 4.49. Now, we can utilize an in-out
bond to swap the 2-port variations of a 0-junction connected between SE and L, a
0-junction connected between L and R_a, and a 0-junction and 1-junction connected
between R_a and both C and R_b. Because of this, it can be deduced that the inductance
L and the resistance R_a are connected directly to the 1-junction of a common flow

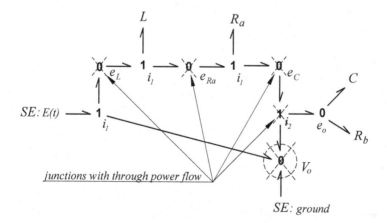

Figure 4.49 Preliminary Bond Graph model of the given circuit

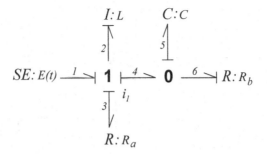

Figure 4.50 Augmented Bond Graph model

i_1. In contrast, the capacitance C and the resistance R_b are connected directly to a 0-junction of a common effort e_o, as shown in Figure 4.50.

Now, let us assign the causality of the Bond Graph model. Since there is only one source of effort, SE, the causal stroke mark should be out of its bond, located on the 1-junction side, as shown in Figure 4.50. Considering the causality restrictions in Section 4.5, apply the causal stork marks to all elements and junctions. It is essential to remember that as we work through the process of assigning causality, we need to consider the integral causality that applies to all associated elements and bonds. We finally number all bonds to get the Augmented Bond Graph model of the circuit shown in Figure 4.50. This figure shows two integral causalities storage elements, L and C represented by bonds 2 and 5.

To derive the dynamic equations of the given circuit, we must first define the Key variables of the model: Input, State variables, and Co-energy (power) variables and their constitutive relations.

Key variables:
- Inputs: $e_1 = SE = E(t)$
- State variables: λ_2, q_5
- Co-energy variables: f_2, e_5

Constitutive relations:
Write equations describing the relationship between co-energy variables and state variables for each energy storage element, as follows:

$$f_2 = \frac{1}{I}\lambda_2, \quad e_5 = \frac{1}{C}q_5 \tag{4.195}$$

Now that we have the connections of flow and effort variables of the model junctions, let us derive the equations that reflect them.

- As shown in Figure 4.50, the flow variables of the 1-junction of a common flow i_1 are given by

$$f_1 = f_2 = f_3 = f_4 \tag{4.196}$$

and the effort variables are given by

$$e_1 = e_2 + e_3 + e_4 \quad \text{or} \quad e_2 = \dot{\lambda}_2 = SE - e_3 - e_4 \tag{4.197}$$

where $\dot{\lambda}_2$ is the rate of change of the flux linkage of the inductance L.
- The effort variables of the 0-junction are given by

$$e_4 = e_5 = e_6 \tag{4.198}$$

and the flow variables are given by

$$f_4 = f_5 + f_6 \tag{4.199}$$

Now, let us derive the equations describing the dynamics of the given circuit.

Considering equation (4.197), to determine the flux linkage change $\dot{\lambda}_2$ of the inductance L, we must determine the effort e_3 and e_4 as follows. The effort e_3 of the resistance R_a is given by

$$e_3 = R_a f_3 \tag{4.200}$$

Equation (4.196) indicates that the flow f_3 equals f_2, given by the constitutive relations indicated in equation (4.195). Considering this, equation (4.200) becomes

$$e_3 = R_a f_2 = \frac{R_a}{I_2} \lambda_2 \tag{4.201}$$

Equation (4.198) indicates that the effort e_4 equals the effort e_5 of the capacitance C, given by the constitutive relations indicated in equation (4.195). Considering this and substituting equation (4.201) with equation (4.197) gives

$$\dot{\lambda}_2 = SE - \frac{R_a}{I} \lambda_2 - \frac{1}{C} q_5 \tag{4.202}$$

Rearranging this equation gives

$$\dot{\lambda}_2 + \frac{R_a}{I} \lambda_2 + \frac{1}{C} q_5 = SE \tag{4.203}$$

To determine the state variable q_5 of the capacitance C, considering equation (4.199) gives

$$\dot{q}_5 = f_5 = f_4 - f_6 \tag{4.204}$$

Equation (4.196) indicates that the flow f_4 equals the flow f_2, given by the constitutive relation indicated in equation (4.195). Considering this, equation (4.204) becomes

$$\dot{q}_5 = f_2 - f_6 = \frac{1}{I} \lambda_2 - f_6 \tag{4.205}$$

The flow f_6 of the resistance R_b is given by

$$f_6 = \frac{1}{R_b} e_6 \tag{4.206}$$

Equation (4.198) indicates that e_6 equals e_5, given by the constitutive relations indicated in equation (4.195). Considering this, equation (4.206) becomes

$$f_6 = \frac{1}{R_b}e_5 = \frac{1}{CR_b}q_5 \qquad (4.207)$$

Substituting equation (4.207) in equation (4.205) gives

$$\dot{q}_5 = \frac{1}{I}\lambda_2 - \frac{1}{CR_b}q_5 \qquad (4.208)$$

So, the state variable q_5 can be determined by integrating equation (4.208). Note that the state variable q_5 equals $(i_1 - i_2)$ as shown in Figure 4.47.
Equations (4.203) and (4.208) are the dynamic equations derived from the Bond Graph model shown in Figure 4.50 for the given circuit shown in Figure 4.47.

Example (4.16):
Using the Bond Graph technique, determine the output voltage e_o as a function of the input voltage $E(t)$ for the transformer circuit shown in Figure 4.51.

Solution:
Let us jump directly to creating the Augmented Bond Graph model of the given circuit. As shown in Figure 4.51, the circuit consists of two separate circuits, primary and secondary, joined together by a transformer with two inductance coils with a turns ratio of n_1/n_2. Since the primary circuit shares the same flow i_1 while the secondary circuit shares the same flow i_2, two different 1-junctions represent the flows i_1 and i_2, respectively, as shown in Figure 4.52. The voltage source $E(t)$ is represented by an effort source SE. Because the inductance L and the source of effort SE share the same flow, they are connected to the 1-junction of common flow i_1. A transformer TF element with a modulus ratio of n_1/n_2 represents the voltage transfer from the primary circuit voltage e_p to the secondary circuit voltage e_s that occurs through the conductor coils, as shown in Figure 4.52. Since the resistance R and the capacitance C share the same flow, they are connected to a 1-junction of common flow i_2. The half-arrow positive power flow directions are selected, as shown in Figure 4.52. It depends on the positive directions of the currents through the circuit elements. Now, let us assign the causality of the Bond Graph model. Since there is only one source of effort, SE, the causal stroke mark should be out of its bond,

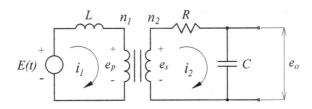

Figure 4.51   Transformer circuit

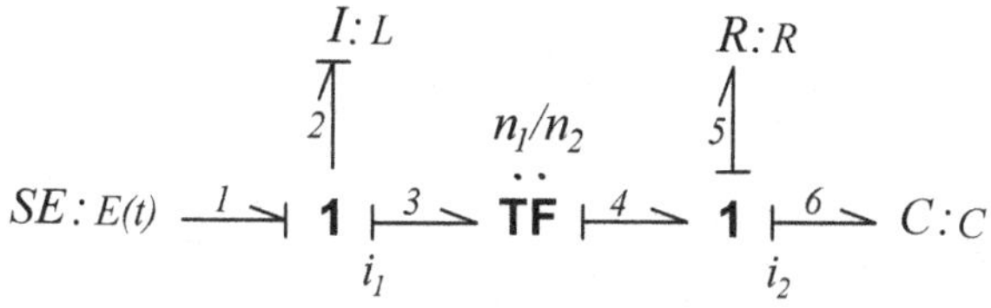

Figure 4.52 Augmented Bond Graph model of the given circuit

located on the 1-junction side, as shown in Figure 4.52. Considering the causality restrictions in Section 4.5, apply the causal stork marks to all elements and junctions. It is essential to remember that as we work through assigning causality, we must consider the integral causality that applies to all associated elements and bonds. We finally number all bonds to get the Augmented Bond Graph model of the circuit shown in Figure 4.51. This figure shows two integral causalities storage elements, L and C represented by bonds 2 and 6.

To derive the dynamic equations of the given circuit, we must first define the key variables of the model: Input, State variables, and Co-energy (power) variables and their constitutive relations.

Key variables:
- Inputs: $e_1 = SE = E(t)$
- State variables: λ_2, q_6
- Co-energy variables: f_2, e_6

Constitutive relations:
Write the constitutive relations that describe the relationship between co-energy variables and state variables for each energy storage element with the integral causality of bonds 2 and 6.

$$f_2 = \frac{1}{I}\lambda_2, \quad e_6 = \frac{1}{C}q_6 \tag{4.209}$$

Now that we have the connections of flow and effort variables of the model junctions let us derive the equations that reflect them.

- As shown in Figure 4.52, the flow variables of the 1-junction of a common flow i_1 are given by

$$f_1 = f_2 = f_3 \tag{4.210}$$

and the effort variables are given by

$$e_1 = e_2 + e_3 \quad \text{or} \quad SE = \lambda_2 + e_3 \tag{4.211}$$

where $\dot{\lambda}_2$ is the rate of change of the magnetic flux linkage of the inductance L.
- Since the effort exerted on one side of the transformer element, TF relates to the effort on the other side, and vice versa, the flow and effort variables of the

transformer TF are given by

$$f_3 = \frac{n_2}{n_1} f_4 \quad \text{and} \quad e_3 = \frac{n_1}{n_2} e_4 \qquad (4.212)$$

Equation (4.212) indicates that the power balance $(e_3 f_3 = e_4 f_4)$ at both sides of the transformer is satisfied.

- The flow variables of the 1-junction of common flow i_2 are given by

$$f_4 = f_5 = f_6 \qquad (4.213)$$

and the effort variables are given by

$$e_4 = e_5 + e_6 \qquad (4.214)$$

Now, let us derive the equations describing the dynamics of the given circuit. Considering equation (4.211), the effort e_3, which is the preliminary circuit voltage e_p, is given by

$$e_3 = SE - \dot{\lambda}_2 \qquad (4.215)$$

It is transferred to the effort e_4, the secondary circuit voltage e_s, via the transformer element TF with a modulus ratio of n_1/n_2, as indicated in the transformer relations in equation (4.212). Considering this, equation (4.215) becomes

$$\frac{n_1}{n_2} e_4 = SE - \dot{\lambda}_2 \qquad (4.216)$$

Substituting equation (4.214) in equation (4.216) gives

$$\frac{n_1}{n_2} (e_5 + e_6) = SE - \dot{\lambda}_2 \qquad (4.217)$$

The effort e_5 of the resistance R is given by

$$e_5 = R f_5 \qquad (4.218)$$

Equation (4.213) indicates that the flow f_5 equals the flow f_4, given by the transformer relations indicated in equation (4.212). Considering this, equation (4.218) becomes

$$e_5 = R f_5 = R f_4 = \frac{n_1}{n_2} R f_3 \qquad (4.219)$$

Equation (4.210) indicates that the flow f_3 equals f_2, given by the constitutive relations indicated in equation (4.209). Considering these, equation (4.219) becomes

$$e_5 = \frac{n_1}{n_2} \frac{R}{I} \lambda_2 \qquad (4.220)$$

The effort e_6 of the capacitor C is given by the constitutive relations indicated in equation (4.209). Substituting equation (4.220) in equation (4.217) and using the constitutive relation gives

$$\frac{n_1}{n_2} \left(\frac{n_1}{n_2} \frac{R}{I} \lambda_2 + \frac{1}{C} q_6 \right) = SE - \dot{\lambda}_2 \qquad (4.221)$$

Rearranging this equation gives

$$\frac{1}{C}q_6 = \frac{n_2}{n_1}[SE - \dot{\lambda}_2] - \frac{n_1}{n_2}(\frac{R}{I}\lambda_2) \qquad (4.222)$$

As shown in Figure 4.51, the transformer output voltage e_o equals the capacitance effort $e_6 = (1/C)q_6$. Considering this, equation (4.222) becomes

$$e_o = \frac{n_2}{n_1}[SE - \dot{\lambda}_2] - \frac{n_1}{n_2}(\frac{R}{I}\lambda_2) \qquad (4.223)$$

Based on this equation, if we want to lower the transformer's output voltage, we need to make the secondary coil have lower turns than the primary coil, meaning that the second term of the right-hand side of equation (4.223) must be increased.

4.6 MULTI-DOMAIN SYSTEMS MODELING THROUGH BOND GRAPH TECHNIQUE

The Bond Graph technique provides the most effective language for describing the dynamics in multi-domain systems because energy is the most challenging physical variable that needs to be considered. In this case, it is a beneficial technique because it is an energy-based description language that monitors energy change across domains. This indicates a valuable technique for modeling multi-domain systems. The following example demonstrates using the Bond Graph technique to model a multi-domain system, such as a DC motor system.

Example (4.17):
Considering the same system presented in Example (4.11), derive the dynamic equations of the electro-mechanical system armature control DC motor shown in Figure 4.53. Check if Kirchhoff's and Newton's laws are embedded in the system's Augmented Bond Graph.

Solution:
Let us jump directly to creating the Augmented Bond Graph model of the given circuit. Figure 4.53 shows that a DC motor transforms electrical power into a mechanical rotating motion. As stated, it operates on electrical and mechanical power,

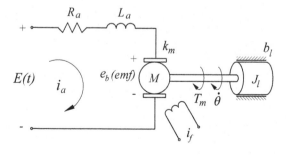

Figure 4.53 DC Motor

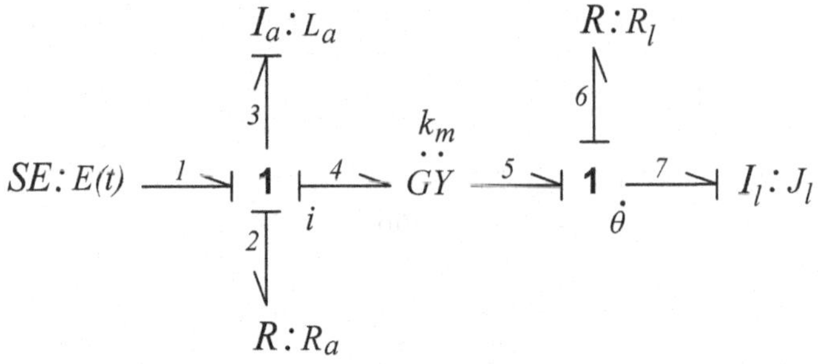

Figure 4.54 Augmented Bond Graph model of the DC motor

making it a typical multi-domain system. In an electrical circuit, the voltage is effort, and the current is flow; hence, the electrical part has constant flow but variable effort, represented by the 1-junction of a common flow i_a, as shown in Figure 4.54. The voltage source $E(t)$ is represented by a source of effort SE, the R_a element represents the armature resistance, and the inertia I_a element represents the armature coil inductance L_a. A Gyrator GY element represents the electro-mechanical transformation, as shown in Figure 4.54. Its gyration ratio k_m equals the motor-torque constant. The mechanical domain covers concepts like rotational inertia, the motor's electro-mechanical conversion, and bearing, as shown in Figure 4.53. The rotational inertia J_l is represented by inertia element I_l, while the bearing friction of a damping factor b_l is represented as an R_l element. Since this combination shares the same flow $\dot{\theta}$, which is the motor speed, it is connected to a 1-junction of common flow $\dot{\theta}$ as shown in Figure 4.54.

As stated previously, the direction of the positive power flow depends on the internal interaction between the system elements. As a result, the half-arrow positive power flow directions are selected, as shown in Figure 4.54. Now, let us assign the causality of the Bond Graph model. Since there is only one source of effort, SE, the causal stroke mark should be out of its bond, located on the junction side, as shown in Figure 4.54. Considering the causality restrictions, apply the causal stork marks to all elements and junctions, as shown in the same figure. We finally number all bonds to get the Augmented Bond Graph model of the circuit shown in Figure 4.54. This figure shows two integral causality storage elements, L_a and L_l represented by bonds 3 and 7.

To derive the dynamic equations of the given circuit, we must first define the key variables of the model: Input, State variables, and Co-energy (power) variables and their constitutive relations.

Key Variables:
- Input: $e_1 = SE = E(t)$
- State variables: λ_3, p_7
- Co-energy variables: f_3, f_7

Constitutive relations:

Write the constitutive relations that describe the relationship between co-energy variables and state variables for each energy storage element with the integral causality of bonds 3 and 7.

$$f_3 = \frac{1}{I_a}\lambda_3 \quad \text{and} \quad f_7 = \frac{1}{I_l}p_7 \tag{4.224}$$

Now that we have the connections between the model junctions' flow and effort variables, let us derive the equations that reflect them.

- As shown in Figure 4.54, the flow variables of the 1-junction of common flow i are given by

$$f_1 = f_2 = f_3 = f_4 \tag{4.225}$$

and the effort variables are given by

$$e_1 = e_2 + e_3 + e_4 \quad \text{and} \quad e_3 = \dot{\lambda}_3 = SE - e_2 - e_4 \tag{4.226}$$

where $\dot{\lambda}_3$ is the magnetic flux linkage change of the inductance L_a.
- Since the effort exerted on one side of the gyrator element GY relates to the flow on the other side and vice versa, and a sense of direction is not required, the effort variables of the Gyrator GY of a gyration factor k_m are given by

$$e_5 = k_m f_4 \quad \text{and} \quad e_4 = k_m f_5 \tag{4.227}$$

where k_m is the motor constant or the back emf constant. Equation (4.227) indicates that the power balance $(e_4 f_4 = e_5 f_5)$ at both sides of the Gyrator is satisfied.
- The flow variables of the 1-junction of a common flow $\dot{\theta}$ are given by

$$f_5 = f_6 = f_7 \tag{4.228}$$

and the effort variables are given by

$$e_5 = e_6 + e_7 \quad \text{and} \quad e_7 = \dot{p}_7 = e_5 - e_6 \tag{4.229}$$

where $\dot{p}_7$ is the momentum change of the rotating mass J_l.

Now, let us deduce the dynamic equations of the given system.

Considering equation (4.226), to determine $\dot{\lambda}_3$, we must determine the efforts e_2 and e_4. The effort e_2 of the resistance R_a is determine as follows

$$e_2 = R_a f_2 \tag{4.230}$$

Equation (4.225) indicates that the flow f_2 equals the flow f_3, given by the constitutive relations indicated in equation (4.224). Considering this, equation (4.230) becomes

$$e_2 = R_a f_3 = \frac{R_a}{I_a}\lambda_3 \tag{4.231}$$

The effort e_4 is given by the gyrator relations indicated in equation (4.227) as follows.

$$e_4 = k_m f_5 \tag{4.232}$$

Equation (4.228) indicates that the flow f_5 equals the flow f_7, given by the constitutive relations indicated in equation (4.224). Considering this, equation (4.232) becomes

$$e_4 = k_m f_7 = \frac{k_m}{I_l} p_7 \tag{4.233}$$

Substituting equations (4.231) and (4.233) in equation (4.226) gives

$$\dot{\lambda}_3 = SE - \frac{R_a}{I_a} \lambda_3 - \frac{k_m}{I_l} p_7 \tag{4.234}$$

Rearranging this equation gives

$$\dot{\lambda}_3 + \frac{R_a}{I_a} \lambda_3 + \frac{k_m}{I_l} p_7 = SE \tag{4.235}$$

Considering equation (4.229), to determine $\dot{p}_7$, we must determine the efforts e_5 and e_6. The effort e_5 is given by the gyrator relations indicated in equation (4.224) as follows.

$$e_5 = k_m f_4 \tag{4.236}$$

Equation (4.225) indicates the flow f_4 equals the flow f_3, given by the constitutive relations indicated in equation (4.224). Considering this, equation (4.236) becomes

$$e_5 = k_m f_3 = \frac{k_m}{I_a} \lambda_3 \tag{4.237}$$

The effort e_6 of the resistance R_l is determined as follows

$$e_6 = R_l f_6 \tag{4.238}$$

Equation (4.228) indicates that the flow f_6 equals the flow f_7, given by the constitutive relations indicated in equation (4.224). Considering this, equation (4.238) becomes

$$e_6 = R_l f_7 = \frac{R_l}{I_l} p_7 \tag{4.239}$$

Substituting equations (4.237) and (4.239) in equation (4.229) gives

$$\dot{p}_7 = \frac{k_m}{I_a} \lambda_3 - \frac{R_l}{I_l} p_7 \tag{4.240}$$

Rearranging this equation yields

$$\dot{p}_7 + \frac{R_l}{I_l} p_7 - \frac{R_a}{I_a} \lambda_3 = 0 \tag{4.241}$$

Equations (4.235) and (4.241) are the differential equations representing the dynamics of the given DC motor system, deduced from the Augmented Bond Graph model shown in Figure 4.54. Now, the question is: How do these equations change if the motor is driven with the current source $i(t)$? (exercise problem).

Now, we will check if Kirchhoff's and Newton's laws are embedded in the Augmented Bond Graph model. The following analyses are what we get when using the constitutive relations indicated in equation (4.224).

- The magnetic flux linkage λ_3 of the inductance L_a can be determined as follows:

$$\lambda_3 = I_a f_3 = I_a i_a \tag{4.242}$$

Taking the derivative of this equation gives

$$\dot{\lambda}_3 = I_a \dot{f}_3 = L_a \frac{di_a}{dt} \tag{4.243}$$

- The momentum p_7 of the rotating mass J_l can be determined as follows:

$$p_7 = I_l f_7 = J_l \dot{\theta} \tag{4.244}$$

Taking the derivative of this equation gives

$$\dot{p}_7 = J_l \dot{f}_7 = J_l \ddot{\theta} \tag{4.245}$$

Substituting equations (4.242) to (4.245) in equations (4.235) and (4.241) and inserting $SE = E(t)$ and $R_l = b_l$ yields

$$L_a \frac{di_a}{dt} + \frac{R_a}{I_a} I_a i_a + \frac{k_m}{I_l} I_l \dot{\theta} = E(t) \tag{4.246}$$

and

$$J_l \ddot{\theta} + \frac{b_l}{I_l} I_l \dot{\theta} - \frac{k_m}{I_a} I_a i_a = 0 \tag{4.247}$$

Rearranging these equations yields

$$L_a \frac{di_a}{dt} + R_a i_a + k_m \dot{\theta} = E(t) \tag{4.248}$$

and

$$J_l \ddot{\theta} + b_l \dot{\theta} = k_m i_a \tag{4.249}$$

By comparing equations (4.248) and (4.249) with equations (4.123) and (4.126), respectively, it is found that they are identical, implying that Kirchhoff's voltage law and Newton's law applied to the given system are satisfied in the Augmented Bond Graph model. As stated previously, the circuit's dynamic equations and equations of motion can be derived from the Bond Graph model without the requirement to create circuit sign conventions and a free body diagram, which is a significant benefit of using the Bond Graph as a modeling technique for the dynamics of multi-domain systems.

This example demonstrates that utilizing a Bond Graph to represent a multi-domain system does not require separately modeling each domain. The key benefit of the Bond Graph technique is that it is sufficient to model the entire system by utilizing it in the same manner.

4.7 DERIVATIVE CAUSALITY IN ELECTRICAL SYSTEMS

The integral relationship, as previously mentioned, should reveal the actual dynamic behavior of the energy storage elements. There are some situations, though, where different options are available. Under these circumstances, we might not have a choice but to reverse the placement of the integral causal stroke mark and apply it to the other end, which indicates the derivative causality. This would be necessary because it would otherwise be impossible to determine which caused what.

Let us now investigate the conditions necessary for the electrical system to have derivative causality. Considering the circuit shown in Figure 4.55a, the flow source *SF* representing the current source $i(t)$ is connected to a 1-junction of a common flow i. Since the causality between the flow source and the 1-junction has been seen in this situation, the causal stroke mark must be located on the source element's side of the bond, as shown in Figure 4.55b. Consequently, when the causality restrictions are considered, the inductance element L indicates a derivative causality, shown in a dotted circle, as shown in the same figure. Now the question is, where did we go wrong? We know that the inductance element does not allow its current to change instantly, implying that the voltage induced will continuously fluctuate. This means that if the current source dictated an instantaneous change, the induced voltage across the inductance would be infinite. This taught us that the current source should not be in series with the inductance element.

An additional example of derivative causality can be seen in an electrical circuit when a voltage source connected in parallel with a combination of resistance, inductance, and capacitance, as shown in Figure 4.56a, meets a similar situation to the one that is observed above. Once again, we have respected the causality between the source and the junction in this situation, as above. The inductance L has integral causality, and the resistance R may carry both causalities. Therefore, it is likewise properly causal; however, the capacitance element now has differential causality, as shown in the Bond Graph model in Figure 4.56b. The explanation for this is the same as before: the capacitance prevents the voltage supplied by the source from fluctuating. If the voltage source generates an output waveform, the instantaneous will experience an infinite amount of current passing through the capacitance. As a result, the capacitance should never be connected in parallel with a voltage source operating independently.

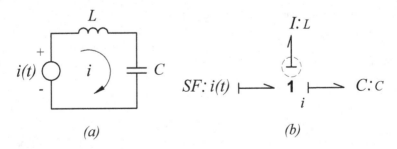

(a)   (b)

Figure 4.55 Electrical circuit and the corresponding Bond Graph model

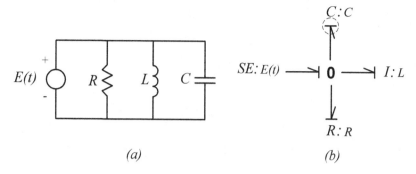

Figure 4.56 Electrical circuit and the corresponding Bond Graph model

When derivative causality arises in an electrical system, the mathematical formulation of the dynamic equations is typically the same as in mechanical systems (see Section 3.3 for further reading). It is essential to remember that the deduced dynamic equation must eliminate all variables of the energy storage elements with derivative causality, meaning that the derived dynamic equation of the circuit should not include the energy variable of the element with derivative causality.

Example (4.18): (Derivative causality)
For the electrical circuit shown in Figure 4.57

1. Create the Augmented Bond Graph model, then determine which storage element has a derivative causality.
2. Derive the equations of motion.
3. Using Kirchhoff's law, derive the equations of motion, and then verify the results with step 2. Explain how Kirchhoff's law estimates the storage element with derivative causality.

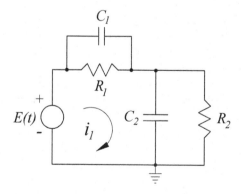

Figure 4.57 Electrical circuit

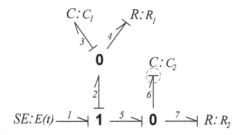

Figure 4.58 Augmented Bond Graph model of the given circuit

Solution:
This example illustrates electrical systems that do not include independent energy storage elements that handle derivative causality. Let us begin the creation of the Augmented Bond Graph model directly. The voltage source $E(t)$ is represented by a source of effort SE connected to a 1-junction of a common flow i, as shown in Figure 4.57. Since the combination of the resistance R_1 and the capacitance C_1 are connected in parallel, they share the same effort (voltage) and are connected to a 0-junction, as shown in Figure 4.58. Since this combination is connected to the current loop, it must connect to the 1-junction of a common flow i. Since the combination of the capacitance C_2 and the resistance R_2 share the same effort (voltage), they are connected to a 0-junction, as shown in Figure 4.58. As stated previously, the direction of the positive power flow depends on the internal interaction between the circuit elements. As a result, the half-arrow positive power flow directions are selected, as shown in Figure 4.58. Considering the causality restrictions, apply the causal stork marks to all elements and junctions. It is essential to remember that as we work through the process of assigning causality, we need to consider the integral causality that applies to all junctions and associated elements. As a result, it is found that the capacitance element C_2 has a derivative causality shown inside a dotted circle, as shown in Figure 4.58. We finally number all bonds to get the Augmented Bond Graph model of the circuit shown in the same figure.

As previously stated, to derive the dynamic equation for a system, considering that its Bond Graph model has derivative causality, we must specify the state variables for the energy storage elements of the integral and the constitutive relations of storage elements with derivative causalities. Note that the energy storage element with derivative causality should be eliminated from the circuit dynamic equations because it is not a state variable that can be considered independent of other energy storage variables.

- Key variables (Integral Causality):
- Inputs: $e_1 = SE$
- State variables: q_3
- Power variables: e_3

Now, write the constitutive relations that describe the relationship between co-energy variables and state variables for each energy storage element with integral causality and the constitutive relations of each energy storage element with derivative causality as follows:

- Constitutive relations (Integral Causality):

$$e_3 = \frac{1}{C_1} q_3 \qquad (4.250)$$

- Constitutive relations (Derivative Causality):

$$e_6 = \frac{1}{C_2} q_6 \qquad (4.251)$$

Now that we have the connections between the flow and effort variables of the model junctions let us derive the equations that reflect them.

- As shown in Figure 4.58, the flow variables of the 1-junction of common flow i are given by

$$f_1 = f_2 = f_5 \qquad (4.252)$$

and the effort variables are given by

$$e_1 = e_2 + e_5 \qquad (4.253)$$

- The effort variables of the 0-junction that connected to both resistance R_1 and capacitance C_1 are given by

$$e_2 = e_3 = e_4 \qquad (4.254)$$

and the flow variables are given by

$$f_2 = f_3 + f_4 \qquad (4.255)$$

- The effort variables of the 0-junction that connected to both resistance R_2 and capacitance C_2 are given by

$$e_5 = e_6 = e_7 \qquad (4.256)$$

and the flow variables are given by

$$f_5 = f_6 + f_7 \qquad (4.257)$$

Let us derive the differential equations describing the given system's dynamics. Considering equation (4.253) and inserting $e_1 = SE$, the effort e_2, is given by

$$e_2 = e_1 - e_5 = SE - e_5 \qquad (4.258)$$

Equation (4.254) indicated that the effort e_2 equals e_3. Considering this and substituting the constitutive relation indicated in equation (4.250) in equation (4.258) gives

$$e_2 = e_3 = \frac{1}{C_2} q_3 = SE - e_5 = E(t) - e_5 \qquad (4.259)$$

Rearranging this equation gives

$$e_5 = E(t) - \frac{1}{C_2}q_3 \tag{4.260}$$

Considering equations (4.252) and (4.257), the flow f_2 is given by

$$f_2 = f_5 = f_6 + f_7 \tag{4.261}$$

The flow f_7 of the resistance R_2 is given by

$$f_7 = \frac{1}{R_2}e_7 \tag{4.262}$$

Note that the flow f_6 equals $\dot{q}_6$. Considering this and substituting equation (4.262) in equation (4.261) gives

$$f_2 = \dot{q}_6 + \frac{1}{R_2}e_7 \tag{4.263}$$

Equation (4.256) indicates that the effort e_7 equals the effort e_5. Considering this and substituting equation (4.260) in equation (4.263) gives

$$f_2 = \dot{q}_6 + \frac{1}{R_2}\left(E(t) - \frac{1}{C_2}q_3\right) = \dot{q}_6 + \frac{1}{R_2}E(t) - \frac{1}{R_2 C_2}q_3 \tag{4.264}$$

The flow f_4 of the resistance R_1 is given by

$$f_4 = \frac{1}{R_1}e_4 \tag{4.265}$$

Equation (4.254) indicates that the effort e_4 equals the effort e_3. Considering this and substituting the constitutive relation indicated in equation (4.250), equation (4.265) gives

$$f_4 = \frac{1}{R_2}e_3 = \frac{1}{R_2 C_1}q_3 \tag{4.266}$$

Using equation (4.255), the flow f_3 of the capacitance C_1 that has integral causality is given by

$$f_3 = f_2 - f_4 \tag{4.267}$$

Note that the flow f_3 equals $\dot{q}_3$. Considering this and substituting equations (4.264) and (4.266) in equation (4.267) gives

$$f_3 = \dot{q}_3 = \dot{q}_6 + \frac{1}{R_2}E(t) - \frac{1}{C_2 R_2}q_3 - \frac{1}{C_1 R_2}q_3 \tag{4.268}$$

Rearranging this equation gives

$$\dot{q}_3 = \dot{q}_6 + \frac{1}{R_2}E(t) - \left(\frac{1}{C_2 R_2} + \frac{1}{C_1 R_2}\right)q_3 \tag{4.269}$$

Let us obtain the flow $\dot{q}_6$ of the capacitance C_2 with a derivative causality. Considering the constitutive relation indicated in equation (4.251) and using equations (4.256) and (4.260) gives

$$e_6 = \frac{1}{C_2}q_6 = e_5 = (E(t) - \frac{1}{C_2}q_3) \tag{4.270}$$

Rearranging this equation yields

$$q_6 = C_2 E(t) - \frac{C_2}{C_1}q_3 \tag{4.271}$$

Taking the derivative of the equation gives

$$\dot{q}_6 = C_2 \dot{E}(t) - \frac{C_2}{C_1}\dot{q}_3 \tag{4.272}$$

Substituting equation (4.272) in equation (4.269) gives

$$(1 + \frac{C_2}{C_1})\dot{q}_3 = C_2 \dot{E}(t) + \frac{1}{R_2}E(t) - (\frac{1}{R_2 C_1} + \frac{1}{R_2 C_1})q_3 \tag{4.273}$$

Rearranging this equation yields

$$\dot{q}_3 = \frac{1}{1 + \frac{C_2}{C_1}}[C_2 \dot{E}(t) + \frac{1}{R_2}E(t) - (\frac{1}{R_2 C_1} + \frac{1}{R_1 C_1})q_3] \tag{4.274}$$

This equation represented the dynamic equation of the given circuit when the derivative causality occurred. It indicates that the flow $\dot{q}_6$ of the capacitance C_2, which has a derivative causality, is not considered in the circuit dynamic equation that satisfies the condition when it has a derivative causality.

4.8 ALGEBRAIC LOOPS IN ELECTRICAL SYSTEMS

As previously discussed in the section, we may have difficulty with the structure of the causal relationships when attempting to assign causalities to particular systems. We may find ourselves in this situation when an electrical circuit has multiple resistances. After the integral causality of the sources and storage elements has been assigned, we must continue arbitrarily assigning the causalities of the resistances and the remaining bonds to finish the causal structure of the Bond Graph model. This is required to complete the model. The causalities of the remaining bonds would also be reversed, creating a whole different causal structure if we gave one of the exciting resistances the opposite causality. This would have been the case if we had given one of the exciting resistances the incorrect causality. If we had given the resistance the incorrect causality, the algebraic loop would have resulted. Section 3.13 provided a detailed clarification and a demonstration of this situation.

4.9 PROBLEMS

Problem (4.1):
Using Kirchhoff's voltage law, derive the differential equations of the electrical circuit shown in Figure 4.59 and put the results in matrix form.

Problem (4.2):
For the electrical circuit shown in Figure 4.60, derive the differential equations. Estimate and draw the corresponding analog mechanical system.

Problem (4.3):
Derive the differential equations of the electrical circuit shown in Figure 4.61 and put the results in matrix form.

Problem (4.4):
Using Kirchhoff's current law, derive the differential equations of the electrical circuit shown in Figure 4.62

Problem (4.5):
Derive the differential equations of the electrical circuit shown in Figure 4.63.

Problem (4.6):
Derive the differential equations of the electrical circuit shown in Figure 4.64.

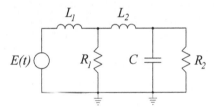

Figure 4.59 Electrical circuit

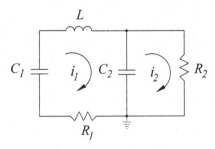

Figure 4.60 Electrical circuit

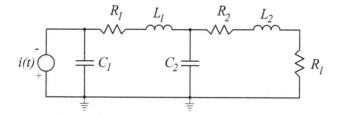

Figure 4.61 Electrical circuit

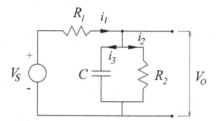

Figure 4.62 Electrical circuit

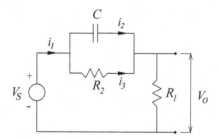

Figure 4.63 Electrical circuit

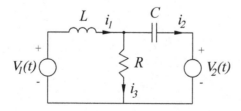

Figure 4.64 Electrical circuit

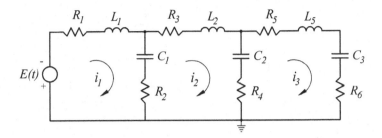

Figure 4.65 Electrical circuit

Problem (4.7):
Derive the differential equations of the electrical circuit shown in Figure 4.65 and put the results in matrix form.

Problem (4.8):
For the electrical circuit shown in Figure 4.66, derive the differential equations. Estimate and draw the corresponding analog mechanical system.

Problem (4.9):
For the electrical circuit shown in Figure 4.67, derive the mathematical model and put the results in matrix form. Estimate and draw the corresponding analog mechanical system.

Problem (4.10):
For the electrical circuit shown in Figure 4.68, derive the differential equations. Estimate and draw the corresponding analog mechanical system.

Problem (4.11):
For the electrical circuit shown in Figure 4.69,

1 Create the preliminary Bond Graph model and simplify it.
2 Create the Augmented Bond Graph model (add the half-arrow positive power flow directions, assign causalities, and number all bonds). Check the derivative causality.

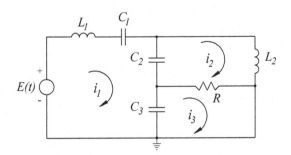

Figure 4.66 Electrical circuit

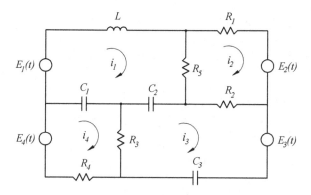

Figure 4.67 Electrical circuit

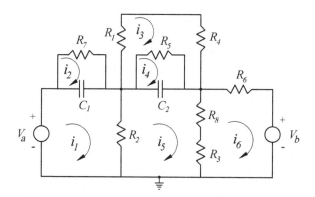

Figure 4.68 Electrical circuit

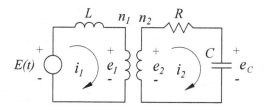

Figure 4.69 Electrical circuit

3 Derive the dynamic equations (write the key variables and their constitutive relations).
4 Check if Kirchhoff's law is embedded in the Augmented Bond Graph.

Problem (4.12):
For the electrical circuit shown in Figure 4.70,

1 Create the preliminary Bond Graph model and simplify it.
2 Create the Augmented Bond Graph model (add the half-arrow positive power flow directions, assign causalities, and number all bonds). Check the derivative causality.
3 Derive the dynamic equations (write the key variables and their constitutive relations).
4 Check if Kirchhoff's law is embedded in the Augmented Bond Graph.

Problem (4.13):
For the electrical circuit shown in Figure 4.65,

1 Create the preliminary Bond Graph model and simplify it.
2 Create the Augmented Bond Graph model (add the half-arrow positive power flow directions, assign causalities, and number all bonds). Check the derivative causality.
3 Derive the dynamic equations (write the key variables and their constitutive relations).
4 Check if Kirchhoff's law is embedded in the Augmented Bond Graph.

Problem (4.14): (Multi-domain system)
For the system shown in Figure 4.71,

1 Create the Augmented Bond Graph model (add the half-arrow positive power flow directions, assign causalities, and number all bonds). Check the derivative causality.

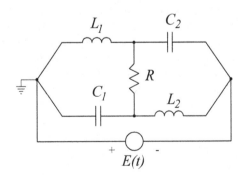

Figure 4.70 Electrical circuit

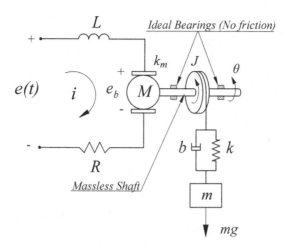

Figure 4.71 Multi-domains system

2 Derive the equations of motion (write the key variables and their constitutive relations).
3 Check if Kirchhoff's and Newton's laws are embedded in the Augmented Bond Graph.

Problem (4.15): (Multi-domain system)
The system shown in Figure 4.72, presents an electrical servomotor connected to a gearbox of ratio N_2/N_1, through inertia load J

1 Create the Augmented Bond Graph model (add the half-arrow positive power flow directions, assign causalities, and number all bonds). Check the derivative causality.
2 Derive the equations of motion (write the key variables and their constitutive relations).
3 Check if Kirchhoff's and Newton's laws are embedded in the Augmented Bond Graph.

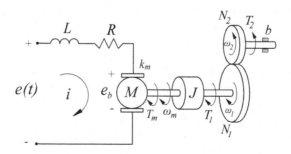

Figure 4.72 Multi-domains system

5 Fluid and Thermal Systems

5.1 INTRODUCTION

As fluids have fewer well-defined shapes (compared to solids) and fluid flow is rarely limited to as clear and accessible pathways as electrical circuit currents, it can be difficult to model these systems using linear terms. So, several assumptions must be considered when dealing with linear fluid systems. The fluid systems are classified into three types: liquid-level, pneumatic, and hydraulic.

5.2 LIQUID-LEVEL SYSTEMS

Two primary variables must be considered when considering liquid flow in a rigid-walled open tank with a constant cross-sectional area: liquid pressure (or liquid level) and volume flow rate. Before beginning to derive the system's dynamic model, it may be necessary to summarize the primary elements of the liquid-level system. It has three main components: Resistance, Compliance, and Inertance (inertia), as well as liquid sources, Pressure, and Flow, which constitute the essential components of a liquid-level system.

The following assumptions must be taken into consideration while modeling the dynamics of the liquid-level system as a linear system:

- In reality, the liquid level fluctuates from point to point throughout the tank cross-section. However, our analyses must utilize a one-dimensional flow model that assumes the liquid level changes uniformly over the tank cross-section area, implying that the liquid will be laminar.
- Since the liquid flow is laminar, it implies that it is steady. When a liquid flow is steady, the velocity and pressure drop at a given location remain constant. As a result, resistance is the only effect that occurs since the inertance and compliance effects prevent them from becoming apparent. The resistance of the bypass valve connected to the tank illustrates this situation.
- There are only minor changes in the liquid density, and its thermal effects are negligible. In this situation, the conservation of mass and Newton's law are the two fundamental physical laws governing our liquid-level system analysis.

It is essential to note that when we derive the dynamic equation of a liquid-level system, we do so under the assumption that the operation of the system reaches steady-state mode and that the change in the liquid level stored in the tank for dt seconds is due to the net inflow to the tank for the same dt seconds. In other words,

DOI: 10.1201/9781032685656-5

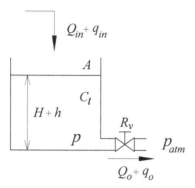

Figure 5.1 Open tank system

if the steady-state value of the input flow rate Q_{in} increases as a result of the net inflow q_{in}, meaning that the input flow rate Q_{in} will be $(Q_{in} + q_{in})$, the liquid level after dt seconds will be $(H + h)$, where H is the value of the liquid level when the system reaches steady-state value, and the output flow rate after dt seconds will be $(Q_o + q_o)$ as shown in Figure 5.1.

5.2.1 LIQUID RESISTANCE

The pure liquid resistance element responds to an applied pressure drop Δp with an instantaneous response of flow rate q_o. A bypass valve of variable orifice area of resistance R_v is the most common device representing the liquid resistance element, as shown in Figure 5.1. This valve's orifice is frequently pre-adjusted before system operation, implying that the orifice is considered a fixed area during system operation. Valves of sharp-edged orifices are considered the virtually pure liquid resistance element because the pressure drop that occurs is so small that the inertance (inertia) and compliance (capacitance) of the valve will be correspondingly negligible, as mentioned above. Most linear flow resistances may be derived using the basic geometrical passageways, assuming laminar flow conditions represent the liquid resistance element. As shown in Figure 5.1, the tank is supplied by a flow rate of q_{in}, the liquid pressure is p, the liquid level is h, the valve resistance is R_v, and the output flow rate passing through the valve is q_o. Usually, the value of the valve resistance is determined experimentally by varying the pressure drop $\Delta p = (p - p_{atm})$ across it and measuring the corresponding flow rate q_o. The following linear form gives it

$$q_o = R_v \Delta p \tag{5.1}$$

This equation indicates that it works similarly to mechanical dampers and electrical resistances, dissipating any liquid power applied to heat. In such tanks, due to the slow vertical motion of liquids, their acceleration and velocity have little effect on the pressure, which is given by $p = \gamma h$, where γ is the specific weight of the liquid,

Figure 5.2 Junction representations of the liquid resistance

which is equal to $\gamma = \rho g$. So, equation (5.1) in terms of liquid level h becomes

$$q_o = R_v \rho g \, \Delta h \tag{5.2}$$

where ρ is the liquid density and g is the gravity.

5.2.2 JUNCTION REPRESENTATIONS OF THE LIQUID RESISTANCE

Since the valve shares the same flow q_o, a 1-junction is used and connected to the resistance element R_v, as shown in Figure 5.2.

5.2.3 LIQUID COMPLIANCE

The tank is generally compliant, as increasing the amount of stored liquid increases the pressure due to gravity. To determine the compliance of a tank, one must force a certain amount of liquid into it while simultaneously measuring the resulting increase in the pressure p or the level h of the liquid. Referring to Figure 5.1 and assuming that the valve is closed ($R_v = 0$), the variation of the liquid pressure is given by

$$p = \frac{1}{C_t} \int q_{in} \, dt \tag{5.3}$$

in terms of the level h, this equation becomes

$$\rho g h = \frac{1}{C_t} \int q_{in} \, dt \tag{5.4}$$

then, the variation of the liquid level h due to the variation of the input flow rate q_{in} is given by

$$h = \frac{1}{\rho g C_t} \int q_{in} \, dt \tag{5.5}$$

where C_t is the tank compliance. When the liquid reaches steady state mode, if we add a volume of the liquid ΔV to the tank, the change of the level Δh goes up an amount $(\Delta V / A)$, and the pressure rises an amount $(\Delta V \gamma / A)$. The compliance is thus

$$C_t = \frac{\text{volume change}}{\text{pressure change}} = \frac{\Delta V}{\Delta V \gamma / A} = \frac{A}{\gamma} = \frac{A}{\rho g} \tag{5.6}$$

Figure 5.3 Junction representations of the tank compliance

where A is the tank cross-section area. Substituting equation (5.6) in equation (5.5) yields

$$h = \frac{\rho g}{\rho g A} \int q_{in}\, dt = \frac{1}{A} \int q_{in}\, dt \qquad (5.7)$$

Taking the derivative of this equation gives

$$A\frac{dh}{dt} = q_{in} \qquad (5.8)$$

Equation (5.8) indicates that the compliance element in the liquid-level system is a storage element and is analogous to the spring element in the mechanical system and the capacitor element in the electrical system.

5.2.4 JUNCTION REPRESENTATIONS OF THE LIQUID COMPLIANCE

Since the thank volume shares the same effort, p or h, a 0-junction is used and connected to the capacitor element C_t, as shown in Figure 5.3.

5.2.5 LIQUID INERTANCE

Usually, the tank is connected to a pipeline with a relatively long length. Consider a laminar liquid flow entering a circular pipeline at a uniform velocity, as shown in Figure 5.4. This figure shows how the velocity gradient develops over the length of the pipeline. The hypothetical boundary surface divides the flow in a pipeline into two essential regions: the core flow region, in which frictional effects are negligible and the velocity remains nearly constant in the radial direction, and the boundary layer region, in which the viscous effects and the velocity changes are significant. Because of the no-slip condition, the liquid particles in contact with the pipe wall come to a complete stop at the boundary layer region, and the velocity profile takes on the shape of a parabolic curve, as shown in Figure 5.4. On the other hand, this is more flattering in turbulent flow than in laminar flow.

Let us consider the laminar one-dimensional flow in a pipeline connected to a tank, as shown in Figure 5.5. These pipes are often called Capillary tubes because their small diameters usually result in laminar flow. Inertance associated with flow in relatively long Capillary pipes is treated analytically, primarily as a liquid property; thus, we will do so here. Any flowing liquid has stored kinetic energy due to its mass and velocity. The inertance of an infinite-sized pipeline is equal to the summation

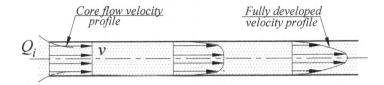

Figure 5.4 Development of the velocity boundary layer in a pipeline

of this kinetic energy throughout the whole volume of the pipeline. The most fundamental assumption in this situation is that all liquid particles have identical velocities
at any given point in time, implying that their velocity profile will behave as the core
flow velocity profile shown in Figure 5.5 along the pipe length, meaning that the velocity v of the liquid will be constant until the liquid reaches the entering port of the
tank, as shown in Figure 5.5. In this situation, the liquid can be treated as a rigid body
of mass m, as shown in this figure. Therefore, the acceleration of the liquid due to
pressure drop Δp across a Capillary tube (inertance element) is defined by Newton's
law as follows

$$\sum F = am \tag{5.9}$$

The force F is given by

$$F = A_p \Delta p \tag{5.10}$$

The mass of the liquid is given by

$$m = \rho A_p L \tag{5.11}$$

where L is the pipeline length and A_p its cross-section area. The liquid velocity is
related to the flow rate q_i by

$$A_p v = q_i \tag{5.12}$$

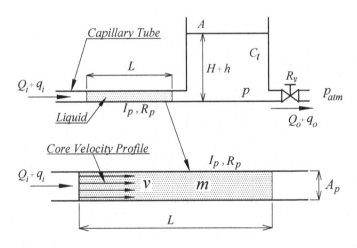

Figure 5.5 Capillary tube

Solving this equation for v gives

$$v = \frac{q_i}{A_p} \tag{5.13}$$

Taking the derivative of equation (5.13) gives

$$\frac{dv}{dt} = a = \frac{1}{A_p} \frac{dq_i}{dt} \tag{5.14}$$

Substituting equations (5.10), (5.11), and (5.14) in equation (5.9) gives

$$A_p \Delta p = \rho A_p L \left(\frac{1}{A_p} \frac{dq_i}{dt} \right) \tag{5.15}$$

Rearranging this equation gives

$$\Delta p = \left(\frac{\rho L}{A_p} \right) \frac{dq_i}{dt} = I_p \dot{q}_i \tag{5.16}$$

where I_p is the liquid inertance (inertia), which is given by

$$I_p = \frac{L}{g A_p} \tag{5.17}$$

Equation (5.17) indicates that the inertance element in the liquid-level system is a storage element and, of course, analogous to the inertia element in the mechanical system and the inductor element in the electrical system.

Since the liquid moves through capillary tubes as a rigid body, the friction between the body and the tube wall will assume a viscous and column friction combination. If the liquid is water, the friction force is relatively small due to its viscosity, resulting in low liquid resistance. In this situation, the viscous friction force is given by

$$F_d = R_{vc} v \tag{5.18}$$

Applying Newton's law gives

$$A_p \Delta p = R_{vc} v \tag{5.19}$$

where R_{vc} is the combination viscous-column friction factor and v is the liquid velocity. From equation (5.19), we obtain

$$\Delta p = F_d = \frac{R_{vc}}{A_p} q_o = R_p q_o \tag{5.20}$$

From this equation, we found that R_p is given by

$$R_p = \frac{R_{vc}}{A_p} \tag{5.21}$$

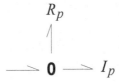

Figure 5.6   Junction representations of the inertance and resistance of the pipeline

where R_p is the liquid resistance in the pipeline. The value of this resistance is arbitrary and established by model analysis, but, in practice, we cannot ignore it.

5.2.6   JUNCTION REPRESENTATIONS OF THE LIQUID INERTANCE

Since the pipeline shares the same flow q, a 1-junction is used and connected to their inertia element I_p and resistance element R_p, as shown in Figure 5.6.

5.2.7   LIQUID SOURCES: PRESSURE AND FLOW

An ideal flow source generates a particular flow regardless of the pressure necessary to produce that flow. Similarly, an ideal pressure source generates a constant pressure at some point in the liquid-level system, regardless of the amount of flow necessary to sustain that pressure. One of the most popular liquid power sources is a pump (positive displacement or centrifugal). Another source of liquid encountered in practice is an elevated tank, in which gravity serves as the power source. In our analysis, we will be concerned with the volume of liquid, or pressure, supplied to the system, regardless of the type of power sources used.

5.2.8   JUNCTION REPRESENTATIONS OF THE LIQUID SOURCES

The source of effort SE or flow SF can be connected to a 1-junction, as shown in Figure 5.7.

As mentioned previously, when we derive the dynamic equation, we do so under the assumption that the operation of the liquid-level system is in steady-state mode and that the change in the liquid level stored in the tank for a time dt is equal to the net inflow to the tank for the same dt. In other words, the liquid level change equals the net inflow in over dt seconds. We are going to consider this in the following examples.

Figure 5.7   Junction representations of the liquid input sources

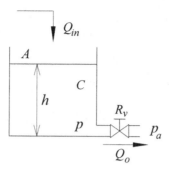

Figure 5.8 Liquid-level system

5.2.9 DYNAMIC MODELING OF LIQUID-LEVEL SYSTEMS

When deriving liquid-level system equations, law of the conservation of mass flow rate (continuity equation) is typically helpful, particularly for systems involving fluid compliance (tanks). So, throughout a period, one is always able to write

Mass in - Mass out = additional mass stored

It is possible to swap volume for mass in this equation when the liquid is considered incompressible (fluid density is constant). So, rewrite the above rule as follows

Volume in - volume out = additional volume stored

Whenever a pipe is in a flow line, the instantaneous sum of flow rates entering and exiting the pipe must be equal. In most cases, it is best to take a straightforward strategy that merely involves equating the pressure difference and the inertance. Considering the liquid-level system shown in Figure 5.8, the elements of our analysis model are the tank compliance C_l, the valve resistance R_v, fluid inertance I_l, and a volume-flow-rate source Q_{in}. The result of applying the principle of conservation of volume with zero initial conditions due to incompressibility is that: So, applying this rule to the system shown in Figure 5.8 gives

$$Q_{in} - A\frac{dh}{dt} = Q_o \tag{5.22}$$

The instantaneous valve output flow rate Q_o is given by

$$Q_o = A_v C_d \sqrt{\frac{2g\Delta p}{\gamma}} = A_v C_d \sqrt{2g} \sqrt{\frac{\gamma h}{\gamma}} = R_v \sqrt{h} \tag{5.23}$$

where R_v $(m/(m^3/sec))$ is the valve resistance, $\gamma = \rho g$ (kg/m^3) is the liquid-specific weight, C_d is the valve orifice discharge coefficient, and A_v (m^2) is the cross-section area of the valve orifice. Substituting equation (5.23) in equation (5.22) gives

$$A\frac{dh}{dt} + R_v \sqrt{h} = Q_{in} \tag{5.24}$$

This is the dynamic equation of the liquid-level system shown in Figure 5.8. To find a solution to this equation, we will use the perturbation method or Taylor's series, which are utilized in most linearization research projects. In this section, we are going to focus on the perturbation method. Let us define the total, operating point, and perturbation values to get things started:

$$h = h_{op} + h_p \tag{5.25}$$

$$Q_{in} = q_{iop} + q_{ip} \tag{5.26}$$

where

- h_{op}=the tank level before the system operation (the level of the tank at the operating pressure).
- h_p=the perturbation quantity of the tank level.
- q_{iop}=the input flow rate to the tank at the operating point.
- q_{ip}=the perturbation quantity of the input flow rate.

The variation of the tank level h_p at the operating point can be linearized as follows

$$\sqrt{h} \approx [\sqrt{h_{op}} + \frac{1}{2\sqrt{h_{op}}}(h - h_{op})] = (\sqrt{h_{op}} + \frac{1}{2\sqrt{h_{op}}}h_p) \tag{5.27}$$

Note that the value of $\sqrt{h_{op}}$ is constant. Now, we will proceed under the assumption that the initial inflow, q_{in}, remained unchanged for an extended period at the value q_{iop}; thus, the level of the tank became stable at h_{op}, is given by

$$q_{iop} = R_v \sqrt{h_{op}} \text{ or } h_{op} = (\frac{q_{iop}}{R_v})^2 \tag{5.28}$$

Substituting equations (5.27) and (5.28) in equation (5.24) and rearranging the results gives

$$A\frac{dh_p}{dt} + R_v(\sqrt{h_{op}} + \frac{1}{2\sqrt{h_{op}}}h_p) = q_{iop} + q_{ip} \tag{5.29}$$

Simplifying this equation gives

$$(\frac{2A\sqrt{h_{op}}}{R_v})\frac{dh_p}{dt} + h_p = (\frac{2\sqrt{h_{op}}}{R_v})q_{ip} \tag{5.30}$$

As discussed above, there is a possibility that the fluid resistances displayed in a fluid system are nonlinear rather than linear. We can establish the system's behavior by linearizing any nonlinear resistances for small changes about a given operating point. Because we know these studies are only approximations, we frequently compare them to simulations of the same nonlinear system to form our opinion regarding the circumstances under which such approximations may be appropriately utilized.

Example (5.1):
For the liquid-level system shown in Figure 5.9, derive the differential equations representing its dynamics, assuming the pipeline has a small cross-section area A_p and a relatively long length L (capillary tube).

Solution:
Figure 5.9 shows that the system consists of a single tank that receives supply from an external flow rate q_{in}. It is linked to a pipeline through a bypass valve. Because it is assumed that the pipeline has a small cross-section area A_p and a long length L (capillary tube), it implies that the liquid mass's inertia and friction as it passes through it will be considered.

- For the tank:
Applying law of the conservation of mass flow rate to the tank gives

$$A\frac{dh}{dt} + q_o = q_{in} \tag{5.31}$$

The flow rate q_o passes through the orifice of the bypass valve of the resistance R_v is given by

$$q_o = \frac{1}{R_v}(p_1 - p_2) \tag{5.32}$$

Considering the pressure p_1 equals ρgh, equation (5.32) becomes

$$q_o = \frac{1}{R_v}(\rho gh - p_2) \tag{5.33}$$

- For the pipeline:
It should be noted that the flow rate q_o will be considered the mass flow rate moving through the pipeline. Applying Newton's law to the pipeline and recalling equations (5.16) and (5.20) gives

$$A_p(p_2 - p_{atm}) = L\dot{q}_o + R_c q_o \tag{5.34}$$

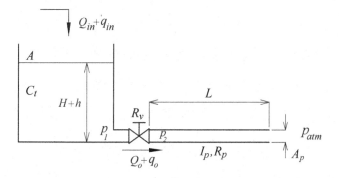

Figure 5.9 Liquid-level system connected to a pipeline at the output

where R_c is the viscous friction factor. Rearranging equation (5.34) gives

$$(p_2 - p_{atm}) = \frac{L}{A_p}\dot{q}_o + \frac{R_c}{A_p}q_o \qquad (5.35)$$

or,

$$(p_2 - p_{atm}) = I_p\dot{q}_o + R_pq_o \qquad (5.36)$$

where I_p is the inertance, which represents the inertia of the mass flow rate, and R_p is the resistance, which represents the friction in the pipeline. Considering absolute pressures ($p_{atm} = 0$), equation (5.36) becomes

$$p_2 = I_p\dot{q}_o + R_pq_o \qquad (5.37)$$

Substituting equation (5.37) in equation (5.36) gives

$$q_o = \frac{1}{R_v}(\rho gh - I_p\dot{q}_o - R_pq_o) \qquad (5.38)$$

Rearranging this equation gives

$$q_o = \frac{1}{(R_p + R_v)}(\rho gh - I_p\dot{q}_o) \qquad (5.39)$$

Taking the derivative of this equation gives

$$\dot{q}_o = \frac{1}{(R_p + R_v)}(\rho g\dot{h} - I_p\ddot{q}_o) \qquad (5.40)$$

Solving this equation for $\dot{h}$ gives

$$\dot{h} = \frac{dh}{dt} = \frac{1}{\rho g}[I_p\ddot{q}_o + (R_v + R_p)\dot{q}_o] \qquad (5.41)$$

Substituting this equation in equation (5.31) gives

$$\frac{A}{\rho g}[(I_p\ddot{q}_o + (R_v + R_p)\dot{q}_o] + q_o = q_{in} \qquad (5.42)$$

Inserting $C_t = (A/\rho g)$ in this equation and rearranging the results gives

$$I_pC_t\ddot{q}_o + C_t(R_v + R_p)\dot{q}_o + q_o = q_{in} \qquad (5.43)$$

At the steady state mode, inserting q_o equals (h/R_v), equation (5.43) becomes

$$\frac{I_pC_t}{R_v}\frac{d^2h}{dt^2} + \frac{C_t(R_v + R_p)}{R_v}\frac{dh}{dt} + \frac{1}{R_v}h = q_{in} \qquad (5.44)$$

This equation is the differential equation that represents the dynamic behavior of the given system. It is essential to note that considering the inertance of the liquid mass as it moves through the pipeline and how it is represented in equation (5.44) indicates that the system will be of the second-order.

Example (5.2):
For the liquid tank system shown in Figure 5.10, derive the differential equations describing its dynamics and put the results in matrix form.

Solution:
The system consists of two tanks, as shown in Figure 5.10. Both tanks are linked to one another through a bypass valve with resistance R_1, while the second tank is connected to the atmosphere through a bypass valve with resistance R_2. The first tank is supplied by an external flow rate q_{in}.

- For the first tank:
Applying the continuity equation on the first tank gives

$$q_{in} = q_{v_1} + C_1 \frac{dh_1}{dt} \qquad (5.45)$$

The flow rate q_{v_1} passing through the first valve of the resistance R_1 is given by

$$q_{v_1} = \frac{h_1 - h_2}{R_1} \qquad (5.46)$$

Substituting equation (5.46) in equation (5.45) gives

$$C_1 \frac{dh_1}{dt} + \frac{h_1 - h_2}{R_1} = q_{in} \qquad (5.47)$$

- For the second tank:
Note that the flow rate q_{v_1} that passes through the valve of the resistance R_1, which is the output of the first tank, will be the input to the second tank. So, applying the continuity equation to the second tank gives

$$q_{v_1} = C_2 \frac{dh_2}{dt} + q_{v_2} \qquad (5.48)$$

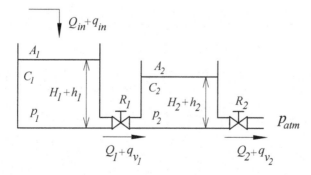

Figure 5.10 Two coupled tank system

The flow rate q_{v_2} passing through the second valve of the resistance R_2 is given by

$$q_{v_2} = \frac{h_2}{R_2} \tag{5.49}$$

Substituting equations (5.46) and (5.49) in equation (5.48) gives

$$\frac{h_1 - h_2}{R_1} = C_2 \frac{dh_2}{dt} + \frac{h_2}{R_2} \tag{5.50}$$

Rearranging this equation gives

$$C_2 \frac{dh_2}{dt} + \frac{h_2 - h_1}{R_1} + \frac{h_2}{R_2} = 0 \tag{5.51}$$

Equations (5.47) and (5.51) are the differential equations representing the dynamics of the given system.

- In matrix form:
Rearranging equations (5.47) and (5.51) gives

$$C_1 \dot{h}_1 + \frac{h_1}{R_1} - \frac{h_2}{R_1} = q_{in} \tag{5.52}$$

$$C_2 \dot{h}_2 + (\frac{h_1}{R_1} + \frac{h_2}{R_2})h_2 - \frac{h_1}{R_1}h_1 = 0 \tag{5.53}$$

So,

$$\begin{bmatrix} C_1 & 0 \\ 0 & C_2 \end{bmatrix} \begin{bmatrix} \dot{h}_1 \\ \dot{h}_2 \end{bmatrix} + \begin{bmatrix} \frac{1}{R_1} & -\frac{1}{R_1} \\ -\frac{1}{R_1} & (\frac{1}{R_1} + \frac{1}{R_2}) \end{bmatrix} \begin{bmatrix} h_1 \\ h_2 \end{bmatrix} = \begin{bmatrix} q_{in} \\ 0 \end{bmatrix} \tag{5.54}$$

Example (5.3):
For the liquid tank system shown in Figure 5.11, derive the differential equations describing its dynamics and put the results in matrix form.

Solution:
As shown in Figure 5.11, the system comprises two tanks. The first tank is supplied by an external flow rate, q_{in}, while the second tank is supplied by the flow rate, q_{v_1}, produced by the first tank. Two throttle valves are located at the exit of each tank. Applying law of the conservation of mass flow rate to the given system gives:

- For the first tank:

$$q_{in} = C_1 \frac{dh_1}{dt} + q_{v_1} \tag{5.55}$$

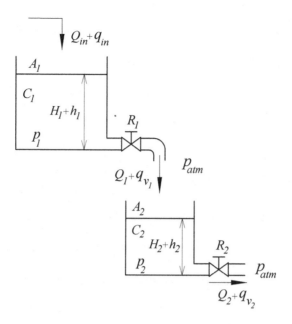

Figure 5.11 Two tank system

The flow rate q_{v_1} passing through the first valve of the resistance R_1 is given by

$$q_{v_1} = \frac{h_1}{R_1} \tag{5.56}$$

Substituting equation (5.56) in equation (5.55) and rearranging the results gives

$$C_1 \frac{dh_1}{dt} + \frac{h_1}{R_1} = q_{in} \tag{5.57}$$

- For the second tank:
Note that the flow rate q_{v_1} passes through the valve of the resistance R_1, which is the output of the first tank and will be the external input of the second tank. Considering this and applying law of the conservation of mass flow rate to the second tank,

$$C_2 \frac{dh_2}{dt} + q_{v_2} = q_{v_1} \tag{5.58}$$

The flow rate q_{v_2} passing through the second valve of the resistance R_2 is given by

$$q_{v_2} = \frac{h_2}{R_2} \tag{5.59}$$

Substituting equations (5.57) and (5.59) in equation (5.58) gives

$$C_2 \frac{dh_2}{dt} + \frac{h_2}{R_2} = \frac{h_1}{R_1} \tag{5.60}$$

Equations (5.57) and (5.60) are the differential equations representing the dynamics of the given system.

- In matrix form:
Considering equations (5.) and (5.) gives

$$\begin{bmatrix} C_1 & 0 \\ 0 & C_2 \end{bmatrix} \begin{bmatrix} \dot{h}_1 \\ \dot{h}_2 \end{bmatrix} + \begin{bmatrix} \frac{1}{R_1} & 0 \\ -\frac{1}{R_1} & \frac{1}{R_2} \end{bmatrix} \begin{bmatrix} h_1 \\ h_2 \end{bmatrix} = \begin{bmatrix} q_{in} \\ 0 \end{bmatrix} \qquad (5.61)$$

5.2.10 BOND GRAPH CONSTRUCTION FOR LIQUID-LEVEL SYSTEMS

Considering the description of the Bond Graph junctions and as previously stated, the simplest way to determine if the junction is a 0-junction or a 1-junction is to check if all pressure or level (efforts) at the junction's elements are identical (0-junction) or if all flow rates (flow) have the same value (1-junction). In other words, if some elements have the same flow rate (flow), they should be connected to a 1-junction, while if they have the same pressure or level (effort), they should be connected to a 0-junction. These situations are explained as follows.

1. For each level or pressure of interest, establish a 0-junction. If using absolute pressure, establish a 0-junction for atmospheric pressure (to be eliminated).
2. For compliance associated with levels or pressures, attach C to appropriate 0-junctions. Attach I, R, and TF elements to 1-junctions (using through power convention) between appropriate 0-junctions.
3. Assign power flow half arrows using the through power flow convention for the 1-junction and two 0-junctions created in parts (1) and (2).
4. If we consider absolute pressures, the atmospheric pressure connection is allocated to a specific node, similar to the ground in mechanical and electrical systems. All bonds related to it must be removed to simplify the model. To further simplify the Bond Graph, any 2-port 1-junctions or 0-junctions with through power flow directions should be swapped by single bonds.
5. Finally, we should add causality stroke marks and bond numbers to finish. As a result, the Augmented Bond Graph is ready to derive the differential equations describing the dynamics of liquid-level systems. It is worth noting that the causality restrictions applied to the Bond Graph model of liquid-level systems will be the same as those applied to mechanical and electrical systems.

The following examples illustrate creating a Bond Graph model of liquid-level systems. We do so under the same assumption considered when the law of mass conservation was applied to obtain the dynamics equation of the liquid-level system, which is the assumption that the system's operation reaches steady-state mode and that the change in the liquid level stored in the tank for a time dt is due to the net inflow to the tank for the same dt. In other words, if the steady-state value of the input flow rate Q_{in} rises as a consequence of the net inflow q_{in}, which indicates that the input flow rate Q_{in} will be $(Q_{in} + q_{in})$, then the liquid level after dt will be $(H + h)$, where

H is the value of the liquid level when the system reaches the steady-state mode. We are going to consider this in the following examples.

Example (5.4):
For the liquid-level system shown in Figure 5.12,

- Create the preliminary Bond Graph model and simplify it.
- Create the Augmented Bond Graph model.
- Write the state and power variables and their constitutive relations.
- Derive the dynamic equations and check if law of the conservation of mass flow rate and Newton's law are embedded in the Augmented Bond Graph model.

Solution:
Figure 5.12 shows that the system comprises a single tank fed by an external flow rate q_{in} and connected to a bypass valve with a resistance of R_v linked to a pipeline with a cross-sectional area of A_p and length of L. As shown in this figure, considering the descriptions of the Bond Graph junctions, we get

- The flow rate q_{in} is represented by a source of flow SF connected to a 1-junction, as shown in Figure 5.13.
- Create a 0-junction for each unique pressure, p_1, p_2 and p_{atm}.
- Since the liquid inside the tank has the same pressure p_1, it is represented by a compliance element C_t connected to the 0-junction of a common effort p_1.
- The resistance of the bypass valve orifice is represented by a resistance element R_v connected to a 1-junction positioned between the two 0-junctions of common efforts p_1 and p_2, respectively.
- Because it is assumed that the pipeline is a capillary tube, it is implied that the inertia of the liquid mass and the friction it encounters while moving through the pipeline will be considered. Therefore, the inertance I_p element, which represents the inertia of mass flow rate, and the resistance R_p element, which represents the friction in the pipeline, are connected to a 1-junction positioned between the two 0-junctions of comment effort p_2 and p_{atm}.

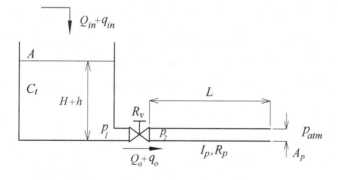

Figure 5.12 Liquid-level system connected to a pipeline at the output

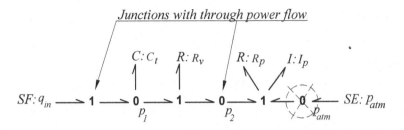

Figure 5.13 Preliminary Bond Graph model of the given system

- When considering absolute pressure, the effect of atmospheric pressure will be similar to that of the effect of the ground in mechanical and electrical systems, as previously stated. Therefore, the atmospheric pressure p_{atm} is represented by a source of effort SE connected to a 0-junction.

At this point, the preliminary Bond Graph model of the given system is constructed, as shown in Figure 5.13. The half-arrows indicating positive power flow directions are selected, as shown in the same figure. Remember that the sources supply power while the elements receive it. To simplify this Bond Graph model, all connections that hold the atmospheric pressure p_{atm} with their associated bonds must be removed, and then any 2-port 1-junctions or 0-junctions with through power flow should be swapped with single bonds. As a result, the 1-junction connected to the source of flow SF, the 0-junction of common effort p_2, and the 0-junction connected to the atmospheric source of effort SE are removed, as shown clearly in Figures 5.13 and 5.14.

Now, let us assign the causality of the Bond Graph model. Since there is only one source of flow, SF, the causal stroke mark should be located on its bond at the element side, as shown in Figure 5.14. Considering the causality restrictions stated in Section 3.3 apply the causal stork marks to all elements and junctions. It is essential to remember that as we work through assigning causality, we must consider the integral causality that applies to all associated elements and bonds. We are finally numbing all bonds to get the Augmented Bond Graph model of the system shown in Figure 5.14. This figure shows two integrally causalities storage elements, I_p and C_t, representing by bonds 2 and 6.

To derive the dynamic equations of the given system, we must first define the Key variables of the model: Input, State variables, and Co-energy (power) variables and their constitutive relations.

Key variables:
- Inputs: $f_1 = SF = q_{in}$.
- State variables: q_2 and p_6.
- Co-energy (Power) variables: e_2 and f_6.

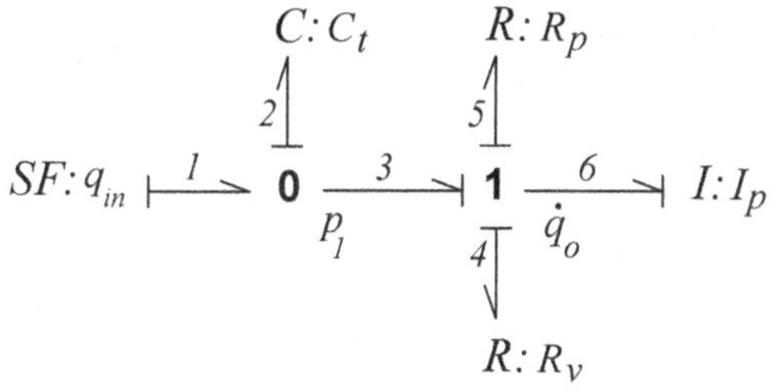

Figure 5.14   Augmented Bond Graph model of the given system.

Constitutive relations:
Write equations describing the relationship between co-energy variables and state variables for each energy storage element, as follows:

$$e_2 = \frac{1}{C_t} q_2 \quad \text{and} \quad f_6 = \frac{1}{I_p} p_6 \tag{5.62}$$

Now that we have the connections between the flow and effort variables of the model junctions shown in Figure 5.14, let us derive the equations that reflect them.

- As shown in Figure 5.14, the effort variables of the 0-junction of common effort p_1 are given by

$$e_1 = e_2 = e_3 \tag{5.63}$$

and the flow variables are given by

$$f_1 = f_2 + f_3 \quad \text{or} \quad f_2 = \dot{q}_2 = f_1 - f_3 \tag{5.64}$$

where $\dot{q}_2$ is the rate of change of the liquid level inside the tank.
- The flow variables of the 1-junction of common flow q_o are given by

$$f_3 = f_4 = f_5 = f_6 \tag{5.65}$$

and the effort variables are given by

$$e_3 = e_4 + e_5 + e_6 \quad \text{or} \quad e_6 = \dot{p}_6 = e_3 - e_4 - e_5 \tag{5.66}$$

where $\dot{p}_6$ is the momentum change of the mass flow rate passing through the pipeline.

Now, let us derive the differential equation describing the dynamics of the given system.

Equation (5.66) indicates that to determine the momentum change $\dot{p}_6$ of the liquid mass, we must first determine the efforts e_3, e_4, and e_5. Equation (5.63) indicates that the effort e_3 equals e_2 given by the constitutive relations indicated in equation (5.66).

Considering this, equation (5.62) becomes

$$\dot{p}_6 = e_2 - e_4 - e_5 = \frac{1}{C_t} q_2 - e_4 - e_5 \tag{5.67}$$

The effort e_4 of the valve orifice resistance R_v is given by

$$e_4 = R_v f_4 \qquad (5.68)$$

The effort e_5 of the pipeline resistance R_p is given by

$$e_5 = R_p f_5 \qquad (5.69)$$

Equation (5.65) indicates that the flows f_4 and f_5 equal f_6 given by the constitutive relations indicated in equation (5.62). Considering these, equations (5.68) and (5.69) become

$$e_4 = R_v f_6 = \frac{R_v}{I_p} p_6 \qquad (5.70)$$

and

$$e_5 = R_p f_6 = \frac{R_p}{I_p} p_6 \qquad (5.71)$$

Substituting equations (5.70) and (5.71) in equation (5.67) and rearranging the results gives

$$\dot{p}_6 + \frac{(R_v + R_p)}{I_p} p_6 = \frac{1}{C_t} q_2 \qquad (5.72)$$

Now let us determine the state variable q_2 of the tank's compliance as a function of the flow source SF. Considering equation (5.64) and inserting $f_1 = SF$ gives

$$\dot{q}_2 = SF - f_3 \qquad (5.73)$$

Equation (5.65) indicates the flow f_3 equals the flow f_6 given by the constitutive relations indicated in equation (5.62). Considering this, equation (5.73) becomes

$$\dot{q}_2 = SF - f_6 = SF - \frac{1}{I_p} p_6 \qquad (5.74)$$

Taking the integration of this equation gives

$$q_2 = \int SF \, dt - \frac{1}{I_p} \int p_6 \, dt \qquad (5.75)$$

Substituting this equation in equation (5.) gives

$$\dot{p}_6 + \frac{(R_v + R_p)}{I_p} p_6 = \frac{1}{C_t} \left(\int SF \, dt - \frac{1}{I_p} \int p_6 \, dt \right) \qquad (5.76)$$

Rearranging this equation gives

$$\dot{p}_6 + \frac{(R_v + R_p)}{I_p} p_6 + \frac{1}{C_t I_p} \int p_6 \, dt = \frac{1}{C_t} \int SF \, dt \qquad (5.77)$$

This equation indicates that the fundamental change in the momentum of the liquid's mass depends on the accumulation of the input flow rate value, $\int SF \, dt$, that fills the tank volume affected by its compliance, as indicated on its right-hand side. Rearranging equation (5.77) gives

$$C_t \dot{p}_6 + C_t \frac{(R_v + R_p)}{I_p} p_6 + \frac{1}{I_p} \int p_6 \, dt = \int SF \, dt \qquad (5.78)$$

This equation represents the dynamic equation of the given liquid-level system shown in Figure 5.12.

Now, we will check if the law of conservation of mass and Newton's law are embedded in the Augmented Bond Graph model shown in Figure 5.14. The following analyses are what we get when using the constitutive relations indicated in equation (5.62).

The momentum of the liquid mass that passes through the pipeline is given by

$$p_6 = I_p f_6 \qquad (5.79)$$

Note that the flow f_6 equals the rate of change of the liquid mass $\dot{q}_o$. Considering this, equation (5.79) becomes

$$p_6 = I_p \dot{q}_o \qquad (5.80)$$

Taking the derivative of this equation gives

$$\dot{p}_6 = I_p \ddot{q}_o \qquad (5.81)$$

Substituting equations (5.79) and (5.81) in equation (5.78) and inserting $SF = \dot{q}_{in}$ gives

$$C_t I_p \ddot{q}_o + C_t \frac{(R_v + R_p)}{I_p} I_p \dot{q}_o + \frac{1}{I_p} I_p \int \dot{q}_o \, dt = \int SF \, dt \qquad (5.82)$$

Rearranging this equation gives

$$C_t I_p \ddot{q}_o + C_t (R_v + R_p) \dot{q}_o + \int \dot{q}_o \, dt = q_{in} \qquad (5.83)$$

Carrying out the integration terms gives

$$C_t I_p \ddot{q}_o + C_t (R_v + R_p) \dot{q}_o + q_o = q_{in} \qquad (5.84)$$

In steady-state mode, the flow rate q_o is given by

$$q_o = \frac{1}{R_v} h \qquad (5.85)$$

Taking the first and second derivatives of this equation, substituting them in equation (5.84), and carrying out the integration terms gives

$$\frac{I_p C_t}{R_v} \frac{d^2 h}{dt^2} + C_t \frac{(R_v + R_p)}{R_v} \frac{dh}{dt} + \frac{1}{R_v} h = q_{in} \qquad (5.86)$$

This equation is the differential equation representing the dynamics of the given system. Comparing this equation with equation (5.44), we found that law of the conservation of mass flow rate and Newton's law are satisfied in the Augmented Bond Graph model of the system shown in Figure 5.14.

Example (5.5):
For the liquid-level system shown in Figure 5.15,

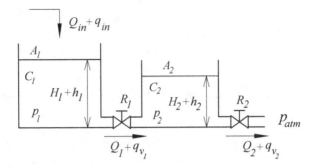

Figure 5.15 Schematic diagram of two tank liquid-level system

a. Create the preliminary Bond Graph model and impose it on the schematic diagram.
b. Simplify it and create the Augmented Bond Graph model.
c. Write the state and power variables and their constitutive relations and then derive the equations of motion.
d. Check if law of the conservation of mass flow rate is embedded in the Augmented Bond Graph of the system.

Solution:
In this example, we will illustrate how to generate a Bond Graph model by superimposing it on the schematic diagram of the given system. The system consists of two tanks, as shown in Figure 5.15. Both tanks are linked to one another through a bypass valve with resistance R_1, while the second tank is connected to the atmosphere through a bypass valve with resistance R_2. The first tank is supplied by an external flow rate of q_{in}. Considering the Bond Graph junction descriptions, the input flow rate q_{in} is represented by a source of flow SF connected to a 1-junction, as shown in Figure 5.16. Since the liquid inside the first tank has the same level (effort) h_1, it is represented by a compliance element C_1 connected to a 0-junction. The resistance of the first valve orifice is represented by a resistance element R_1 connected to a 1-junction. Since the liquid inside the second tank has the same level of effort (effort) h_2, it is represented by a compliance element C_2 connected to a 0-junction. The resistance of the second valve orifice is represented by the resistance R_2 connected to a 1-junction. As previously stated, when considering the absolute pressure, the effect of the atmospheric pressure will be similar to that of the ground in mechanical and electrical systems. Therefore, the atmospheric pressure p_{atm} is represented by a source of effort SE connected to a 0-junction, as shown in Figure 5.16.

At this point, the preliminary Bond Graph model of the given system is created and superimposed on the system schematic diagram, as shown in Figure 5.16. The half-arrow positive power flow directions are selected, as shown in the same figure. Remember that the sources supply power while the elements receive it. To simplify this Bond Graph model, all connections that hold the atmospheric pressure p_{atm} with their associated bonds must be removed, and then any 2-port 1-junctions or

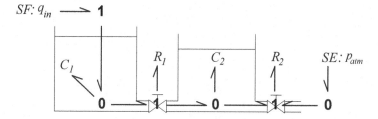

Figure 5.16 Preliminary Bond Graph model imposed on the schematic diagram of the system

0-junctions with through power flow should be swapped with single bonds. As a result, the 0-junction connected to $SE : p_{atm}$, as well as both 1-junctions connected to the flow source SF and the 1-junction connected to the resistance R_2, have all been eliminated, as shown in Figure 5.17. The resulting simplified Bond Graph of the given system is shown in Figure 5.18.

Now, let us assign the causality of the Bond Graph model. There is only one source of flow; the causal stroke mark should be located on its bond at the sides of the element, as shown in Figure 5.18. Considering the causality restrictions, apply the causal stork marks to all elements and junctions. It is essential to remember that as we work through assigning causality, we must consider the integral causality that applies to all associated elements and bonds. We are finally numbing all bonds to get the Augmented Bond Graph model of the given system shown in Figure 5.18. This figure shows two integrally causal storage elements, C_1 and C_2.

To derive the dynamic equations of the given system, we must first define the key variables of the model: Input, State variables, and Co-energy (power) variables and their constitutive relations.

Key variables:
- Inputs: $f_1 = SF = q_{in}$.
- State variables: q_2 and q_7.
- Co-energy (Power) variables: e_2 and e_7.

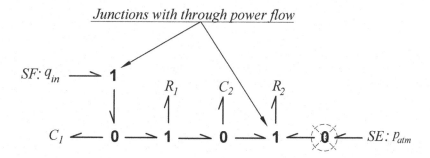

Figure 5.17 Preliminary Bond Graph model of the system

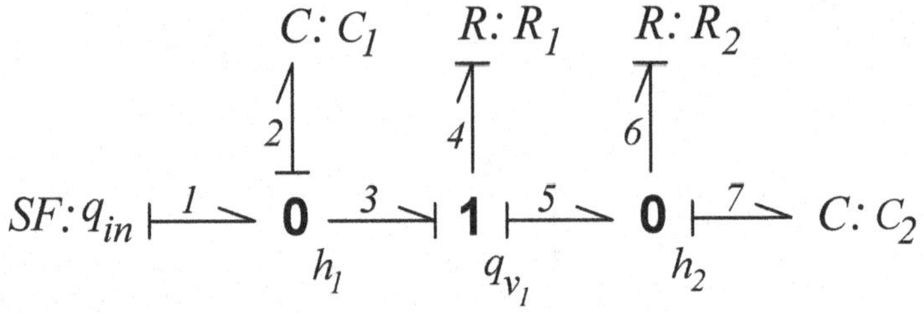

$$C\!:\!C_1 \qquad R\!:\!R_1 \qquad R\!:\!R_2$$

Figure 5.18 Augmented Bond Graph model of the system

Constitutive relations:
Write equations describing the relationship between co-energy variables and state variables for each energy storage element, as follows:

$$e_2 = \frac{1}{C_1}q_2 \quad \text{and} \quad e_7 = \frac{1}{C_2}q_7 \tag{5.87}$$

Now that we have the connections between the flow and effort variables of the model junctions shown in Figure 5.18, let us derive the equations that reflect them.

- As shown in Figure 5.18, the effort variables of the 0-junction of a common effort h_1 that is connected to the capacitor C_1 are given by

$$e_1 = e_2 = e_3 \tag{5.88}$$

and the flow variables are given by

$$f_1 = f_2 + f_3 \quad \text{or} \quad f_2 = \dot{q}_2 = f_1 - f_3 \tag{5.89}$$

where $\dot{q}_2$ is the rate of change of the liquid level inside the first tank.
- The flow variables of the 1-junction of a common flow q_{v_1} are given by

$$f_3 = f_4 = f_5 \tag{5.90}$$

and the effort variables are given by

$$e_3 = e_4 + e_5 \tag{5.91}$$

- The effort variables of the 0-junction of a common effort h_2 that is connected to both the capacitance C_2 and the resistance R_2 are given by

$$e_5 = e_6 = e_7 \tag{5.92}$$

and the flow variables are given by

$$f_5 = f_6 + f_7 \quad \text{and} \quad f_7 = \dot{q}_7 = f_5 - f_6 \tag{5.93}$$

Now, let us derive the differential equation describing the dynamics of the given system.

- For the first tank:

Equation (5.89) indicates that to determine $\dot{q}_2$, we must determine the flow f_3. Equation (5.90) indicates that the flow f_3 equals f_4. Considering this and inserting $f_1 = SF$, equation (5.89) becomes

$$f_2 = \dot{q}_2 = SF - f_4 \tag{5.94}$$

The flow f_4 of the resistance R_1 is given by

$$f_4 = \frac{e_4}{R_1} \tag{5.95}$$

Equation (5.91) indicates that the effort e_4 equals $(e_3 - e_5)$. Considering this, equation (5.95) becomes

$$f_4 = \frac{1}{R_1}(e_3 - e_5) \tag{5.96}$$

Equations (5.88) and (5.92) indicate that the effort e_3 equals e_2 and the effort e_5 equals e_7, respectively. Considering these, equation (5.96) becomes

$$f_4 = \frac{1}{R_1}(e_2 - e_7) \tag{5.97}$$

Using the constitutive relations indicated in equation (5.87), equation (5.97) becomes

$$f_4 = \frac{1}{R_1}\left(\frac{1}{C_1}q_2 - \frac{1}{C_2}q_7\right) \tag{5.98}$$

Substituting equation (5.98) in equation (5.94) and rearranging the results gives

$$\dot{q}_2 + \frac{1}{R_1}\left(\frac{1}{C_1}q_2 - \frac{1}{C_2}q_7\right) = SF \tag{5.99}$$

This equation is the differential equation representing the dynamics of the first tank of the given system.

- For the second tank:

Equation (5.93) indicates that to determine $\dot{q}_7$, we must determine the flows f_5 and f_6. Equation (5.90) indicates that the flow f_5 equals f_4, which is given by equation (5.98). Considering this, the flow f_5

$$f_5 = \frac{1}{R_1}\left(\frac{1}{C_1}q_2 - \frac{1}{C_2}q_7\right) \tag{5.100}$$

The flow f_6 of the resistance R_2 is given by

$$f_6 = \frac{e_6}{R_2} \tag{5.101}$$

Equation (5.92) indicates that e_6 equals e_7. Considering this, Using the constitutive relations indicated in equation (5.87), equation (5.101) becomes

$$f_6 = \frac{1}{R_2}\frac{1}{C_2}q_7 \tag{5.102}$$

Substituting equations (5.100) and (5.101) in equation (5.93) gives

$$\dot{q}_7 = \frac{1}{R_1}(\frac{1}{C_1}q_2 - \frac{1}{C_2}q_7) - \frac{1}{R_2C_2}q_7 \tag{5.103}$$

Rearranging this equation yields

$$\dot{q}_7 + \frac{1}{R_1}(\frac{1}{C_2}q_7 - \frac{1}{C_1}q_2) + \frac{1}{R_2C_2}q_7 = 0 \tag{5.104}$$

This equation is the differential equation representing the dynamics of the second tank of the given system.

Now, we will check if law of the conservation of mass flow rate is embedded in the Augmented Bond Graph model shown in Figure 5.18. The following analyses are what we get when using the constitutive relations indicated in equation (5.87). Note that the efforts e_2 and e_7 equal the liquid levels h_1 and h_2, respectively. The state variable q_2 of the compliance C_1 is given by

$$q_2 = C_1 e_2 = C_1 h_1 \tag{5.105}$$

Taking the derivative of this equation gives

$$\dot{q}_2 = C_1 \frac{dh_1}{dt} \tag{5.106}$$

The state variable q_7 of the compliance C_2 is given by

$$q_7 = C_2 e_7 = C_2 h_2 \tag{5.107}$$

Taking the derivative of this equation gives

$$\dot{q}_7 = C_2 \frac{dh_2}{dt} \tag{5.108}$$

Substituting equations (5.105), (5.106), and (5.107) in equation (5.99) and inserting $SF = q_{in}$ gives

$$C_1 \frac{dh_1}{dt} + \frac{1}{R_1}(\frac{1}{C_1}C_1 h_1 - \frac{1}{C_2}C_2 h_2) = q_{in} \tag{5.109}$$

Rearranging this equation gives

$$C_1 \frac{dh_1}{dt} + \frac{(h_1 - h_2)}{R_1} = q_{in} \tag{5.110}$$

Substituting equations (5.105), (5.107), and (5.108) in equation (5.99) gives

$$C_2 \frac{dh_2}{dt} + \frac{1}{R_1} \left(\frac{1}{C_2} C_2 h_2 - \frac{1}{C_1} C_1 h_1 \right) + \frac{1}{R_2 C_2} C_2 h_2 = 0 \qquad (5.111)$$

Rearranging this equation gives

$$C_2 \frac{dh_2}{dt} + \frac{(h_2 - h_1)}{R_1} + \frac{h_2}{R_2} = 0 \qquad (5.112)$$

By comparing equations (5.110) and (5.112) with equations (5.47) and (5.51), we found that they are identical, implying that law of the conservation of mass flow rate is satisfied in the Augmented Bond Graph model of the given system shown in Figure 5.18.

Example (5.6):
For the liquid-level system shown in Figure 5.19,

- Create the preliminary Bond Graph model and simplify it.
- Create the Augmented Bond Graph model.
- Write the state and power variables and their constitutive relations.
- Derive the dynamic equations and check if law of the conservation of mass flow rate is embedded in the Augmented Bond Graph model.

Solution:
As shown in Figure 5.19, the system comprises two tanks. The first tank is supplied by an external flow rate, q_{in}, while the second tank is supplied by the output flow

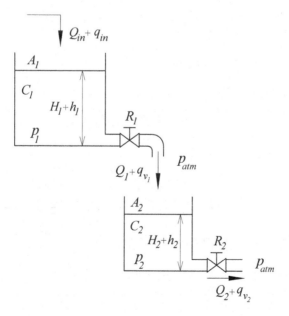

Figure 5.19 Two separated tanks liquid-level system

rate, q_{v_1}, produced by the first tank. Two throttle valves are located at the exit of each tank.

The flow rate q_{in} is represented by a source of flow SF_1 connected to a 1-junction, as shown in Figure 5.20. Since the liquid inside the first tank has the same pressure p_1 (same level h_1), it is represented by a compliance element C_1 connected to a 0-junction. The resistance of the orifice of the first bypass valve is represented by the resistance R_1 connected to a 1-junction. When considering the absolute pressure, the effect of the atmospheric pressure will be similar to that of the effect of the ground in mechanical and electrical systems, as previously stated. Therefore, the atmospheric pressure p_{atm} is represented by a source of effort SE connected to two 0-junctions for both tanks, as shown in Figure 5.20. Note that the output flow q_{v_1} of the first tank will be the supply flow rate to the second tank. So, the supply flow rate to the second tank is represented by a source of flow SF_2. Since the liquid inside the second tank has the same pressure p_2 (same level h_2), it is represented by a compliance element C_2 connected to a 0-junction. The resistance of the orifice of the second throttle valve is represented by the resistance R_2 connected to a 1-junction, as shown in Figure 5.20.

At this point, the preliminary Bond Graph model of the given system is created. The half-arrows positive power flow directions are selected, as shown in the same figure. Remember that the sources supply power while the elements receive it. To simplify this Bond Graph model, all connections that hold the atmospheric pressure p_{atm} for both tanks with their associated bonds must be removed, and then any 2-port 1-junctions or 0-junctions with through power flow should be swapped with single bonds. As a result, the 0-junctions connected to both atmospheric source of effort SE and the 1-junction connected to the resistance R_2 are removed, as shown in Figure 5.20. Note that although the 1-junction that is connected to the resistance R_1 is a through power flow junction, we cannot remove it during the process of simplifying the Bond Graph model since the flow rate q_{v_1} that it will serve as the supply flow

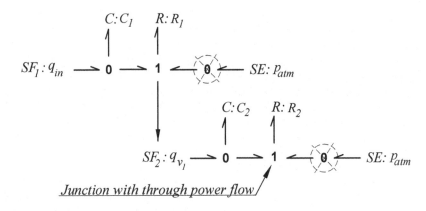

Figure 5.20 Preliminary Bond Graph model of the system

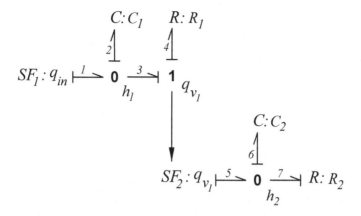

Figure 5.21 Augmented Bond Graph model

rate for the second tank. The resulting simplified Bond Graph of the given system is shown in Figure 5.21.

Now, let us assign the causality of the Bond Graph model. There are two sources of flow; the causal stroke marks should be located on their bonds at the sides of the elements, as shown in Figure 5.21. Considering the causality restrictions stated in Section (3.3), apply the causal stork marks to all elements and junctions. It is essential to remember that as we work through assigning causality, we must consider the integral causality that applies to all associated elements and bonds. We are finally numbing all bonds to get the Augmented Bond Graph model of the system shown in Figure 5.21. This figure shows two integrally causal storage elements, C_1 and C_2.

To derive the dynamic equations of the given system, we must first define the key variables of the model: Input, State variables, and Co-energy (power) variables and their constitutive relations.

Key variables:
- Inputs: $f_1 = SF_1 = q_{in}$ and $f_5 = SF_2 = q_{v_1}$.
- State variables: q_2 and q_6 .
- Co-energy (Power) variables: e_2 and e_6.

Constitutive relations:
Write equations describing the relationship between co-energy variables and state variables for each energy storage element, as follows:

$$e_2 = \frac{1}{C_1} q_2 \quad \text{and} \quad e_6 = \frac{1}{C_2} q_6 \tag{5.113}$$

Now that we have the connections between the flow and effort variables of the model junctions shown in Figure 5.21, let us derive the equations that reflect them.

- As shown in Figure 5.21, the effort variables of the 0-junction that is connected to the capacitor C_1 are given by

$$e_1 = e_2 = e_3 \qquad (5.114)$$

and the flow variables are given by

$$f_1 = f_2 + f_3 \quad \text{or} \quad f_2 = \dot{q}_2 = f_1 - f_3 \qquad (5.115)$$

where $\dot{q}_2$ is the rate of change of the liquid level inside the first tank.
- As previously stated, the 1-junction connected to the resistance R_1 is a through power flow junction that cannot be removed during the process of simplifying the Bond Graph model since the flow rate q_{v_1} that passes through it serves as the supply flow rate for the second tank. So, the flow variables of this 1-junction are given by

$$f_3 = f_4 \qquad (5.116)$$

and the effort variables are given by

$$e_3 = e_4 \qquad (5.117)$$

So, the flow f_5 equals the supply flow rate q_{v_1} to the second tank.
- The effort variables of the 1-junction that is connected to both the capacitance C_2 and the resistance R_2 are given by

$$e_5 = e_6 = e_7 \qquad (5.118)$$

and the flow variables are given by

$$f_5 = f_6 + f_7 \quad \text{or} \quad f_6 = \dot{q}_6 = f_5 - f_7 \qquad (5.119)$$

where $\dot{q}_6$ is the rate of change of the liquid level inside the first tank. Now, let us derive the differential equations describing the dynamics of the given system.

- For the first tank:
Considering equation (5.115) indicates that to determine $\dot{q}_2$, we must obtain the flow f_3. Equation (5.116) indicates that the flow f_3 equals the flow f_4, which is the flow of the resistance R_1 and it is given by

$$f_3 = f_4 = \frac{e_4}{R_1} \qquad (5.120)$$

Equation (5.117) indicates that the effort e_4 equals e_3, which equals e_2 as indicated in equation (5.114). Considering this, equation (5.120) becomes

$$f_4 = \frac{e_2}{R_1} \qquad (5.121)$$

Note that the effort e_2 is given by the constitutive relation indicated in equation (5.113). Considering this, inserting $f_1 = SF_1$ equation (5.115) becomes

$$\frac{1}{R_1 C_1} q_2 = SF_1 - \dot{q}_2 \tag{5.122}$$

Rearranging this equation gives

$$\dot{q}_2 + \frac{1}{R_1 C_1} q_2 = SF_1 \tag{5.123}$$

This equation is the differential equation representing the dynamics of the first tank of the given system.

- For the second tank:
Equation (5.119) indicates that to determine $\dot{q}_6$ we must obtain the flow f_7 of the resistance R_2 which is given by

$$f_7 = \frac{e_7}{R_2} \tag{5.124}$$

Equation (5.118) indicates that the effort e_7 equals e_6. Considering this and using the constitutive relations indicated in equation (5.113), equation (5.124) becomes

$$f_7 = \frac{1}{R_2 C_2} q_6 \tag{5.125}$$

Substituting this equation in equation (5.119) and inserting $f_5 = SF_2$ gives

$$\dot{q}_6 = SF_2 - \frac{1}{R_2 C_2} q_6 \tag{5.126}$$

Rearranging this equation gives

$$\dot{q}_6 + \frac{1}{R_2 C_2} q_6 = SF_2 \tag{5.127}$$

This equation is the differential equation representing the dynamics of the second tank of the given system. Equations (5.123) and (5.127) are the differential equations representing the dynamics of the given liquid-level system shown in Figure 5.19

Now, we will check if law of the conservation of mass flow rate is embedded in the Augmented Bond Graph model shown in Figure 5.21. The following analyses are what we get when using the constitutive relations indicated in equation (5.113). Note that the efforts e_2 and e_6 equal the liquid levels h_1 and h_2, respectively.

- For the first tank:
The state variable q_2 of the compliance C_1 is given by

$$q_2 = C_1 e_2 = C_1 h_1 \tag{5.128}$$

Taking the derivative of this equation gives

$$\dot{q}_2 = C_1 \frac{dh_1}{dt} \tag{5.129}$$

Substituting equations (5.128) and (5.129) in equation (5.123) and inserting $SF_1 = q_{in}$ gives

$$C_1 \frac{dh_1}{dt} + \frac{1}{R_1 C_1} C_1 h_1 = q_{in} \tag{5.130}$$

Rearranging this equation gives

$$C_1 \frac{dh_1}{dt} + \frac{h_1}{R_1} = q_{in} \tag{5.131}$$

- For the second tank:
The state variable q_6 of the compliance C_2 is given by

$$q_7 = C_2 e_6 = C_2 h_2 \tag{5.132}$$

Taking the derivative of this equation gives

$$\dot{q}_6 = C_2 \frac{dh_2}{dt} \tag{5.133}$$

Substituting equations (5.132) and (5.133) in equation (5.127) and inserting $SF_2 = q_{v_1}$ gives

$$C_2 \frac{dh_2}{dt} + \frac{1}{R_2 C_2} C_2 h_2 = q_{v_1} \tag{5.134}$$

Rearranging this equation gives

$$C_2 \frac{dh_2}{dt} + \frac{h_2}{R_2} = q_{v_1} \tag{5.135}$$

By comparing equations (5.131) and (5.135) with equations (5.57) and (5.60), respectively, we found that they are identical, implying that the law of the conservation of mass is satisfied in the Augmented Bond Graph model of the given system shown in Figure 5.21.

5.3 PNEUMATIC SYSTEMS

Pneumatic systems are significantly utilized in the automation of industrial machines. Pneumatic circuits, for example, which transform the energy of compressed air into mechanical energy, are famous, and many types of pneumatic controllers are found in use across a wide range of commercial and industrial domains. Air movement through interconnected pipelines and pressure vessels is essential to various industrial processes. Considering the pneumatic system shown in Figure 5.22, when the

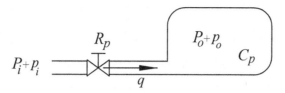

Figure 5.22 Pneumatic system

upstream pressure P_i has reached a steady-state value, assume that it shifts a small value to $P_i + p_i$, the downstream pressure (the pressure inside the vessel) will also shift to $P_o + p_o$. If the airflow state is subsonic and $P_i \gg p_i$ as well as $P_o \gg p_o$, then the airflow rate through the restriction of resistance R_p will become proportional to the pressure difference $(p_i - p_o)$. This will be the case if all of the conditions above are satisfied. Whenever we deal with pneumatic systems, we assume that the flow state is subsonic, implying that they can be characterized by resistance R_p and capacitance C_p.

5.3.1 PNEUMATIC RESISTANCE

When we talk about airflow resistance in restricting devices such as pipelines, orifices, valves, and anything else along these lines, we refer to the change in differential pressure between the upstream and downstream regions of the device that limits the flow of air. In this case, the pressure required to change a unit in the mass flow rate (kg/sec) can be expressed as

$$R_p = \frac{\text{change in differential pressure}}{\text{change in mass flow rate}} = \frac{d(\Delta p)}{dq} \tag{5.136}$$

where R_p ((N m/kg kg) is the pneumatic resistance and $d(\Delta p)$ is a change in the differential pressure, and dq is a change in the mass flow rate. However, this may be easily verified experimentally using Figure 5.23, which is the relation between the pressure difference $d\Delta p$ and the flow rate q, by simply determining the slope of the

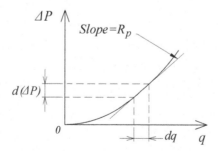

Figure 5.23 Pressure difference versus flow rate

curve at any given set of operating conditions. Note that the resistance value R_p is not constant but varies as operating conditions change.

5.3.2 JUNCTION REPRESENTATION OF THE PNEUMATIC RESISTANCE

Since the valve or orifice shares the same flow q, a 1-junction is used and connected to the resistance element R_p, as shown in Figure 5.24.

Figure 5.24 Junction representations of the pneumatic resistance

5.3.3 PNEUMATIC CAPACITANCE

Capacitance can be defined as the change in the air mass in a pneumatic vessel necessary to make a unit change in pressure. This change in mass is required to make a unit change in pressure, which can be expressed as

$$C_p = \frac{\text{change in of air or gas}}{\text{change in pressure}} = \frac{dm}{dp} \tag{5.137}$$

where m (kg) is the mass of air in vessel and p (N/m^2) is the absolute pressure of air and C_p(kg/(Nm^2)) is the pneumatic capacitance, which can be expressed as

$$C_p = \frac{dm}{dp} = V\frac{d\rho}{dp} \tag{5.138}$$

where V (m^3) is the volume of vessel and ρ (kg/m^3) is the mass density of air. The use of the perfect gas law may be calculated as follows. For air, we have

$$pv = \frac{p}{\rho} = \frac{R}{M}T = R_{pa}T \tag{5.139}$$

where R (N m/kg-mole K^o) is the universal gas constant, M (kg/kg-mole) is the molecular gas constant, R_{pa} (N m/kg K^o) is the gas constant of air, T (K^o) is the absolute temperature of air, and ρ (N/m^2) is the density of air. The expansion process can be expressed as polytropic if the change in the state of the air is between isothermal and adiabatic, as follows.

$$\frac{p}{\rho^n} = \text{constant} \tag{5.140}$$

where n is the polytropic exponent. Since $d\rho/dp$ can be obtained from equation (5.140) as follows

$$\frac{d\rho}{dp} = \frac{\rho}{np} \qquad (5.141)$$

Substituting equation (5.141) in equation (5.139) gives

$$\frac{d\rho}{dp} = \frac{1}{nR_{pa}T} \qquad (5.142)$$

Then, the capacitance C_{pa} of a vessel is found by substituting equation (5.142) in equations (5.138) as follows:

$$C_{pa} = \frac{V}{nR_{pa}T} \qquad (5.143)$$

where C_{pa} is the capacitance of the air and R_{pa} is the gas constant for the particular air involved. Note that if gas other than air is used in a pressure vessel, capacitance C_{pg} is given by

$$C_{pg} = \frac{V}{nR_{pg}T} \qquad (5.144)$$

where C_{pg} is the capacitance of the gas and R_{pg} is the gas constant for particular gas involved.

The above analysis makes it readily apparent that the capacitances C_{pa} and C_{pg} of a pressure vessel are not constant. They vary depending on the expansion process, the kind of gas contained inside the vessel (air, nitrogen, hydrogen, etc.), and the temperature of the gas within the vessel. The polytropic exponent n is roughly constant for gases in an insulated metal vessel and ranges from 1 to 1.2.

5.3.4 JUNCTION REPRESENTATIONS OF THE PNEUMATIC CAPACITANCE

Since the vessel volume shares the same effort, p_o, a 0-junction is used and connected to the capacitor element C_p, as shown in Figure 5.25.

5.3.5 PNEUMATIC SOURCES: PRESSURE AND FLOW

An ideal flow source generates a particular flow regardless of the pressure necessary. Similarly, an ideal pressure source generates a constant pressure at some point in

Figure 5.25 Junction representation of the pneumatic capacitance

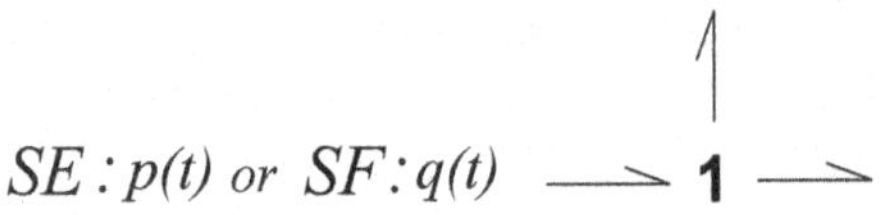

$$SE : p(t) \ or \ SF : q(t) \longrightarrow 1 \longrightarrow$$

Figure 5.26 Junction representation of the pneumatic sources

the pneumatic system, regardless of the amount of flow necessary to sustain that pressure. One of the most popular pneumatic power sources is the air compressor. In our analysis, we will be concerned with the volume of air (gas), or pressure supplied to the system, regardless of the type of power sources used.

5.3.6 JUNCTION REPRESENTATIONS OF THE PNEUMATIC SOURCES

The source of effort SE or flow SF can be connected to a 1-junction, as shown in Figure 5.26. As mentioned previously, when we derive the dynamic equation, we do so under the assumption that the operation of the pneumatic system is in steady-state mode and that the change in the flow rate stored in the vessel for dt seconds is equal to the net inflow to the vessel for the same dt seconds. In other words, the pressure change equals the net inflow in over dt seconds. We are going to consider this in the following analysis.

5.3.7 DYNAMIC MODELING OF PNEUMATIC SYSTEMS

Consider the pneumatic system in Figure 5.22. It comprises a pressure vessel and a valve-equipped connecting pipe. We may deal with a linear system if the system variables only slightly deviate from their respective steady-state values. Now, let us derive the dynamic equations of the given system.

Assume that the system operates at a subsonic flow condition throughout the operation. As a result, we must use an average resistance throughout the working zone of the system. As shown in Figure 5.27, the average resistance R_p, which is the value of the slope at the operating point, is given by

$$R_p = \frac{(p_i - p_o)}{q} \quad or \quad q = \frac{(p_i - p_o)}{R_p} \tag{5.145}$$

Remember that the resistance R_p is not constant. Considering equation (5.137), the capacitance C_p of the vessel is given by

$$C_p = \frac{dm}{dp_o} \quad or \quad C_p dp_o = dm \tag{5.146}$$

The change in mass dm equals the change in the airflow rate qdt passing through the valve to the vessel. Hence,

$$C_p dp_o = qdt \quad or \quad q = C_p \frac{dp_o}{dt} \tag{5.147}$$

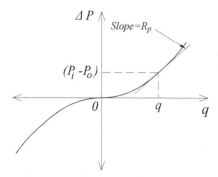

Figure 5.27 Pressure difference versus flow rate

Substituting equation (5.147) in equation (5.145) gives

$$C_p \, dp_o = \frac{(p_i - p_o)}{R_p} dt \qquad (5.148)$$

Rearranging this equation gives

$$R_p C_p \frac{dp_o}{dt} + p_o = p_i \qquad (5.149)$$

Equation (5.149) is the dynamic equation of the given pneumatic system shown in Figure 5.22. This equation indicated that the given pneumatic system is analogous to a first-order electrical, mechanical, or liquid-level system.

5.3.8 BOND GRAPH CONSTRUCTION FOR PNEUMATIC SYSTEMS

As stated previously, if some elements have the same flow rate (flow), they should be connected to a 1-junction, while if they have the same pressure (effort), they should be connected to a 0-junction. These situations are explained as follows.

1. For each pressure of interest, establish a 0-junction. If we consider absolute pressures, the atmospheric pressure connection is allocated to a specific node, similar to the ground in mechanical and electrical systems (to be eliminated).
2. Attach capacitance elements C to appropriate 0-junctions for pneumatic capacitance associated with pressures. Attach resistance elements R and transformer element TF to 1-junctions (using the power convention) between appropriate 0-junctions.
3. Assign power flow half arrows using the through power flow convention for the 1-junction and two 0-junctions.
4. All bonds related to the atmospheric pressure must be removed to simplify the model. To further simplify, any 2-port 1-junctions or 0-junctions with through power flow directions should be swapped by single bonds.

5. Finally, we should add causality stroke marks and bond numbers to finish. Consider the causality restrictions applied to the Bond Graph model, which will be the same as those applied to previous systems (Section 3.3). As a result, the Augmented Bond Graph is ready to derive the differential equations describing the dynamics of pneumatic systems.

Example (5.7):
For the pneumatic system shown in Figure 5.22,

a. Create the preliminary Bond Graph and then the Augmented Bond Graph model and derive the dynamic equation(s).
b. Check if the law of conservation of mass flow is embedded in the Augmented Bond Graph.

Solution:
Consider the pneumatic system shown in Figure 5.22; as we stated in Section 5.3, when the upstream pressure P_i has reached a steady-state value and shifts a small value to $P_i + p_i$, the downstream pressure (the pressure inside the vessel) will also shift to $P_o + p_o$. In this case, the Bond Graph model is created as follows (see Figure 5.28):

- The change in the upstream pressure p_i is represented by a source of effort SE connected to a 0-junction of a common effort.
- Assume that the airflow state is subsonic, then the airflow rate through the restriction will become proportional to the pressure difference $(p_i - p_o)$. Since the valve's orifice shares the same flow q, it is represented by a resistance element R_p connected to the same 1-junction of a common flow q.
- Since the airflow stored in the vessel shares the same pressure p_o, it is represented by a capacitance element C_p connected to a 0-junction of common effort.

At this point, the preliminary Bond Graph model of the given system is constructed, as shown in Figure 5.28. The half-arrows indicating positive power flow directions

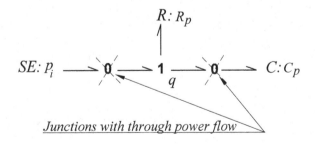

Figure 5.28 Preliminary Bond Graph model of the given system

are selected, as shown in the same figure. In order to simplify this Bond Graph model, any 2-port 1-junctions or 0-junctions with through power flow should be swapped with single bonds. As a result, the 0-junction connected to the source of effort SE and the 0-junction connected to the capacitance C_p are removed, as shown in Figures 5.28 and 5.29.

Now, let us assign the causality of the Bond Graph model. Since there is only one source of effort, SE, the causal stroke mark should be located on its bond at the junction side, as shown in Figure 5.29. Considering the causality restrictions in Section (3.3), apply the causal stork marks to all elements and junctions. We are finally numbing all bonds to get the Augmented Bond Graph model of the system shown in Figure 5.29. This figure shows one integral causality storage element, C_p represented by bond 3.

To derive the dynamic equations of the given system, we must first define the key variables of the model: Input, State variables, and Co-energy (power) variables and their constitutive relations.

Key variables:
- Inputs: $e_1 = SE = p_i$.
- State variables: q_3.
- Co-energy (Power) variables: e_3.

Constitutive relations:
Write equations describing the relationship between the co-energy variable and state variable for the capacitance C_p storage element as follows:

$$e_3 = \frac{1}{C_p} q_3 \tag{5.150}$$

Now that we have the connections of flow and effort variables of the model junction shown in Figure 5.29, let us derive the equations that reflect them. As shown in this figure, the flow variables of the 1-junction of common flow q are given by

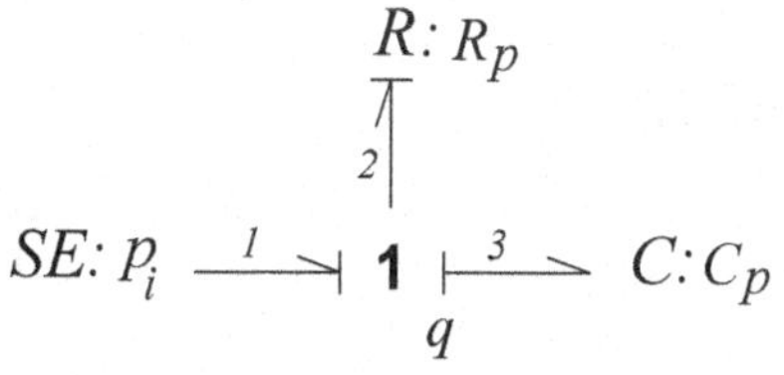

Figure 5.29   Augmented Bond Graph model of the given system

$$f_1 = f_2 = f_3 \tag{5.151}$$

and the effort variables are given by

$$e_1 = e_2 + e_3 \quad \text{or} \quad SE = e_2 + e_3 \tag{5.152}$$

Considering this equation, to derive the differential equation describing the dynamics of the given system, we have to determine e_2 and e_3. The effort e_2 of the resistance R_p is determined as follows:

$$e_2 = R_p f_2 \qquad (5.153)$$

Equation (5.152) indicates that the flow f_2 equals the flow f_3, which equals $\dot{q}_3$. Considering this and taking the derivative of the constitutive relation indicated in equation (5.150) gives

$$f_3 = \dot{q}_3 = C_p \dot{e}_3 \qquad (5.154)$$

Substituting this equation in equation (5.153) gives

$$e_2 = R_p f_3 = R_p \dot{q}_3 = R_p C_p \dot{e}_3 \qquad (5.155)$$

Substituting this equation in equation (5.152) gives

$$R_p C_p \dot{e}_3 + e_3 = SE \qquad (5.156)$$

This is the dynamic equation of the given pneumatic system deduced from its Augmented Bond Graph.

Now, we will check if law of the conservation of mass flow rate is embedded in the Augmented Bond Graph model shown in Figure 5.29. Note that the effort e_3 equals the pressure p_o and the effort e_1 equals SE which is equal the upstream pressure p_i. Considering these, equation (5.156) becomes

$$C_p R_p \dot{p}_o + p_o = p_i \qquad (5.157)$$

This equation represents the dynamics of the pneumatic system as deduced from the Augmented Bond Graph model shown in Figure 5.29. Comparing equation (5.157) with equation (5.149) it was found that they are identical, implying that the law of conservation of mass is satisfied in the Augmented Bond Graph model shown in Figure 5.29.

5.4 HYDRAULIC SYSTEMS

Hydraulic systems have widespread applications in aircraft control, heavy construction machinery, machine tool applications, and other uses. Their high horsepower-to-weight ratio plays a significant role in their widespread use. High-pressure, low-velocity flows characterize hydraulic systems, and the initial models we are developing will be based on the idea that any fluid power is the product of pressure and a volume flow rate. Many hydraulic components, including pumps, valves, cylinders, motors, and pipelines, have valuable models based on this concept.

In this section, within the limits of our study, a simple hydraulic circuit and a hydraulic servo system will be dealt with. We will begin by providing some fundamental concepts regarding a straightforward hydraulic circuit, and a dynamic model of the hydraulic servo system will then be obtained. Since the hydraulic servo system is

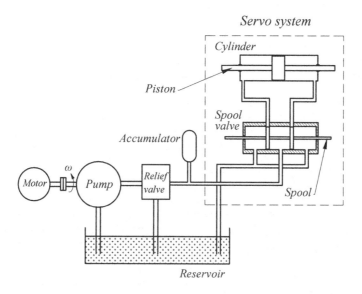

Figure 5.30 Hydraulic circuit

nonlinear, the nonlinear equation that characterizes its dynamics must be linearized. We use a linearization strategy that is either based on the perturbation method that was covered in Section 5.2.9 or on the Taylor series expansion. Let us first begin with the hydraulic circuits.

Building hydraulic circuits can provide a wide range of motion and force combinations. However, regardless of the application, they are all basically the same. Four essential parts of such circuits are a reservoir to suck and retain the hydraulic oil, a pump to force the oil through the circuit, valves to control oil pressure and flow rate, and an actuator (hydraulic motor or cylinder) to transform hydraulic power into mechanical power to perform the required work. Figure 5.30 shows a simple hydraulic circuit consisting of a reservoir, an electrical motor, a pump, an accumulator, a relief valve, and a servo system (spool valve and cylinder).

When working with the hydraulic system, we worked under the assumption that the oil was incompressible. Because while the system is running, oil flows through orifices and is stored in the cylinder control cavities, we can characterize it as having hydraulic resistance and capacitance. In addition, if we assume that the interconnecting pipelines between the hydraulic components are relatively long, we will consider the inertance and resistance of the oil passing through them.

5.4.1 HYDRAULIC RESISTANCE

When defining hydraulic resistance in restricting devices such as orifices and valves, we refer to the change in the pressure difference between the upstream and downstream regions of the orifice that limits the oil flow rate. In this case, the pressure

Figure 5.31 Junction representation of the hydraulic resistance

required to change a unit in the flow rate q can be expressed as

$$R_h = \frac{\text{change in differential pressure}}{\text{change in flow rate}} = \frac{\Delta p}{dq} \qquad (5.158)$$

where R_h ((N m/kg kg) is the hydraulic resistance, Δp (N/m^2) is the change in the pressure difference across the orifice, and dq (m^3/sec) is the change in the flow rate that passes through the orifice. Referring to Figure 5.31, rewrite equation (5.158) as follows:

$$R_h = \frac{\Delta p}{dq_s} = \frac{(p_s - p_a)}{dq_s} \qquad (5.159)$$

where p_s and p_a are the upstream and downstream pressures, respectively. Note that the resistance value R_h is not constant but varies as operating conditions change.

5.4.2 JUNCTION REPRESENTATION OF THE HYDRAULIC RESISTANCE

Since the valve orifice shares the same flow q_s, a 1-junction is used and connected to the resistance element R_h, as shown in Figure 5.37.

5.4.3 HYDRAULIC CAPACITANCE

The compressibility of the hydraulic oil, a function of the oil's Bulk modulus, is the most essential factor in determining the capacitance of the oil contained within a cavity. It offers a relationship between the pressure that occurs and the volume of the oil contained within the cavity. In other words, it is a term that refers to the change in oil pressure that occurs within a cavity and controls the rate at which the oil's volume changes. So, the Bulk modulus is defined as

$$B = -\frac{\text{change in oil volume}}{\text{change in the pressure}} = -\frac{dV/V}{dp} \qquad (5.160)$$

or

$$dp = -B\frac{dV}{V} \qquad (5.161)$$

Considering this rule, the oil pressure increases when the volume is compressed. Let ΔV be the decrease in oil volume $\Delta V = -dv$. So,

$$\Delta p = \frac{B}{V_o}\Delta V \quad \text{or} \quad \frac{\Delta p}{\Delta V} = \frac{B}{V_o} \qquad (5.162)$$

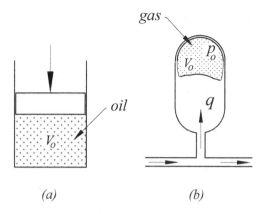

Figure 5.32 Oil control cavity and accumulator

where V_o is the volume of the compressed oil in the cavity, as shown in Figure 5.32a. Considering equation (5.162), the hydraulic capacitance C_h, which represents the oil volume capacitance, is given by

$$C_h = \frac{V_o}{B} \tag{5.163}$$

Similarly, for a gas pressure accumulator shown in Figure 5.32b, we found that the hydraulic capacitance C_{hc} is given by

$$C_{hc} = \frac{V_o}{\gamma p_o} \tag{5.164}$$

where γ is the ratio of specific heat at constant pressure to that at constant volume; its value is 1.4 at 1 atm pressure.

5.4.4 JUNCTION REPRESENTATION OF THE HYDRAULIC CAPACITANCE

Since the control volume shares the same effort, a 0-junction is used and connected to the capacitor element C_h, as shown in Figure 5.33.

$$C_h$$

Figure 5.33 Junction representation of the hydraulic capacitance

5.4.5 HYDRAULIC INERTANCE

The hydraulic components are linked to the circuit via hydraulic pipelines. If these pipelines are relatively long, we must recognize the inertance and resistance of the oil mass flow as it passes through. In this case, the quantity of oil that flows through them may significantly affect the system's operation. As a result, it is necessary to consider the oil's inertance and resistance (flow losses). Let us again perform the analysis presented in Section 5.2.5 for hydraulic pipelines.

Consider a laminar flow of oil entering a circular pipeline at a uniform velocity, as shown in Figure 5.34. This figure shows how the velocity gradient develops over the length of the pipeline. The hypothetical boundary surface divides the flow in a pipeline into two essential regions: the core flow region, in which frictional effects are negligible and the velocity remains nearly constant in the radial direction, and the boundary layer region, in which the viscous effects and the velocity changes are significant. Because of the no-slip condition, the oil particles in contact with the pipe wall come to a complete stop at the boundary layer region, and the velocity profile takes on the shape of a parabolic curve, as shown in Figure 5.34. On the other hand, this is more flattering in turbulent flow than in laminar flow.

Let us consider the laminar one-dimensional flow in a pipeline, as shown in Figure 5.34a. Inertance associated with mass flow in relatively long hydraulic pipelines is treated analytically, primarily as an oil property; thus, we will do so here. Any flowing oil has stored kinetic energy due to its mass and velocity. The inertance of an infinite-sized pipeline is equal to the summation of this kinetic energy throughout the whole volume of the pipeline. The most fundamental assumption in this situation is that all oil particles have identical velocities at any given point in time, implying that

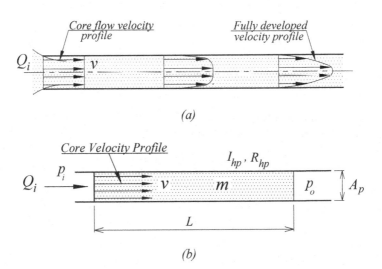

(a)

(b)

Figure 5.34 Development of the velocity boundary layer in a hydraulic tube

their velocity profile will behave as the core flow velocity profile shown in Figure 5.34b along the pipe length, meaning that the velocity v of the oil will be constant until it reaches the outlet port of the pipelines. Considering this, the acceleration of the oil due to a pressure drop $(p_i - p_o)$ across a pipeline is defined by Newton's law as follows:

$$\sum F = am \tag{5.165}$$

The force F is given by

$$F = A_p(p_i - p_o) \tag{5.166}$$

The mass of the oil is given by

$$m = \rho A_p L \tag{5.167}$$

The oil velocity is related to the flow rate Q_i by

$$v = \frac{Q_i}{A_p} \tag{5.168}$$

and the acceleration is given by

$$a = \frac{\dot{Q}_i}{A_p} \tag{5.169}$$

where L is the pipeline length and A_p its cross-section area. Substituting equations (5.166), (5.167) and (5.169) in equation (5.165) gives

$$(p_i - p_o) = \frac{\rho L}{A_p}\dot{Q}_i = I_h\dot{Q}_i \tag{5.170}$$

where I_{ph} is the hydraulic inertance which is given by

$$I_{ph} = \frac{\rho L}{A_p} \tag{5.171}$$

Equation (5.171) indicates that the inertance element in the hydraulic system is a storage element analogous to the inertia element in the mechanical system and the inductor element in the electrical system.

Now, let us determine the hydraulic resistance of the pipelines. Since the oil moves through pipelines as a rigid body, the friction between the body and the tube wall will assume a viscous and column friction combination. In this situation, the viscous and column friction forces F_d are given by

$$F_d = R_d v \tag{5.172}$$

Applying Newton's law gives

$$A_p\Delta p = R_d v \tag{5.173}$$

where R_d is the viscous and column friction factor. Substituting equation (5.168) in equation (5.173) and rearranging the results, we get

$$\Delta p = F_d = \frac{R_d}{A_p^2}Q_i = R_{hp}Q_i \tag{5.174}$$

where R_{hp} is the oil resistance in the hydraulic pipelines, which is given by

$$R_{hp} = \frac{R_d}{A_p^2} \qquad (5.175)$$

The value of this resistance is arbitrary and established by model analysis.
In most hydraulic systems, the pipelines are considered short, implying that the inertance and resistance of the mass flow rate passing through them are ignored.

5.4.6 JUNCTION REPRESENTATIONS OF THE OIL INERTANCE AND RESISTANCE

Since the hydraulic pipeline shares the same flow, a 1-junction is used and connected to their inertia element I_{hp} and resistance element R_{hp}, as shown in Figure 5.35.

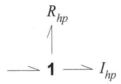

Figure 5.35 Junction representations of the inertance and resistance of the hydraulic tube

5.4.7 HYDRAULIC SOURCES: PRESSURE AND FLOW

As mentioned, an ideal flow source generates a particular flow regardless of the pressure necessary. Similarly, an ideal pressure source generates a constant pressure at some point in the hydraulic system, regardless of the amount of flow necessary to sustain that pressure. One of the most popular hydraulic power sources is the hydraulic pump. In our analysis, we will be concerned with the oil flow rate, or pressure, supplied to the system, regardless of the type of power sources used.

5.4.8 JUNCTION REPRESENTATIONS OF THE HYDRAULIC SOURCES

The source of effort SE or flow SF can be connected to a 1-junction, as shown in Figure 5.36.

$$SE : p(t) \ or \ SF : q(t) \longrightarrow 1 \longrightarrow$$

Figure 5.36 Junction representations of hydraulic sources

5.4.9 DYNAMIC MODELING OF A HYDRAULIC SERVO SYSTEM

The hydraulic servo system comprises a spool valve that controls the piston of a hydraulic cylinder, also called a servo motor, as shown in Figure 5.37. Assume that the spool valve is symmetrical and has no overlapping and that the area of the valve orifices is proportional to the displacement of the valve x_s. As a result, the orifice coefficient and the pressure drop across the orifice are constant and unaffected by the valve's location. Because the valve allows high-pressure oil to pass through a cylinder that already includes a piston, a significant amount of hydraulic force is created, which can then be used to move a load. Also, let us say that the return pressure p_r in the return line is low and can be ignored. The hydraulic oil cannot be compressed, the piston's inertia force and the load's reactive forces are not significant compared to the piston's hydraulic force, and there are not any leakage flows around the spool valve from the supply pressure side of pressure p_s to the return pressure side of pressure p_r. Let us begin by deriving the linearized model of the spool valve.

If we push the spool valve with a displacement of x_s, both orifices open, as shown in Figure 5.37. As a result, the flow rate q_a flows from the high-pressure cavity of pressure p_s to the right side cavity of the cylinder of pressure p_a. At the same time, the flow rate q_r flows from the left side cavity of the cylinder of pressure p_b to the return line of pressure p_r. As a result, the resulting difference in the pressure force applied on both sides of the piston will cause it to move a displacement x_s to the left, as shown in Figure 5.37. Therefore, the flow rates q_a and q_b passing through both

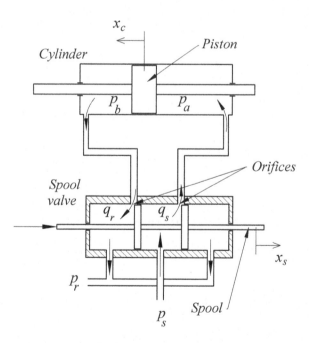

Figure 5.37 Hydraulic servo system

orifices are given by

$$q_a = C_s x_s \sqrt{(p_s - p_a)} \quad \text{so} \quad q_b = C_s x_s \sqrt{(p_b - p_r)} \tag{5.176}$$

where C_s is the orifice coefficient, which is proportionally constant. Note that we assumed that the spool valve has no overlaps and that both orifices are of the same dimensions and unaffected by the spool's location, implying that both q_a and q_b are equal. Considering equations (5.176) and assuming that the return pressure p_r equals 0, we get

$$q_a = q_b \quad \text{and} \quad (p_s - p_a) = p_b \tag{5.177}$$

Note that the pressure difference $\Delta p = (p_a - p_b)$ is the resultant differential pressure applied to the piston during system operation. Then, p_a and p_b can be written

$$p_a = \frac{p_s + \Delta p}{2} \quad \text{and} \quad p_b = \frac{p_s - \Delta p}{2} \tag{5.178}$$

Since we assume that $q_a = q_b$, we will only do the following analysis for q_a. Substituting equation (5.178) in equation (5.176) gives

$$q_a = C_s x_s \sqrt{\frac{p_s - \Delta p}{2}} \tag{5.179}$$

This equation indicates that the flow rate q_a is a function of both spool displacement x_s and the differential pressure Δp, and it is a nonlinear equation. Now let us linearize this equation using Taylor series expansion as follows:

$$q_a - Q_a = c_1(x_s - X_s) + c_2(\Delta p - \Delta P) \tag{5.180}$$

where Q_a, x_s and ΔP are the system parameters at the operating point. Using the Taylor series expansion to obtain c_1 and c_2 as follows.

$$c_1 = \left. \frac{\partial q_a}{\partial x_s} \right|_{x_s = X_s, \Delta p = \Delta P} = C_s \sqrt{\frac{p_s - \Delta p}{2}} \tag{5.181}$$

and

$$c_2 = \left. \frac{\partial q_a}{\partial \Delta p} \right|_{x_s = X_s, \Delta p = \Delta P} = -X_s \frac{C_s}{2\sqrt{2}\sqrt{p_s - \Delta P}} \leq 0 \tag{5.182}$$

Substituting equations (5.181) and (5.182) in equation (5.180) gives

$$q_a - Q_a = C_s \sqrt{\frac{p_s - \Delta p}{2}}(x_s - X_s) - X_s \frac{C_s}{2\sqrt{2}\sqrt{p_s - \Delta P}}(\Delta p - \Delta P) \tag{5.183}$$

Assume that the operating point is near the origin. So, $X_s \approx 0, \Delta P \approx 0, Q_a \approx 0$. Considering these, equation (5.180) become

$$q_a = c_1 x_s - c_2 \Delta p \tag{5.184}$$

where

$$c_1 = C_s \sqrt{\frac{p_s - \Delta P}{2}}\bigg|_{X_s \approx 0, \Delta P \approx 0} = C_s \sqrt{\frac{p_s}{2}} \tag{5.185}$$

and

$$c_2 = \frac{C_s X_s}{2\sqrt{2}\sqrt{p_s - \Delta P}}\bigg|_{X_s \approx 0, \Delta P \approx 0} = 0 \tag{5.186}$$

Substituting equations (5.185) and (5.186) in equation (5.184) gives

$$q_a = c_1 x_s = C_s x_s \sqrt{\frac{p_s}{2}} \tag{5.187}$$

Equation (5.187) is the linearized dynamic equation of the spool valve.

Now, let us derive the linearized model of the hydraulic cylinder. Note that the flow rate q_a times dt equals the piston displacement dx_s times the piston area A_c times the density of oil ρ. Therefore,

$$A_c \rho\, dx_s = q_a\, dt \tag{5.188}$$

Substituting equation (5.187) in equation (5.188) and rearranging the results gives

$$A_c \rho \frac{dx_s}{dt} = C_s x_s \sqrt{\frac{p_s}{2}} \tag{5.189}$$

This equation is the linearized dynamic equation of the hydraulic cylinder. So, equations (5.187) and (5.189) are the linear dynamic models of the given hydraulic servo system shown in Figure 5.37.

5.4.10 BOND GRAPH CONSTRUCTIONS FOR HYDRAULIC SYSTEMS

When modeling the hydraulic system with the Bond Graph technique, we employ the same procedures applied to the pneumatic systems.

1. For each pressure of interest, establish a 0-junction. The return pressure connection is allocated to a specific node, similar to the ground in mechanical and electrical systems. All bonds related to it must be removed to simplify the model. To further simplify the Bond Graph, any 2-port 1-junctions or 0-junctions with through power flow directions should be swapped by single bonds.
2. Attach capacitance element C to appropriate 0-junctions for hydraulic capacitance elements associated with pressures. Attach the resistance element R and transformer element TF to 1-junctions (using the power convention) between appropriate 0-junctions.
3. Assign power flow half arrows using the through power flow convention.
4. Finally, we should add causality stroke marks and bond numbers to finish. As a result, the Augmented Bond Graph is ready to derive the differential equations describing the dynamics of hydraulic systems.

Example (5.8):
For the hydraulic servo system shown in Figure 5.37, create the preliminary Bond Graph and then the Augmented Bond Graph model and derive the dynamic equation(s).

Solution:
As shown in Figure 5.37, the system includes four operating pressures: the supply pressure p_s, the cylinder pressures p_a and p_b, and the return pressure p_r. Considering these, four 0-junctions are created, as shown in Figure 5.38. The supply pressure p_s is represented by a source of effort SE connected to a 0-junction of a common effort p_s. In contrast, the return pressure p_r is represented by a source of effort p_r $SE : p_r$ connected to the 0-junction of a common effort p_r (to be eliminated). Two 0-junctions are created for pressures p_a and p_b. The mass m_p of the piston is represented by an inertia element I connected to a 1-junction of common flow $\dot{x}_c$. When the valve opens, the flow rates q_a and q_b pass through the valve orifices. Both orifices are represented by two resistance elements, R_a and R_b, and each is connected to a 1-junction, as shown in Figure 5.38. We assume that the spool valve has no overlaps, which means that the flow rate is proportional to the spool valve displacement x_s, and the orifice coefficients and pressure drop across them are both constant and unaffected by the position of the spool. As a result, both q_a and q_b are equal, implying that the resistance R_a equals R_b.

At this point, the preliminary Bond Graph model of the given system is constructed. The half-arrows indicating positive power flow directions are selected, as shown in Figure 5.38. Since we assumed that the return pressure p_r equals 0, to simplify this bond graph model, all connections that hold the return pressure p_r with their associated bonds must be removed, and then any 2-port 1-junctions or 0-junctions with through power flow should be swapped with single bonds. As a result, the 0-junction connected to the source of flow SE, the 0-junction connected to the source of flow $SE : p_r$, and the 1-junction connected to resistance R_b are removed, as shown in Figures 5.38 and 5.39.

Now, let us assign the causality of the Bond Graph model. Since there is only one source of effort, SE, the causal stroke mark should be located on its bond at

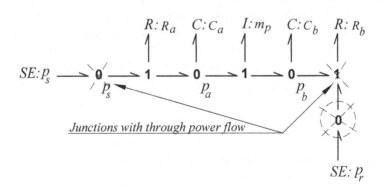

Figure 5.38 Preliminary Bond Graph model of a hydraulic servo system

the junction side, as shown in Figure 5.39. Considering the causality restrictions in Section 3.3, apply the causal stork marks to all elements and junctions. We finally number all bonds to get the Augmented Bond Graph model of the system shown in Figure 5.39. This figure shows three integral causalities storage elements, C_a, I and C_b representing by bond 4, 6 and 8.

To derive the dynamic equations of the given system, we must first define the key variables of the model: Input, State variables, and Co-energy (power) variables and their constitutive relations.

- Key variables:
- Inputs: $e_1 = SE = p_s$.
- State variables: q_4, p_6 and q_8.
- Co-energy (Power) variables: e_4, f_6 and e_8.

- Constitutive relations: Write equations describing the relationship between co-energy variables and state variables for energy storage elements, C_a, I, and C_b, as follows:

$$e_4 = \frac{1}{C_a}q_4, \quad f_6 = \frac{1}{I}p_6 \quad \text{and} \quad e_8 = \frac{1}{C_b}q_8 \tag{5.190}$$

Now that we have the connections between the flow and effort variables of the model junctions shown in Figure 5.39, let us derive the equations that reflect them.

- As shown in Figure 5.39, the flow variables of the 1-junction of common flow $\dot{x}_s$ are given by

$$f_1 = f_2 = f_3 \tag{5.191}$$

and the effort variables are given by

$$e_1 = e_2 + e_3 \tag{5.192}$$

- The effort variables of the 0-junction of common effort p_a are given by

$$e_3 = e_4 = e_5 \tag{5.193}$$

and the flow variables are given by

$$f_3 = f_4 + f_5 \tag{5.194}$$

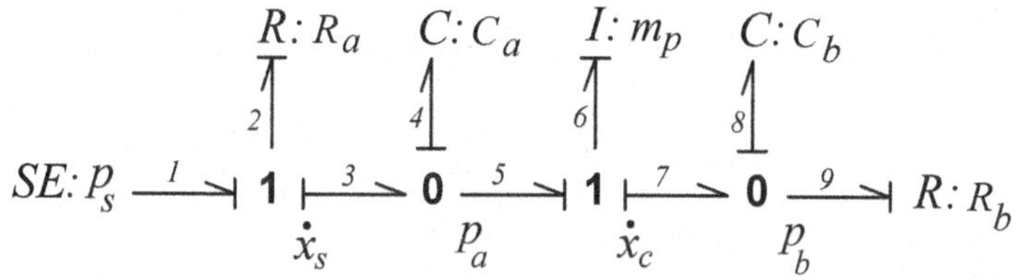

Figure 5.39 Augmented Bond Graph model of a hydraulic servo system

- The flow variables of the 1-junction of common flow $\dot{x}_c$ are given by

$$f_5 = f_6 = e_7 \tag{5.195}$$

and the effort variables are given by

$$e_5 = e_6 + e_7 \quad \text{or} \quad e_6 = \dot{p}_6 = e_5 - e_7 \tag{5.196}$$

where $\dot{p}_6$ is the momentum change of the piston.
- The effort variables of the 0-junction of common effort p_b are given by

$$e_7 = e_8 = e_9 \tag{5.197}$$

and the flow variables are given by

$$f_7 = f_8 + f_9 \tag{5.198}$$

Now, let us derive the differential equation describing the dynamics of the given system.

- **For the spool valve:**
  Considering equation (5.192) and inserting $e_1 = SE$, the effort e_3 is given by

$$e_3 = e_1 - e_2 = SE - e_2 \tag{5.199}$$

The effort e_2 of the resistance R_a is given by

$$e_2 = R_a f_2 \tag{5.200}$$

Equation (5.191) indicates that the flow f_2 equals f_3, which is equal to $(f_4 + f_5)$ as indicated in equation (5.194). Equation (5.195) indicates that the flow f_5 equals the flow f_6, given by the constitutive relations indicated in equation (5.190). Note that the flow f_4 is equal to $\dot{q}_4$, which is the derivative of the state variable q_4 of the capacitance C_a, representing the velocity $\dot{x}_s$ of the spool element. Considering these, equation (5.200) becomes

$$e_2 = R_a(f_4 + f_6) = R_a\left(\dot{q}_4 + \frac{1}{I}p_6\right) \tag{5.201}$$

Substituting this equation in equation (5.199) gives

$$e_3 = SE - R_a\left(\dot{q}_4 + \frac{1}{I}p_6\right) \tag{5.202}$$

Equation (5.196) indicates that to determine the piston's momentum change $\dot{p}_6$, we must first determine the efforts e_5 and e_7. Equation (5.193) indicates that the effort e_5 equals e_3 given by equation (5.202), while equation (5.179) indicates that the effort e_7 equals e_8, given by the constitutive relations indicated in equation (5.190). Considering these, equation (5.196) becomes

$$\dot{p}_6 = e_3 - e_8 = SE - R_a \dot{q}_4 - \frac{R_a}{I} p_6 - \frac{1}{C_b} q_8 \tag{5.203}$$

Rearranging this equation gives

$$\dot{p}_6 + \frac{R_a}{I} p_6 + R_a \dot{q}_4 + \frac{1}{C_b} q_8 = SE \tag{5.204}$$

This equation represents the dynamic equation of the spool valve.

- **For the hydraulic cylinder:**
Equation (5.198) indicates that the flow f_9 equals $(f_7 - f_8)$, while equation (5.195) indicates that the flow f_7 equals f_6, given by the constitutive relations indicated in equation (5.190). Note that the flow f_8 is equal to $\dot{q}_8$, which is the derivative of the state variable q_8 of the capacitance C_b, representing the velocity $\dot{x}_c$ of the piston. Considering these, equation (5.198) becomes

$$f_9 = f_6 - f_8 = \frac{1}{I} p_6 - \dot{q}_8 \tag{5.205}$$

The effort e_9 of the resistance R_b is given by

$$e_9 = R_b f_9 \quad \text{or} \quad f_9 = \frac{e_9}{R_b} \tag{5.206}$$

Equation (5.197) indicates that the effort e_9 equals the effort e_8, given by the constitutive relations indicated in equation (5.190). Considering this, equation (5.206) becomes

$$f_9 = \frac{1}{C_b R_b} q_8 \tag{5.207}$$

Substituting this equation in equation (5.205) and rearranging the results gives

$$\frac{1}{I} p_6 = \dot{q}_8 + \frac{1}{C_b R_b} q_8 \tag{5.208}$$

This equation represents the dynamic equation of the hydraulic cylinder. So, equations (5.204) and (5.208) are the dynamic equations of the given hydraulic servo system. As an exercise, check if Newton's law and law of the conservation of mass flow rate are embedded in the Augmented Bond Graph model shown in Figure 5.39.

5.5 THERMAL SYSTEMS

When the temperatures of two bodies differ, heat is transferred from the hotter body to the colder body to bring the temperatures back to equality. There are three heat transfer modes: conduction, convection, and radiation. These modes govern both the type and quantity of heat transferred. In practice, the radiation mode is a highly non-linear system. Since we are working with linear systems in this book, heat transfer modeling in the radiation mode will be ignored. It may be necessary to describe the essential parameters of a thermal system before starting dynamic modeling. It is important to know that heat is transferred through flux, meaning there is no mass movement and no inertia in thermal systems. So, the fundamental parameters of a thermal system are only resistance and capacitance, as well as thermal sources, temperature, and heat flow, which all constitute the essential parameters of a thermal system.

5.5.1 THERMAL RESISTANCE

5.5.1.1 Conduction mode

The temperature difference between the two bodies and the thermal resistance of the material connecting them affect the heat transmission rate. Figure 5.40 shows two bodies with temperatures T_1 and T_2 ($T_1 > T_2$) connected by a solid rod with a constant cross-section area A and length L. The rod is constructed from a thermally conductive material with a thermal conductivity of k. Although, in theory, k is constant, in practice, it can vary depending on weather, the temperature, the location of the body, and the direction of the heat flux q. As a means of describing thermal resistance, Fourier's law of heat conduction may be expressed as follows:

$$\text{Heat transfer rate} = q = \frac{kA}{L}(T_1 - T_2) = \frac{kA}{L}\Delta T \qquad (5.209)$$

This equation indicates an instantaneous relation between ΔT and q. In many real applications, the assumption of constant k is adequate and required for a linear model. If the rod has no thermal capacitance, this equation would be true. Since a real rod would have thermal capacitance, we must use equation (5.209) to define the resistance element of the rod. Its thermal capacitance will be dealt with separately. The

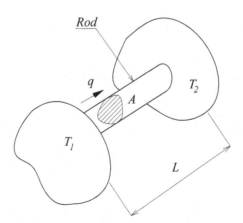

Figure 5.40   Heat transfer between bodies, Doebelin [1]

pure and ideal thermal resistance for conduction heat transfer may be defined as follows.

$$q = \frac{1}{R_c}\Delta T \tag{5.210}$$

Substituting equation (5.210) in equation (5.209) and rearranging the results gives

$$R_c = \frac{1}{q}\Delta T = \frac{L}{kA} \tag{5.211}$$

where R_c is the thermal resistance associated with the conduction mode. Equation (5.211) indicates that the analogy to a mechanical damper or electrical resistance is obvious if we consider ΔT to be the mechanical force (voltage) and heat flux q to be the velocity (current).

5.5.1.2 Convection mode

Heat transfer through the fluid-solid interface by convection mode occurs in many practical cases. Heat is conducted through a thin layer of fluid (referred to as the boundary layer) that adheres to the solid wall, as shown in Figure 5.41. Whenever a fluid particle moves through an interface between a boundary layer and the mainstream, heat is transferred away from that interface into the mainstream. This process is called convection heat transfer, as shown in Figure 5.41. The following equation may describe this process:

$$q = hA(T_1 - T_2) = hA\Delta T \tag{5.212}$$

where h is the film coefficient of heat transfer and A is the surface area of the body wall. Equation (5.212) is a linear representation of reality, which is realistic since the

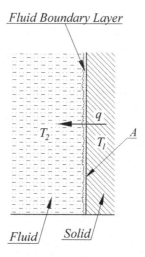

Figure 5.41 Heat transfer by convection, Doebelin [1]

coefficient h changes with temperature. Therefore, the thermal resistance R_v associated with the convection mode is given by

$$R_v = \frac{1}{q}\Delta T \tag{5.213}$$

Substituting equation (5.213) in equation (5.212) gives

$$R_v = \frac{1}{hA} \tag{5.214}$$

5.5.1.3 Conduction and convection modes

Usually, conduction and convection modes are combined in thermal systems. We must define an overall heat transfer coefficient and thermal resistance in this case. For example, Figure 5.42 shows a cross-section of a car radiator system, showing how heat is transferred from the heated internal water to the solid radiator wall and then to the air flowing over the radiator by the fan. Since the same amount of heat flux q flows through all three resistances (water, radiator wall, and air), we may write

$$q(\frac{1}{h_wA}) + q(\frac{L}{kA}) + q(\frac{1}{h_aA}) = T_w - T_a \tag{5.215}$$

where:

- h_a = film coefficient of heat transferred to the air.
- h_w = film coefficient of heat transferred to the water.
- L = thickness of the solid body.
- T_a = temperature of the air.
- T_w = temperature of the water.
- T_{ya} = temperature of the air-side boundary layer.
- T_{yw} = temperature of the water-side boundary layer.

Assuming that the boundary layers on both the air and the water sides are very thin, this would imply that the temperatures of both layers sides, T_{ya} and T_{yw}, are the same as T_a and T_w, respectively. Considering these, equation (5.215) becomes

$$q = \frac{T_w - T_a}{\frac{1}{h_wA} + \frac{L}{kA} + \frac{1}{h_aA}} = \frac{1}{R_{cv}}\Delta T \tag{5.216}$$

where R_cv is the overall resistance, which is

$$R_{cv} = (\frac{1}{h_wA} + \frac{L}{kA} + \frac{1}{h_aA}) \tag{5.217}$$

As we might have expected from the electrical or mechanical system, we observe that the total thermal resistance, indicated in equation (5.217), is simply the sum of the individual resistances (parallel connections).

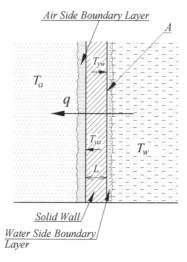

Figure 5.42 Overall heat transfer of a combination between conduction and convection modes, Doebelin [1]

5.5.2 THERMAL CAPACITANCE

Heat flowing into a body of material can manifest in various forms, such as mechanical work or changes in the kinetic energy of a moving fluid, depending on how the heat will transfer. Bodies of material for which thermal energy does not cause significant mechanical work or kinetic energy changes are those whose energy can be stored without causing a significant temperature change. When energy is added, it will appear as stored internal energy and as a rise in temperature. In this instance, the rise in temperature is directly proportional to the entire amount of heat transmitted through the body at an optimal thermal capacitance. When heat is present, the body's temperature T_s rises, as shown in Figure 5.43, which is given by

$$(T_f - T_s) = \frac{1}{C_t} \int q \, dt \qquad (5.218)$$

where C_t is the thermal capacitance. We assume that the body's temperature is uniform across its volume at any time. If the fluid is thoroughly and constantly mixed, this ideal state is very close to being achieved. Thermal conductivity k plays an essential role in heating the body uniformly. If a body has infinite thermal conductivity, then regardless of the heat flow rate, the temperature would be uniform. A body's temperature will never be perfectly uniform during transient temperature changes since no actual material has an infinite k value. Referring to Figure 5.43, the boundary layer fluid film has negligible thermal capacitance as the film is very thin, resulting in pure convection resistance. As a result, the amount of heat added when the body's temperature rises is

$$Q = \int q \, dt = m c_p (T_f - T_s) \qquad (5.219)$$

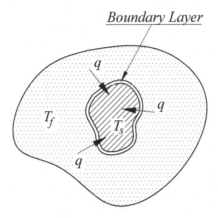

Figure 5.43 Heat transfer by conduction, Doebelin [1]

where m is the mass flow rate and c_p is the specific heat. Thus,

$$C_t = \frac{\text{amount of heat added}}{\text{temperature rise}} = \frac{m\,c_p(T_f - T_s)}{(T_f - T_s)} = mc_p \qquad (5.220)$$

Although the specific heat c_p of actual materials changes somewhat with temperature, it is common to assume a constant mean value for this parameter. When heat is applied to or removed from the material in the process of changing phase, the thermal capacitance of the material is almost infinite, as it is possible to apply heat without causing any temperature rise.

5.5.3 THERMAL SOURCES: TEMPERATURE AND HEAT FLOW

An ideal heat flow source generates a predetermined heat flow regardless of the temperature needed. In contrast, an ideal temperature source maintains a predetermined temperature regardless of the heat flow it must supply. By utilizing phase change materials, constant-temperature sources may become relatively effective frequently. Electrical resistance heating may be the most valuable heat source for many applications. An example would be using an electrically resisting heating coil immersed in a fluid to serve as a heat transfer source, as shown in Figure 5.44. When the coil is heated, a portion of the electrically generated heat transfers to its metal, and the remainder is released as intended into the fluid. Unless the thermal capacitance of the heating coil metal is small enough, almost all of the electrically generated heat goes into the fluid, and we have a good approximation of a heat flow source.

5.5.4 DYNAMIC MODELING OF THERMAL SYSTEMS

Thermal devices and processes are widespread and significant in the global economy. This is because there are only two thermal elements in thermal systems (capacitor and resistance) instead of three elements in other systems (inertia, capacitor, and

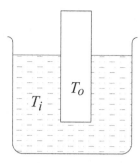

Figure 5.44 First order thermal system

resistance), meaning that thermal systems do not have an inertia element. As a result, there are fewer conceivable ways for these two elements to operate together. Regarding the fundamental rules used to set up equations of thermal systems, the energy conservation law is the most applicable. We can always write the equation that describes every thermal system over any given time interval as follows:

$$\text{Energy in - Energy out = additional energy stored} \qquad (5.221)$$

Consider the thermal system shown in Figure 5.44, a solid body with a temperature of T_o submerged in a fluid with a temperature of T_i. This configuration provides an adequate model for a wide variety of real-world thermal systems situations, such as the temperature-sensing instruments (thermometers, thermocouples, resistance temperature detectors, etc.), the heat treating of metal parts, the heating and cooling of food, and many other cases. We assume that the temperature of the solid body T_o can be uniform throughout at any instant and that the surface coefficient of convective heat transfer is uniform over the surface even with time. Because all of the heat that is transferred from the fluid to the solid is stored in (or removed from) the solid, we may write the following over a time dt:

$$\text{Energy in (or out) = additional energy stored (or removed)} \qquad (5.222)$$

Applying this rule to the thermal system shown in Figure 5.44 gives

$$hA_s(T_i - T_o)dt = MC\,dT_o \qquad (5.223)$$

where:

- A_s=surface area for heat transfer.
- h=surface heat transfer coefficient.
- M=mass of the solid body.
- C=specific heat of the solid body.

Rearranging equation (5.223) yields

$$MC\frac{dT_o}{dt} + hA_sT_o = hA_sT_i \qquad (5.224)$$

or,

$$\left(\frac{MC}{hA_s}\right)\frac{dT_o}{dt} + T_o = T_i \tag{5.225}$$

This is the dynamic equation of the heat transferred to the solid body shown in Figure 5.44.

Example (5.9):
A water-boiling system is shown in Figure 5.45. The process aims to heat the water inside the tank to a boiling temperature. Considering a thin tank wall, derive the differential equation describing the heat transfer dynamics.

Solution:
The water in the tank is heated by an electrical-resistance heating coil positioned inside the tank, as shown in Figure 5.45. This process is called the batch process. The heater supplies a constant heat input at a rate of q_i. We assume that the temperature of the water T_w is uniform throughout the water at any time. Since the wall of the tank is assumed to be thin, its capacitor is neglected. As a result, only the capacitance and resistance of the water are considered. Applying the conservation of energy law to the system gives

$$q_i - hA(T_w - T_a) = mc_p T_w \tag{5.226}$$

where h, A, m, and c_p are the surface heat transfer coefficient, surface area for heat transfer, the water's mass, and the water's specific heat, respectively. We may rewrite equation (5.226) over a time interval dt as

$$q_i\,dt - hA(T_w - T_a)\,dt = mc_p\,dT_w \tag{5.227}$$

or,

$$q_i - hA(T_w - T_a) = mc_p\frac{dT_w}{dt} \tag{5.228}$$

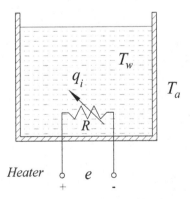

Figure 5.45 Boiling water system

If we assume that the capacitance of the heater coil is not significant, then a constant or time-varying input voltage e to the resistance heating coil will result in a heat generation rate of

$$q_i = \frac{1}{R}e^2 \tag{5.229}$$

where R is the resistance of the heater coil. Substituting equation (5.229) in equation (5.228) and rearranging the results gives

$$mc_p\frac{dT_w}{dt} + hA(T_w - T_a) = \frac{1}{R}e^2 \tag{5.230}$$

This differential equation represents the heat transfer dynamics in the boiling water system shown in Figure 5.45.

5.5.5   BOND GRAPH CONSTRUCTIONS FOR THERMAL SYSTEMS

In general, thermal systems have been studied in terms of how they are like electrical circuits, with temperature analogous to voltage and heat flux analogous to current. In this analogy, thermal resistances and capacitors, series 1-junctions and parallel 0-junctions, and thermal sources are the same as voltage and current in electrical systems. Thermal inertia, however, does not exist, as previously stated.

This analogy, in which temperature is related to effort and heat flux is related to flow, has proven helpful, which is why we provide it here. In thermal systems, the product of heat flux and temperature is not power! The power is inherent in the heat flux itself. Pseudo-Bond Graphs are those in which e and f are not power variables; we will refer to these as such from this point forward. The pseudo-Bond Graph cannot be represented using power variables as a regular Bond Graph. Suppose the fundamental components of the pseudo-Bond Graph accurately relate the variables e, f, p, and q. In this case, Bond Graph techniques can be successfully applied to the pseudo-Bond Graph. The following section demonstrates that it is possible to construct an adequate Bond Graph using temperature and heat flux as the effort and flow variables. On the other hand, the thermodynamic models are more difficult to understand than those necessary to demonstrate the pseudo-Bond Graph. Two situations can arise when investigating thermal systems, as shown in Figures 5.46 and 5.47. In both cases, we will assume that temperature gradients and heat movement exclusively take place along the x-axis.

Let us now establish the two most essential parameters, thermal capacitance and resistance, that will be used to represent the thermal behavior of the system in the pseudo-Bond Graph model and its associated junction.

5.5.5.1   Thermal resistance element and 1-junction

The situation shown in Figure 5.46 is an example of complete resistance. If a surface of a material of area A has temperatures of T_1 and T_3, we can assume that the heat flux through the material, $\dot{Q}_2$, is a function of $T_2 = T_1 - T_3$. In the linear case, we

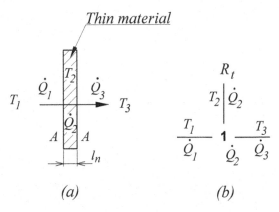

Figure 5.46 Thermal resistance element in the conduction mode and the corresponding 1-junction of common flow, Karnopp [2]

have

$$\dot{Q}_2 = \frac{1}{R_t}T_2 = \frac{1}{R_t}(T_1 - T_3) \tag{5.231}$$

where $\dot{Q}_2$ can be measured in power units (Btu/sec) or (cal/sec) or any other unit. Since this is a pseudo-Bond Graph, it does not matter whether power units are used as long as they are used consistently for the properties of the two elements. For a thin material with a thickness, l_n, area, A, and thermal conductivity, k, as shown in Figure 5.46a, the thermal resistance in conduction mode is given by

$$R_t = \frac{l_n}{kA} \tag{5.232}$$

In this case, the heat transferred through this element is represented by a 1-junction of common flow $\dot{Q}_2$ as shown in Figure 5.46b. Now, let us estimate the equations that reflect both the flow and effort of this junction. The flow variables are given by

$$f_1 = f_2 = f_3 \quad \text{or} \quad \dot{Q}_1 = \dot{Q}_2 = \dot{Q}_3 \tag{5.233}$$

and the effort variables are given by

$$e_1 + e_2 + e_3 = 0 \quad \text{or} \quad T_1 + T_2 + T_3 = 0 \tag{5.234}$$

Notice that the thermal resistance indicates a relation between T_2 and $\dot{Q}_2$ as indicated in equation (5.231) and that the 1-junction implies that the flow relation indicated in equation (5.233) and the effort variables indicated in equation (5.234). We have shown how the resistance R_t and the 1-junction may be combined to limit the common flow as a function of the difference between two efforts $(T_1 - T_3)$, as indicated in equation (5.231).

5.5.5.2 Thermal capacitor element and 0-junction:

A thick material that varies in temperature due to the overall amount of heat energy stored in it is shown in Figure 5.47a. This situation is an example of a complete capacitor. Therefore, T_2 is a function of,

$$T_2 = \frac{1}{C_t} Q_2 \tag{5.235}$$

and Q_2 can be determined by

$$Q_2 = \int \dot{Q}_2 \, dt = \int (\dot{Q}_1 - \dot{Q}_3) \, dt \tag{5.236}$$

The element is a capacitor, given that T_2 represents an effort and Q_2 represents an integral of heat flux. Similarly, the Bond Graph shown in Figure 5.47b connects a C_t element with a 0-junction to show that $T_1 = T_2 = T_3$ and that $\dot{Q}_2 = (\dot{Q}_1 - \dot{Q}_3)$. Again, this indicates that $T_1 = T_2 = T_3$. When considering a linear system, the thermal capacitance C_t can be defined as follows. By assuming that the temperature at $t = t_o$ is the same as that at T_1, it is possible to determine the capacitance C_t by assuming that the element performs little work when expanding or contracting, which is negligible. Hence, changes in its internal energy are only the results of $\dot{Q}_2$. Then, if c_p is a specific heat, we have

$$c_p = \frac{\partial u}{\partial T} \tag{5.237}$$

where u is the internal energy per unit mass, then

$$C_t = m c_p \tag{5.238}$$

Combining equations (5.235), (5.236) and (5.238) gives

$$T_2 = \frac{1}{m c_p} \int \dot{Q}_2 \, dt = \frac{1}{m c_p} Q_2 \quad \text{or} \quad Q_2 = m c_p T_2 = m c_p (T_1 - T_3) \tag{5.239}$$

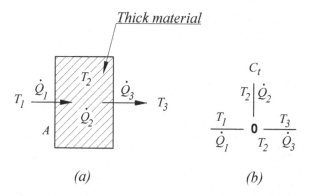

(a)   *(b)*

Figure 5.47 Thermal capacitance element in the conduction mode and the corresponding 0-junction of common effort, Karnopp [2]

where m is the mass of the substance. Note that c_p is the specific heat at constant volume, but for most solids and fluids, the work performed by expansion is negligible compared to Q_2, as stated above. As a result, even if the material is allowed to expand, c_p will not change significantly.

Example (5.10):
Considering the thin tank wall for the water-boiling system shown in Figure 5.45, create the Augmented pseudo-Bond Graph model and derive the dynamic equations.

Solution:
As shown in Figure 5.45, the fluid stored in the tank is heated by an electrical resistance coil. Since the tank wall is thin, its capacitance will be neglected. In this case, we will be concerned only with the fluid's capacitance C_w and resistance R_w. Based on the descriptions of the elements in Figures 5.46 and 5.47, C_w is connected to a 0-junction, and R_w is connected to a 1-junction, as well as the source of flow; SF represents the input heat flow q_i and the ambient temperature T_a is represented by a source of effort SE as shown in Figure 5.48. Keep in mind that, in heating systems, the ambient temperature reflects the lowest temperature of the system. So, because $(T_w - T_a)$ should be of a positive value, the ambient temperature T_a must be of a negative sign, as shown in Figure 5.48. This implies that the source of effort in the thermal system that represents the ambient temperature should have a minus sign. The half-arrows indicating the direction of the positive power flow are selected, as shown in the same figure.

Now, let us assign the causality of the pseudo-Bond Graph model. There are two sources: flow SF and effort SE. The causal stroke mark of SF should be located on its bond at the element side, while SE should be located on its bond at the junction side, as shown in Figure 5.48. Considering the causality restrictions in Section (3.3), apply the causal stork marks to all elements and junctions. We finally number all bonds to get the Augmented pseudo-Bond Graph model of the system shown in Figure 5.48. This figure shows only one integral causal storage element, C_w, represented by bond 2.

To derive the dynamic equations of the given system, we must first define the key variables of the model: Input, State variables, and Co-energy (power) variables and their constitutive relations.

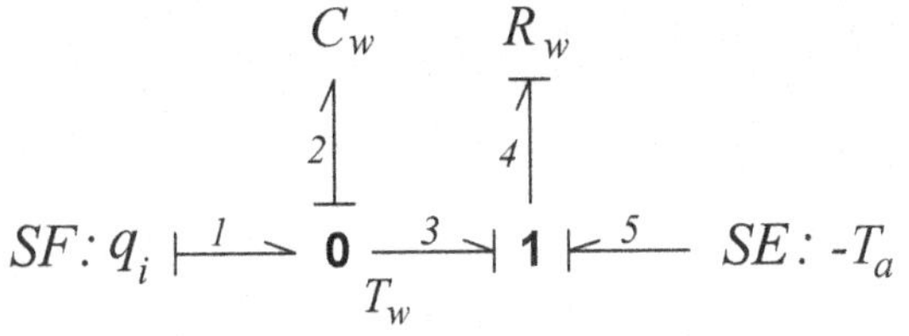

Figure 5.48 Augmented pseudo-Bond Graph model of the boiling system

Key variables:
- Inputs: $f_1 = SF = q_i$ and $e_5 = SE = -T_a$.
- State variables: q_2.
- Co-energy (Power) variables: e_2.

Constitutive relations:
Write equations describing the relationship between co-energy variables and state variables for the energy storage element C_w, as follows:

$$e_2 = \frac{1}{C_w} q_2 \tag{5.240}$$

Now that we have the connections between flow and effort variables at the model junctions, let us derive the equations that reflect them.

- As shown in Figure 5.48, the effort variables of the 0-junction that is connected to the capacitance C_w are given by

$$e_1 = e_2 = e_3 \tag{5.241}$$

and the flow variables are given by

$$f_1 = f_2 + f_3 \quad \text{or} \quad f_2 = \dot{q}_2 = f_1 - f_3 \tag{5.242}$$

- the flow variables of the 1-junction that is connected to the resistance R_w are given by

$$f_3 = f_4 = f_5 \tag{5.243}$$

and the effort variables are given by

$$e_3 = e_4 + e_5 \tag{5.244}$$

Now, let us derive the differential equations of the given system. Considering equation (5.242) and inserting $f_1 = SF$ gives

$$f_2 = \dot{q}_2 = f_1 - f_3 = SF - f_3 \tag{5.245}$$

Equation (5.243) indicates that the flow f_3 equals f_4. Considering this, equation (5.245) becomes

$$\dot{q}_2 = SF - f_4 \tag{5.246}$$

The flow f_4 of the resistance R_w is given by

$$f_4 = \frac{1}{R_w} e_4 \tag{5.247}$$

Equation (5.244) indicates that the effort e_4 equals $(e_3 - e_5)$. Considering this, equation (5.247) becomes

$$f_4 = \frac{1}{R_w}(e_3 - e_5) \tag{5.248}$$

Equation (5.241) indicates that the effort e_3 equals e_2, given by the constitutive relation indicated in equation (5.240). Considering this and inserting $e_5 = SE$, equation (5.248) becomes

$$f_4 = \frac{1}{R_w}(\frac{1}{C_w}q_2 - SE) \tag{5.249}$$

Substituting equation (5.249) in equation (5.246) gives

$$\dot{q}_2 = SF - \frac{1}{R_w}(\frac{1}{C_w}q_2 - SE) \tag{5.250}$$

Rearranging this equation gives

$$\dot{q}_2 + \frac{1}{R_wC_w}q_2 + \frac{1}{R_w}SE = SF \tag{5.251}$$

This equation represents heat transfer dynamics in the water-boiling system (Boiler).

We will now check that the Augmented pseudo-Bond Graph model shown in Figure 5.48 satisfies the law of conservation of energy. It is essential to remember that in the pseudo-Bond Graph model, multiplying effort and flow is not a power. The following analyses are what we get when using the constitutive relations indicated in equation (5.240). The effort e_2 equals T_w while the effort e_5 equals $SE = -T_a$. Considering these, we have

$$q_2 = C_we_2 = C_wT_w \tag{5.252}$$

Taking the derivative of this equation gives

$$\dot{q}_2 = C_w\frac{dT_w}{dt} \tag{5.253}$$

Substituting equations (5.252) and (5.253) in equation (5.251) and inserting $SF = q_i$ and $SE = -T_a$ yields

$$C_w\frac{dT_w}{dt} + \frac{1}{R_wC_w}C_w T_w - \frac{1}{R_w}T_a = q_i \tag{5.254}$$

Rearranging this equation gives

$$C_w\frac{dT_w}{dt} + \frac{1}{R_w}(T_w - T_a) = q_i \tag{5.255}$$

Referring to equations (5.214) and (5.220), the capacitance C_w and resistance R_w are given by

$$C_w = mc_p \quad \text{and} \quad R_w = \frac{1}{hA} \tag{5.256}$$

Substituting this equation in equation (5.256) and inserting $q_i = (e^2/R)$, where e and R are the input voltage and resistance of the heating coil, gives

$$mc_p\frac{dT_w}{dt} + hA(T_w - T_a) = \frac{1}{R}e^2 \tag{5.257}$$

By comparing equation (5.257) with equation (5.230), we found that they are identical, implying that the law of energy conservation is satisfied in the Augmented pseudo-Bond Graph model of the given system shown in Figure 4.48.

Finally, as previously discussed in this chapter, the fluid and thermal systems are assumed to be in steady-state mode when we develop their dynamic equations, and the change in the system parameters over a time dt is due to the net change in the input over the same time dt. Therefore, we should consider these assumptions when solving the problems below.

REFERENCES

1. Doebelin, E.O., "System Dynamics, Modeling, Analysis, Simulation, Design" Marcel Dekker, Columbus, Ohio, 1998.
2. Karnopp, D., and Rosenberg, R., "Systems Dynamics: A Unified Approach," A Wiley-Interscince Publication, USA, 1975.

5.6 PROBLEMS

Problem (5.1):
For the system shown in Figure 5.49, derive the differential equation describing the dynamics of the level h.

Problem (5.2):
For the system shown in Figure 5.50, derive the differential equation describing the dynamics of the level h. The fluid inertance is I_l, and the pipe resistance is R_l, lumped at the tank's inlet port.

Problem (5.3):
For the system shown in Figure 5.51, derive the differential equation describing the dynamics of the level h. The fluid inertance is I_l, and the pipe resistance is R_l, lumped at the outlet port of the tank.

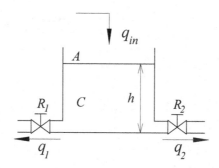

Figure 5.49 Liquid-level system

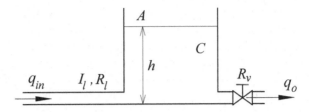

Figure 5.50 Liquid-level system

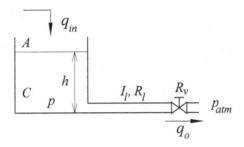

Figure 5.51 Liquid-level system

Problem (5.4):
For the system shown in Figure 5.52, derive the differential equation describing the dynamics of the level h. Assume that h is always greater than h_o ($h > h_o$).

Problem (5.5):
For the system shown in Figure 5.53, derive the differential equation describing the dynamics of the level h as a function of θ.

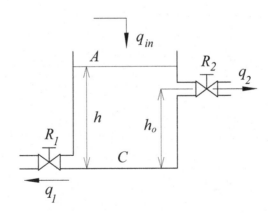

Figure 5.52 Liquid-level system

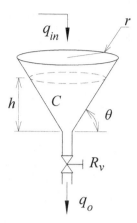

Figure 5.53 Liquid-level system

Problem (5.6):
For the system shown in Figure 5.54, derive the differential equation describing the dynamics of the level h. Assume that h is always greater than h_o ($h > h_o$).

Problem (5.7):
For the system shown in Figure 5.55,

1. Apply law of the conservation of mass flow rate to derive the differential equations that describe the dynamics of the levels h_1 and h_2. Put the results in matrix form.
2. Create the preliminary Bond Graph model and simplify it.
3. Create the Augmented Bond Graph (add the half-arrow positive power flow directions, assign causalities, and number all bonds).

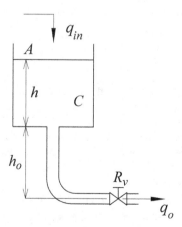

Figure 5.54 Liquid-level system

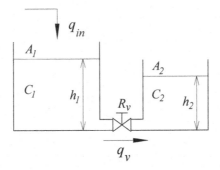

Figure 5.55 Liquid-level system

4. Derive the dynamic equations (write the key variables and their constitutive relations). Put the results in matrix form.
5. Check if law of the conservation of mass flow rate is embedded in the Augmented Bond Graph model.

Problem (5.8):
In the system shown in Figure 5.56, the fluid inertance is I_l, and the pipe resistance is R_l, lumped between the two tanks.

1. Apply law of the conservation of mass flow rate to derive the differential equations that describe the dynamics of the levels h_1 and h_2. Put the results in matrix form.
2. Create the preliminary Bond Graph model (consider the effect of the atmospheric pressure p_{atm}), impose it on the system diagram, and simplify it.
3. Create the Augmented Bond Graph (add the half-arrow positive power flow directions, assign causalities, and number all bonds). Check the derivative causalities.
4. Derive the dynamic equations (write the key variables and their constitutive relations). Put the results in matrix form.
5. Check if the law of conservation of mass flow rate is embedded in the augmented bond graph model.

Problem (5.9):
In the system shown in Figure 5.57, the fluid inertance is I_l, and the pipe resistance is R_l, lumped at the outlet port of the second tank.

1. Apply law of the conservation of mass flow rate to derive the differential equations that describe the dynamics of the levels h_1 and h_2. Put the results in matrix form.
2. Create the preliminary Bond Graph model (consider the effect of the atmospheric pressure p_{atm}) and simplify it.
3. Create the Augmented Bond Graph (add the half-arrow positive power flow directions, assign causalities, and number all bonds). Check the derivative causalities.
4. Derive the dynamic equations (write the key variables and their constitutive relations). Put the results in matrix form.

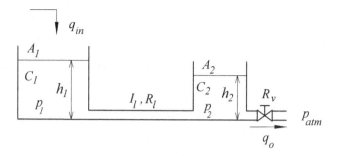

Figure 5.56 Liquid-level system

5. Check if law of the conservation of mass flow rate is embedded in the Augmented Bond Graph model.

Problem (5.10):
In the system shown in Figure 5.58, the fluid inertance is I_l, and the pipe resistance is R_l, lumped between the two tanks.

1. Apply law of the conservation of mass flow rate to derive the differential equations that describe the dynamics of the levels h_1 and h_2. Put the results in matrix form.
2. Create the preliminary Bond Graph model (consider the effect of the atmospheric pressure p_{atm}) and simplify it.
3. Create the Augmented Bond Graph (add the half-arrow positive power flow directions, assign causalities, and number all bonds). Check the derivative causalities.
4. Derive the dynamic equations (write the key variables and their constitutive relations). Put the results in matrix form.
5. Check if law of the conservation of mass flow rate is embedded in the Augmented Bond Graph model.

Problem (5.11):
For the system shown in Figure 5.59,

1. Apply law of the conservation of mass flow rate to derive the differential equation that describes the dynamics of the levels h_1, h_2, and h_3. Put the results in matrix form.

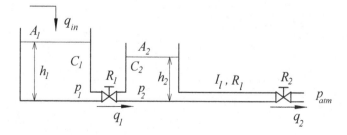

Figure 5.57 Liquid-level system

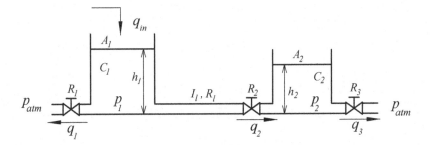

Figure 5.58 Liquid-level system

2. Create the preliminary Bond Graph model (consider the effect of the atmospheric pressure p_{atm}) and simplify it.
3. Create the Augmented Bond Graph (add the half-arrow positive power flow directions, assign causalities, and number all bonds). Check the derivative causalities.
4. Derive the equations of motion (write the key variables and their constitutive relations). Put the results in matrix form.
5. Check if law of the conservation of mass flow rate is embedded in the Augmented Bond Graph model.

Problem (5.12):
Consider the hydraulic control valve shown in Figure 5.60, assuming that the area A_p, subjected to the valve pressure p_v, and the resistance R_v of the valve orifice are linear. The valve flow rate is given by $q_v = R_v \sqrt{p_{vo}}$, where p_{vo} is the valve pressure at the operating point. The inlet cavity is of linear compliance C_v, and the compliance of the return cavity is neglected. Assuming also that the return pressure P_t is equal to the atmospheric pressure,

1. Apply Newton's law and law of the conservation of mass flow rate to derive the dynamic equations of the given system.
2. Create the preliminary Bond Graph model (consider the effect of the return pressure p_t) and simplify it.

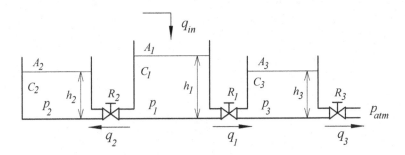

Figure 5.59 Liquid-level system

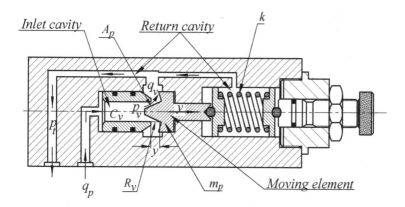

Figure 5.60 Hydraulic pressure control valve

3. Create the Augmented Bond Graph (add the half-arrow positive power flow directions, assign causalities, and number all bonds). Check the derivative causalities.
4. Derive the dynamic equations (write the key variables and their constitutive relations).
5. Check if Newton's law and law of the conservation of mass flow rate law are embedded in the Augmented Bond Graph model.

Problem (5.13):
For the system shown in Figure 2.61,

1. Apply law of the conservation of mass flow rate to derive the dynamic equations for the given system.
2. Create the preliminary Bond Graph model (consider the effect of the atmospheric pressure p_{atm}) and simplify it.
3. Create the Augmented Bond Graph (add the half-arrow positive power flow directions, assign causalities, and number all bonds). Check the derivative causalities.
4. Derive the dynamic equations (write the key variables and their constitutive relations).
5. Check if law of the conservation of mass flow rate is embedded in the Augmented Bond Graph model.

Problem (5.14):
For the system shown in Figure 5.62,

1. Apply law of the conservation of mass heat flow to derive the differential equations of the given system.
2. Create the pseudo-Bond Graph model.
3. Create the augmented pseudo-Bond Graph (add the half-arrow positive power flow directions, assign causalities, and number all bonds). Check the derivative causalities.

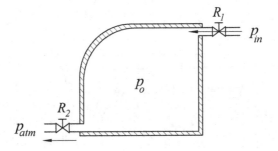

Figure 5.61 Pneumatic vessel system

4. Derive the dynamic equations (write the key variables and their constitutive relations).
5. Check if law of the conservation of mass heat flow is embedded in the Augmented Bond Graph model.

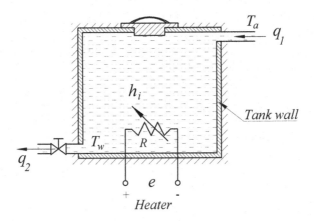

Figure 5.62 Thermal tank system

6 Lagrange Technique

6.1 INTRODUCTION

Lagrange's method is an alternative technique that can derive equations of motion for dynamical systems. The application of work and energy is essential for this method. The energy terms are responsible for the account of conservative forces, whereas the work terms are responsible for the account of non-conservative forces.

To fully understand the ideas behind the Lagrangian technique, it is necessary to have a strong foundation in the concepts detailed below.

- Kinetic energy can be defined as the quantity of energy in a system due to the masses' motion.
- The quantity of energy that an element of a system possesses as a result of any work done to it is the system's potential energy. In most cases, it comes from the gravitational cause of its vertical movement, a spring-related displacement element, or both.
- A momentum term is created by taking the derivative of Lagrangian's kinetic energy expressions concerning a generalized velocity.
- The time derivative of a momentum term produces an inertial force term, which accounts for the accelerations in Newton's second law.
- A conservative force term is produced by taking the derivative of the potential energy terms concerning a generalized displacement. This term is used in Newton's second law to account for the forces caused by springs and gravity. Usually, the vertical movement of the mass from rest is what generates gravitational force.
- The non-conservative forces represent a variety of forces, including viscous damping and friction, among others.

The following is a brief overview of the benefits that come from utilizing Lagrange's formulation:

- Lagrange's equations are force equations, like those deduced from Newton's second law. However, unlike Newton's equations, Lagrange's equations can be formulated by examining the system as a whole rather than breaking it down into its component masses.
- It can be applied to problems involving particles and rigid bodies, such as electrodynamics.
- It simplifies every conceivable process into a single method, following the same fundamental stages.
- This technique can be used in many different coordinate systems, producing the equations of motion in the selected coordinate system. These coordinates must be inertial, even though they cannot be inertial.
- Constraining forces are automatically removed from the equation.

DOI: 10.1201/9781032685656-6

- Instead of vector quantities, the method is based on scalar quantities.
- Finally, free-body diagrams are unnecessary for this method, which is a consider-
  able benefit in modeling dynamic systems using the Lagrange method.

6.2 LAGRANGE'S EQUATIONS OF MOTION OF MECHANICAL SYSTEMS

Lagrange's equations are derived from Newton's laws, and as a result, they are based
on the same restrictions as those laws. Consider Newton's second law as follows:

$$\sum F = m\ddot{x} \tag{6.1}$$

where F are the conservative forces presented in the system. Suppose that

$$x = x(q) \tag{6.2}$$

where q is a generalized coordinate. So, we may rewrite equation (6.1) along a generalized coordinate q as follows:

$$m(\ddot{x}\frac{dx}{dq}) = F\frac{dx}{dq} \tag{6.3}$$

Now let us determine the acceleration $\ddot{x}$ along the generalized coordinated q. Considering the acceleration is the time derivative of the velocity in generalized coordi-
nation, we obtain

$$\frac{d}{dt}(\dot{x}\frac{dx}{dq}) = \ddot{x}\frac{dx}{dq} + \dot{x}\frac{d}{dt}(\frac{dx}{dq}) \tag{6.4}$$

Rearranging this equation gives

$$\ddot{x}\frac{dx}{dq} = \frac{d}{dt}(\dot{x}\frac{dx}{dq}) - \dot{x}\frac{d}{dt}(\frac{dx}{dq}) \tag{6.5}$$

Substituting equation (6.5) in equation (6.3) gives

$$m\frac{d}{dt}(\dot{x}\frac{dx}{dq}) - m\dot{x}\frac{d}{dt}(\frac{dx}{dq}) = F\frac{dx}{dq} \tag{6.6}$$

We know that

$$\frac{dx}{dt} = \frac{dx}{dq}\cdot\frac{dq}{dt} \tag{6.7}$$

So,

$$\dot{x} = \frac{dx}{dq}\dot{q} \quad \text{or} \quad \frac{\dot{x}}{\dot{q}} = \frac{dx}{dq} \tag{6.8}$$

Taking into account the concepts behind partial derivatives of differential equations,
we get

$$\frac{\partial\dot{x}}{\partial\dot{q}} = \frac{dx}{dq} \tag{6.9}$$

where $\dot{x}$ is a function of q and $\dot{q}$. Substituting equation (6.9) in equation (6.6) gives

$$m\ddot{x}\frac{dx}{dq} = m\frac{d}{dt}(\dot{x}\frac{\partial \dot{x}}{\partial \dot{q}}) - m\dot{x}\frac{\partial \dot{x}}{\partial q} = F\frac{dx}{dq} \tag{6.10}$$

Rewriting this equation gives

$$m\ddot{x}\frac{dx}{dq} = \frac{d}{dt}(\frac{\partial (\frac{1}{2}m\dot{x}^2)}{\partial \dot{q}}) - \frac{\partial (\frac{1}{2}m\dot{x}^2)}{\partial q} = F_q \tag{6.11}$$

Thus,

$$\frac{d}{dt}(\frac{\partial T}{\partial \dot{q}}) - \frac{\partial T}{\partial q} = F_q \tag{6.12}$$

where $T = \frac{1}{2}m\dot{x}^2$, which is the kinetic energy, and F_q is the summation of the potential forces along a generalized coordinate q.

A force can be described in a conservative system, where the work done in changing locations depends only on the starting and ending coordinates and not on the path. So,

$$F = -\frac{dU}{dx} \tag{6.13}$$

where U is potential energy. So, for a generalized coordinate q we can now write

$$F_q = -\frac{\partial U}{\partial q} \tag{6.14}$$

Substituting equation (6.14) in equation (6.12) gives

$$\frac{d}{dt}(\frac{\partial T}{\partial \dot{q}}) - \frac{\partial (T-U)}{\partial q} = 0 \tag{6.15}$$

Note that U does not depend on $\dot{q}$ in a conservative system. So, we can introduce the Lagrangian function as follows:

$$L = T - U \tag{6.16}$$

which is just the difference between kinetic and potential energies. It is important to note that U represents the work done by the forces of constraints in moving a particle from a generalized position to a reference position or the work that we would have to perform against the forces of constraints to move it from a reference position to a general position. Therefore, we now have

$$\frac{d}{dt}(\frac{\partial (L)}{\partial \dot{q}}) - \frac{\partial (L)}{\partial q} = 0 \tag{6.17}$$

This equation is Lagrange's equation of motion. In general, if $x = x(q_1, q_2)$ and $y = y(q_1, q_2)$,

$$\frac{d}{dt}(\frac{\partial (T-U)}{\partial \dot{q}_i}) - \frac{\partial (T-U)}{\partial q_i} = 0 \quad \text{for} \quad i = 1, 2 \tag{6.18}$$

where the generalized coordinates q_i are independent, there are just enough to describe the dynamic system, and time is not counted among them.

6.3 SYSTEMS WITH NON-CONSERVATIVE ELEMENTS

We have only thought about conservative systems so far. In these systems, deriving the generalized forces from a scalar potential is possible. The force exerted across a damping element is non-conservative because it is proportional to the velocity at which the element is being moved. This is a problem with damping elements (non-conservative elements). It should go without saying that one cannot derive the force from a function based on the position coordinates. However, we would prefer to keep the benefits of Newton-Lagrange formalism. While doing so, let us reflect on the fundamental expression of the Lagrange equation found in equation (6.15). So, we can rewrite this equation as follows:

$$\frac{d}{dt}\left(\frac{\partial T}{\partial \dot{q}}\right) - \frac{\partial T}{\partial q} = Q_i \qquad (6.19)$$

where Q_i is the generalized force. Starting from this point, we incorporated all potential conservative force configurations and produced equation (6.19) involving the Lagrangian function.

To extend the formulation to non-conservative elements, we must include the friction (damping) forces in the generalized force term Q_i. This will extend the formulation to include dissipation (non-conservative) elements. Since this force depends on velocity $\dot{q}$, we can accomplish this goal by introducing a potential dependent on velocity. When differentiated by $\dot{q}$, this potential would produce a non-conservative force.

6.4 RAYLEIGH DISSIPATION FUNCTION

The Rayleigh dissipation function is a function that is utilized to deal with the effects of velocity-proportional forces in Lagrange dynamics. This function, known as the "Rayleigh potential," represents half of the system's energy loss due to friction.

$$D_i = \frac{1}{2}\sum R_i \dot{q}_i^2 \qquad (6.20)$$

where D_i is the dissipation energy and R_i is the resistance along the coordinates q_i. The friction force is negative for the velocity gradient of the dissipation. It is important to note that friction is a non-conservative element, and as such, it needs to be included in the Q_i term indicated in equation (6.19). Therefore, the generalized force term Q_i would consist of both conservative and non-conservative components as follows:

$$Q_i = -\frac{\partial U}{\partial q_i} - \frac{\partial D_i}{\partial \dot{q}_i} \qquad (6.21)$$

Substituting this equation in equation (6.19) would be modified as

$$\frac{d}{dt}\left(\frac{\partial T}{\partial \dot{q}}\right) - \frac{\partial T}{\partial q_i} + \frac{\partial U}{\partial q_i} + \frac{\partial D_i}{\partial \dot{q}_i} = 0 \qquad (6.22)$$

Rearranging this equation gives

$$\frac{d}{dt}\left(\frac{\partial (T-U)}{\partial \dot{q}}\right) - \frac{\partial (T-U)}{\partial q_i} + \frac{\partial D_i}{\partial \dot{q}_i} = 0 \qquad (6.23)$$

Even though it depends on the term $\dot{q}_i$, the term $\partial D_i/\partial \dot{q}_i$ should not be included in the time derivative term. It is added as a generalized force term for Lagrange's formula for the dissipation (non-conservative) element, reaching its formal determination. The following examples illustrate how the Lagrange method can be used to derive the equations of motion for specific systems. When deriving the Lagrange equations of motion, keep the following points in mind:

- It is usually essential to substitute variables into the Lagrange function so that only the set of generalized displacements and velocities shows up in the formula for the Lagrange function.
- The selection of generalized displacements is arbitrary and should be convenient for subsequent analysis, but the number of generalized displacements must equal the number of system degrees of freedom.
- When forming the Lagrangian function and Lagrange's equations of motion, it is essential to observe the negative signs.

Example (6.1):
For the system shown in Figure 6.1, derive the equations of motion using the Lagrange method,

Solution:
As shown in Figure 6.1, the system is a two-degrees-of-freedom system; since it includes two masses m_1 and m_2, the total kinetic energy is given by

$$T = \frac{1}{2}m_1\dot{x}_1^2 + \frac{1}{2}m_2\dot{x}_2^2 \tag{6.24}$$

Given that the system has two springs, the total potential energy is

$$U = \frac{1}{2}k_1(x_1 - x_2)^2 + \frac{1}{2}k_2x_2^2 \tag{6.25}$$

Given that the system has two dampers, the non-conservative dissipation energy is

$$D = \frac{1}{2}b_1(\dot{x}_1 - \dot{x}_2)^2 + \frac{1}{2}b_2\dot{x}_2^2 \tag{6.26}$$

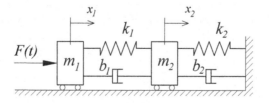

Figure 6.1 Two-degrees mass-spring-damper system

Then, the non-conservative forces are obtained using the Rayleigh potential term indicated in equation (6.26) along the generalized coordinates as follows:

$$\frac{\partial D}{\partial \dot{x}_1} = b_1(\dot{x}_1 - \dot{x}_2) \tag{6.27}$$

and

$$\frac{\partial D}{\partial \dot{x}_2} = -b_1(\dot{x}_1 - \dot{x}_2) + b_2\dot{x}_2 \tag{6.28}$$

The Lagrange's function in x_1 and x_2 coordinates is given by

$$L = T - U = \frac{1}{2}m_1\dot{x}_1^2 + \frac{1}{2}m_2\dot{x}_2^2 - \frac{1}{2}k_1(x_1 - x_2)^2 - \frac{1}{2}k_2 x_2^2 \tag{6.29}$$

Considering equation (6.23), Lagrange's equations of motion for each generalized coordinate are given by

$$\frac{d}{dt}\left(\frac{\partial(T-U)}{\partial \dot{x}_1}\right) - \frac{\partial(T-U)}{\partial x_1} = \frac{d}{dt}[m_1\dot{x}_1] - [-k_1(x_1 - x_2)] = Q_{x_1} \tag{6.30}$$

and

$$\frac{d}{dt}\left(\frac{\partial(T-U)}{\partial \dot{x}_2}\right) - \frac{\partial(T-U)}{\partial x_2} = \frac{d}{dt}[m_2\dot{x}_2] - [-k_1(x_1 - x_2) * (-1) - k_2 x_2] = Q_{x_2} \tag{6.31}$$

As previously determined, the non-conservative forces should be incorporated into the system's generalized forces terms. Since the external force $F(t)$ is applied to the mass m_2, the generalized forces of the system are given by

$$Q_{x_1} = -b_1(\dot{x}_1 - \dot{x}_2) \quad \text{and} \quad Q_{x_2} = F(t) + b_1(\dot{x}_1 - \dot{x}_2) - b_2\dot{x}_2 \tag{6.32}$$

Substituting equation (6.32) in equations (6.22) and (6.31) gives

$$\frac{d}{dt}[m_1\dot{x}_1] - [-k_1 x_1 + k_2(x_2 - x_1)] = -b_1(\dot{x}_1 - \dot{x}_2) \tag{6.33}$$

and

$$\frac{d}{dt}[m_2\dot{x}_2] - [-k_2(x_2 - x_1)] = F(t) + b_1(\dot{x}_1 - \dot{x}_2) - b_2\dot{x}_2 \tag{6.34}$$

Carrying out the time derivative terms in equations (6.32) and (6.33) and rearranging the results gives

$$m_1\ddot{x}_1 + b_2(\dot{x}_1 - \dot{x}_2) + k_2(x_1 - x_2) = F(t) \tag{6.35}$$

and

$$m_2\ddot{x}_2 + b_1(\dot{x}_2 - \dot{x}_1) + k_1(x_2 - x_1) + b_2\dot{x}_2 + k_2 x_2 = 0 \tag{6.36}$$

Equations (6.34) and (6.35) are the equations of motion derived from the Lagrange method for the given system shown in Figure 6.1.

Example (6.2):
Using the Lagrange method, derive the equations of motion for the pendulum gantry system of a massless rod of length l shown in Figure 6.2, assuming that the angular displacement θ is of a small value.

Solution:
Let the generalized coordinates be x and θ, as shown in Figure 6.3. When the system is at rest, the angle θ corresponds to a completely vertical, downward-pointing pendulum rod of zero value. When facing the system, we assumed that the rotation of the pendulum rod is considered to be positive direction is counter-clockwise (CCW). In addition, the positive direction of the cart displacement x_c is to the right, as indicated by the Cartesian frame of coordinates (x, y) shown in Figure 6.3 This figure shows that when the cart moves a displacement x_c, the pendulum rod rotates an angular displacement θ, and according to the reference frame definition, the displacements x_p and y_p of the pendulum bob mass m_p are given by

$$x_p = x_c + l\sin\theta \quad \text{and} \quad y_p = -l\cos\theta \tag{6.37}$$

Taking the time derivative of this equation gives

$$\dot{x}_p = \dot{x}_c + l\cos\theta\,\dot{\theta} \quad \text{and} \quad \dot{y}_p = l\sin\theta\,\dot{\theta} \tag{6.38}$$

Now, let us determine the total kinetic energy of the system. It is the sum of the translational and rotational kinetic energies arising from the cart mass m_c and pendulum bob mass m_p. Note that the cart mass considered only translational kinetic energy, while the bob mass considered both translational and rotational kinetic energies. So, the translational kinetic energy of the cart mass m_c is given by

$$T_c = \frac{1}{2}m_c\dot{x}_c^2 \tag{6.39}$$

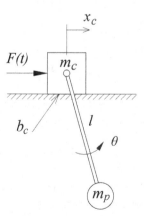

Figure 6.2 Pendulum gantry system

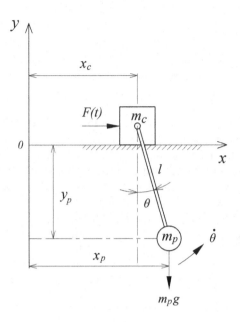

Figure 6.3 Motion of pendulum gantry system

and the translational kinetic energy of the bob mass m_p can be expressed as a function
of its velocity components as follows

$$T_{pt} = \frac{1}{2}m_p(\sqrt{\dot{x}_p^2 + \dot{y}_p^2})^2 \tag{6.40}$$

Substituting equation (6.39) in equation (6.41) and rearranging the results gives

$$T_{pt} = \frac{1}{2}m_p[(\dot{x}_c + l\cos\theta\,\dot{\theta})^2 + (l\sin\theta\dot{\theta})^2] \tag{6.41}$$

The rotational kinetic energy of the bob mass m_p is given by

$$T_{pr} = \frac{1}{2}J_p\dot{\theta}^2 \tag{6.42}$$

Thus, the total kinetic energy of the system is the sum of the three individual kinetic
energies indicated in equations (6.38), (6.40), and (6.42). It is given by

$$T = \frac{1}{2}m_c\dot{x}_c^2 + \frac{1}{2}m_p[(\dot{x}_c + l\cos\theta\,\dot{\theta})^2 + (l\sin\theta\dot{\theta})^2] + \frac{1}{2}J_p\dot{\theta}^2 \tag{6.43}$$

This equation indicated that the total kinetic energy regarding the generalized coordinates and their time derivative could be expressed.

Since the system does not include a spring, the potential energy is only due to
gravity. It is usually caused by the mass's vertical movement from rest. So, the total

potential energy in the system is due to the gravitational potential energy of the bob mass m_p, which is given by

$$U = m_p g y_p \tag{6.44}$$

Substituting equation (6.38) in equation (6.45) gives

$$U = m_p g l \cos \theta \tag{6.45}$$

This equation indicates that potential energy can only be expressed in generalized coordinates. So, the Lagrange function in x_c and θ coordinates is given by

$$L = T - U = \frac{1}{2} m_c \dot{x}_c^2 + \frac{1}{2} m_p [(\dot{x}_c + l \cos \theta \, \dot{\theta})^2 + (l \sin \theta \dot{\theta})^2] + \frac{1}{2} J_p \dot{\theta}^2 + m_p g l \cos \theta \tag{6.46}$$

Considering the friction that is provided to the cart, as shown in Figure 6.3, the non-conservative dissipation energy is given by Rayleigh potential as,

$$D_c = -\frac{1}{2} b_c \dot{x}_c^2 \tag{6.47}$$

So,

$$\frac{\partial D_c}{\partial \dot{x}_c} = -b_c \dot{x}_c \tag{6.48}$$

Let us now derive Lagrange's equations of motion of the system. Applying the Lagrange equation indicated in equation (6.47) along the generalized coordinates x_c and θ gives

- For the generalized coordinate x_c, we have

$$\frac{d}{dt} \left(\frac{\partial L}{\partial \dot{x}_c} \right) - \frac{\partial L}{\partial x_c} = \frac{d}{dt} [m_c \dot{x}_c + m_p (\dot{x}_c + l \cos \theta \, \dot{\theta})] - 0 = Q_c \tag{6.49}$$

- For the generalized coordinate θ we have

$$\frac{d}{dt} \left(\frac{\partial L}{\partial \dot{\theta}} \right) - \frac{\partial L}{\partial \theta} = \frac{d}{dt} [m_p (\dot{x}_c + l \cos \theta \, \dot{\theta}) + l \sin \theta \dot{\theta} + J_p \dot{\theta}] - \frac{\partial}{\partial \theta} (m_p g l \cos \theta) = Q_p \tag{6.50}$$

Since the external force $F(t)$ and friction forces are applied only to the cart, so the generalized forces of the system are given by

$$Q_c = F(t) - b_c \dot{x}_c \quad \text{and} \quad Q_p = 0 \tag{6.51}$$

Substituting equation (6.52) in equations (6.50) and (6.51) and carrying out the partial derivative $(\partial / \partial \theta)$ gives

$$\frac{d}{dt} [m_c \dot{x}_c + m_p (\dot{x}_c + l \cos \theta \, \dot{\theta}) + m_p l \sin \theta \dot{\theta}] = F(t) - b_c \dot{x}_c \tag{6.52}$$

and

$$\frac{d}{dt}[m_p(\dot{x}_c + l\cos\theta\,\dot{\theta}) + m_p l\sin\theta\dot{\theta} + J_p\dot{\theta}] + m_p g\,l\sin\theta = 0 \qquad (6.53)$$

Now, carrying out the time derivative terms presented in equations (6.52) and (6.53) and rearranging the results gives

$$(m_p + m_c)\ddot{x}_c + m_p l\cos\theta\,\ddot{\theta} - m_p l\sin\theta\dot{\theta}^2 + m_p l\cos\theta\dot{\theta}^2 + b_c\dot{x}_c = F(t) \qquad (6.54)$$

and

$$m_p l\cos\theta\ddot{x}_c + J_p\ddot{\theta} + m_p l^2\cos\theta\ddot{\theta} + m_p g\,l\sin\theta - m_p l\sin\theta\dot{\theta}\dot{x}_c = 0 \qquad (6.55)$$

Assuming small θ, so, $\sin\theta \approx \theta$, $\cos\theta \approx 1$, $\dot{\theta}^2 \approx 0$ and $\dot{x}_c\dot{\theta} \approx 0$. Considering these, equations (6.54) and (6.55) become

$$(m_p + m_c)\ddot{x}_c + m_p l\ddot{\theta} + b_c\dot{x}_c = F(t) \qquad (6.56)$$

and

$$(J_p + m_p l^2)\ddot{\theta} + m_p l\ddot{x}_c + m_p g l\theta = 0 \qquad (6.57)$$

These equations are the equations of motion derived from the Lagrange method for the pendulum gantry system shown in Figure 6.2.

6.5 LAGRANGIAN METHOD APPLIED TO ELECTRICAL SYSTEMS

The Lagrangian method was designed to reduce the constraints on obtaining differential equations for mechanical systems. It can also be utilized to acquire the dynamic equations for electrical circuits. A unified framework can be provided for electro-mechanical systems with electrical and mechanical components when the same technique is used for both electrical and mechanical systems. One particular benefit of employing the same technique for both systems is similar to the Bond Graph modeling technique discussed in Chapters 3 and 4.

The position variable represents the generalized coordinate in mechanical systems. The limitations in an electrical circuit that are connected are given by the circuit components, and the charge variable is the electrical equivalent to the position variable in mechanical systems. Therefore, our choice of a variable will be the charge q, which represents the generalized coordinate in electrical circuits. So, the variables following in the meshes must also be defined as the charge variable. This implies that the charges moving through each mesh are defined appropriately in terms of the variables that have been specified. Once the electrical circuit configuration coordinates q_i have been defined, the $\dot{q}_i$ will be the mesh current. As previously determined, comparing Kirchhoff's and Newton's laws, we found that the input voltage, inductance, capacitor, and resistance in electrical circuits are analogous to the input force, mass, spring, and damper in mechanical systems. This will lead us to apply the Lagrange method to electrical circuits. The following example will demonstrate how the Lagrangian method is equivalent in mechanical and electrical systems.

Example (6.3):
For the electrical circuit shown in Figure 6.4, derive the dynamic equations using the Lagrange method.

Solution:
There is only one mesh, as shown in Figure 6.4, and the coordinates are specified as the charge that follows in the mesh. This implies the circuit has two generalized coordinates, q_1 and q_2. Notably, these coordinates do not correspond to the charges q_1 and q_2. Applying the Lagrange equation to the given electrical circuit shown in Figure 6.4, we obtain

$$T = \frac{1}{2}L_1\dot{q}_1^2 + \frac{1}{2}L_2\dot{q}_2^2 \tag{6.58}$$

$$U = \frac{1}{2}\frac{1}{C_1}(q_1 - q_2)^2 + \frac{1}{2}\frac{1}{C_2}q_2^2 \tag{6.59}$$

The voltages across the resistances, representing the damping in mechanical systems, are $R_1(\dot{q}_1 - \dot{q}_2)$ and $R_2\dot{q}_2$ in the given electrical circuit. So, the non-conservative dissipation energies of both resistances are given by

$$R_e = \frac{1}{2}R_1(\dot{q}_1 - \dot{q}_2)^2 + \frac{1}{2}R_2\dot{q}_2^2 \tag{6.60}$$

where R_e is the total dissipation energy. Therefore, the Rayleigh potential term in equation (6.21) along the generalized coordinates gives the non-conservative terms of both resistances as follows:

$$\frac{\partial R_{e_1}}{\partial \dot{q}_1} = R_1(\dot{q}_1 - \dot{q}_2) \tag{6.61}$$

and

$$\frac{\partial R_{e_2}}{\partial \dot{q}_2} = -R_1(\dot{q}_1 - \dot{q}_2) + R_2\dot{q}_2 \tag{6.62}$$

The Lagrange's function in q_1 and q_2 coordinates is given by

$$L = T - U = \frac{1}{2}L_2\dot{q}_2^2 + \frac{1}{2}L_1\dot{q}_1^2 - \frac{1}{2}\frac{1}{C_1}(q_1 - q_2)^2 - \frac{1}{2}\frac{1}{C_2}q_2^2 \tag{6.63}$$

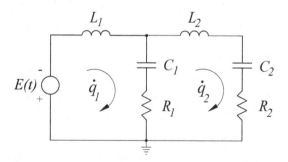

Figure 6.4 Electrical circuit

Considering equation (6.22), Lagrange's dynamic equation for each generalized co-ordinate is given by

$$\frac{d}{dt}(\frac{\partial(T-U)}{\partial \dot{q}_1}) - \frac{\partial(T-U)}{\partial q_1} = \frac{d}{dt}[L_1\dot{q}_1] - [-\frac{1}{C_1}(q_1-q_2)] = Q_{e_1} \qquad (6.64)$$

and

$$\frac{d}{dt}(\frac{\partial(T-U)}{\partial \dot{q}_2}) - \frac{\partial(T-U)}{\partial q_2} = \frac{d}{dt}(L_2\dot{q}_2) - [-\frac{1}{C_1}(q_1-q_2)*(-1) - \frac{1}{C_2}q_2] = Q_{e_2}$$

$$(6.65)$$

As previously determined, the non-conservative forces should be incorporated into the system's generalized forces terms. Since the external voltage $e(t)$, is applied to the loop of charge $\dot{q}_1$, the generalized forces of the system are given by

$$Q_{e_1} = e(t) - R_1(\dot{q}_1 - \dot{q}_2) \quad \text{and} \quad Q_{e_2} = R_1(\dot{q}_1 - \dot{q}_2) - R_2\dot{q}_2 \qquad (6.66)$$

Substituting equation (6.66) in equations (6.64) and (6.65) gives

$$\frac{d}{dt}(\frac{\partial(T-U)}{\partial \dot{q}_1}) - \frac{\partial(T-U)}{\partial q_1} = \frac{d}{dt}(L_1\dot{q}_1) - [-\frac{1}{C_1}q_1 + \frac{1}{C_2}(q_2-q_1)] = e(t) - R_1(\dot{q}_1 - \dot{q}_2)$$

$$(6.67)$$

and

$$\frac{d}{dt}(\frac{\partial(T-U)}{\partial \dot{q}_2}) - \frac{\partial(T-U)}{\partial q_2} = \frac{d}{dt}(L_2\dot{q}_2) - [-\frac{1}{C_2}(q_2-q_1)] = R_1(\dot{q}_1 - \dot{q}_2) - R_2\dot{q}_2$$

$$(6.68)$$

These equations are then rearranged with all variable terms on the left side and everything else on the right side, as follows:

$$L_1\ddot{q}_1 + R_1(\dot{q}_1 - \dot{q}_2) = e(t) \qquad (6.69)$$

and

$$L_2\ddot{q}_2 + R_1(\dot{q}_2 - \dot{q}_1) + \frac{1}{C_2}(q_2 - q_1) = 0 \qquad (6.70)$$

These two equations are the dynamic equations of the given circuit using the Lagrange method.

6.6 PROBLEMS

Problem (6.1):
For the system shown in Figure 6.5, derive the equations of motion of the masses m_1, m_2, and m_3 using the Lagrange method.

Problem (6.2):
For the system shown in Figure 6.6, assuming the angular displacements θ_1 and θ_2 are small, derive the equation(s) of motion using the Lagrange method.

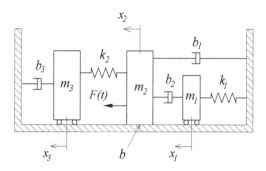

Figure 6.5 Multi-degree of freedom mechanical translational system

Problem (6.3):
For the system shown in Figure 6.7, assuming the inclination angle θ is large enough so that the system moving down, derive the equations of motion using the Lagrange method.

Problem (6.4):
For the system shown in Figure 6.8, derive the equation(s) of motion using the Lagrange method. Assume the rotating angle θ of the pinion is small.

Problem (6.5):
For the system shown in Figure 6.9, assuming the rod is massless $J_o \approx 0$, and its angular displacement θ is small, derive the equations of motion using the Lagrange method.

Problem (6.6):
For the system shown in Figure 6.10, assuming the pendulum angle θ is small, derive the equation of motion using the Lagrange method.

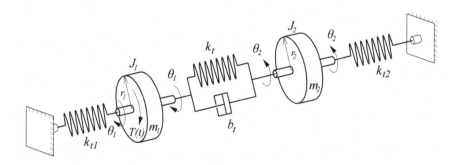

Figure 6.6 Mechanical rotational system

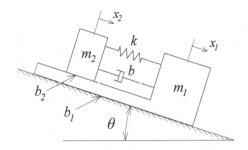

Figure 6.7 Mechanical system

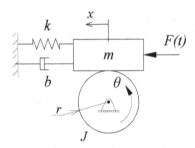

Figure 6.8 Rack and pinion system

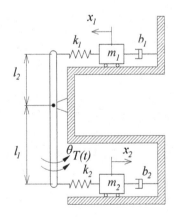

Figure 6.9 Mechanical system

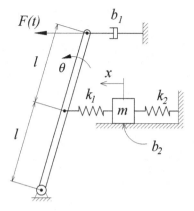

Figure 6.10   Mechanical system

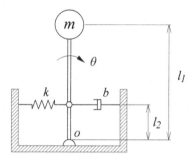

Figure 6.11   Inverted pendulum system

Problem (6.7):
For the system shown in Figure 6.11, assuming the pendulum angle θ is small, derive
the differential equations using the Lagrange method.

Problem (6.8):
For the electrical circuit shown in Figure 6.12, derive the differential equations using
the Lagrange method.

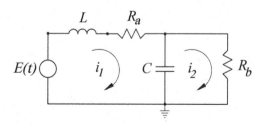

Figure 6.12   Electrical circuit

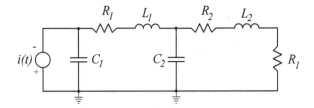

Figure 6.13 Electrical circuit

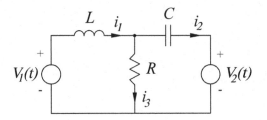

Figure 6.14 Electrical circuit

Problem (6.9):
For the electrical circuit shown in Figure 6.13, derive the differential equations using the Lagrange method.

Problem (6.10):
Obtain the mathematical model of the circuit shown in Figure 6.14 using the Lagrange method.

Problem (6.11):
Obtain the mathematical model of the circuit shown in Figure 6.15 using the Lagrange method.

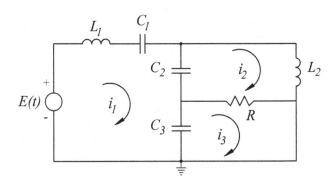

Figure 6.15 Electrical circuit

7 System Differential Equations Solution

7.1 INTRODUCTION

The differential equations representing the motion of a dynamic system can be solved by analytical and software simulation to determine the behavior of a dynamic system. The solution obtained through software simulation can numerically solve the equations, allowing system parameter effects to be revealed after many cases have to be run. In contrast, analytical techniques provide general solutions that directly demonstrate the effects of system parameters. In this chapter, we will explain how these two techniques are implemented. The analytical solution techniques are only applicable to linear equations with constant coefficients. We will briefly review the classical method for finding general solutions to differential equations, which is taught in detail in the majority of first-year differential equations courses. This chapter will also detail an alternate analytical solution technique, the Laplace transform.

7.2 CLASSICAL METHOD (CONSTANT-COEFFICIENTS METHOD)

The general form of the differential equation, which will be dealing with, is given by

$$a_n \frac{d^n x}{dt^n} + a_{n-1} \frac{d^{(n-1)} x}{dt^{(n-1)}} + \cdots + a_1 \frac{dx}{dt} + a_o x = F(t) \qquad (7.1)$$

where a_n are the constants and $F(t)$ is a known function of time. Three steps are required to solve equation (7.1), which are outlined below

1. Find the complementary function part of the general solution, x_c.
2. Find the particular function part of the general solution, x_p.
3. x_c and x_p should be combined to obtain the general solution $x(t)$, and initial conditions should be applied to determine constants of integration.

7.2.1 COMPLEMENTARY FUNCTION PART (HOMOGENEOUS SOLUTION)

Using the operator $\lambda = d/dt$, we can rewrite equation (7.1) as follows:

$$(a_n \lambda^n + a_{n-1} \lambda^{(n-1)} + \cdots + a_1 \lambda + a_o) x = F(t) \qquad (7.2)$$

and

$$a_n \lambda^n + a_{n-1} \lambda^{(n-1)} + \cdots + a_1 \lambda + a_o = 0 \qquad (7.3)$$

DOI: 10.1201/9781032685656-7

Equation (7.3) is the characteristic equation of the generalized differential equation (7.1). It is considered an algebraic equation in the unknown λ, and we must solve it for its n roots. Once these roots are identified, we can immediately write down the complementary solution x_c. This solution depends on the number of roots of the equation, which can be classified as:

1. The characteristic equation has a real root, s_1, the solution is

$$x_c = C_1 e^{s_1 t} \tag{7.4}$$

2. The characteristic equation has two real roots, s_1 and s_2, which are not repeated; the solution is

$$x_c = C_1 e^{s_1 t} + C_2 e^{s_2 t} \tag{7.5}$$

3. The characteristic equation has two real repeated roots, s_1 and s_1; the solution is

$$x_c = C_1 e^{s_1 t} + C_2 t e^{s_1 t} \tag{7.6}$$

4. The characteristic equation has three real roots (no repeated roots), s_1, s_2 and s_3; the solution is

$$x_c = C_1 e^{s_1 t} + C_2 e^{s_2 t} + C_3 e^{s_3 t} \tag{7.7}$$

5. The characteristic equation has three real roots (two repeated roots), s_1, s_1, and s_2; the solution is

$$x_c = C_1 e^{s_1 t} + C_2 t e^{s_1 t} + C_3 e^{s_2 t} \tag{7.8}$$

6. The characteristic equation has three repeated real roots, s_1, s_1 and s_1; the solution is

$$x_c = C_1 e^{s_1 t} + C_2 t e^{s_1 t} + C_3 t^2 e^{s_1 t} \tag{7.9}$$

7. The characteristic equation has complex roots.
   When the complex roots arise, they always come in pairs of the form $(a \pm bi)$, and the solution in general for such a pair is given by

$$x_c = C e^{at} \sin(bt + \phi) \tag{7.10}$$

where C and ϕ are constants of integration. If the complex pair is repeated $(a \pm bi)$ and $(a \pm bi)$, the solution is given by

$$x_c = C_1 e^{at} \sin(bt + \phi_1) + C_2 t e^{at} \sin(bt + \phi_2) \tag{7.11}$$

where $C's$ and $\phi's$ are constants of integration. Note that the values of a usually have negative values.

After the above analysis, once the roots of the characteristic equation are established, the complementary solution is immediately obtained.

7.2.2 PARTICULAR FUNCTION PART (NON-HOMOGENEOUS SOLUTION)

Whereas the approach described above is universal for obtaining the complementary solution x_c, there are no universal methods for obtaining the particular solution x_p because it depends on the form of the forcing function $f(t)$ indicated in equation (7.1). In fact, for most $f(t)'s$ forms, a simple procedure, the approach of indeterminate coefficients, is enough. If consecutive derivatives of the $f(t)$ eventually become zero or repeat themselves, the solution is satisfactory; otherwise, there are no solutions for the specific differential equation.

The forms of the forcing input function can be classified into:

1. The forcing function is of the form $f(t) = A$. In this case, the first derivative will equal zero, meaning the solution will satisfy.
2. The forcing function is of the form $f(t) = At^b$. In this case, all derivatives beyond the $b's$ will equal zero, and the solution will satisfy. For example, assuming that $f(t) = 5t^3$, the particular solution will be in the form

$$x_p = At^3 \qquad (7.12)$$

Taking the first, second, and third derivatives of this equation yields

$$\dot{x}_p = 3At^2, \quad \ddot{x}_p = 6At \quad \text{and} \quad \dddot{x}_p = 6A \qquad (7.13)$$

This equation indicates that the fourth derivative will be zero, meaning the solution will satisfy.
3. The forcing function is of the form $f(t) = A \sin bt$. In this situation, successive derivatives yield only $sin\,(bt)$ or $\cos\,(bt)$ functions, and the solution is satisfied. For example, assuming that $f(t) = 2sin(5t)$ the particular solution will be in the form

$$x_p = A sin(5t) \qquad (7.14)$$

Taking the first, second, and third derivatives of this equation yields

$$\dot{x}_p = 5A\cos(5t), \quad \ddot{x}_p = -25A sin(5t) \quad \text{and} \quad \dddot{x}_p = -125A\cos(5t) \qquad (7.15)$$

This equation indicated that the fourth derivative would repeat the second derivative with a different coefficient, the fifth derivative would repeat the third derivative, and so on, meaning the solution would be satisfied.
4. The forcing function is of the form $f(t) = e^{bt^a}$. Regardless of how far the successive differentiation process progresses, it will continue to generate new functional forms, rendering the solution unsatisfactory.

In general, where the particular solution x_p is satisfied, it is written as a sum of terms consisting of every different kind of function found in $f(t)$ and its derivative, each multiplied by an undetermined coefficient. By substituting x_p and its derivatives into the differential equation, it is possible to obtain these coefficients immediately. The general solution of the differential equation is simply the sum of both complimentary (homogeneous) and particular (non-homogeneous) solutions, as follows:

$$x(t) = x_c + x_p \qquad (7.16)$$

7.2.3 GENERATING THE CLASSICAL SOLUTION OF A DIFFERENTIAL EQUATION USING MATLAB®

MATLAB code, text, equations, and graphics can merge into a single document in a Live Editor environment. These documents are termed MATLAB live scripts and MATLAB live functions. These interactive scripts are adaptable to various domains and can be used in various applications. In addition, the output of live scripts is saved and displayed with the code that prompted it to be generated; see Appendix (1).

The MATLAB command known as *syms* executes the symbolic scalar variables and functions. The differential equation provided should be written down first, and then the *dsolve* command should be used to create the general solution that includes the integral constants. To obtain the values of the integral constants and create the final solution, write the initial conditions and again use the *dsolve* command for both the differential equation and their initial conditions. Type "help syms" and "help dsolve" and see the results in the command window for further details.

Let us now look at some basic examples that demonstrate how to use the classical approaches explained previously to obtain the complementary and particular solutions of the differential equation. These solutions may be verified using MATLAB.

Example (7.1) (Repeated roots):
Solve the following differential equation using the classical method and verify the results using MATLAB.

$$5\ddot{x} + 20\dot{x} + 20x = 28 \quad \text{if} \quad x(0) = 5 \quad \text{and} \quad \dot{x}(0) = 8 \tag{7.17}$$

Solution:
- Complementary function part
The characteristic equation is given by:

$$5\lambda^2 + 20\lambda + 20 = 0 \quad \text{or} \quad \lambda^2 + 4\lambda + 4 = 0 \tag{7.18}$$

Obtaining the roots of this equation gives

$$(\lambda + 2)^2 = 0 \quad \text{and} \quad \lambda = -2, -2 \quad \text{(repeated roots)} \tag{7.19}$$

Therefore, the complementary solution is

$$x_c = C_1 e^{\lambda_1 t} + C_2 t e^{\lambda_2 t} = C_1 e^{-2t} + C_2 t e^{-2t} \tag{7.20}$$

- Particular function part
Because the particular function part must take the form of the input forcing function, we assume that

$$x_p = A \tag{7.21}$$

Since the given differential equation is of second-order, taking the first and second derivatives of this equation yields

$$\dot{x}_p = \ddot{x}_p = 0 \tag{7.22}$$

Substituting equations (7.21) and (7.22) in equation (7.17) and solving for A gives

$$5*0+20*0+20A = 28 \text{ so } A = \frac{7}{5} \tag{7.23}$$

Substituting the value of A in equation (7.21), the particular solution is

$$x_p = \frac{7}{5} \tag{7.24}$$

- General Solution
The general solution of the given differential equation, which is the summation of the complementary function indicated in equation (7.20) and the particular function indicated in equation (7.24), is given by

$$x(t) = x_c + x_p = C_1 e^{-2t} + C_2 t e^{-2t} + \frac{7}{5} \tag{7.25}$$

We must now apply initial conditions to find the constants C_1 and C_2. To accomplish that, we must first differentiate equation (7.25) as follows:

$$\dot{x}(t) = -2C_1 e^{-2t} + C_2 e^{-2t} + C_2 *(-2)*t e^{-2t} = (-2C_1 + C_2 - 2C_2 t)e^{-2t} \tag{7.26}$$

```
% Solution of the Ordinary Differential Equation using classical method.
% Use the syms command to create symbolic scalar variables and functions.
% For more details about this command, on the command window, write
% help syms and see the results.
  syms x(t) a
% Specify the second and first order derivative of x by
% using diff(x,t,2) and diff(x,t,1), respectively and write the given
% differential equation using the operator ==.
  eqn=5*diff(x,t,2)+20*diff(x,t,1)+20*x==28;
% Using dsolve command to obtain the general solution which contains
% both complementary and particular solutions.
  xSol(t)=dsolve(eqn)
```

xSol(t) =

$$C_1 e^{-2t} + C_2 t e^{-2t} + \frac{7}{5}$$

```
% Now, we have to consider the initial conditions I.C. to eliminate the
% constants C1 and C2 from the solution.
% Specify the second initial condition by assigning diff(x,t) to Dx.
  Dx=diff(x,t);
% Insert the initial by assiging x(0)==5 Dx(0)==8 to cond command
  cond=[x(0)==5,Dx(0)==8];
% Now, Considering I.C. cond as the second input to the dsolve command
% and obtaining the general solution (response equation).
  x(t)=dsolve(eqn,cond)
```

x(t) =

$$\frac{18 e^{-2t}}{5} + \frac{76 t e^{-2t}}{5} + \frac{7}{5}$$

Figure 7.1 MATLAB live script file generating the solution of the given differential equation $5\ddot{x} + 20\dot{x} + 20x = 28$

Substituting $x(0) = 5$ in equation (7.25) and solving for C_1 gives

$$C_1 + C_2 * 0 + 1.4 = 5 \quad \text{gives} \quad C_1 = \frac{5}{1.4} = 3.6 \tag{7.27}$$

Substituting $\dot{x}(0) = 8$ in equation (7.26) yields

$$\dot{x}(0) = (-2C_1 + C_2 - 0) * 1 = 8 \tag{7.28}$$

Solving this equation for C_2 yields

$$C_2 = 8 + 2C_1 \quad \text{and} \quad C_1 = 3.6 \quad \text{so} \quad C_2 = 15.2 \tag{7.29}$$

Substituting values of C_1 and C_2 in equations (7.25) gives

$$x(t) = 3.6e^{-2t} + 15.2te^{-2t} + 1.4 \tag{7.30}$$

This equation is the time-domain solution of the given differential equation indicated in equation (7.17).

Now, let us generate the solution using MATLAB. Figure 7.1 shows the MATLAB live script file and the generated results. Note that the script provides an explanation for each step it contains. Comparing equation (7.30) with the generated $x(t)$ presented in Figure 7.1, we found they are the same.

Example (7.2) (Real roots):
Using the classical method, find the solution of the following differential equation.

$$\frac{d^2x}{dt^2} + 3\frac{dx}{dt} + 2x = 4e^{-5t} \quad \text{if} \quad x(0) = 2 \quad \text{and} \quad \dot{x}(0) = 0 \tag{7.31}$$

Solution:
- Complementary function part:
The characteristic equation is

$$\lambda^2 + 3\lambda + 2 = 0 \tag{7.32}$$

This equation has two real roots: $\lambda_1 = -2$ and $\lambda_2 = -1$. Then, the complementary function is given by

$$x_c = C_1 e^{\lambda_1 t} + C_2 e^{\lambda_2 t} = C_1 e^{-2t} + C_2 e^{-t} \tag{7.33}$$

- Particular function part:
Considering the definition of the forcing function $f(t) = 4e^{-5t}$, clearly give the particular solution of the form

$$x_p = Ae^{-5t} \tag{7.34}$$

Taking the first and second derivatives of equation (7.34) yields

$$\dot{x}_p = -5Ae^{-5t} \tag{7.35}$$

and

$$\ddot{x}_p = 25Ae^{-5t} \tag{7.36}$$

Rewrite equation (7.31) as follows:

$$\ddot{x}_p + 3\dot{x}_p + 2x_p = 4e^{-5t} \tag{7.37}$$

Substituting equations (7.34), (7.35) and (7.36) in equation (7.37) gives

$$25Ae^{-5t} - 15Ae^{-5t} + 2Ae^{-5t} = 4e^{-5t} \tag{7.38}$$

Rearranging this equation gives

$$12Ae^{-5t} = 4e^{-5t} \tag{7.39}$$

Solving this equation for A gives

$$A = \frac{1}{3} \tag{7.40}$$

Substituting this equation in equation (7.34) yields

$$x_p = \frac{1}{3}e^{-5t} \tag{7.41}$$

- **General solution:** The general solution is given by

$$x(t) = x_c + x_p = C_1 e^{-2t} + C_2 e^{-t} + \frac{1}{3}e^{-5t} \tag{7.42}$$

Applying the initial conditions to this equation gives

$$x(0) = 2 = C_1 + C_2 + \frac{1}{3} \tag{7.43}$$

Take the first derivative of equation (7.42) and apply the initial conditions as follows,

$$\dot{x}(0) = 0 = -2C_1 - C_2 - \frac{5}{3} \tag{7.44}$$

Solving equations (7.43) and (7.44) for C_1 and C_2 gives

$$C_1 = -\frac{10}{3} \quad \text{and} \quad C_2 = 5 \tag{7.45}$$

Substituting equation (7.45) in equation (7.42) gives

$$x(t) = 5e^{-t} - \frac{10}{3}e^{-2t} + \frac{1}{3}e^{-5t} \tag{7.46}$$

This equation is the time-domain solution of the given differential equation.

Example (7.3) (Complex roots):
Using the classical method, find the solution of the following differential equation and verify the results using MATLAB.

$$\frac{d^2x}{dt^2} + 8\frac{dx}{dt} + 65x = 130 \quad \text{if} \quad x(0) = 10 \quad \text{and} \quad \dot{x}(0) = 4 \tag{7.47}$$

Solution:
- Complementary function part
The characteristic equation is

$$\lambda^2 + 8\lambda + 6 = 0 \tag{7.48}$$

This equation has two distinct complex roots, which are given by

$$\lambda_{1,2} = \frac{-8 \pm \sqrt{8^2 - 4(1)(65)}}{2} = \frac{-8 \pm \sqrt{64 - 260}}{2} \tag{7.49}$$

or,

$$\lambda_{1,2} = \frac{-8 \pm 14i}{2} = -4 \pm 7i \tag{7.50}$$

Then, the complementary function is given by

$$x_c = C_1 e^{-4t} \cos(7t) + C_2 e^{-4t} \sin(7t) \tag{7.51}$$

or

$$x_c = e^{-4t}[C_1 \cos(7t) + C_2 \sin(7t)] \tag{7.52}$$

- Particular function part:
The particular function is of the form $x_p = A$. Taking the first and second derivatives of this function yields

$$\dot{x}_p = \ddot{x}_p = 0 \tag{7.53}$$

Substituting equation (7.53) and the value of $x_p = A$ in equation (7.47) and solving for A gives

$$65A = 130 \quad \text{gives} \quad A = 2 \tag{7.54}$$

Then, the particular solution is given by

$$x_p = 2 \tag{7.55}$$

- General Solution:
The general solution is given by

$$x(t) = x_c + x_p = e^{-4t}[C_1 \cos(7t) + C_2 \sin(7t)] + 2 \tag{7.56}$$

We must apply initial conditions to find the constants C_1 and C_2. To accomplish that, we must first differentiate equation (7.56) as follows:

$$\dot{x}(t) = -4e^{-4t}[C_1 \cos(7t) + C_2 \sin(7t)] + e^{-4t}[-7C_1 \sin(7t) + 7C_2 \cos(7t)] \tag{7.57}$$

Applying initial conditions and solving equations (7.56) and (7.57) for C_1 and C_2 gives:

$$x(0) = 10 \quad \text{yields} \quad C_1 + 2 = 10 \quad \text{and gives} \quad C_1 = 8 \tag{7.58}$$

and

$$\dot{x}(0) = 4 \ \text{yields} \ -4C_1 + 7C_2 = 4 \ \text{and gives} \ C_2 = \frac{4C_1 + 4}{7} = \frac{36}{7} \qquad (7.59)$$

Substituting the values of C_1 and C_2 indicated in equations (7.58) and (7.59) in equation (7.56) gives

$$x(t) = [8\cos(7t) + \frac{36}{7}\sin(7t)]e^{-4t} + 2 \qquad (7.60)$$

This equation is the time-domain solution of the given differential equation.

Now, let us generate the solution using MATLAB. Figure 7.2 shows the MATLAB live script file and the generated results. The script provides an explanation for each step it contains. Comparing equation (7.60) with the generated $x(t)$ presented in Figure 7.2, we found that they are the same.

```
% Solution of the Ordinary Differential Equation using classical method.
% Use the syms command to create symbolic scalar variables and functions.
% For more details about this command, on the command window, write
% help syms and see the results.
  syms a x(t)
% Specify the first and second derivatives of x by
% using diff(x,t,1) and diff(x,t,2) and write the given
% differential equation using the operator ==.
  eqn=diff(x,t,2)+8*diff(x,t,1)+65*x==130;
% Using dsolve command to obtain the general solution which contains
% both complementary and particular solutions.
  xSol(t)=dsolve(eqn)
```

xSol(t) = $C_1 \cos(7\,t)\,e^{-4t} - C_2 \sin(7\,t)\,e^{-4t} + 2$

```
% Now, we have to consider the initial conditions I.C. to eliminate the
% constants C1 and C2 from the solution.
% Specify the second initial condition by assigning diff(x,t) to Dx.
  Dx=diff(x,t);
% Insert the initial by assiging x(0)==10 Dx(0)==4 to cond.
  cond=[x(0)==10,Dx(0)==4];
% Now, Considering I.C. cond as the second input to the dsolve command
% and obtaining the general solution (response equation).
  x(t)=dsolve(eqn,cond)
```

x(t) =

$8\cos(7\,t)\,e^{-4t} + \dfrac{36\sin(7\,t)\,e^{-4t}}{7} + 2$

```
% This equation is the solution of the given differential equation
% generated from MATLAB using classical method.
```

Figure 7.2 MATLAB live script file generating the solution of the given differential equation $\ddot{x} + 8\dot{x} + 65x = 130$

7.3 LAPLACE TRANSFORM METHOD

The Laplace transform method provides an alternative analytical method for solving ordinary linear differential equations. This section will provide an overview of this approach and how it differs from the classical method. It allows us to transform the differential equation into an algebraic equation that efficiently solves the unknown $X(s)$. This algebraic equation is known as a *Transfer function*. It is an analytical expression that describes the ratio between the system's output and input. We then inverse Laplace, transforming the $X(s)$ back into $x(t)$ using the tables to get our final solution.

Before we start outlining the analysis of the Laplace transform method, it is essential to go over the algebra of complex conjugate numbers, which are often one of the roots of the transfer function characteristic equation.

The complex conjugate number can be expressed in rectangular form as

$$x = a \pm ib \tag{7.61}$$

where a and b are the real and imaginary parts, respectively. Note that i is equal to $\sqrt{-1}$. Equation (7.61) indicates that both numbers have the same real, negative, and positive imaginary parts, as shown in Figure 7.3. We can also express the complex number in polar form as

$$x = |x| \angle \pm \phi \tag{7.62}$$

and

$$x = |x| = \sqrt{a^2 + b^2} \quad \text{and} \quad \phi = \tan^{-1} \frac{b}{a} \tag{7.63}$$

where $|x|$ and ϕ are the magnitude and the angle of x. We can express the complex conjugate number indicated in equation (7.61) as

$$x_1 = a + ib = |x| \angle \phi = |x|(\cos \phi + i \sin \phi) \quad \text{and}$$
$$x_2 = a - ib = |x| \angle -\phi = |x|(\cos \phi - i \sin \phi) \tag{7.64}$$

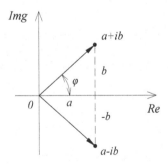

Figure 7.3 Complex numbers

We can express sine and cosine in complex form using Euler's theorem. As a result, we can express $\cos\phi$ and $\sin\phi$, presented in equation (7.64) as

$$(\cos\phi + i\sin\phi) = e^{i\phi} \quad \text{and}$$
$$(\cos\phi - i\sin\phi) = e^{-i\phi} \tag{7.65}$$

Solving this equation for $\cos\phi$ and $\sin\phi$ gives

$$\cos\phi = \frac{e^{i\phi} + e^{-i\phi}}{2} \quad \text{and}$$
$$\sin\phi = \frac{e^{i\phi} - e^{-i\phi}}{2i} \tag{7.66}$$

Now, let us start the analysis of solving the differential equation using the Laplace transform. It is defined by the linear transformation as follows:

$$\mathscr{L}[f(t)] = F(s) = \int_0^\infty f(t)e^{-st}dt \quad \text{for } t > 0 \tag{7.67}$$

where s is an arbitrary complex number $(a + bi)$ and $f(t)$ is of exponential order for $t > 0$. On the other hand, the inverse Laplace transform is defined as

$$\mathscr{L}^{-1}[F(s)] = f(t) = \frac{1}{2\pi i}\int_{b-i\infty}^{b+i\infty} F(s)e^{st}dt \quad \text{for } t > 0 \tag{7.68}$$

These are the formal definitions of the Laplace and inverse Laplace transforms. In practice, conducting a contour integration in the complex plane is only sometimes required, as is the case with the classical technique, unless the situation requires it. Rather than that, a dictionary of Laplace transform pairs is created, and some simple rules are used to convert between the time-domain $f(t)$ and the s-domain (frequency domain) $F(s)$.

The following analysis will find the Laplace transform of some simple functions for output $x(t)$ or input $f(t)$; both are the same.

1. Impulse function $f(t) = I\delta(t)$:

$$F(s) = \int_0^\infty I\delta(t)e^{-st}dt = I \tag{7.69}$$

2. Step function $f(t) = A$:

$$F(s) = \int_0^\infty Ae^{-st}dt = -A\frac{e^{-st}}{s}\bigg|_0^\infty = A(-\frac{e^{-\infty s}}{s} + \frac{e^{-0s}}{s}) \quad \text{for } t > 0 \tag{7.70}$$

So,

$$F(s) = A(-\frac{0}{s} + \frac{1}{s}) = \frac{A}{s} \tag{7.71}$$

3. Ramp function $f(t) = Rt$:

$$F(s) = \int_0^\infty Rt\, e^{-st} dt = R[\frac{e^{-st}}{s^2}(-st-1)]\Big|_0^\infty = \frac{R}{s^2} \qquad (7.72)$$

4. Exponential function $f(t) = e^{-at}$:

$$F(s) = \int_0^\infty e^{-at} e^{-st} dt = \int_0^\infty e^{-(s+a)t} dt = \frac{-1}{(s+a)} e^{-(s+a)t}\Big|_0^\infty = \frac{1}{(s+a)} \qquad (7.73)$$

5. Sinusoidal function $f(t) = \sin(\omega t)$:
Applying superposition gives

$$f(t) = af_1(t) + bf_2(t) \quad \text{and} \quad F(s) = aF_1(s) + bF_2(s) \qquad (7.74)$$

Let us use this relation to obtain $\mathscr{L}\sin(\omega t)$. Using equation (7.66), we get

$$\sin(\omega t) = \frac{e^{i\omega t} - e^{-i\omega t}}{2i} = \frac{e^{i\omega t}}{2i} - \frac{e^{-i\omega t}}{2i} \qquad (7.75)$$

So,

$$f(t) = f_1(t) + f_2(t) = \frac{e^{i\omega t}}{2i} - \frac{e^{-i\omega t}}{2i} \qquad (7.76)$$

Taking the Laplace transform gives

$$F_1(s) = \frac{1}{2i}[\frac{1}{s-i\omega}] \quad \text{and} \quad F_2(s) = \frac{-1}{2i}[\frac{1}{s+i\omega}] \qquad (7.77)$$

and

$$F(s) = \frac{1}{2i}[\frac{1}{s-i\omega} - \frac{1}{s+i\omega}] = \frac{1}{2i}[\frac{s+i\omega-s+i\omega}{s^2+\omega^2}] = \frac{\omega^2}{s^2+\omega^2} \qquad (7.78)$$

6. Time delay function $f(t) = f_1(t-t_o)u(t-t_o)$:

$$F(s) = \int_0^\infty f_1(t-t_o)u(t-t_o)e^{-st} dt = \int_{t_o}^\infty f_1(t-t_o)e^{-st} dt \qquad (7.79)$$

Letting $\tau = (t-t_o)$ and $dt = d\tau$, gives

$$F(s) = \int_0^\infty f_1(\tau)e^{-s(\tau+t_o)} d\tau = e^{-st_o}\int_0^\infty f_1(\tau)e^{-s\tau} d\tau = F_1(s)e^{-st_o} \qquad (7.80)$$

Note that the time delay is given formally using the delayed unit step function, $u(t-t_o)$. Suppose the causal time function is defined as identically zero for a negative argument (i.e., $f(t-t_o) = 0$ for $t < t_o$). In this case, the unit step notation $u(t-t_o)$ is unnecessary. Therefore, we denote such a function as $f(t-t_o)$ and its Laplace transform is $F(s)e^{-st_o}$ where $F(s) = \mathscr{L}[f(t)]$.

7. Differentiation:
- For the first derivative, $f(t) = \dot{f}_1$

$$F(s) = \int_0^\infty \frac{df_1(t)}{dt} e^{-st} dt \tag{7.81}$$

Integrating this equation by parts gives,

$$\text{Let } u = e^{-st}, \; dv = \frac{df_1(t)}{dt} dt \; \text{ and } \; du = -se^{-st} dt \quad v = f_1(t) \tag{7.82}$$

Then,

$$F(s) = f_1(t)e^{-st}\Big|_0^\infty - \int_0^\infty (-se^{-st})f_1(t)dt = -f(0) + sF_1(s) \tag{7.83}$$

- For the second derivative, $f(t) = \ddot{f}_1$:
In the same way, we get,

$$F(s) = s^2 F_1(s) - sf_1(t)\Big|_0 - \dot{f}_1(t)\Big|_0 \tag{7.84}$$

- In general, for n^{th} derivative, we have

$$F(s) = s^n F_1(s) - s^{n-1} f_1(t)\Big|_0 - s^{n-2}\frac{df_1(t)}{dt}\Big|_0 \cdots - \frac{d^{n-1}f_1(t)}{dt^{n-1}}\Big|_0 \tag{7.85}$$

The transformation of a derivative of any order is possible using equation (7.85). It also automatically creates the initial conditions that are required. That is, $f(0)$ and $\dot{f}(0)$, etc., are the conditions of $f(t)$ and its derivatives, whereas $f(t)$ and its derivatives are evaluated at a time just before the deriving input is applied.

8. Integration:

$$f(t) = \int_0^t f_1(\tau)d\tau \tag{7.86}$$

The Laplace transform of this equation gives

$$F(s) = \int_0^\infty [\int_0^t f_1(\tau)d\tau]e^{-st} dt \tag{7.87}$$

Integration by parts gives

$$\text{Let } u = \int_0^t f_1(\tau)d\tau, \; dv = e^{-st}dt \; \text{ and } \; du = f_1(t)dt, \; v = \frac{-1}{s}e^{-st} \tag{7.88}$$

So,

$$F(s) = \frac{-1}{s}e^{-st}\int_0^t f_1(\tau)d\tau\Big|_0^\infty - \int_0^\infty \frac{-1}{s}e^{-st}f_1(t)dt = \frac{1}{s}F_1(s) \tag{7.89}$$

Hundreds of such pairs have been worked out and published as tables: one of the advantages of the Laplace transform method is that we rarely need to use the Table 7.1 and instead look it up in a table instead of using the transform method. As a result, if we know what $F(s)$ is, we can use this table to figure out what $f(t)$ is because the uniqueness of the function guarantees that there is only one $F(s)$ for each $f(t)$.

7.3.1 TRANSFORM AND FREQUENCY DOMAIN METHODS

7.3.1.1 Inverse transforms

Laplace transform solutions to n^{th} order linear time-invariant systems are typically of the form

$$X(s) = \frac{(s-a_1)(s-a_b).....(s-a_m)}{(s-b_1)(s-b_2)....(s-b_n)} = \frac{Z(s)}{P(s)} \qquad (7.90)$$

where a_m and b_n are referred to as zeros and poles of $X(s)$, respectively. It represents the ratio of two polynomials numerator $Z(s)$ (zeros) and denominator $P(s)$ (poles), as indicated in equation (7.90). The denominator $P(s)$ is usually called the characteristic equation of the transfer function $X(s)$.

There are two principal methods for finding the inverse Laplace transform $\mathcal{L}^{-1}X(s) = x(t)$: Partial fraction expansion and the Method of Residues.

7.3.1.2 Inverse Laplace transforms using partial fraction expansion

The following is a breakdown of the parts that comprise the Laplace transform of $f(t)$.

$$X(s) = X_1(s) + X_2(s) + \cdots + X_n(s) \qquad (7.91)$$

Moreover, if the inverse Laplace transform of $X_1(s), X_2(s),..., X_n(s)$ is already available, then

$$\mathcal{L}^{-1}X(s) = \mathcal{L}^{-1}X_1(s) + \mathcal{L}^{-1}X_2(s) + \cdots + \mathcal{L}^{-1}X_n(s) = x_1(t) + x_2(t) + \cdots + x_n(t) \qquad (7.92)$$

where $x_1(t), x_2(t),..., x_n(t)$ are the inverse Laplace transforms of $X_1(s), X_2(s),..., X_n(s)$, respectively. The obtained inverse Laplace transform of $X(s)$ is unique, except possibly at discontinuous locations in the time function. When the time function is continuous, the $x(t)$ and its Laplace transform $X(s)$ have a one-to-one relationship. In general, the form $X(s)$ is when frequent occurrences of the variable occur, as given by equation (7.90), and thus

$$X(s) = \frac{Z(s)}{P(s)} \qquad (7.93)$$

where $Z(s)$ and $P(s)$ are polynomials in s-domain and the degree $P(s)$ is always higher than that of $Z(s)$.

Table 7.1
Inverse Laplace transform table

No	$f(t)$ time-domain	$F(s)$ s-domain
1	Unite impulse $\delta(t)$	1
2	Unite step $1(t)$	$\frac{1}{s}$
3	t	$\frac{1}{s^2}$
4	$\frac{t^{n-1}}{(n-1)!}$ $(n=1,2,3,\dots)$	$\frac{1}{s^n}$
5	t^n $(n=1,2,3,\dots)$	$\frac{n!}{s^{n+1}}$
6	e^{-at}	$\frac{1}{s+a}$
7	te^{-at}	$\frac{1}{(s+a)^2}$
8	$\frac{1}{(n-1)!}t^{n-1}e^{-at}$ $(n=1,2,3,\dots)$	$\frac{1}{(s+a)^n}$
9	$t^n e^{-at}$ $(n=1,2,3,\dots)$	$\frac{n!}{(s+a)^{n+1}}$
10	$\sin\omega t$	$\frac{\omega}{(s^2+\omega^2)}$
11	$\cos\omega t$	$\frac{s}{(s^2+\omega^2)}$
12	$\sinh\omega t$	$\frac{\omega}{(s^2-\omega^2)}$
13	$\cosh\omega t$	$\frac{s}{(s^2-\omega^2)}$
14	$\frac{1}{a}(1-e^{-at})$	$\frac{1}{s(s+a)}$
15	$\frac{1}{b-a}(e^{-at}-e^{-bt})$	$\frac{1}{(s+a)(s+b)}$
16	$\frac{1}{b-a}(be^{-bt}-ae^{-at})$	$\frac{s}{(s+a)(s+b)}$
17	$\frac{1}{ab}[1+\frac{1}{a-b}(be^{-at}-ae^{-bt})]$	$\frac{1}{s(s+a)(s+b)}$
18	$\frac{1}{a^2}(1-e^{-at}-ate^{-at})$	$\frac{1}{s(s+a)^2}$
19	$\frac{1}{a^2}(at-1+e^{-at})$	$\frac{1}{s^2(s+a)}$
20	$e^{-at}\sin\omega t$	$\frac{\omega}{(s+a)^2+\omega^2}$
21	$e^{-at}\cos\omega t$	$\frac{s+a}{(s+a)^2+\omega^2}$
22	$\omega t\sin\omega t$	$\frac{2\omega^2 s}{(s^2+\omega^2)^2}$
23	$\omega t\cos\omega t$	$\frac{\omega(s^2-\omega^2)}{(s^2+\omega^2)^2}$
24	$\frac{\omega_n}{\sqrt{1-\zeta^2}}e^{-\zeta\omega_n t}\sin\omega_n\sqrt{1-\zeta^2}t$	$\frac{\omega_n^2}{s^2+2\zeta\omega_n s+\omega_n^2}$
25	$-\frac{1}{\sqrt{1-\zeta^2}}e^{-\zeta\omega_n t}\sin(\omega_n\sqrt{1-\zeta^2}t-\phi),\ \phi=\tan^{-1}(\frac{\sqrt{1-\zeta^2}}{\zeta})$	$\frac{s}{s^2+2\zeta\omega_n s+\omega_n^2}$
26	$1-\frac{1}{\sqrt{1-\zeta^2}}e^{-\zeta\omega_n t}\sin(\omega_n\sqrt{1-\zeta^2}t+\phi),\ \phi=\tan^{-1}(\frac{\sqrt{1-\zeta^2}}{\zeta})$	$\frac{\omega_n^2}{s(s^2+2\zeta\omega_n s+\omega_n^2)}$
27	$1-\cos\omega t$	$\frac{\omega^2}{s(s^2+\omega^2)}$
28	$\omega t-\sin\omega t$	$\frac{\omega^3}{s^2(s^2+\omega^2)}$
29	$\sin\omega t-\omega t\cos\omega t$	$\frac{2\omega^3}{(s^2+\omega^2)^2}$
30	$\frac{1}{2\omega}t\sin\omega t$	$\frac{s}{(s^2+\omega^2)^2}$
31	$t\cos\omega t$	$\frac{s^2-\omega^2}{(s^2+\omega^2)^2}$
32	$\frac{1}{\omega_2^2-\omega_1^2}(\cos\omega_1 t-\cos\omega_2 t),\ (\omega_1^2\neq\omega_1^2)$	$\frac{s}{(s^2+\omega_1^2)(s^2+\omega_2^2)}$
33	$\frac{1}{2\omega}(\sin\omega_1 t+\omega t\cos\omega t)$	$\frac{s^2}{(s^2+\omega^2)^2}$

7.3.1.3 Laplace and inverse Laplace transforms using MATLAB

The Laplace and inverse Laplace transforms of the transfer functions are carried out with the support of the two primary MATLAB commands known as *laplace* and *ilaplace*. A symbolic expression, function, vector, or matrix must be provided as the input functions for both commands. To get further information, type "help laplace" or "help ilaplace" and see the results on the command window.

The following examples show how to use partial fraction expansions to find the time-domain solution (inverse Laplace transform) to transfer functions with various types of characteristic equation roots. To be comparable, we will use the same transfer functions given in Examples (7.1), (7.2), and (7.3).

Example (7.4) (Repeated roots):
Considering the same differential equation presented in Example (7.1), equation (7.17),

$$5\ddot{x} + 20\dot{x} + 20x = 28 \quad \text{if } x(0) = 5 \text{ and } \dot{x}(0) = 8 \tag{7.94}$$

Obtain the transfer function $X(s)$ and then solve for the time domain solution $x(t)$. Compare the results with equation (7.30) (classical method) and verify them using MATLAB.

Solution:
Applying the Laplace transform to the given differential equation gives

$$5[s^2X(s) - x(0)s - \dot{x}(0)] + 20[sX(s) - x(0)] + 20X(s) = \frac{28}{s} \tag{7.95}$$

Considering the initial conditions, equation (7.95) becomes

$$5[s^2X(s) - 5*s - 8] + 20[sX(s) - 5] + 20X(s) = \frac{28}{s} \tag{7.96}$$

Rearranging this equation gives

$$s(5s^2 + 20s + 20)X(s) = 25s^2 + 140s + 28 \tag{7.97}$$

So,

$$X(s) = \frac{25s^2 + 140s + 28}{s(5s^2 + 20s + 20)} = \frac{1}{5}\frac{25s^2 + 140s + 28}{s(s^2 + 4s + 4)} \tag{7.98}$$

The characteristic equation of this transfer function has one root equal to zero and two repeated roots: $s_1 = 0$, $s_2 = -2$, and $s_3 = -2$. Considering these, equation (7.98) becomes

$$X(s) = \frac{1}{5}\frac{25s^2 + 140s + 28}{s(s+2)^2} \tag{7.99}$$

Taking the partial fraction expansions of this equation and clearing fractions gives

$$X(s) = \frac{1}{5}[\frac{A}{s} + \frac{B}{(s+2)} + \frac{C}{(s+2)^2}] = \frac{(A+B)s^2 + (4A+2B+C)s + 4A}{s(s+2)^2} \tag{7.100}$$

```
% Solution of the differential equation using Laplace and inverse Laplace transforms.
% Use syms command to create symbolic scalar variables and functions. For more details
% about this command, on the command window, write help syms and see the results.
syms s t x(t) X(s)
% Specify the first and second derivatives of x using D1x=diff(x) and D2x=diff(D1x).
D1x=diff(x);
D2x=diff(D1x);
% write the given differential equation
Eqn=5*D2x+20*D1x+20*x==28
```

Eqn(t) =

$$5\frac{\partial^2}{\partial t^2}x(t) + 20\frac{\partial}{\partial t}x(t) + 20\,x(t) = 28$$

```
% Using laplace command to obtain the laplace transform of the above equation.
LEqn=laplace(Eqn)
```

LEqn =

$$20\,s\,\text{laplace}(x(t),t,s) - 20\,x(0) - 5\,s\,x(0) + 5\,s^2\,\text{laplace}(x(t),t,s) - 5\left(\left(\frac{\partial}{\partial t}x(t)\right)\Big|_{t=0}\right) + 20\,\text{laplace}(x(t),t,\ldots$$

```
% Use subs command and consider the initial conditions to replace the above equation to
% a symbolic expression only in s.
LEqn=subs(LEqn, {laplace(x(t),t,s), x(0), D1x(0)}, {X(s), 5, 8})
```

LEqn =

$$20\,X(s) - 25\,s + 20\,s\,X(s) + 5\,s^2\,X(s) - 140 = \frac{28}{s}$$

```
% Use isolate command to obtain X(s)
LEqn=isolate(LEqn, X(s))
```

LEqn =

$$X(s) = \frac{25\,s + \frac{28}{s} + 140}{5\,s^2 + 20\,s + 20}$$

```
% Use simplifyFraction command to simplify X(s)
LEqn1=simplifyFraction(LEqn)
```

LEqn1 =

$$X(s) = \frac{25\,s^2 + 140\,s + 28}{5\,s\,(s^2 + 4\,s + 4)}$$

```
% Now, use ilaplace command to obtain the response equation x(t)
x(t)=ilaplace(rhs(LEqn1))
```

x(t) =

$$\frac{18\,e^{-2t}}{5} + \frac{76\,t\,e^{-2t}}{5} + \frac{7}{5}$$

Figure 7.4 MATLAB live script file generating the solution of the given differential equation $5\ddot{x} + 20\dot{x} + 20x = 28$

Solve this equation for A, B, and C as follows. Equating like terms of equation (7.100) give

$$s^2 : \quad A + B = 25 \tag{7.101}$$

$$s : \quad 4A + 2B + C = 140 \tag{7.102}$$

$$0 : \quad 4A = 28 \text{ so } A = \frac{28}{4} = 7 \tag{7.103}$$

Substituting the value of A indicated in equation (7.103) in equations (7.101) and (7.102) and solving for B and C gives

$$B = 18 \text{ and } C = 76 \tag{7.104}$$

Substituting the values of A, B and C in equation (7.100) gives

$$X(s) = \frac{1}{5}[\frac{7}{s} + \frac{18}{(s+2)} + \frac{76}{(s+2)^2}] \tag{7.105}$$

From Table 7.1, items (2), (5), and (6), taking the inverse Laplace transform of equation (7.105) gives

$$x(t) = \frac{7}{5} + \frac{18}{5}e^{-2t} + \frac{76}{5}te^{-2t} \tag{7.106}$$

This equation is the time-domain solution of the given differential equation. Comparing this equation with equation (7.30), we found they are the same.

Let us generate the solution using MATLAB. Figure 7.4 shows the MATLAB live script file and the generated results. It generates the transfer function $X(s)$ and the solution $x(t)$. The script provides an explanation for each step it contains. Comparing equation (7.106) with the generated $x(t)$ presented in Figure 7.4, we found they are the same.

Example (7.5) (Real roots):
Considering the same differential equation presented in Example (7.2), equation (7.31),

$$\frac{d^2x}{dt^2} + 3\frac{dx}{dt} + 2x = 4e^{-5t} \text{ when } x(0) = 2 \text{ and } \dot{x}(0) = 0 \tag{7.107}$$

obtain the transfer function $X(s)$ and then solve for $x(t)$ and compare the results with equation (7.46) (classical method).

Solution:
Applying the Laplace transform to the given differential equation gives

$$[s^2X(s) - x(0)s - \dot{x}(0)] + 3[sX(s) - x(0)] + 2X(s) = \frac{4}{s+5} \tag{7.108}$$

Considering the initial conditions, equation (7.108) becomes

$$s^2X(s) - 2*s - 0 + 3sX(s) - 3*2 + 2X(s) = \frac{4}{s+5} \tag{7.109}$$

So,

$$(s+5)(s^2 + 3s + 2)X(s) = 2s^2 + 16s + 34 \tag{7.110}$$

Rearranging this equation gives

$$X(s) = \frac{2s^2 + 16s + 34}{(s+5)(s^2 + 3s + 2)} \tag{7.111}$$

The characteristic equation of this equation has three real roots: $s_1 = -5$, $s_2 = -2$, and $s_3 = -1$. Considering these, equation (7.111) becomes

$$X(s) = \frac{2s^2 + 16s + 34}{(s+5)(s+2)(s+1)} \tag{7.112}$$

Taking the partial fraction expansions of this equation gives

$$X(s) = \frac{2s^2 + 16s + 34}{(s+5)(s^2 + 3s + 2)} = \frac{A}{(s+5)} + \frac{B}{(s+2)} + \frac{C}{(s+1)} \tag{7.113}$$

Solving this equation for A, B, and C gives

$$A = \frac{2s^2 + 16s + 34}{(s+2)(s+1)}\bigg|_{s=-5} = \frac{4}{12} = \frac{1}{3} \tag{7.114}$$

$$B = \frac{2s^2 + 16s + 34}{(s+5)(s+1)}\bigg|_{s=-2} = -\frac{10}{3} \tag{7.115}$$

$$C = \frac{2s^2 + 16s + 34}{(s+5)(s+2)}\bigg|_{s=-1} = \frac{20}{4} = 5 \tag{7.116}$$

Substituting the values of A, B and C in equation (7.113) gives

$$X(s) = \frac{1}{3}\frac{1}{(s+5)} - \frac{10}{3}\frac{1}{(s+2)} + 5\frac{1}{(s+1)} \tag{7.117}$$

Using table (7.1), item (6), taking the inverse Laplace transform of this equation gives

$$x(t) = \frac{1}{3}e^{-5t} - \frac{10}{3}e^{-2t} + 5e^{-t} \tag{7.118}$$

This equation is the time-domain solution of the given differential equation. Comparing this equation with equation (7.46), we found they are the same.

Example (7.6) (Complex roots):
Considering the same differential equation presented in Example (7.3), equation (7.47),

$$\frac{d^2x}{dt^2} + 8\frac{dx}{dt} + 65x = 130 \quad \text{when } x(0) = 10 \text{ and } \dot{x}(0) = 4 \tag{7.119}$$

Obtain the transfer function $X(s)$ and then solve for $x(t)$. Compare the results with equation (7.60) (classical method) and verify the results using MATLAB.

Solution:
Applying the Laplace transform to the given differential equation gives

$$[s^2X(s) - x(0)s - \dot{x}(0)] + 8[X(s) - x(0)] + 65X(s) = \frac{130}{s} \tag{7.120}$$

Considering the initial conditions, equation (7.120) becomes

$$s^2X(s) - 10s - 4 + 8X(s) - 8*10 + 65X(s) = \frac{130}{s} \tag{7.121}$$

Rearranging this equation gives

$$s(s^2 + 8s + 65)X(s) = 10s^2 + 84s + 130 \tag{7.122}$$

So,

$$X(s) = \frac{10s^2 + 84s + 130}{s(s^2 + 8s + 65)} \tag{7.123}$$

Taking the partial fraction expansions of this equation gives

$$\frac{10s^2 + 84s + 130}{s(s^2 + 8s + 65)} = \frac{A}{s} + \frac{Bs + C}{(s^2 + 8s + 65)} \tag{7.124}$$

Clearing fractions gives

$$X(s) = \frac{As^2 + 8As + 65A + Bs^2 + Cs}{s(s^2 + 8s + 65)} = \frac{(A+B)s^2 + (8A+C)s + 65A}{s(s^2 + 8s + 65)} \tag{7.125}$$

Solving this equation for A, B, and C is as follows. Equating like terms gives

$$s^2 : A + B = 10 \tag{7.126}$$

$$s : 8A + C = 84 \tag{7.127}$$

$$0 : 65A = 130 \text{ so } A = \frac{130}{65} = 2 \tag{7.128}$$

Substituting the value of A in equations (7.126) and (7.127) and solving for B and C, we get

$$B = 10 - 2 = 8 \text{ and } C = 84 - 16 = 68 \tag{7.129}$$

Substituting values of A, B and C in equation (7.124) gives

$$X(s) = \frac{2}{s} + 8\frac{s}{s^2 + 8s + 65} + 68\frac{1}{s^2 + 8s + 65} \tag{7.130}$$

In the characteristic equation of the given transfer function, there are two distinct complex roots that can be determined as follows:

$$s_{1,2} = \frac{-8 \pm \sqrt{8^2 - 4(1)(65)}}{2} = \frac{-8 \pm \sqrt{64 - 260}}{2} \tag{7.131}$$

$$s_{1,2} = \frac{-8 \pm 14i}{2} = -a \pm bi = -4 \pm 7i \tag{7.132}$$

Considering these roots, equation (7.130) becomes

$$X(s) = \frac{2}{s} + 8\frac{s}{(s+4)^2 + 7^2} + 68\frac{1}{(s+4)^2 + 7^2} \tag{7.133}$$

Using Table 7.1, items (2), (10), and (11), take the inverse Laplace transform of this equation. For the second and third terms, we need to rearrange them to use the inverse Laplace transform table, items (10) and (11) so that they look like this instead:

$$X(s) = \frac{2}{s} + 8\frac{s+a-a}{(s+4)^2 + 7^2} + 68\frac{1}{b}\frac{b}{(s+4)^2 + 7^2} = \frac{2}{s} + 8\frac{s+a}{(s+4)^2 + 7^2} +$$
$$(68 - 8a)\frac{1}{b}\frac{b}{(s+4)^2 + 7^2} \tag{7.134}$$

```
% Solution of the differential equation using Laplace and inverse Laplace transform.
% Use syms command to create symbolic scalar variables and functions. For more details
% about this command, on the command window, write help syms and see the results.
  syms s t x(t) X(s)
% Specify the first and order order derivative of x using D1x=diff(x) and D2x=diff(D1x).
  D1x=diff(x);
  D2x=diff(D1x);
% write the given differential equation
  Eqn=D2x+8*D1x+65*x==130
```

Eqn(t) =

$$\frac{\partial^2}{\partial t^2} x(t) + 8 \frac{\partial}{\partial t} x(t) + 65 x(t) = 130$$

```
% Using laplace command to obtain the laplace transform of the above equation.
LEqn=laplace(Eqn)
```

LEqn =

$$8 s \, \mathrm{laplace}(x(t), t, s) - 8 x(0) - s x(0) + s^2 \, \mathrm{laplace}(x(t), t, s) - \left(\left(\frac{\partial}{\partial t} x(t) \right)\Big|_{t=0} \right) + 65 \, \mathrm{laplace}(x(t), t, s) = \frac{130}{s}$$

```
%  Use subs command and consider the initial conditions to replace the above equation to
% a symbolic expression only in S.
LEqn=subs(LEqn, {laplace(x(t),t,s), x(0), D1x(0)}, {X(s), 10, 4})
```

LEqn =

$$65 X(s) - 10 s + 8 s X(s) + s^2 X(s) - 84 = \frac{130}{s}$$

```
%  Use isolate command to obtain X(s)
LEqn=isolate(LEqn, X(s))
```

LEqn =

$$X(s) = \frac{10 s + \frac{130}{s} + 84}{s^2 + 8 s + 65}$$

```
%  Use simplifyFraction command to simplify X(s)
LEqn1=simplifyFraction(LEqn)
```

LEqn1 =

$$X(s) = \frac{2 \, (5 s^2 + 42 s + 65)}{s \, (s^2 + 8 s + 65)}$$

```
% Now, use ilaplace command to obtain the response equation x(t)
x(t)=ilaplace(rhs(LEqn1))
```

x(t) =

$$8 e^{-4t} \left(\cos(7 t) + \frac{9 \sin(7 t)}{14} \right) + 2$$

Figure 7.5 MATLAB live script file generating the solution of the given differential equation $\frac{d^2 x}{dt^2} + 8 \frac{dx}{dt} + 65 x = 130$

where $a = 4$ and $b = 7$. Substituting these values in equation (7.134) and rearranging the results gives

$$X(s) = \frac{2}{s} + 8 \frac{s+4}{(s+4)^2 + 7^2} + \frac{36}{7} \frac{7}{(s+4)^2 + 7^2} \qquad (7.135)$$

Using Table 7.1, items (10) and (11), taking the inverse Laplace transform of this equation gives

$$x(t) = 2 + 8 e^{-4t} \cos(7t) + \frac{36}{7} e^{-4t} \sin(7t) \qquad (7.136)$$

This equation is the time-domain solution of the given differential equation. Compare this equation with equation (7.60), classical method; we found they are the same.

Let us generate the solution using MATLAB. Figure (7.5) shows the MATLAB live script file and the generated results. It generates the transfer function $X(s)$ and the solution $x(t)$. Comparing equation (7.136) with the generated $x(t)$ presented in Figure 7.5, we found they are the same.

Example (7.7) (Non-repeated roots):
Considering the following transfer function,

$$X(s) = \frac{(s^2+1)}{(s+1)(s+3)(s+4)} \tag{7.137}$$

Find the time-domain solution $x(t)$ and verify the results using MATLAB.

Solution:
The partial fraction expansion of equation (7.137) is given by

$$X(s) = \frac{(s^2+1)}{(s+1)(s+3)(s+4)} = \frac{A}{(s+1)} + \frac{B}{(s+3)} + \frac{C}{(s+4)} \tag{7.138}$$

Solving this equation for A, B, and C gives

$$A = \frac{(s^2+1)}{(s+3)(s+4)}\bigg|_{s=-1} = \frac{2}{(2*3)} = \frac{1}{3} \tag{7.139}$$

$$B = \frac{(s^2+1)}{(s+1)(s+4)}\bigg|_{s=-3} = \frac{10}{(-2)*(1)} = -5 \tag{7.140}$$

$$C = \frac{(s^2+1)}{(s+1)(s+3)}\bigg|_{s=-4} = \frac{17}{(-3)*(-1)} = \frac{17}{3} \tag{7.141}$$

Substituting values of A, B and C in equation (7.138) gives

$$X(s) = \frac{(s^2+1)}{(s+1)(s+3)(s+4)} = \frac{1}{3}\frac{1}{(s+1)} - 5\frac{1}{(s+3)} + \frac{17}{3}\frac{1}{(s+4)} \tag{7.142}$$

Using Table 7.1 item (6), we can obtain the inverse Laplace transform of equation (7.142) as follows.

$$x(t) = \frac{1}{3}e^{-t} - 5e^{-3t} + \frac{17}{3}e^{-4t} \tag{7.143}$$

This equation is the time-domain solution of the transfer function indicated in equation (7.137). Now, let us verify equation (7.143) using MATLAB. The MATLAB live script and the generated results are shown in Figure 7.6. Comparing equation (7.143) with the MATLAB results shown in this Figure 7.6, we found they are the same.

Example (7.8) (Repeated roots):
Considering the following transfer function,

$$X(s) = \frac{s^2}{(s+1)^2(s+2)} \tag{7.144}$$

Find the time-domain solution $x(t)$ and verify the results using MATLAB.

```
% Write the symbolic form of the given transfer function as follows
syms s;
Xs=(s^2+1)/((s^2+4*s+3)*(s+4))
```

Xs =

$$\frac{s^2+1}{(s+4)\,(s^2+4\,s+3)}$$

```
% Use ilaplace command to generate the time-domain solution.
xt=ilaplace(Xs)
```

xt =

$$\frac{e^{-t}}{3} - 5\,e^{-3t} + \frac{17\,e^{-4t}}{3}$$

Figure 7.6 MATLAB live script file for solving the transfer function $X(s) = \frac{(s^2+1)}{(s+1)(s+3)(s+4)}$

Solution:
Applying the partial fraction expansion of the given transfer function gives

$$X(s) = \frac{s^2}{(s+1)^2(s+2)} = \frac{A}{(s+1)} + \frac{B}{(s+1)^2} + \frac{C}{(s+2)} \qquad (7.145)$$

Solving this equation for A, B, and C gives

$$B = \frac{s^2}{(s+2)}\bigg|_{s=-1} = \frac{1}{(1)} = 1 \qquad (7.146)$$

$$C = \frac{s^2}{(s+1)^2}\bigg|_{s=-2} = \frac{4}{(1)} = 4 \qquad (7.147)$$

$$A = \frac{s^2}{(s+1)(s+2)}\bigg|_{s=-1} = \frac{1}{(0*1)} = \frac{1}{0} \text{ we cannot do this} \qquad (7.148)$$

Instead, multiplying both sides of equation (7.145) by $(s+1)^2$ and rearranging the results gives

$$(s+1)^2 X(s) = \frac{s^2}{(s+2)} = A(s+1) + B + C\frac{(s+1)^2}{(s+2)} \qquad (7.149)$$

Taking the derivative of this equation and evaluating it at $s = -1$ gives

$$\frac{d}{ds}[(s+1)^2 X(s)]\bigg|_{s=-1} = [\frac{2s}{(s+2)} - \frac{s^2}{(s+2)^2}]\bigg|_{s=-1} = A + 0 + C[\frac{2(s+1)}{(s+2)} - \frac{(s+1)^2}{(s+2)^2}]\bigg|_{s=-1}$$
$$(7.150)$$

Then,

$$A = \frac{2s(s+2) - s^2}{(s+1)^2}\Big|_{s=-1} = \frac{s^2 + 4s}{(s+2)^2}\Big|_{s=-1} = \frac{1-4}{1} = -3 \qquad (7.151)$$

Substituting values of A, B, and C in equation (7.145), yields

$$X(s) = \frac{s^2}{(s+1)^2(s+2)} = \frac{-3}{(s+1)} + \frac{1}{(s+1)^2} + \frac{4}{(s+2)} \qquad (7.152)$$

Taking the inverse Laplace transform of this equation using Table 7.1 items (6) and (7), the time-domain solution of the given transfer function is given by

$$x(t) = -3e^{-t} + te^{-t} + 4e^{-2t} \qquad (7.153)$$

Since the characteristic equation of the given transfer function contains two repeated roots, let us verify equation (7.153) using an alternate method for obtaining the factors A, B, and C. Rearranging equation (7.145) as follows

$$X(s) = \frac{s^2}{(s+1)^2(s+2)} = \frac{A(s+1)(s+2) + B(s+2) + C(s+2)^2}{(s+1)^2(s+1)} \qquad (7.154)$$

Clearing fractions gives

$$X(s) = \frac{A(s^2 + 3s + 2) + B(s+2) + C(s^2 + 4s + 4)}{(s+1)^2(s+1)} \qquad (7.155)$$

From this equation, we have;

$$\frac{s^2}{(s+1)^2(s+2)} = \frac{(A+C)s^2 + (3A + B + 4C)s + (2A + 2B + 4C)}{(s+1)^2(s+1)} \qquad (7.156)$$

Equating the factors of both sides of this equation yields,

$$s^2 : A + C = 1 \qquad (7.157)$$

$$s : 3A + B + 4C = 0 \qquad (7.158)$$

$$0 : 2A + 2B + 4C = 0 \qquad (7.159)$$

Solving these equations for A, B, and C yields

$$A = -3, \ B = 1 \ \text{and} \ C = 4 \qquad (7.160)$$

Substituting these values in equation (7.145) and taking the inverse Laplace transform using Table 7.1 items (6) and (7), the time-domain solution of the given transfer function is given by

$$x(t) = -3e^{-t} + te^{-t} + 4e^{-2t} \qquad (7.161)$$

We get the same results comparing equations (7.161) and (7.153). Let us verify equation (7.161) using MATLAB. The MATLAB live script and the generated results

```
% Write the symbolic form of the given transfer function as follows
syms s;
Xs=(s^2)/((s^2+2*s+1)*(s+2))
```

Xs =

$$\frac{s^2}{(s+2)\,(s^2+2\,s+1)}$$

```
% Use ilaplace command to generate the time-domain solution.
xt=ilaplace(Xs)
```

xt = $4\,e^{-2t} - 3\,e^{-t} + t\,e^{-t}$

Figure 7.7 MATLAB live script file for solving the transfer function $X(s) = \frac{s^2}{(s+1)^2(s+2)}$

are shown in Figure 7.7. Comparing equation (7.161) and the MATLAB results, we found they are the same.

Example (7.9) (Complex roots):
Considering the following transfer function,

$$X(s) = \frac{2s^2 + 6s + 6}{(s+2)(s^2 + 2s + 2)} \tag{7.162}$$

Find the time-domain solution $x(t)$ and verify the results using MATLAB.

Solution:
The partial fraction expansion of the given transfer function is given by

$$X(s) = \frac{2s^2 + 6s + 6}{(s+2)(s^2 + 2s + 2)} = \frac{A}{(s+2)} + \frac{Bs + C}{(s^2 + 2s + 2)} \tag{7.163}$$

Rearranging this equation gives

$$X(s) = \frac{2s^2 + 6s + 6}{(s+2)(s^2 + 2s + 2)} = \frac{A(2s^2 + 6s + 6) + (Bs + C)(s+2)}{(s+2)(s^2 + 2s + 2)} \tag{7.164}$$

Clearing fractions gives

$$X(s) = \frac{2s^2 + 6s + 6}{(s+2)(s^2 + 2s + 2)} = \frac{(A+B)s^2 + (2A + 2B + C)s + (2A + 2C)}{(s+2)(s^2 + 2s + 2)} \tag{7.165}$$

Equating the factors of both sides of this equation yields,

$$s^2 : A + B = 2 \tag{7.166}$$

$$s : 2A + 2B + C = 6 \tag{7.167}$$

$$0 : 2A + 2B = 6 \tag{7.168}$$

```
% Write the symbolic form of the given transfer function as follows
syms s;
Xs=(2*s^2+6*s+6)/((s+2)*(s^2+2*s+2))
```

Xs =

$$\frac{2s^2 + 6s + 6}{(s+2)(s^2 + 2s + 2)}$$

```
% Generate the time-domain solution using ilaplace command
xt=ilaplace(Xs)
```

xt = $e^{-2t} + e^{-t}(\cos(t) + \sin(t))$

Figure 7.8 MATLAB live script file for solving the transfer function $X(s) = \frac{2s^2+6s+6}{(s+2)(s^2+2s+2)}$

Solving these equations for A, B, and C yields

$$A = 1, \quad B = 1 \quad \text{and} \quad C = 2 \tag{7.169}$$

Substituting these values in equation (7.163) gives

$$X(s) = \frac{1}{(s+2)} + \frac{s+2}{(s^2+2s+2)} \tag{7.170}$$

The roots of the denominator of the second term of this equation are complex conjugate, which are given by

$$s_{1,2} = -1 \pm i \tag{7.171}$$

Now, we can rewrite the denominator of the second term of equation (7.170) as follows

$$(s^2 + 2s + 2) = (s+1)^2 + (1)^2 \tag{7.172}$$

Considering this, equation (7.170) becomes

$$X(s) = \frac{1}{(s+2)} + \frac{s+2}{(s+1)^2 + (1)^2} \tag{7.173}$$

Rearranging the second term of this equation to be able to use the inverse Laplace transform table (7.1) items (20) and (21) as follows

$$X(s) = \frac{1}{s+2} + \frac{s+2}{(s+1)^2 + 1^2} = \frac{1}{s+2} + \frac{(s+1)}{(s+1)^2 + 1^2} + \frac{1}{(s+1)^2 + 1^2} \tag{7.174}$$

Taking the inverse Laplace transform of this using Table 7.1 items (6), (10), and (11), the time-domain solution of the given transfer function is given by

$$x(t) = e^{-2t} + e^{-t}\cos(t) + e^{-t}\sin(t) \tag{7.175}$$

Let us verify this equation using MATLAB. The MATLAB live script and the generated results are shown in Figure 7.8. Comparing equation (7.175) with the MATLAB results, we found they are the same.

7.3.1.4   Inverse Laplace transforms using residue partial fraction expansion (method of residues)

If $X(s)$ is a ratio of polynomials in s, then

$$\mathscr{L}^{-1}[X(s)] = \sum \text{residues of } (X(s)e^{st}) \tag{7.176}$$

where the residue of an nth order pole at $s = s_1$ is given by

$$R_{s_1} = \frac{1}{(n-1)!}[\frac{d^{n-1}}{ds^{n-1}}(s-s_1)^n X(s)e^{st}]\Big|_{s=s_1} \tag{7.177}$$

- First-order pole:

$$R_{s_1} = [(s-s_1)X(s)e^{st}]\Big|_{s=s_1} \tag{7.178}$$

- Second-order pole:

$$R_{s_2} = [\frac{d}{ds}(s-s_2)^2 X(s)e^{st}]\Big|_{s=s_2} \tag{7.179}$$

7.3.1.5   Method of residues using MATLAB

The residue function of MATLAB can be used to compute the partial fraction expansion of a ratio of two polynomials. It provides three types of data:

1. The residues are given in the output vector r.
2. The poles are given in the output vector p, the roots of the characteristic equation (the denominator).
3. The so-called direct terms are given in the output vector k.

Three cases of the direct term k are provided depending on the order of the numerator polynomial:

- If it is less than the order of the denominator polynomial, there will be no direct terms.
- If it equals the order of the denominator, there will be one direct term.
- If there is one greater than the order of the denominator, there will be two direct terms.

The MATLAB script file generates the partial fraction expansion using the Method of Residues. The MATLAB begins by dividing the numerator by the denominator, creating a polynomial in s (the transfer function), denoted as a row vector k. After MATLAB extends the residual into partial fractions, they are returned as column vector r, and the pole positions are replaced as column p. The next step, outside

MATLAB, is to use the generated values for r, p, and k to represent partial fraction expansion of the given transfer function and then take the inverse Laplace transform to obtain the solution. Write "help residue" on the MATLAB command window and see the results for more information.

The following examples demonstrate how to use the residue method to find the time-domain solution to transfer functions with different kinds of characteristic equation roots. To be comparable, we will use the same transfer functions given in Examples (7.7), (7.8), and (7.9).

Example (7.10) (Non-repeated roots):
Considering the same transfer function presented in Example (7.7), equation (7.137), which is

$$X(s) = \frac{(s^2+1)}{(s+1)(s+3)(s+4)} \tag{7.180}$$

Expand the transfer function using the Method of Residues and determine the time-domain solution $x(t)$. Verify the results using MATLAB.

Solution:
Applying the Method of Residues, equation (7.176), to equation (7.180) gives

$$x(t) = \left.\frac{(s^2+1)}{(s+3)(s+4)}e^{st}\right|_{s=-1} + \left.\frac{(s^2+1)}{(s+1)(s+4)}e^{st}\right|_{s=-3} + \left.\frac{(s^2+1)}{(s+1)(s+3)}e^{st}\right|_{s=-4} \tag{7.181}$$

By solving this equation for $x(t)$, the time-domain solution of the given transfer function is given by

$$x(t) = \frac{1}{3}e^{-t} - 5e^{-3t} + \frac{17}{3}e^{-4t} \tag{7.182}$$

The MATLAB live script file that generates the results of the partial fraction expansion of the given transfer function using the Method of Residues is shown in Figure 7.9. The results indicated that the roots p of the characteristic equation are three real roots and no direct term in this partial fraction expansion since the raw vector k is empty. Considering these, we get the following equation:

$$X(s) = \frac{r_1}{(s+p_1)} - \frac{r_2}{(s+p_2)} + \frac{r_3}{(s+p_3)} = \frac{5.667}{(s+4)} - \frac{5}{(s+3)} + \frac{0.333}{(s+1)} \tag{7.183}$$

Using table (7.1), item (6), taking the inverse Laplace transform of this equation gives

$$x(t) = 5.667e^{-4t} - 5e^{-3t} + 0.333e^{-t} \tag{7.184}$$

This equation is the time-domain solution of the given transfer function using the Method of Residues. Comparing equation (7.184) with equation (7.182), we found they are the same.

```
% Partial fraction expansion using Method of Residues.
% Write the given transfer function as follows
Z=[1 0 1];
P=conv([1 4 3],[1 4]);
xs=tf(Z,P)
```

```
xs =

        s^2 + 1
  -------------------------
  s^3 + 8 s^2 + 19 s + 12
```

Continuous-time transfer function.

```
% Use Residue command
[r,p,k]=residue(Z,P)
```

```
r = 3×1
   5.6667
  -5.0000
   0.3333
p = 3×1
  -4.0000
  -3.0000
  -1.0000
k =

  []
```

Figure 7.9 MATLAB live script file generates the partial fraction expansion for the transfer function, $X(s) = \frac{(s^2+1)}{(s+1)(s+3)(s+4)}$, using the Method of Residues

Example (7.11) (Repeated roots):
Considering the same transfer function presented in Example (7.8), equation (7.144), which is

$$X(s) = \frac{s^2}{(s+1)^2(s+2)} \tag{7.185}$$

Expand the transfer function using the Method of Residues and determine the time-domain solution. Verify the results using MATLAB.

Solution:
Applying the Method of Residues, equation (7.176), to equation (7.185) gives

$$x(t) = \frac{s^2}{(s+1)^2} e^{st} \Big|_{s=-2} + \frac{d}{ds} \left[\frac{s^2}{(s+2)} e^{st} \right] \Big|_{s=-1} \tag{7.186}$$

$$x(t) = \frac{s^2}{(s+1)^2} e^{st} \Big|_{s=-2} + \left(\left[\frac{2s}{(s+2)} - \frac{s^2}{(s+2)^2} \right] e^{st} + \frac{s^2}{(s+2)} t e^{st} \right) \Big|_{s=-1} \tag{7.187}$$

Evaluating this equation and using table (7.1), items (6) and (7), give

$$x(t) = 4e^{-2t} - 3e^{-t} + te^{-t} \tag{7.188}$$

This equation is the time-domain solution of the given transfer function using the Method of Residues. Figure 7.10 shows the MATLAB live script file that generates the results of the partial fraction expansion of the given transfer function using the

```
% Partial fraction expansion using Method of Residues.
% Write the given transfer function as follows
Z=[1 0 0];
P=conv([1 2 1],[1 2]);
xs=tf(Z,P)
```

xs =

```
          s^2
    --------------------
    s^3 + 4 s^2 + 5 s + 2
```

Continuous-time transfer function.

```
% Use Residue command
[r,p,k]=residue(Z,P)
```

r = 3×1
    4.0000
   -3.0000
    1.0000
p = 3×1
   -2.0000
   -1.0000
   -1.0000
k =

    []

Figure 7.10 MATLAB live script file generates the partial fraction expansion for the transfer function, $X(s) = \frac{s^2}{(s+1)^2(s+2)}$, using the Method of Residues

Method of Residues. The results indicated that the roots p of the characteristic equation have two repeated roots and one real root, and there is no direct term in this partial fraction expansion since the raw vector k is empty. Considering these, we get

$$X(s) = \frac{r_1}{(s+p_1)} - \frac{r_2}{(s+p_2)} + \frac{r_3}{(s+p_2)^2} = \frac{4}{(s+2)} - \frac{3}{(s+1)} + \frac{1}{(s+1)^2} \quad (7.189)$$

Using table (7.1), items (6) and (7), taking the inverse Laplace transform of this equation gives

$$x(t) = 4e^{-2t} - 3e^{-t} + te^{-t} \quad (7.190)$$

Comparing equation (7.190) with equation (7.188), we found they are the same.

Example (7.12) (Complex roots):
Considering the same transfer function presented in Example (7.9), which is

$$X(s) = \frac{2s^2 + 6s + 6}{(s+2)(s^2 + 2s + 2)} \quad (7.191)$$

Expand the transfer function using the Method of Residues and determine the time-domain solution. Verify the results using MATLAB.

Solution:

The roots of the second term of the denominator of equation (7.191) are

$$s_{1,2} = -1 \pm i \tag{7.192}$$

Therefore, the given transfer function $X(s)$ can be written in terms of linear factors as follows:

$$X(s) = \frac{2s^2 + 6s + 6}{(s+2)[s-(-1+i)][s-(-1-i)]} \tag{7.193}$$

Apply the Method of Residues to this equation as follows

$$x(t) = \frac{2s^2 + 6s + 6}{(s^2 + 2s + 2)} e^{st}\Big|_{s=-2} + \frac{2s^2 + 6s + 6}{(s+2)[s-(-1-i)]} e^{st}\Big|_{s=-1+i}$$

$$+ \frac{2s^2 + 6s + 6}{(s+2)[s-(-1+i)]} e^{st}\Big|_{s=-1-i} \tag{7.194}$$

Expand the first term of this equation, which gives

$$x(t) = \frac{2s^2 + 6s + 6}{(s^2 + 2s + 2)} e^{st}\Big|_{s=-2} = e^{-2t} \tag{7.195}$$

Expand the second term as follows

$$\frac{2s^2 + 6s + 6}{(s+2)[s-(-1-i)]} e^{st}\Big|_{s=-1+i} = \frac{2(-1+i)^2 + 6(-1+i) + 6}{(1+i)[-1+i+1+i]} e^{(-1+i)t} =$$

$$\frac{2(1-2i-1) - 6 + 6i + 6}{2i-2} e^{(-1+i)t} = \frac{2i}{2i-2} e^{(-1+i)t} = \frac{i(-1-i)}{-1+i(-1-i)} e^{(-1+i)t} =$$

$$\frac{1-i}{1+1} e^{(-1+i)t} = \frac{1}{2}(1-i)e^{(-1+i)t} \tag{7.196}$$

Expand the third term in the same manner gives

$$\frac{2s^2 + 6s + 6}{(s+2)[s-(-1+i)]} e^{st}\Big|_{s=-1-i} = \frac{1}{2}(1+i)e^{(-1-i)t} \tag{7.197}$$

Substituting equations (7.189), (7.190) and (7.191) gives

$$x(t) = e^{-2t} + \frac{1}{2}(1-i)e^{(-1+i)t} + \frac{1}{2}(1+i)e^{(-1-i)t} =$$

$$e^{-2t} + \frac{1}{2}e^{-t}(e^{it} + e^{-it}) + \frac{i}{2}e^{-t}(e^{-it} - e^{it}) =$$

$$e^{-2t} + \frac{1}{2}e^{-t}(e^{it} + e^{-it}) + \frac{1}{2i}e^{-t}(e^{it} - e^{-it}) \tag{7.198}$$

Now, apply Euler's theorem, equation (7.65), to the exponential terms e^{2t} and e^{-2t} indicated in equation (7.198). By rearranging the results, we get the time-domain solution of the given transfer function as

$$x(t) = e^{-2t} + e^{-t}\cos(t) + e^{-t}\sin(t) \tag{7.199}$$

```
% Partial fraction expansion using Method of Residues.
% Write the given transfer function as follows
Z=[2 6 6];
P=conv([1 2],[1 2 2]);
xs=tf(Z,P)
```

xs =

```
   2 s^2 + 6 s + 6
  ---------------------
  s^3 + 4 s^2 + 6 s + 4
```

Continuous-time transfer function.

```
% Use Residue command
[r,p,k]=residue(Z,P)
```

```
r = 3×1 complex
   1.0000 + 0.0000i
   0.5000 - 0.5000i
   0.5000 + 0.5000i
p = 3×1 complex
  -2.0000 + 0.0000i
  -1.0000 + 1.0000i
  -1.0000 - 1.0000i
k =

   []
```

Figure 7.11 MATLAB live script file generates the partial fraction expansion for the transfer function, $X(s) = \frac{2s^2+6s+6}{(s+2)(s^2+2s+2)}$, using the Method of Residues

Figure 7.11 shows the MATLAB live script file that generates the results of the partial fraction expansion of the given transfer function using the Method of Residues. The results indicated that the roots of the characteristic equation have one real and complex root, and there is no direct term in this partial fraction expansion since the raw vector k is empty. Considering these, we get

$$X(s) = \frac{1}{(s+2)} + \frac{(0.5-0.5i)}{(s+1-i)} + \frac{(0.5+0.5i)}{(s+1+i)} = \frac{1}{(s+2)} + \frac{(0.5-0.5i)(s+1+i)+(0.5+0.5i)(s+1-i)}{(s+1-i)(s+1+i)} \tag{7.200}$$

Evaluating this equation gives

$$X(s) = \frac{1}{(s+2)} + \frac{s+2}{s^2+2s+2} \tag{7.201}$$

By rearranging the second term of this equation in a manner that is similar to what we did in equation (7.173) and using Table 7.1, items (6), (10), and (11), the inverse Laplace transform of equation (7.201) is

$$x(t) = e^{-2t} + e^{-t}\cos(t) + e^{-t}\sin(t) \tag{7.202}$$

This equation is the time-domain solution of the given transfer function. Comparing equation (7.202) with equation (7.199), we found they are the same.

Finally, from the above analysis, we found that the partial fraction expansion approach and the Method of Residues are highly effective and can be used in various cases of interest. Both methods can be challenging to understand algebraically, but conceptually, finding $x(t)$ for the given $X(s)$ and vice versa is easy.

7.3.2 SOLUTION OF DIFFERENTIAL EQUATIONS

Three basic steps are required to perform the Laplace transform for solving ordinary differential equations (equations of motion),

1. Each term in the differential equation should be transformed using the Laplace method. It is then transformed into an algebraic equation.
2. Find the unknown dependent variable by solving the algebraic equation for different input functions.
3. Using the inverse Laplace transform, find the time-domain solution.

Let us illustrate these procedures in the following analysis. Considering the following differential equation, the goal is to obtain the time-domain solution $x(t)$ for different input functions $f(t)$.

$$m\ddot{x} + b\dot{x} + kx = f(t) \quad \text{with} \quad x(0) = \dot{x}(0) = 0 \qquad (7.203)$$

Taking the Laplace transform of each term of equation (7.203) gives

$$m[s^2 X(s) - sx(0) - \dot{x}(0)] + b[sX(s) - x(0)] + kX(s) = F(s) \qquad (7.204)$$

or

$$X(s) = \frac{F(s)}{ms^2 + bs + k} + \frac{msx(0) + m\dot{x}(0) + bx(0)}{ms^2 + bs + k} \qquad (7.205)$$

and with zero initial conditions, we have

$$X(s) = \frac{F(s)}{ms^2 + bs + k} \qquad (7.206)$$

Given the characteristic equation $(ms^2 + bs + k)$, the roots are

$$s_{1,2} = \frac{-b \pm \sqrt{b^2 - 4mk}}{2m} \qquad (7.207)$$

Rearranging this equation gives

$$s_{1,2} = \left(\frac{-b}{2m}\right) \pm \frac{1}{2m}(\sqrt{b^2 - 4mk})i = a_1 \pm a_2 i \qquad (7.208)$$

Considering these, the transfer function indicated in equation (7.206) becomes

$$X(s) = \frac{F(s)}{(s + a_1)^2 + (a_2)^2} \qquad (7.209)$$

where

$$a_1 = (\frac{b}{2m}) \quad \text{and} \quad a_2 = \frac{\sqrt{b^2 - 4mk}}{2m} \tag{7.210}$$

As a result, for any input forcing function, $f(t)$, we may determine $F(s)$ and $X(s)$ by applying the Laplace transform. Then, by applying the inverse Laplace transform, we can obtain $x(t)$, which represents the time-domain solution of the given differential equation. The standard input functions $f(t)$ are classified into the following categories:

1. Unit impulse input, $f(t) = \delta(t)$,
The Laplace transform of the unit impulse input is given by,

$$F(s) = \mathscr{L}[f(t)] = \mathscr{L}[\delta(t)] = 1 \tag{7.211}$$

Substituting this equation in equation (7.206) gives

$$X(s) = \frac{1}{(s + a_1)^2 + (a_2)^2} \tag{7.212}$$

We need to rearrange this equation to use the inverse Laplace transform table, item (20), so that it looks like this instead:

$$X(s) = \frac{1}{a_2} \frac{a_2}{(s + a_1)^2 + (a_2)^2} \tag{7.213}$$

Taking the inverse Laplace transform of this equation gives

$$x(t) = \frac{1}{a_2} e^{a_1 t} \sin(a_2 t) \tag{7.214}$$

Substituting the values of a_1 and a_2 indicated in equation (7.210) in equation (7.214) gives

$$x(t) = \frac{2m}{\sqrt{b^2 - 4mk}} e^{-(b/2m)t} \sin(\frac{\sqrt{b^2 - 4mk}}{2m}) t \tag{7.215}$$

This equation is the time-domain solution of equation (7.209) for a unit impulse input.

2. Unit step input, $f(t) = u(t)$,
The Laplace transform of the unit step input is given by,

$$F(s) = \mathscr{L}[f(t)] = \mathscr{L}[u(t)] = \frac{1}{s} \tag{7.216}$$

Substituting this equation in equation (7.206) gives

$$X(s) = \frac{1}{s(ms^2 + bs + k)} \tag{7.217}$$

Applying partial fraction gives

$$X(s) = \frac{A}{s} + \frac{Bs + C}{ms^2 + bs + k} \tag{7.218}$$

We can determine the coefficient associated with the linear factor as

$$A = \frac{1}{ms^2 + bs + k}\Big|_{s=0} = \frac{1}{k} \tag{7.219}$$

Substituting the value of A in equation (7.218) and clearing fractions gives

$$1 = \frac{1}{k}(ms^2 + bs + k) + Bs^2 + Cs = (\frac{m}{k} + B)s^2 + (\frac{b}{k} + C)s + 1 \tag{7.220}$$

Equating coefficients of like terms and solving for B and C, we get

$$B = -\frac{m}{k} \quad \text{and} \quad C = -\frac{b}{k} \tag{7.221}$$

Substituting equations (7.219) and (7.221) in equation (7.218) gives

$$X(s) = \frac{1}{k}\frac{1}{s} - \frac{m}{k}\frac{s}{(ms^2 + bs + k)} - \frac{b}{k}\frac{1}{(ms^2 + bs + k)} \tag{7.222}$$

Considering equation (7.212), equation (7.222) becomes

$$X(s) = \frac{1}{k}\frac{1}{s} - \frac{m}{k}\frac{s}{(s + a_1)^2 + (a_2)^2} - \frac{b}{k}\frac{1}{(s + a_1)^2 + (a_2)^2} \tag{7.223}$$

where a_1 and a_2 are given by equation (7.210). To use the Laplace transform Table 7.1 items (1), (20), and (21), we need to rearrange the second and third terms of equation (7.223) so that it looks like this instead:

$$X(s) = \frac{1}{k}\frac{1}{s} - \frac{m}{k}\frac{s + a_1 - a_1}{(s + a_1)^2 + (a_2)^2} - \frac{b}{k}\frac{1}{a_2}\frac{a_2}{(s + a_1)^2 + (a_2)^2} \tag{7.224}$$

or,

$$X(s) = \frac{1}{k}\frac{1}{s} - \frac{m}{k}\frac{s + a_1}{(s + a_1)^2 + (a_2)^2} + \frac{m}{k}\frac{a_1}{a_2}\frac{a_2}{(s + a_1)^2 + (a_2)^2} - \frac{b}{k}\frac{1}{a_2}\frac{a_2}{(s + a_1)^2 + (a_2)^2} \tag{7.225}$$

Rearranging this equation gives

$$X(s) = \frac{1}{k}\frac{1}{s} - \frac{m}{k}\frac{s + a_1}{(s + a_1)^2 + (a_2)^2} + [\frac{m}{k}\frac{a_1}{a_2} - \frac{b}{k}\frac{1}{a_2}]\frac{a_2}{(s + a_1)^2 + (a_2)^2} \tag{7.226}$$

Taking the inverse Laplace transform of this equation gives

$$x(t) = \frac{1}{k} - \frac{m}{k}e^{a_1 t}\cos(a_2 t) + [\frac{m}{k}\frac{a_1}{a_2} - \frac{b}{k}\frac{1}{a_2}]e^{a_1 t}\sin(a_2 t) \tag{7.227}$$

Substituting equation (7.210) in this equation gives

$$x(t) = \frac{1}{k} - \frac{m}{k}e^{-(b/2m)t}\cos(\frac{\sqrt{b^2-4mk}}{2m})t + [\frac{m}{k}\frac{(-b/2m)}{((1/2m)\sqrt{b^2-4mk})} -$$

$$\frac{b}{k}\frac{1}{((1/2m)\sqrt{b^2-4mk})}]e^{-(b/2m)t}\sin(\frac{\sqrt{b^2-4mk}}{2m})t \qquad (7.228)$$

This equation is the time-domain solution of equation (7.217) for a unit step input.

3. Unit ramp input, $f(t) = t$,
The Laplace transform of the unit ramp input is given by

$$X(s) = \mathscr{L}[f(t)] = \mathscr{L}[t] = \frac{1}{s^2} \qquad (7.229)$$

Thus, rewriting equation (7.206) as,

$$X(s) = \frac{1}{s^2}\frac{1}{ms^2+bs+k} \qquad (7.230)$$

Applying partial fraction expansions to this equation gives

$$X(s) = \frac{1}{s^2}\frac{1}{ms^2+bs+k} = \frac{A_1}{s^2} + \frac{A_2}{s} + \frac{Bs+C}{ms^2+bs+k} \qquad (7.231)$$

Clearing fractions gives

$$X(s) = \frac{(mA_2+B)s^3 + (mA_1+bA_2+C)s^2 + (bA_1+kA_2)s + kA_1}{s^2(ms^2+bs+k)} \qquad (7.232)$$

Equating the factors of both sides of this equation yields,

$$s^3 : mA_2 + B = 0 \qquad (7.233)$$

$$s^2 : mA_1 + bA_2C = 0 \qquad (7.234)$$

$$s : bA_1 + kA_2 = 1 \qquad (7.235)$$

$$0 : kA_1 = 1 \qquad (7.236)$$

Solving these equations for A_1, A_2, B, and C yields

$$A_1 = -\frac{1}{k}, \ A_2 = -\frac{b}{k^2}, \ B = \frac{mb}{k^2}, \ \text{and} \ C = (\frac{m}{k} + \frac{b^2}{k^2}) \qquad (7.237)$$

Substituting these values in equation (7.231) gives

$$X(s) = \frac{1}{k}\frac{1}{s^2} - \frac{b}{k^2}\frac{1}{s} + \frac{(mb/k^2)s + [(m/k)+(b^2/k^2)]}{ms^2+bs+k} \qquad (7.238)$$

Using equation (7.209), equation (7.238) becomes

$$X(s) = \frac{1}{k}\frac{1}{s^2} - \frac{b}{k^2}\frac{1}{s} + \frac{(mb/k^2)s + [(m/k) + (b^2/k^2)]}{(s+a_1)^2 + (a_2)^2} \qquad (7.239)$$

where a_1 and a_2 are given by equation (7.210). Considering these and rearranging the third part of equation (7.239) so that it looks like items (10) and (11) in Table 7.1, and then taking the inverse Laplace transform of this equation, the time-domain solution of equation (7.239) is provided by

$$x(t) = \frac{1}{k}t - \frac{b}{k^2} + \frac{mb}{k^2}e^{-(b/2m)t}\cos(\frac{\sqrt{b^2-4mk}}{2m})t + \frac{2m}{\sqrt{b^2-4mk}}$$

$$[(\frac{m}{k}) + \frac{b^2}{k^2}) - (\frac{(b/2m)}{(mb/k^2)}]e^{-(b/2m)t}\sin(\frac{\sqrt{b^2-4mk}}{2m})t \qquad (7.240)$$

This equation is the time-domain solution of the differential equation indicated in equation (7.230) for a unit ramp input.

Now, let us look at the Laplace transform for a specific function. The following two examples illustrate how to accomplish this.

Example (7.13):
Find the Laplace transform $F(s)$ of the function $f(t)$ shown in Figure 7.12.

Solution:
The function $f(t)$, shown in Figure 7.12, can be written as

$$f(t) = \frac{1}{b}1(t) - \frac{2}{b}1(t-a) + \frac{1}{b}1(t-2a) \qquad (7.241)$$

Taking the Laplace transform of this equation gives

$$F(s) = \mathscr{L}[f(t)] = \frac{1}{b}\mathscr{L}1(t) - \frac{2}{b}\mathscr{L}[1(t-a)] + \frac{1}{b}\mathscr{L}[1(t-2a)] \qquad (7.242)$$

$$F(s) = \frac{1}{b}\frac{1}{s} - \frac{2}{b}\frac{1}{s}e^{-at} + \frac{1}{b}\frac{1}{s}e^{-2at} = \frac{1}{bs}(1 - 2e^{-at} + e^{-2at}) \qquad (7.243)$$

This equation is the transfer function of the given function.

Example (7.14):
Find the Laplace transform $F(s)$ of the function $f(t)$ shown in Figure 7.13.

Solution:
The function $f(t)$, shown in Figure 7.13, can be written as

$$f(t) = (t-a)1(t-a) \qquad (7.244)$$

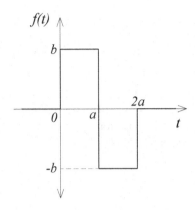

Figure 7.12 Function $f(t)$

Taking the Laplace transform of this equation gives

$$F(s) = \mathcal{L}[f(t)] = \mathcal{L}[(t-a)1(t-a)] = \frac{1}{s^2}e^{-as} \qquad (7.245)$$

As a result of the previous analysis, some features that distinguish the Laplace transform from the classical technique are as follows:

1. The Laplace transform method does not require separate steps to be carried out to determine the complementary solution, the particular solution, and the integration constant.
2. The final solution, including the initial conditions, is obtained simultaneously in the Laplace transform. In other words, no debate is needed to define the necessary initial conditions because the method always includes them.
3. The classical technique demands a piecewise solution that precisely matches one piece's end conditions and the next's initial conditions for inputs that cannot be described by a single formula for their whole path but must be defined over time. This means the Laplace transform method is good at dealing with such discontinuous input.

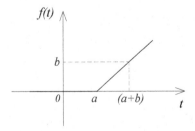

Figure 7.13 Function $f(t)$

7.4   PROBLEMS

Problem (7.1):
Find the time-domain solution $x(t)$ for the following differential equations using the classical method and verify the results using MATLAB

$$1. \quad \ddot{x} + 2x = 0, \quad x(0) = 5 \quad \text{and} \quad \dot{x}(0) = 0$$

$$2. \quad \ddot{x} + 5x = t, \quad x(0) = 0 \quad \text{and} \quad \dot{x}(0) = 0$$

$$3. \quad 4\ddot{x} + 2\dot{x} + x = 4, \quad x(0) = 0 \quad \text{and} \quad \dot{x}(0) = 2$$

$$4. \quad 2\ddot{x} + 4\dot{x} + x = 5e^{-2t}, \quad x(0) = 3 \quad \text{and} \quad \dot{x}(0) = 2$$

$$5. \quad \ddot{x} + \dot{x} + 2x = 4t, \quad x(0) = 0 \quad \text{and} \quad \dot{x}(0) = 2$$

Problem (7.2):
Derive the Laplace transform of the following functions, considering each is $f(t) = 0$ at $t \le 0$.

$$1. \quad f(t) = 5te^{-2s} \quad \text{at} \quad t > 0$$

$$2. \quad f(t) = 5\sin(3t) \quad \text{at} \quad t > 0$$

$$3. \quad f(t) = bt^2 e^{-as} \quad \text{at} \quad t > 0$$

Problem (7.3):
Using the inverse Laplace transform, obtain the response equations of the same differential equations presented in Problem (7.1) and verify the results using MATLAB.

Problem (7.4):
Find the inverse Laplace transform $x(t)$ for the following transfer functions.

$$1. \quad X(s) = \frac{2s+1}{(s+1)(s+2)^2}$$

$$2. \quad X(s) = \frac{2s^2+6s+6}{s(s+2)}$$

$$3. \quad X(s) = \frac{s^2+4s+5}{(s+2)(s^2+2s+2)}$$

$$4. \quad X(s) = \frac{s}{s^2+2s+5}$$

$$5. \quad X(s) = \frac{s^2+2s+4}{s^2(s+2)}$$

Problem (7.5):
Expand the following transfer functions into partial fraction expansions using the Method of Residues and determine the time-domain solution $x(t)$. Verify the results using MATLAB.

$$1. \quad X(s) = \frac{1}{s^2(s^2+8)}$$

$$2. \quad X(s) = \frac{1}{s^4 + 2s^3 + 2s^2 + 4}$$

$$3. \quad X(s) = \frac{3s^2 + 4s + 1}{s^4 + 2s^3 + 5s^2 + 8s + 10}$$

Problem (7.6):

For the following figures, what are the Laplace transforms $F(s)$ of the function $f(t)$.

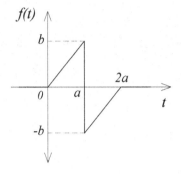

Figure 7.14   Function $f(t)$

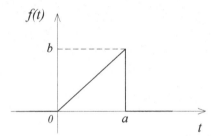

Figure 7.15   Function $f(t)$

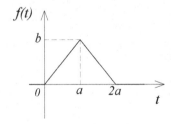

Figure 7.16   Function $f(t)$

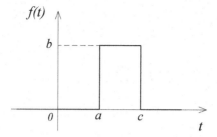

Figure 7.17   Function $f(t)$

8 Dynamic System Responses

8.1 INTRODUCTION

This chapter will investigate various combinations of fundamental system elements from the mechanical, electrical, liquid, and thermal domains. Because elements often behave similarly, we should expect similar behavior in system implementations. Understanding the features of a specific class of systems, regardless of their physical nature, allows us to analyze system responses more effectively due to this similarity. We find two classes of systems of critical importance: first-order and second-order systems. Since most practical systems will fall into one of these two categories, they are significant in and of themselves.

8.2 FIRST-ORDER SYSTEMS

8.2.1 FIRST-ORDER MECHANICAL SYSTEMS

There are two basic configurations of first-order mechanical systems: those that include mass elements and those that do not. The system with a mass element should be connected only to the damper, as shown in Figure 8.1, with mass velocity as the independent coordinate; otherwise, the system will be second-order. If the system does not have a mass element, the independent coordinate will be the displacement of the massless element, and it must include a spring and damper, as shown in Figure 8.2.

8.2.1.1 First-order system includes a mass element

Applying Newton's second law on the systems shown in Figure 8.1 gives

1. For translational system (fFigure 8.1a)

$$m\frac{dv}{dt} + bv = f(t) \tag{8.1}$$

2. For rotational system (Figure 8.1b)

$$J\frac{d\omega}{dt} + b_t\omega = T(t) \tag{8.2}$$

According to these two equations, the mass velocity will be the independent coordinate of a first-order system that includes a mass element.

DOI: 10.1201/9781032685656-8

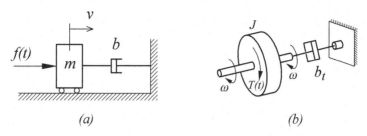

Figure 8.1 A mechanical first-order system includes a mass element: (a) translational system; (b) rotational system

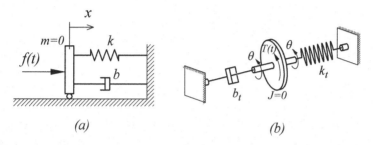

Figure 8.2 A mechanical first-order system without mass element: (a) translational system; (b) rotational system

8.2.1.2 First-order system without mass element

Applying Newton's second law on the systems shown in Figure 8.2 gives

1. For the translational system (Figure 8.2a)

$$b\frac{dx}{dt} + kx = f(t) \tag{8.3}$$

2. For the rotational system (Figure 8.2b)

$$b_t\frac{d\theta}{dt} + k_t\theta = T(t) \tag{8.4}$$

Equations (8.3) and (8.4) indicate that the displacement of system elements is the independent coordinate of a first-order system that does not include a mass element.

8.2.2 FIRST-ORDER SYSTEM ANALYSES

8.2.2.1 System includes a mass element

It is worth noting that any system whose equations follow equations (8.1) to (8.4), by definition, is a first-order system. In the following analysis, we will only consider equations (8.1), representing the first-order translational system with a mass element shown in Figure 8.1a. Rewriting equation (8.1) gives

$$\frac{m}{b}\dot{v} + v = \frac{1}{b}f(t) \tag{8.5}$$

If we define

$$\tau = \frac{m}{b} \quad \text{is the system time constant}$$

and

$$K = \frac{1}{b} \quad \text{is the system steady-state gain}$$

then, the equation representing the general standard form of a first-order system is given by

$$\tau\dot{v} + v = Kf(t) \tag{8.6}$$

This equation indicated that only two parameters, τ and K, are required to fully characterize the first-order system. The following analysis will reveal the physical relevance of these two parameters and the reasons for their definitions. Now, it is necessary to investigate how such systems respond to various types of inputs. Taking the Laplace transform of equation (8.6) and rearranging the results gives

$$\frac{V(s)}{F(s)} = \frac{K}{(\tau s + 1)} \tag{8.7}$$

This equation represents the standard form of the transfer function of a first-order system.

8.2.2.2 First-order system response to step input

The step input assumes that the system is initially in equilibrium, with both input and output at zero, until the input jumps to a certain value and continues at this value for an extended time, as shown in Figure 8.3a. It is necessary to figure out how the mass moves due to this motion and precisely how the velocity v of the mass changes over time. Assume the step input is $f(t) = A_i u(t)$, where $u(t)$ is the symbol of a unit step function. Rewriting equation (8.6) as follows

$$\tau\dot{v} + v = KA_i u(t) \tag{8.8}$$

Assuming zero initial conditions, taking the Laplace transform of this equation and rearranging the results gives

$$V(s) = \frac{KA_i}{s(\tau s + 1)} \tag{8.9}$$

This equation represents the transfer function of a first-order system applied to the step input force. Rearrange this equation gives

$$V(s) = \frac{(KA_i/\tau)}{s(s+(1/\tau))} \tag{8.10}$$

Assume $a = (1/\tau)$, applying partial fraction expansion to equation (8.10) yields

$$V(s) = \frac{(aKA_i)}{s(s+a)} = \frac{A}{s} + \frac{B}{s+a} \tag{8.11}$$

Solving for A and B gives

$$A = \frac{(aKA_i)}{(s+a)}\bigg|_{s=0} \quad \text{so} \quad A = KA_i \tag{8.12}$$

$$B = \frac{(aKA_i)}{s}\bigg|_{s=-a} \quad \text{so} \quad B = -KA_i \tag{8.13}$$

Substituting these values in equation (8.11) yields

$$V(s) = \frac{KA_i}{s} - \frac{KA_i}{s+a} \tag{8.14}$$

Using Table 7.1, items (2) and (6), the inverse Laplace transform of equation (8.14) is given by

$$v(t) = KA_i - KA_i e^{-at} \tag{8.15}$$

Substituting $a = (1/\tau)$ in this equation and rearranging it gives

$$v(t) = KA_i(1 - e^{-t/\tau}) \tag{8.16}$$

This response equation represents the dynamic motion of a first-order system that includes a mass element applied to step input. In contrast, its response is shown in Figure 8.3a, which is the plot of mass velocity v versus time. This figure can be used for any first-order system that includes a mass element with any values of K and τ, as well as any size of the step input A_i.

Based on the final value theorem, to obtain the steady-state value of the system response, we will use two different ways to get it: either from the transfer function, equation (8.10), or from the system response, equation (8.16); however, the results should be the same in both cases, as shown below.

1. From the transfer function $V(s)$, equation (8.14):

$$V_{ss} = \lim_{s \to 0} sV(s) = \lim_{s \to 0} s\frac{(KA_i/\tau)}{s(s+(1/\tau))} = KA_i \tag{8.17}$$

2. From the response equation $v(t)$, equation (8.16):

$$V_{ss} = \lim_{t \to \infty} v(t) = \lim_{t \to \infty} KA_i(1 - e^{-t/\tau}) = KA_i \tag{8.18}$$

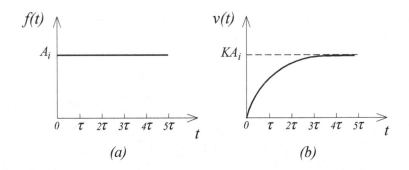

Figure 8.3 Generalized step response of first-order systems: (a) step input; (b) step response

From the above analysis, it is possible to identify several characteristics and features of the step response of a first-order system as follows.

- The value of the mass velocity v approaches approximately its steady state value KA_i as the time t approaches infinity ∞, as shown in Figure 8.3b. Achieving this value would take an infinite amount of time; in practice, it would take 4 to 5 times the time constant τ. At this point, we would say that the system response has reached a steady state for the particular input applied.
- For the steady-state condition, the output velocity to input force ratio is $KA_i/A_i = K$, which is the steady-state gain or static sensitivity. As a result, we consider K the amount of steady-state output produced for each unit of the input provided to the system. It is important to note that the numerical value of K does not affect how quickly the steady-state output will reach.
- In general, the speed of a response in a linear system determines how long it takes for the system to reach a certain percentage of its steady-state value. This is possible because the output is directly proportional to the input. As a result, the time required to reach the same actual value of the output may differ depending on whether the step input is large or small, but the time required to reach the same percentage of the steady state value will be the same for all step sizes. According to this definition, the speed of a first-order system response is determined by the numerical value of its time constant τ, since the percentage of steady-state response is given by

$$\frac{v}{KA_i} = 1 - e^{-t/\tau} \tag{8.19}$$

- As we can see from this equation, if τ, for example, is cut in half, similar values of v/KA_i will be obtained in one-half the time; as a result, the speed of the response is inversely proportional to the value of the time constant τ.
- If we define the response time as the time it takes for the system response to reach the steady state, we can only increase system speed by decreasing the time constant $\tau = (m/b)$. Now let us consider changing m and b. In this case, the steady-state gain $K = (1/b)$ is unchanged by changing m; however, changing b will directly affect K. As a result, a system with small b and/or large K implies that a small value of τ indicates a fast response.

Now let us define two additional parameters of a first-order system, which are the rise time T_r and the settling time T_s as follows:

1. Rise time:
It is defined as the response time going from 0.1 to 0.9 or 10% to 90% of its steady-state, as shown in Figure 8.4. To obtain an expression for rise time, we put 0.1 and 0.9 in the general first-order response equation as follows:

- For $t = 0.1$:

$$T_{r_1} = 1 - e^{-at} = 1 - e^{-0.1a} = \frac{0.11}{a} \tag{8.20}$$

- For $t = 0.9$:

$$T_{r_2} = 1 - e^{-0.9a} = \frac{2.31}{a} \tag{8.21}$$

Taking the difference between T_{r_2} and T_{r_1} gives

$$T_r = \frac{2.31}{a} - \frac{0.11}{a} = \frac{2.2}{a} \tag{8.22}$$

2. Settling time:
It is when the response reaches and stays within ±2% of its steady-state, as shown in Figure 8.4. It is given by:

$$T_s = \frac{4}{a} \tag{8.23}$$

Inserting $a = 1/\tau$ in equations (8.22) and (8.23) gives

$$T_r = 2.2\tau \quad \text{and} \quad T_s = 4\tau \tag{8.24}$$

This equation indicated that the rise time T_r and the settling time T_s are functions of the time constant τ.

Finally, we may summarize the relevance of the two fundamental first-order system parameters by stating that K indicates how much steady-state output will be created for each input unit and that τ indicates how quickly the steady-state will be achieved. Now, let us compute the initial slope of the curve of v versus time t shown in Figure 8.4. Taking the derivative of equation (8.16) yields

$$\frac{dv}{dt}\bigg|_{t=0} = \frac{KA_i}{\tau} e^{-t/\tau}\bigg|_{t=0} = \frac{KA_i}{\tau} \tag{8.25}$$

This equation indicated that the initial slope of the velocity response is KA_i/τ, as shown in Figure 8.4. From this figure, we noted that the steady state would be attained in τ seconds if the slope KA_i/τ remained constant at the starting initial value rather than decreasing according to $(KA_i/\tau)e^{-t/\tau}$. This fact is helpful when sketching the response curves of first-order systems applied to step inputs. So, for $t = \tau$ we have

$$v = KA_i(1 - e^{-1}) = 0.632\,(KA_i) \tag{8.26}$$

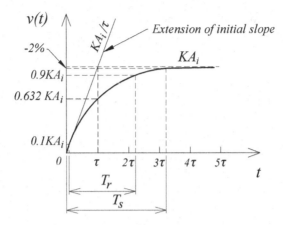

Figure 8.4 First-order system parameters characteristics

Thus, 63.2% of the steady state value KA_i is achieved in one-time constant, as shown in Figure 8.4.

To determine the displacement response of the mass, the integration of equation (8.16) should be carried out. Assuming zero initial conditions and carrying out this integration gives

$$x(t) = \int v(t)dt = \int KA_i(1 - e^{-t/\tau})dt \tag{8.27}$$

So,

$$x(t) = KA_i t + kA_i \tau(e^{-t/\tau} - 1) \tag{8.28}$$

Based on this equation, Figure 8.5 shows the response of the mass displacement $x(t)$, which is the plot of the displacement x versus time. As shown in this figure, after the

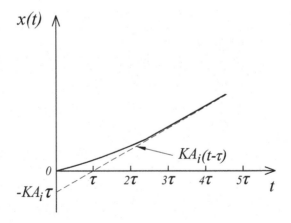

Figure 8.5 The generalized displacement response of a first-order system includes a mass element

transient period during which $e^{-t/\tau}$ passes out, we observed that the displacement becomes asymptotic to the straight line $KA_i(t - \tau)$.

8.2.2.3 First-order system response to impulse input

The impulse input has infinite height and infinitesimal duration but a finite area, as shown in Figure 8.6a. Although no real physical variable behaves precisely like this, an approximation near enough for many physical applications is frequently possible. Considering again the translational system shown in Figure 8.1a, which is applying to the impulse input $f(t) = I_i\delta(t)$, where $\delta(t)$ is the symbol of a unit impulse function, the equation of motion of the system is given by

$$m\dot{v} + bv = I_i\delta(t) \tag{8.29}$$

or

$$\frac{m}{b}\dot{v} + v = \frac{I_i}{b}\delta(t) \tag{8.30}$$

Assuming zero initial conditions and taking the Laplace transform of this equation,

$$\frac{m}{b}sV(s) + V(s) = I_i\frac{1}{b} \tag{8.31}$$

Rearranging this equation gives

$$V(s) = \frac{I_i/b}{(m/b)s + 1} = \frac{I_iK}{\tau s + 1} = \frac{I_iK/\tau}{s + (1/\tau)} \tag{8.32}$$

where $K = (1/b)$ and $\tau = (m/b)$. Rewrite equation (8.32) and put $a = (1/\tau)$ yields

$$V(s) = \frac{I_iK}{\tau}\frac{1}{(s + a)} \tag{8.33}$$

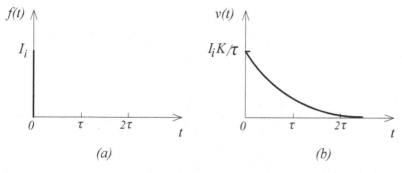

(a)   (b)

Figure 8.6 Generalized impulse response of first-order systems: (a) impulse input; (b) impulse response

Using Table 7.1, item (6), the inverse Laplace transform of this equation is given by

$$v(t) = \frac{I_i K}{\tau} e^{-at} \tag{8.34}$$

Substituting $a = (1/\tau)$ in this equation yields

$$v(t) = \frac{I_i K}{\tau} e^{-t/\tau} \tag{8.35}$$

This response equation represents the dynamic motion of the velocity of a first-order system applied to impulse input. Based on this equation, the velocity response of the first-order system due to the impulse input is shown in Figure 8.6b. This figure can be used for any first-order system with any values of K and τ and any size of impulse input I_i. From this figure we noted that the mass velocity $v(t)$ is an exponentially decaying signal over the response time until it approaches zero.

To determine the displacement response of the mass due to impulse input, the integration of equation (8.35) should be carried out. Assuming zero initial conditions and carrying out this integration gives

$$x(t) = \int v(t) dt = \int \frac{I_i K}{\tau} e^{-t/\tau} dt \tag{8.36}$$

So,

$$x(t) = I_i K e^{-t/\tau} \tag{8.37}$$

The steady-state value of the system response can be obtained as follows.

1. From the transfer function $V(s)$, equation (8.33):

$$V_{ss} = \lim_{s \to 0} sV(s) = \lim_{s \to 0} s \frac{(I_i K/\tau)}{(s + (1/\tau))} = 0 \tag{8.38}$$

2. From the response equation $v(t)$, equation (8.35):

$$V_{ss} = \lim_{t \to \infty} v(t) = \lim_{t \to \infty} I_i K e^{-t/\tau} = 0 \tag{8.39}$$

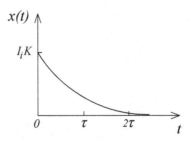

Figure 8.7 Generalized displacement response of first-order systems

Based on equation (8.37), Figure 8.7 shows the response of the mass displacement, which is the plot of the displacement x versus time. This figure shows that the mass displacement $x(t)$ is an exponentially decaying signal over the response time until approaching zero, implying that there is no steady-state value, which is indicated clearly in equation (8.39) and Figure 8.7.

8.2.2.4 First-order system response to ramp input

The ramp input changes uniformly throughout time, as shown in Figure 8.8a. Considering again the translational system shown in Figure 8.1a, which is applying to the ramp input $f(t) = r(t) = R_i t$; the equation of motion of the system is given by

$$m\dot{v} + bv = f(t) = R_i t \qquad (8.40)$$

Rearranging this equation gives

$$\frac{m}{b}\dot{v} + v = \frac{1}{b}R_i t = KR_i t \qquad (8.41)$$

Assuming zero initial conditions and taking the Laplace transform of this equation gives

$$V(s) = \frac{KR_i}{s^2((m/b)s + 1)} = \frac{KR_i}{s^2(\tau s + 1)} = \frac{KR_i a}{s^2(s + a)} \qquad (8.42)$$

where $\tau = (m/b)$, $a = (1/\tau)$ and $K = (1/b)$. Applying partial fraction expansion to equation (8.42) yields

$$V(s) = \frac{R_i K a}{s^2(s + a)} = \frac{A}{s} + \frac{B}{s^2} + \frac{C}{(s + a)} \qquad (8.43)$$

Rearranging this equation gives

$$V(s) = \frac{R_i K a}{s^2(s + a)} = \frac{As(s + a) + B(s + a) + Cs^2}{s^2(s + a)} \qquad (8.44)$$

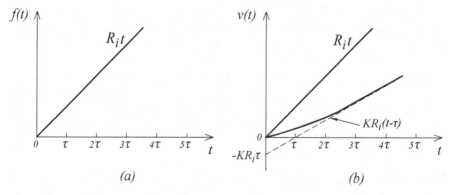

(a) (b)

Figure 8.8 Generalized velocity response of a first-order system does not include mass element due to ramp input

or,

$$V(s) = \frac{R_i Ka}{s^2(s+a)} = \frac{(A+C)s^2 + (Aa+B)s + Ba}{s^2(s+a)} \qquad (8.45)$$

Equating the factors of both sides of this equation gives

$$s^2 : A + C = 0 \qquad (8.46)$$

$$s : Aa + B = 0 \qquad (8.47)$$

$$0 : Ba = R_i Ka \quad \text{so} \quad B = R_i K \qquad (8.48)$$

Substituting equation (8.48) in equation (8.47) gives

$$Aa + R_i K = 0 \quad \text{so} \quad A = \frac{-R_i K}{a} \qquad (8.49)$$

Substituting equation (8.49) in equation (8.46) gives

$$\frac{-R_i K}{a} + C = 0 \quad \text{so} \quad C = \frac{R_i K}{a} \qquad (8.50)$$

Substituting values of A, B, and C, equations (8.48), (8.49), and (8.50), in equation (8.43) gives

$$V(s) = \frac{(-R_i K/a)}{s} + \frac{R_i K}{s^2} + \frac{(R_i K/a)}{s+a} \qquad (8.51)$$

Using Table 7.1, items (2), (3), and (6), taking the inverse Laplace transform of this equation and putting $a = (1/\tau)$ yields

$$v(t) = R_i Kt + R_i K\tau(e^{-t/\tau} - 1) \qquad (8.52)$$

This response equation represents the dynamic motion of a first-order system applied to ramp input. Based on this equation, the velocity response of the first-order system due to the ramp input is shown in Figure 8.8b. This figure can be used for any first-order system with any K and τ values, as well as any size ramp input R_i. From this figure we noted that, after an initial transient period, the velocity v rises at a constant rate, i.e., the motion is one of constant acceleration. This is true for first-order mass-damper systems as well.

The steady-state value of the system response can be obtained as follows.

1. From the transfer function $V(s)$, equation (8.42):

$$V_{ss} = \lim_{s \to 0} sV(s) = \lim_{s \to 0} s \frac{(R_i K/\tau)}{s^2(s + (1/\tau))} = \infty \qquad (8.53)$$

2. From the response equation $v(t)$, equation (8.52):

$$V_{ss} = \lim_{t \to \infty} [(\frac{R_i K}{2})t^2 - (R_i K\tau)t - (R_i K\tau^2)e^{-t/\tau}] = \infty \qquad (8.54)$$

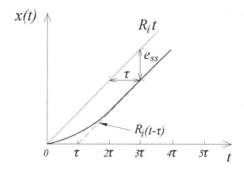

Figure 8.9 Generalized displacement response of a first-order system does not include mass element due to ramp input

To determine the displacement response of the mass due to ramp input, integration of equation (8.52) should be carried out. Assuming zero initial conditions and carrying out this integration gives

$$x(t) = \int v(t)dt = \int R_i Kt + R_i K\tau(e^{-t/\tau} - 1)dt \qquad (8.55)$$

So,

$$x(t) = (\frac{R_i K}{2})t^2 - (R_i K\tau)t - (R_i K\tau^2)e^{-t/\tau} \qquad (8.56)$$

Based on this equation, the displacement response $x(t)$ of the first order includes a mass element applied to ramp input, which is shown in Figure 8.9. As shown in this figure, which is the response of the mass displacement for a translational system, after the transient period during which $e^{-t/\tau}$ passes out, we observe that the displacement becomes asymptotic to the straight line $KA_i(t - \tau)$. We also observed that if the ramp input force is kept applied, the translational damper element must encounter a mechanical stop to cause the mass to stall; otherwise, the system will become unbalanced and fall entirely. As a result, we cannot sustain a ramp input in a real system for unlimited time intervals since the input force and output displacement approach infinity; however, a ramp input may be an acceptable model for actual input over a bounded time interval. On the other hand, the rotating system allows for contentious, unhindered motion, with the result that the output angular displacement may reach infinity.

Finally, when analyzing the response of a first-order system subjected to a variety of input-forcing functions, we found the following observations:

- When the step and impulse inputs are applied, the system remains balanced because their responses have bounded out and become more predictable.
- The impulse response has no steady-state output.
- The system response is unbalanced when the ramp input is applied since it increases indefinitely.

As a result, the step input is frequently employed to analyze the dynamics of first-order systems in the time domain.

8.2.2.5 First-order system does not include a mass element

Since the previous analysis was performed on a mechanical first-order system containing a mass element, we will now determine whether or not we will receive the same results when considering a first-order system that does not have a mass element. In the following analysis, we will focus on the translational system shown in Figure 8.2a, which is applied only to step input force $f(t) = A_i u(t)$. Applying Newton's second law gives

$$b\frac{dx}{dt} + kx = f(t) \tag{8.57}$$

Rearranging this equation gives

$$\frac{b}{k}\frac{dx}{dt} + x = \frac{1}{k}f(t) \tag{8.58}$$

Then, the equation representing the general form of a first-order system that does not include a mass element is given by

$$\tau \dot{x} + x = Kf(t) \tag{8.59}$$

where

$$K = \frac{1}{k} \text{ is the steady-state gain}$$

and

$$\tau = \frac{b}{k} \text{ is the time constant}$$

Substituting $f(t) = A_i u(t)$ in equation (8.89) yields

$$\tau \dot{x} + x = KA_i u(t) \tag{8.60}$$

Assuming zero initial conditions, taking the Laplace transform of this equation gives

$$\tau s X(s) + X(s) = \frac{KA_i}{s} \tag{8.61}$$

or,

$$X(s) = \frac{KA_i}{s(\tau s + 1)} \tag{8.62}$$

Rearrange this equation to give

$$X(s) = \frac{(KA_i/\tau)}{s(s + (1/\tau))} \tag{8.63}$$

Assuming $a = (1/\tau)$ and applying partial fraction expansion to equation (8.63) yields

$$X(s) = \frac{(aKA_i)}{s(s+a)} = \frac{A}{s} + \frac{B}{s+a} \tag{8.64}$$

Solving for A and B gives

$$A = KA_i \quad \text{and} \quad B = -KA_i \tag{8.65}$$

Substituting these values in equation (8.64) gives

$$X(s) = \frac{KA_i}{s} - \frac{KA_i}{s+a} \tag{8.66}$$

Using Table 7.1, items (2) and (6), the inverse Laplace transform of this equation yields

$$x(t) = KA_i(1 - e^{-t/\tau}) \tag{8.67}$$

This equation represents the dynamic response of the displacement of a first-order system that does not include a mass element. Its response, which is the plot of the displacement x versus time t, is shown in Figure 8.10. As a comparison, this equation is similar to equation (8.16), which represents the velocity response of a first-order system that includes a mass element. This is because the velocity and the displacement are the independent coordinates of both systems, based on whether or not they have a mass element. This implies that the displacement response $x(t)$ of the system without mass element is equivalent to the velocity response $v(t)$ of the system includes mass elements, as shown clearly in Figures 8.4 and 8.10. However, the difference is only observed in the time constant values τ and the steady-state system gain K.

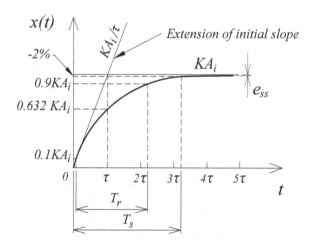

Figure 8.10 Step response of first-order systems

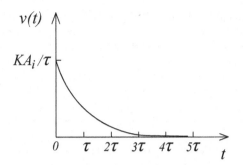

Figure 8.11   Velocity response of a first-order system for step response

To determine the velocity response $v(t)$ of the given system shown in Figure 8.2a, the derivative of equation (8.67) should be carried out. As a result, it is given by

$$v(t) = \frac{KA_i}{\tau} e^{-t/\tau} \tag{8.68}$$

This equation represents the dynamic response of the velocity of a first-order system that does not include a mass element. Its response, which is the plot of velocity v versus time t, is shown in Figure 8.11. This figure shows that the velocity $v(t)$ is an exponentially decaying signal over the response time until it approaches zero.

Considering the definitions of the time constant and steady-state gain in both systems, the characteristics of the velocity step response of the system include a mass element that can be applied to the displacement response of the system that does not include a mass element. Note that the first-order system, which includes a mass element, cannot be linked to a spring element; otherwise, the system will be second-order. Although we have strongly emphasized the fundamental significance of K and τ, we must maintain sight of the reality that modifying the design of a particular first-order system can only be done by changing the values of these physical parameters. In general, when the system induces a mass element, the time constant is $\tau = m/b$, and the steady-state gain is $K = 1/b$, whereas they are $\tau = b/k$ and $K = 1/k$ when the system does not include a mass element.

The following outlined comments are what we observed when we compared the corresponding responses of a first-order mechanical system based on whether or not it has a mass element.

- When the system includes a mass element, changing the value of b affects both the speed of response and the steady-state value, whereas changing the value of m will affect only the speed of response. If we consider the rate at which the system reaches a steady state to be the speed of response, then the only way the system speed can be increased is by decreasing the value of the time constant $\tau = m/b$. Thus, we might reduce m, raise b, or do both. On the other hand, the steady-state gain of the system, $K = 1/b$, is unaffected by changes in m, while changes in b change it.

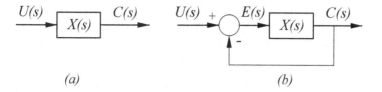

(a)   *(b)*

Figure 8.12   Block diagram of a transfer function $X(s)$: (a) open loop; (b) closed loop

- When the system does not include a mass element if we need to increase the system speed, we must decrease the time constant $\tau = (b/k)$ by increasing k or decreasing b, or doing both. If we consider changing in b and k, the steady-state gain $K = (1/k)$ is unchanged by changing b; however, changing k will directly affect K. This means that the system with a large k, or small K, implies a value of τ will decrease, indicating a fast response.

We are now aware of how the values of K and τ affect the system's response, and as a result, we can make adjustments to m, b, and k based on their definitions.

As a result of the above analyses, we found that all parameters and their characteristic analyses for a first-order system that includes a mass element, in which the mass velocity $v(t)$ is the independent coordinate, are considered to be the same for a system that does not include a mass element, in which the displacement $x(t)$ is the independent coordinate, as shown in Figure 8.4 and 8.10.

8.2.2.6   Steady-state error

The steady-state error e_{ss} is defined as the difference between the actual output $C(s)$ and the input $U(s)$ at the time approaches ∞, as shown in Figures 8.9 and 8.10. As shown in Figure 8.12, the system error $E(s)$ is given by

$$E(s) = U(s) - C(s) \tag{8.69}$$

First, considering the transfer function $X(s)$ in open loop, as shown in Figure 8.12a, the output $C(s)$ is given

$$C(s) = U(s)X(s) \tag{8.70}$$

Substituting this equation in equation (8.69) gives

$$E(s) = U(s)[1 - X(s)] \tag{8.71}$$

Applying the final value theorem, the steady state error e_{ss} is given by

$$e_{ss} = \lim_{s \to 0} s E(s) = \lim_{s \to 0} s U(s)[1 - X(s)] \tag{8.72}$$

So, considering this equation, the steady-state error e_{ss} is given by

1. For step input:

$$e_{ss} = \lim_{s \to 0} s\,E(s) = \lim_{s \to 0} s\,\frac{A_i}{s}[1 - X(s)] \tag{8.73}$$

2. For impulse input:

$$e_{ss} = \lim_{s \to 0} s\,E(s) = \lim_{s \to 0} s\,I_i[1 - X(s)] \tag{8.74}$$

3. For ramp input:

$$e_{ss} = \lim_{s \to 0} s\,E(s) = \lim_{s \to 0} s\,\frac{R_i}{s^2}[1 - X(s)] \tag{8.75}$$

Now, considering the transfer function $X(s)$ in closed loop, as shown in Figure 8.12a, the output $C(s)$ is given

$$C(s) = U(s)\frac{X(s)}{1 + X(s)}$$

Substituting this equation in equation (8.69) and rearranging the results gives

$$E(s) = U(s)[\frac{1}{1 + X(s)}]$$

Applying the final value theorem, the steady state error e_{ss} is given by

$$e_{ss} = \lim_{s \to 0} s\,E(s) = \lim_{s \to 0} s\,U(s)[1 + \frac{1}{1 + X(s)}]$$

So, considering this equation, the steady-state error e_{ss} is given by

1. For step input:

$$e_{ss} = \lim_{s \to 0} s\,E(s) = \lim_{s \to 0} s\,\frac{A_i}{s}[\frac{1}{1 + X(s)}]$$

2. For impulse input:

$$e_{ss} = \lim_{s \to 0} s\,E(s) = \lim_{s \to 0} s\,I_i[\frac{1}{1 + X(s)}]$$

3. For ramp input:

$$e_{ss} = \lim_{s \to 0} s\,E(s) = \lim_{s \to 0} s\,\frac{R_i}{s^2}[\frac{1}{1 + X(s)}]$$

Example (8.1):
For the rotational mechanical system shown in Figure 8.13, assuming that $k_t = 50$
(N/m), $b_t = 10$ (N.s/m) and $T(t) = 20\,\delta(t)$ (N m), (impulse input), determine

1. The transfer function $\Theta(s)/T(s)$.
2. The steady-state gain K and time constant τ.
3. The angular displacement $\theta(t)$ and angular velocity $\omega(t)$ response equations and
   plot the responses over a time interval ranging from $t = 0$ to 5τ.
4. The steady-state value Θ_{ss} and the steady-state error e_{ss}.

Solution: The given system does not include a mass element, as shown in Figure
8.13. Applying Newton's second law gives

$$b_t\,\dot{\theta} + k_t\,\theta = T(t) \tag{8.76}$$

Substituting the values of k_t and b_t in this equation yields

$$10\dot{\theta} + 50\theta = T(t) \tag{8.77}$$

This differential equation represents the dynamic motion of the given system.

1. Transfer function $\Theta(s)/T(s)$:
Assuming zero initial conditions, taking the Laplace transform of equation (8.77)
gives
$$10s\,\Theta(s) + 50\Theta(s) = T(s) \tag{8.78}$$
Rearranging this equation gives

$$\frac{\Theta(s)}{T(s)} = \frac{1}{(10s + 50)} \tag{8.79}$$

This equation is the transfer function of angular displacement θ of the given system.

2. Steady-state gain K and time constant τ:
Rearranging equation (8.79) as follows:

$$\frac{\Theta(s)}{T(s)} = \frac{0.02}{(0.2s + 1)} \tag{8.80}$$

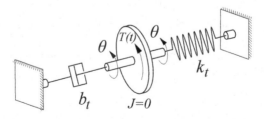

Figure 8.13 First-order rotational mechanical system

Table 8.1

Values of the responses, $\theta(t)$ and $\omega(t)$, calculated from equations (8.83) and (8.84) due to impulse input

	time (sec)	$\theta(t)$ (rad.)	$\omega(t)$ (rad./sec)
0	0	2	-10
0.5τ	0.1	1.21	-6.06
τ	0.2	0.73	-3.69
1.5τ	0.3	0.44	-2.23
2τ	0.4	0.27	-1.35
2.5τ	0.5	0.16	-0.82
3τ	0.6	0.099	-0.49
3.5τ	0.7	0.06	-0.3
4τ	0.8	0.036	-0.18
4.5τ	0.9	0.022	-0.11
5τ	1	0.013	-0.067

Comparing this equation with the general form of the transfer function of a first-order system indicated in equation (8.7), we found that,

$$K = 0.02 \ (\frac{1}{\text{N/m}}) \quad \text{and} \quad \tau = 0.2 \ \text{(sec)} \tag{8.81}$$

3. System responses $\theta(t)$ and $\omega(t)$:

To determine the angular displacement response equation $\theta(t)$, taking the Laplace transform of the impulse input, $T(s) = 20$, inserting it in equation (8.79), and rearranging the results gives

$$\Theta(s) = \frac{20}{(10s + 50)} = \frac{2}{(s + 5)} \tag{8.82}$$

Using Table 7.1, item (6), taking the inverse Laplace transform of equation (8.82), gives

$$\theta(t) = 2e^{-5t} \tag{8.83}$$

This equation is the angular displacement response equation representing the dynamic motion of the first-order system, shown in Figure 8.13, applied to the impulse input. To determine the angular velocity $\omega(t)$ response equation, take the first derivative of equation (8.83) as follows:

$$\omega(t) = \dot{\theta}(t) = -10e^{-5t} \tag{8.84}$$

To plot the responses $\theta(t)$ and $\omega(t)$ due to the impulse input, carry out the calculation of equations (8.83) and (8.84) over a time interval ranging from $t = 0$ to 5τ. The

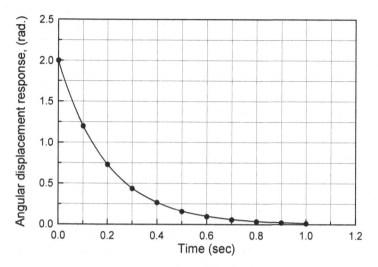

Figure 8.14 The plot of the angular displacement response $\theta(t)$ of the given system applied to the impulse input

obtained results are tabulated in Table 8.1. Insert the calculated results on sketch paper and draw running average curves throughout each point, as illustrated in Figures 8.14 and 8.15. These figures show the plot of the responses $\theta(t)$ and $\omega(t)$.

4. Steady-state value Θ_{ss} and steady-state error e_{ss}:
Considering the final value theorem, the steady-state value Θ_{ss} can be calculated from the transfer function indicated in equation (8.82) as follows:

$$\Theta_{ss} = \lim_{s \to 0} s\, D(s)\, \Theta(s) = \lim_{s \to 0} s * 20\frac{1}{(10s + 50)} = 0 \quad \text{(rad.)} \qquad (8.85)$$

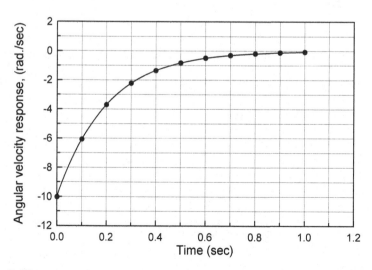

Figure 8.15 Plot of the angular velocity response $\omega(t)$ of the given system applied to the impulse input

We can also calculate the steady-state value from the response equation. Considering the response equation indicated in equation (8.83) and inserting $t = \infty$, the steady-state value Θ_{ss} is given by

$$\Theta_{ss} = \theta(\infty) = \lim_{t \to \infty}(2e^{-5t}) = 0 \quad (\text{rad.}) \tag{8.86}$$

The steady-state error e_{ss} can be calculated as follows. Recalling equation (8.74) gives

$$e_{ss} = \lim_{s \to 0} s * D(s)[1 - \Theta(s)] = \lim_{s \to 0} s * 20[1 - \frac{1}{(10s + 50)}] = 0 \quad (\text{rad.}) \tag{8.87}$$

Example (8.2):
For the translational mechanical system shown in Figure 8.16, assuming that $k = 100$ (N/m), $b = 10$ (N.s/m) and $v_i = 0.5\,u(t)$ (m/sec), (step input), determine

1. The transfer function $X(s)/V_i(s)$.
2. The steady-state gain K and the time constant τ.
3. The displacement $x(t)$ and the velocity $v(t)$ response equations and plot them over a time interval ranging from $t = 0$ to 5τ.
4. The steady-state value X_{ss} and the steady-state error e_{ss}.
5. Use MATLAB to generate the $x(t)$ and $v(t)$ plots from the response equations. Based on the transfer function $X(s)/V_i(s)$, generate the plot of the displacement $x(t)$ using Simulink.

Solution:
1. The transfer function $X(s)/V_i(s)$:
Note that the given system does not include a mass element, as shown in Figure 8.16. Applying Newton's law gives

$$b(\dot{x} - v_i) + kx = 0 \quad \text{or} \quad b\dot{x} + kx = bv_i \tag{8.88}$$

Substituting the values of b and k in this equation gives

$$10\dot{x} + 100x = 10v_i \tag{8.89}$$

Assuming zero initial conditions, taking the Laplace transform of this equation gives

$$10sX(s) + 100X(s) = 10V_i(s) \tag{8.90}$$

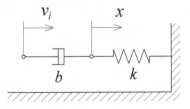

Figure 8.16 First-order translational mechanical system

Rearranging this equation gives

$$\frac{X(s)}{V_i(s)} = \frac{10}{(10s + 100)} \tag{8.91}$$

This equation is the transfer function of the displacement x of the given system.

2. Steady-state gain K and time constant τ:
Rearranging equation (8.91) as follows:

$$\frac{X(s)}{V(s)} = \frac{0.1}{(0.1s + 1)} \tag{8.92}$$

By comparing this equation with the general form of the transfer function of a first-order system indicated in equation (8.7), we found that

$$K = 0.1 \quad (\frac{\text{m/sec}}{\text{N/m}}) \quad \text{and} \quad \tau = 0.1 \text{ (sec)} \tag{8.93}$$

3. System responses $x(t)$ and $v(t)$:
To determine the displacement response equation $x(t)$, taking the Laplace transform of the step input $T(s) = 0.5/s$, inserting it in equation (8.91), and rearranging the results gives

$$X(s) = \frac{0.5 * 10}{s(10s + 100)} = \frac{0.5}{s(s + 10)} \tag{8.94}$$

To determine the system response equation $x(t)$, applying partial fraction expansion on the transfer function indicated in equation (8.94) gives

$$X(s) = \frac{0.5}{s(s + 10)} = \frac{A}{s} + \frac{B}{s + 10} \tag{8.95}$$

Solving this equation for A and B as follows

$$A = \frac{0.5}{(s + 10)}\Big|_{s=0} = \frac{0.5}{10} \tag{8.96}$$

$$B = \frac{0.5}{s}\Big|_{s=-10} = -\frac{0.5}{10} \tag{8.97}$$

Substituting values of A and B in equation (8.84) yields

$$X(s) = \frac{0.5}{10}\frac{1}{s} - \frac{0.5}{10}\frac{1}{s + 10} \tag{8.98}$$

Using Table 7.1, items (2) and (6), taking the inverse Laplace transform of equation (8.98) and rearranging the results gives

$$x(t) = 0.05(1 - e^{-10t}) \tag{8.99}$$

Table 8.2

The values of the responses, $x(t)$ and $v(t)$, calculated from equations (8.99) and (8.100) due to step input

time (sec)	$x(t)$ (m)	$v(t)$ (m/sec)
0	0	0	0.5
0.5τ	0.05	0.0197	0.3033
τ	0.1	.0316	0.1839
1.5τ	0.15	0.0388	0.1116
2τ	0.2	.0432	0.0677
2.5τ	0.25	0.0459	0.41
3τ	0.3	0.0475	0.0249
3.5τ	0.35	0.0485	0.0151
4τ	0.4	0.0491	0.0092
4.5τ	0.45	0.0494	0.0056
5τ	0.5	0.0497	0.0034

This is the displacement response equation of the given system applied to step input. Since Table 7.1, item (14) where $a = 1/\tau = 10$, is similar to the system transfer function term indicated in equation (8.94), we may not need to carry out the analysis above (determining A and B), which means that we may extract the response equation directly from the table.

To determine the velocity $v(t)$ response equation, take the first derivative of equation (8.99) as follows:

$$v(t) = \frac{d}{dt}x(t) = 0.5e^{-10t} \tag{8.100}$$

To plot the responses $x(t)$ and $v(t)$ due to the step input, carry out the calculation of equations (8.99) and (8.100) over a time interval ranging from $t = 0$ to 5τ. The obtained results are tabulated in Table 8.2. Insert the calculated results on sketch paper and draw running average curves throughout each point, as illustrated in Figures 8.17 and 8.18. These figures show the plot of the responses $x(t)$ and $v(t)$.

4. Steady-state value X_{ss} and steady-state error e_{ss}:
Considering the final value theorem, the steady-state value X_{ss} can be calculated as follows:

$$X_{ss} = \lim_{s \to 0} s\, U(s)\, X(s) = \lim_{s \to 0} s\frac{0.5}{s}[\frac{10}{(10s + 100)}] = 0.05 \quad \text{(m)} \tag{8.101}$$

We can also calculate the steady-state value from the response equation. Considering the response equation indicated in equation (8.98) and inserting $t = \infty$, the steady-state value X_{ss} is given by

$$X_{ss} = x(\infty) = \lim_{t \to \infty} 0.05(1 - e^{-10t}) = 0.05 \quad \text{(m)} \tag{8.102}$$

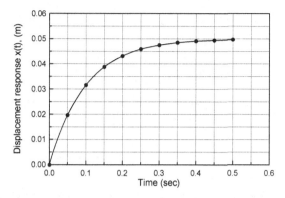

Figure 8.17 The plot of the displacement response $x(t)$ of the given first-order system applied to the step input

The steady-state error e_{ss} can be calculated as follows. Recalling equation (8.73) gives

$$e_{ss} = \lim_{s \to 0} s * \frac{0.5}{s}[1 - \frac{10}{(10s + 100)}] = 0.45 \quad (m) \tag{8.103}$$

This equation indicates that the value of the steady-state error is too high. This may be attributed to the fact that the damper and spring are connected in series.

5. Generate $x(t)$ and $v(t)$ responses using MATLAB and Simulink:
Utilizing a computer program, such as Matlab, will generate the responses immediately by carrying out an extensive computation, varying the time from 0 to 5τ, applying to the response equations indicated in equations (8.99) and (8.100), through the program's toolboxes. The Matlab live script file and the generated response plots of $x(t)$ and $v(t)$ are shown in Figure 8.19.

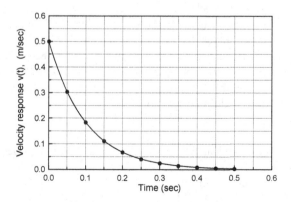

Figure 8.18 Plot of the velocity response $v(t)$ of the given first-order system applied to the step input

```
% Generating the time response of a first-order system applied to step input.
% Define the simulation time interval
t=[0:0.01:0.6];;
% Write the response equation of the displacement x(t).
xt=(0.5/10)*(1-exp(-10*t));
% Write the response equation of the velocity v(t).
vt=0.5*exp(-10*t);
% Now, Let us observe the time response of our system.
% Use the subplot command to place the figures side by side.
fig_raw_resp = figure('position', [0, 0, 3000, 1500]);
subplot(1,2,1); plot(t, xt); title('Displacement response'); xlabel('Time (sec)');
ylabel('Displacement x(t)')
subplot(1,2,2); plot(t, vt); title('Velocity response'); xlabel('Time (sec)');
ylabel('Velocity v(t)')
```

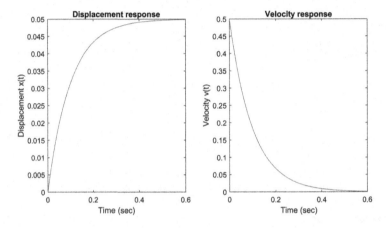

Figure 8.19 MATLAB live script file generates the system responses $x(t)$ and $v(t)$ due to step input

We may create a Simulink model using the Simulink software, as shown in Figure 8.20. As shown in this figure, two alternative Simulink models can be created. The first one, the Simulink block diagram, is created using the system differential equation (8.89), as shown in Figure 8.20a. On the other hand, the second one, the Simulink transfer function block diagram, is created directly using the system transfer function indicated in equation (8.91), as shown in Figure 8.20b. The parameter values, $b = 10$ (N m/sec), $k = 50$ (N/m), and $v_i = 0.5$ (m/sec), are set to their respective default values in Simulink models. Figure 8.21 shows the displacement response $x(t)$ plot generated from both Simulink models.

Example (8.3):
For the translational mechanical system shown in Figure 8.22, consider the values of the following parameters: $m = 2$ (kg), $b = 4$ (N.s/m) and $f(t) = r(t)$ (N) where $r(t) = t$ (unit Ramp input). Determine

1. The transfer function $V(s)/F(s)$.
2. The steady-state gain K and the time constant τ,

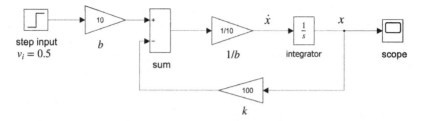

(a) Simulink block diagram model

- To create the Simulink block diagram model, rearrange the equation of motion of the given system, indicated in equation (8.89), as follows:

$$\dot{x} = \frac{1}{b}[bv_i - kx] = \frac{1}{10}[10v_i - 100x]$$

- Use the Simulink Library Browser to build the following Simulink model: see Appendix (2).

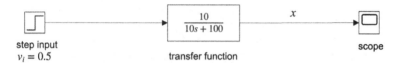

(b) Simulink transfer function block diagram model

- To create the Simulink transfer function block diagram, consider the transfer function indicated in equation (8.91) and use the Simulink Library Browser to build the

Figure 8.20 Simulink models of the given system

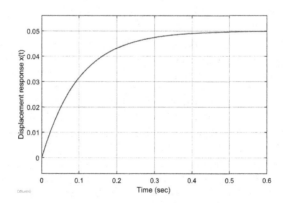

Figure 8.21 Displacement response $x(t)$ generated from Simulink block diagrams

3. The mass velocity $v(t)$ and displacement $x(t)$ response equations.
4. plot $v(t)$ and $x(t)$ over a time interval ranging from $t = 0$ to 6τ.

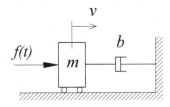

Figure 8.22 First-order mechanical system includes mass element

5. The steady-state value, V_{ss}, and the steady-state error, e_{ss}.
6. Use MATLAB to generate the plot of $v(t)$ and $x(t)$ responses.

Solution:
Note that the given system includes a mass element, as shown in Figure 8.22. When a first-order system includes a mass element, the independent coordinate must be the mass velocity; otherwise, the system will be a second-order system. So, applying Newton's second law to the systems shown in Figure 8.22 gives

$$m\dot{v} + bv = f(t) \tag{8.104}$$

Substituting the values of m, b and $f(t)$ in this equation yields

$$2\dot{v} + 4v = f(t) \tag{8.105}$$

1. Transfer function $V(s)/F(s)$:
Assuming zero initial conditions, taking the Laplace transform of equation (8.105) and rearranging the results gives

$$\frac{V(s)}{F(s)} = \frac{1}{(2s+4)} \tag{8.106}$$

This equation is the transfer function of the velocity response v of the given system.
2. Steady-state gain K and time constant τ:
Rearranging equation (8.106) as follows:

$$\frac{V(s)}{F(s)} = \frac{0.25}{(0.5s+1)} \tag{8.107}$$

Comparing this equation with the general form of a first-order indicated in equation (8.7), we found that

$$K = 0.25 \ (\frac{1}{N\,m}) \quad \text{and} \quad \tau = 0.5 \ (\text{sec})$$

3. System responses $v(t)$ and $x(t)$:
To determine the velocity response equation $x(t)$, taking the Laplace transform of the unit ramp input, $F(s) = 1/s^2$, inserting it in equation (8.106), and rearranging the results gives

$$V(s) = \frac{0.5}{s^2(s+2)} \tag{8.108}$$

Now, apply partial fraction expansion on this equation as follows.

$$V(s) = \frac{0.5}{s^2(s+2)} = \frac{A}{s} + \frac{B}{s^2} + \frac{C}{s+2} \tag{8.109}$$

Rearranging this equation gives

$$V(s) = \frac{0.5}{s^2(s+2)} = \frac{As(s+2) + B(s+2) + Cs^2}{s^2(s+2)} \tag{8.110}$$

The clearing fraction gives

$$V(s) = \frac{0.5}{s^2(s+2)} = \frac{(A+C)s^2 + (2A+B)s + 2B}{s^2(s+2)} \tag{8.111}$$

Solving this equation for A, B, and C is as follows. Equating like terms gives

$$s^2 : A + C = 0 \tag{8.112}$$

$$s : 2A + B = 0 \tag{8.113}$$

$$0 : 2B = 0.5 \tag{8.114}$$

Solving these equations gives

$$A = -\frac{0.5}{4}, \quad B = \frac{0.5}{2} \text{ and } C = \frac{0.5}{4} \tag{8.115}$$

Substituting these values in equation (8.109) gives

$$V(s) = -\frac{0.5}{4}\frac{1}{s} + \frac{0.5}{2}\frac{1}{s^2} + \frac{0.5}{4}\frac{1}{(s+2)} \tag{8.116}$$

Using Table 7.1, items (2), (3), and (6), taking the inverse Laplace transform of this equation give

$$v(t) = -\frac{0.5}{4} + \frac{0.5}{2}t + \frac{0.5}{4}e^{-2t} \tag{8.117}$$

This equation is the velocity response equation of the given system applied to the ramp input. To determine the displacement $x(t)$ response equation, take the integration of equation (8.117) as follows:

$$x(t) = \int \left(-\frac{0.5}{4} + \frac{0.5}{2}t + \frac{0.5}{4}e^{-2t}\right) dt = \left(-\frac{0.5}{4}t + \frac{0.5}{4}t^2 - \frac{0.5}{8}e^{-2t}\right) + C_1 \tag{8.118}$$

Assuming zero initial conditions and solving this equation for C_1 gives

$$x(0) = 0 = 0 + 0 - \frac{0.5}{8} + C_1 \quad \text{so} \quad C_1 = \frac{0.5}{8} \tag{8.119}$$

Substituting the value of C_1 in equation (8.118) gives

$$x(t) = -\frac{0.5}{4}t + \frac{0.5}{4}t^2 - \frac{0.5}{8}e^{-2t} + \frac{0.5}{8} \tag{8.120}$$

Table 8.3

Values of the responses, $x(t)$ and $v(t)$, calculated from equations (8.117) and (8.120) due to ramp input

	time (sec)	$x(t)$ (m)	$v(t)$ (m/sec)
0	0	0	0
0.5τ	0.25	0133	0.0012
τ	0.5	0.046	0.0083
1.5τ	0.75	0.0904	0.0251
2τ	1.0	0.1419	0.054
2.5τ	1.25	0.1978	0.0964
3τ	1.5	0.2562	0.1531
3.5τ	1.75	0.3163	0.2247
4τ	2.0	0.3773	0.3114
4.5τ	2.25	0.438	0.4134
5τ	2.5	0.5008	0.5308
5.5τ	2.75	0.563	0.6638
6τ	3	0.6253	0.8123

This is the displacement response equation of the given system applied to unit ramp input.

4. Plot $v(t)$ and $x(t)$:
To plot the responses $v(t)$ and $x(t)$, carry out the calculation of equations (8.117) and (8.120) over a time interval ranging from $t = 0$ to 5τ. The obtained results are tabulated in Table 8.3. Insert the calculated results on sketch paper and draw running average curves throughout each point, as illustrated in Figure 8.23. This figure shows the plot of the responses $x(t)$ and $v(t)$.

5. Steady-state value X_{ss} and steady-state error e_{ss}: Considering the final value theorem, the steady-state value V_{ss} can be calculated as follows,

$$V_{ss} = \lim_{s \to 0} s\, U(s)V(s) = \lim_{s \to 0} s \, \frac{1}{s^2} \frac{1}{(2s+4)} = \infty \quad \text{(m/sec)} \qquad (8.121)$$

Recalling equation (8.75), the steady state error e_{ss} due to the ramp input is given by

$$e_{ss} = \lim_{s \to 0} s * \frac{1}{s^2}[1 - \frac{1}{(2s+4)}] = \infty \quad \text{(m/sec)} \qquad (8.122)$$

6. Generate $v(t)$ and $x(t)$ using MATLAB:
Figure 8.24 shows the MATLAB live script file that generates the system responses from equations (8.117) and (8.120). Note that we extend the simulation stop time in this script so that we may observe the whole behavior of the system's response.

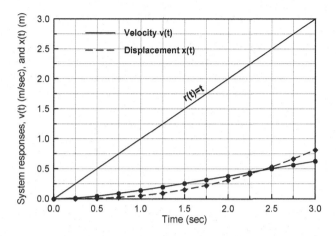

Figure 8.23   The plot of the velocity and displacement responses, $x(t)$ and $v(t)$, of the given system applied to the unit ramp input

```
% Generating the time response of a first-order system applied to ramp input.
% Define the simulation time interval
t=[0:0.25:8];
% Define the ramp input.
u=t;
% Write the response equations v(t) and x(t)
vt=-(0.5/4)+(0.5/2)*t+(0.5/4)*exp(-2*t);
xt=-(0.5/4)*t+(0.5/4)*t.*t-(0.5/8)*exp(-2*t)+(0.5/8);
% Use the subplot command to place the figures side by side as follows.
fig_raw_resp = figure('position', [0, 1000, 4000, 1500]);
subplot(1,2,1); plot(t, vt); hold on, plot(t,u)
title('Velocity response'); xlabel('Time (sec)');
ylabel('Velocity v(t)')
subplot(1,2,2); plot(t, xt);  hold on, plot(t,u)
title('Displacement response'); xlabel('Time (sec)');
ylabel('Displacement x(t)')
```

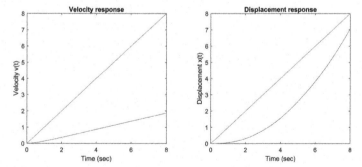

Figure 8.24   MATLAB live script generating the system responses, $x(t)$ and $v(t)$, due to unit ramp input

8.2.3 FIRST-ORDER ELECTRICAL SYSTEMS

Consider the electrical circuit shown in Figure 8.25, and assume that the output terminals are not experiencing any current flow (no load circuit). As a result, there is only one current in the system, and the capacitor voltage e_C equals the output voltage e_o. Applying Kirchhoff's voltage law gives

$$Ri + \frac{1}{C}\int i\,dt = e_i(t) \tag{8.123}$$

and e_o is given by

$$e_o = \frac{1}{C}\int i\,dt \tag{8.124}$$

These two equations are the dynamic equations of the given first-order electrical circuit shown in Figure 8.25. Assuming zero initial conditions, I.C.=0, taking the Laplace transform of equations (8.123) and (8.124), gives

$$RI(s) + \frac{1}{Cs}I(s) = E_i(s) \tag{8.125}$$

and

$$E_o(s) = \frac{1}{Cs}I(s) \tag{8.126}$$

Solving equation (8.125) for $I(s)$ gives

$$I(s) = \frac{Cs}{RCs + 1}E_i(s) \tag{8.127}$$

Substituting equation (8.127) in equation (8.126) and rearranging the results gives

$$\frac{E_o(s)}{E_i(s)} = \frac{1}{RCs + 1} \tag{8.128}$$

This equation represents the transfer function of the first-order electrical system shown in Figure 8.25. Comparing this equation with the general form of a first-order transfer function, we found that the steady-state gain K and the time constant τ are given by

$$K = 1 \quad \text{and} \quad \tau = RC \quad \text{(sec)} \tag{8.129}$$

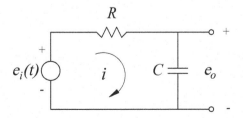

Figure 8.25 First order electrical system

Since equations (8.128) and (8.129) satisfy our description of a first-order electrical system, all standard responses from previous studies on the generalized first-order mechanical system discussed previously in Section 8.2.2 are readily available; there is no need for any extra analyses.

Example (8.4):
Considering the same electrical circuit shown in Figure 8.25, and assuming the values of the following parameters: $R = 5\,(\Omega)$, $C = 0.01$ (F) and $e_i(t) = 5\,u(t)$ (V) (step input), determine

1. The transfer function $E_o(s)/E_i(s)$.
2. The system gain K and the time constant τ.
3. The output voltage response equation $e_o(t)$ and plot the response over a time interval ranging from $t = 0$ to 5τ.
4. The steady-state value E_{ss} and the steady-state error e_{ss}.

Solution:
1. Transfer function $E_o(s)/E_i(s)$:
Considering the transfer function indicated in equation (8.128) and substituting values of $R = 5\,(\Omega)$, $C = 0.01$ (F) gives

$$\frac{E_o(s)}{E_i(s)} = \frac{1}{(0.05s + 1)} \tag{8.130}$$

Rearranging this equation gives

$$\frac{E_o(s)}{E_i(s)} = \frac{20}{(s + 20)} \tag{8.131}$$

This equation is the transfer function of the given circuit shown in Figure 8.25.

2. Steady-state gain K and time constant τ:
Comparing equation (8.130) with the general form of the transfer function of a first-order system indicated in equation (8.7), we found that

$$K = 1 \quad \text{and} \quad \tau = 0.05 \ \text{(sec)}$$

3. Output voltage response equation $e_o(t)$:
To determine the voltage response equation $e_o(t)$, taking the Laplace transform of the step input, $E_i(s) = 5/s$, inserting it in equation (8.131), and rearranging the results gives

$$E_o(s) = \frac{100}{s(s + 20)} \tag{8.132}$$

Since Table 7.1, item (14), where $a = 20$, is similar to the transfer function terms indicated in equation (8.132), we may extract the response equation directly from

---

Table 8.4

Values of the output voltage $e_o(t)$ response, calculated from equation (8.133) due to step input

	time (sec)	$e_o(t)$
0	0	0
0.5τ	0.025	1.976
τ	0.05	3.16
1.5τ	0.075	3.88
2τ	0.1	4.32
2.5τ	0.125	4.59
3τ	0.15	4.75
3.5τ	0.175	4.85
4τ	0.2	4.91
4.5τ	0.225	4.94
5τ	0.25	4.96

---

the table as follows. Considering this, insert $a = 20$ and taking the inverse Laplace transform of equation (8.132) gives

$$e_o(t) = 5(1 - e^{-20t}) \tag{8.133}$$

This response equation represents the dynamics of the given circuit applied to the step input.

To plot the response $e_o(t)$ due to the step input, calculate equation (8.133) over a time interval ranging from $t = 0$ to 5τ. The obtained results are tabulated in Table 8.4. Insert the calculated results on a sketch paper and draw running average curves throughout each point, as illustrated in Figure 8.26. This figure shows the response $e_o(t)$ plot.

4. Steady-state value E_{ss} and the steady state error e_{ss}:
Based on the transfer function indicated in equation (8.132), applying the final value theorem, the steady-state value E_{ss} is given by

$$E_{ss} = \lim_{s \to 0} s\, U(s)\, E_o(s) = \lim_{s \to 0} s \frac{5}{s} \frac{20}{(s+20)}) = 5 \ (V) \tag{8.134}$$

or we can calculate it from the response equation as follows. Considering the response equation indicated in equation (8.133) and inserting $t = \infty$, the steady-state value E_{ss} is given by

$$E_{ss} = e_o(\infty) = \lim_{t \to \infty} 5(1 - e^{-20t}) = 5 \ (V) \tag{8.135}$$

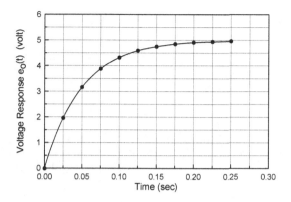

Figure 8.26   The plot of the output voltage $e_o(t)$ response of the given circuit applied to the step input

The steady-state error e_{ss} can be calculated as follows. Recalling equation (8.73) gives

$$e_{ss} = \lim_{s \to 0} s \frac{5}{s} [1 - \frac{20}{s+20}] = 0 \quad (V) \tag{8.136}$$

8.2.4   FIRST-ORDER LIQUID-LEVEL SYSTEMS

Considering the liquid-level system shown in Figure 8.27, the variation of the liquid level at the operating point is given by

$$A\frac{dh}{dt} + \frac{h}{R_v} = q_{in} \tag{8.137}$$

Rearranging this equation gives

$$AR_v\frac{dh}{dt} + h = R_v q_{in} \tag{8.138}$$

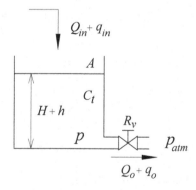

Figure 8.27   First-order liquid level system

Taking the Laplace transform of this equation yields

$$AR_v sH(s) + H(s) = R_v q_{in}(s) \qquad (8.139)$$

Rearranging this equation gives

$$\frac{H(s)}{q_{in}(s)} = \frac{R_v}{AR_v s + 1} \qquad (8.140)$$

This equation represents the transfer function of a first-order liquid-level system. Comparing this equation to the general form of the first-order system indicated in equation (8.7), we found that the steady-state gain K and the time constant τ are given by

$$K = R_v \; \left(\frac{m}{m^3/sec}\right) \quad \text{and} \quad \tau = AR_v \; (sec) \qquad (8.141)$$

Since equations (8.140) and (8.141) satisfy our description of a first-order liquid level system, all the standard responses from previous studies on the generalized first-order mechanical system presented in Section 8.2.2 are readily available; no additional analyses are required.

Example (8.5):
Considering the same liquid-level system shown in Figure 8.27, assuming the values of the following parameters: A=0.02 (m^2), $R_v = 0.1 * 10^4$ $(m/(m^3/sec))$, and $q_{in} = 1.8 * 10^{-4} \delta(t)$ (m^3/sec) (impulse input), determine

1. The transfer function $H(s)/q_{in}(s)$.
2. The steady-state gain K and time constant τ.
3. The liquid-level $h(t)$ response equation and plot the response over a time interval ranging from $t = 0$ to 5τ.
4. The steady-state value H_{ss} and the steady-state error e_{ss}.

Solution:
First, we assume that the system is at a steady state, then the inflow Q_{in} is added by an amount of $q_{in} = 1.8 * 10^{-4}$ (m^3/sec) (impulse input), as shown in Figure 8.27.

- Transfer function $H(s)/q_{in}(s)$:
Considering the transfer function indicated in equation (8.140) and inserting parameter values gives

$$\frac{H(s)}{q_{in}(s)} = \frac{0.1 * 10^4}{0.02 * 0.1 * 10^4 s + 1} = \frac{0.1 * 10^4}{(20s + 1)} \qquad (8.142)$$

This equation is the transfer function of the given system.

- Steady-state gain K and time constant τ:
Comparing the transfer function indicated in equation (8.142) with the general form of a first-order, we found that the steady-state gain K and the time constant τ are given by

$$K = 0.1 * 10^4 \; \left(\frac{m}{m^3/sec}\right) \quad \text{and} \quad \tau = 20 \; (sec)$$

Table 8.5

Values of the liquid level response $h(t)$, calculated from equation (8.145) due to impulse input

	time (sec)	$h(t)$ (m)
0	0	0.009
0.5τ	10	0.0055
τ	20	0.0033
1.5τ	30	0.002
2τ	40	0.0012
2.5τ	50	0.0007
3τ	60	0.0004
3.5τ	70	0.00027
4τ	80	0.00018
4.5τ	90	0.00012
5τ	100	.0001

- Liquid level response equation $h(t)$:

Taking the Laplace transform of the inflow $q_{in}(s) = 1.8 * 10^{-4}$ (m^3/sec) and substituting it in equation (8.142), gives

$$H(s) = \frac{(0.1 * 10^4) * (1.8 * 10^{-4})}{20s + 1} = \frac{0.18}{20s + 1} \tag{8.143}$$

Rearranging this equation gives

$$H(s) = \frac{0.009}{s + 0.05} \tag{8.144}$$

To obtain the response equation of the liquid level, we must take the inverse Laplace transform of this equation. Since Table 7.1 item (6) is similar to the system transfer function indicated in equation (8.144), we may extract the response equation directly from this table as follows.

$$h(t) = 0.009e^{-0.05t} \tag{8.145}$$

This equation is the liquid-level response equation representing the dynamics of the first-order system applied to impulse input, shown in Figure 8.27.

To plot the response $h(t)$ due to the impulse input, calculate equation (8.145) over a time interval ranging from $t = 0$ to 5τ. The obtained results are tabulated in Table 8.5. Insert these results on a piece of sketch paper and draw running average curves throughout each point, as illustrated in Figure 8.28. This figure shows the response $h(t)$ plot.

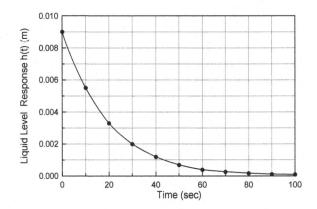

Figure 8.28 The plot of the liquid level $h(t)$ due to impulse input

3. Steady-state value H_{ss} and the steady state error e_{ss}:

The steady-state value H_{ss} can be obtained by applying the final value theorem to the transfer function indicated in the equation (8.144).

$$H_{ss} = \lim_{s \to 0} s \, D(s) H(s) = \lim_{s \to 0} s * (1.8 * 10^{-4}) \frac{0.1 * 10^4}{20s + 1} = 0 \quad \text{(m)} \qquad (8.146)$$

Alternatively, we can calculate it from the response equation. Considering the response equation indicated in equation (8.145) and inserting $t = \infty$, the steady-state value H_{ss} is given by

$$H_{ss} = h(\infty) = \lim_{t \to \infty} 0.009 e^{-0.05t} = 0 \quad \text{(m)} \qquad (8.147)$$

The steady-state error e_{ss} can be calculated as follows. Recalling equation (8.74) gives

$$e_{ss} = \lim_{s \to 0} s * (1.8 * 10^{-4})[1 + \frac{0.1 * 10^4}{20s + 1}] = 0 \quad \text{(m)} \qquad (8.148)$$

8.2.5 FIRST-ORDER PNEUMATIC SYSTEMS

Consider the pneumatic system in Figure 8.29 and assume that the system operates at a subsonic flow condition throughout the operations. Recalling equation (5.150) as follows

$$R_p C_p \frac{dp_o}{dt} + p_o = p_i \qquad (8.149)$$

Taking the Laplace transform of this equation as follows:

$$R_p C_p s \, p_o(s) + p_o(s) = p_i(s) \qquad (8.150)$$

Rearranging this equation gives

$$\frac{p_o(s)}{p_i(s)} = \frac{1}{R_p C_p s + 1} \qquad (8.151)$$

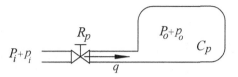

Figure 8.29 First order pneumatic system

This equation represents the transfer function of the first-order pneumatic system shown in Figure 8.29. Compare this equation with the general form of the first-order system as follows:

$$\frac{p_o(s)}{p_i(s)} = \frac{1}{(R_pC_p)s + 1} = \frac{K}{\tau s + 1} \tag{8.152}$$

From this equation, the steady-state gain K and the time constant τ are given by

$$K = 1 \quad \text{and} \quad \tau = R_pC_p \text{ (sec)} \tag{8.153}$$

where R_p and C_p are the pneumatic resistance and capacitance, respectively.

8.2.6 FIRST-ORDER THERMAL SYSTEMS

The variety of a simple first-order thermal system that may display is more restricted than the various mechanical, electrical, and liquid-level systems that can display because it has only two elements (capacitance and resistance). Considering a thermal system shown in Figure 8.30 and recalling equation (5.239) as follows:

$$\left(\frac{MC}{hA_s}\right)\frac{dT_o}{dt} + T_o = T_i \tag{8.154}$$

Taking the Laplace transform of this equation gives:

$$\left(\frac{MC}{hA_s}\right)sT_o(s) + T_o(s) = T_i(s) \tag{8.155}$$

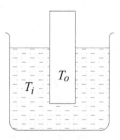

Figure 8.30 First-order thermal system

Rearranging this equation gives

$$\frac{T_o(s)}{T_i(s)} = \frac{1}{(\frac{MC}{hA_s})s + 1} \tag{8.156}$$

This equation represents the transfer function of the first-order thermal system shown in Figure 8.30. Comparing this equation with the general form of the first-order system, we found that,

$$K = 1 \text{ and } \tau = (\frac{1}{hA_s})(MC) = (R_t)(C_t) \text{ (sec)} \tag{8.157}$$

where R_t and C_t are thermal resistance and capacitance, respectively, which are given by

$$R_t = \frac{1}{hA_s} \text{ and } C_t = MC \tag{8.158}$$

From equation (8.156), we can see that for T_o to have a quick response (small τ) to T_i, the values h and A_s need to be quite large, while the values M and C need to be relatively small. According to the definition of fast response in terms of thermal resistance R_t and thermal capacitance C_t, both must be small, exactly like R and C in first-order electrical systems.

8.3   SECOND-ORDER SYSTEMS

Since many useful machines and processes are made by connecting different parts or subsystems, we will look at how this can lead to second-order systems. Second-order system models can show up independently when a whole system is seen as a group of separate parts. To make the whole system, we can also make these models by connecting two subsystems, each of which can be modeled as a first-order system.

There are two approaches to analyzing such a system. The first approach is to interact with both subsystems; the first subsystem is entirely separate from the second and is connected in a cascade way. The second approach is to analyze the system when the two subsystems initially interact as a combined system. These approaches apply to any dynamic system (mechanical, electrical, liquid, or thermal).

8.3.1   SECOND-ORDER MECHANICAL SYSTEMS

8.3.1.1   Second-order mechanical system formed from two first-order subsystems

In this case, the system consists of two first-order subsystems that are initially separated. However, the separation is split after the connection between the two subsystems occurs. Consider the mechanical system shown in Figure 8.31, which is a combination of the two first-order systems shown in Figures 8.16 and 8.22 presented in examples (8.2) and (8.3). When the two subsystems come into contact, the

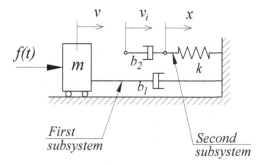

Figure 8.31 Second-order system formed by connecting two first-order subsystems

first subsystem's output will be the second subsystem's input. This is what we mean when we talk about a cascade connection. In this case, the second subsystem acts as a loading effect on the first subsystem. Applying Newton's law on the first subsystem shown in Figure 8.31 gives

$$m\dot{v} + b_1 v = f(t) \tag{8.159}$$

Taking the Laplace transform of this equation and rearranging the results gives

$$\frac{V(s)}{F(s)} = \frac{(1/b_1)}{(m/b_1)s + 1} \tag{8.160}$$

Similarly, applying Newton's law on the second subsystem gives,

$$b_2(\dot{x} - v_i) + kx = 0 \tag{8.161}$$

Taking the Laplace transform of this equation and rearranging the results to gives

$$\frac{X(s)}{V_i(s)} = \frac{(b_2/k)}{(b_2/k)s + 1} \tag{8.162}$$

It is to be noted that whenever the first subsystem moves and comes into contact with the second subsystem, the velocity $V_i(s)$, which is the input of the second subsystem, will be equal to the velocity $V(s)$, which is the output of the first subsystem. Considering this fact, we get the total transfer function if we multiply equations (8.160) by (8.162) as a cascade connection, which is divided into two first-order transfer functions, as shown in Figure 8.32. As a result,

$$\frac{X(s)}{F(s)} = \frac{(b_2/(b_1 k))}{(mb_2/kb_1)s^2 + ((mk + b_1 b_2)/b_1 k)s + 1} \tag{8.163}$$

This equation represents the transfer function of a second-order system when both first-order subsystems are initially separated.

Let us examine the system without applying the separated first-order models and compare the results with equation (8.163). Since both analyses are performed with

$$\xrightarrow{F(s)} \boxed{\dfrac{(1/b_1)}{(m/b_1)\,s+1}} \xrightarrow{V(s)} \boxed{\dfrac{(b_2/k)}{(b_2/k)\,s+1}} \xrightarrow{X(s)}$$

Figure 8.32 Block diagram of a second-order system formed from two first-order systems

the same system, the results are expected to be identical. Assuming that the two subsystems are in contact, as shown in Figure 8.33, applying Newton's second law gives

$$m\dot{v} + b_1 v + b_2(v - \dot{x}) = f(t) \tag{8.164}$$

and

$$b_2(\dot{x} - v) + kx = 0 \tag{8.165}$$

Taking the Laplace transform of equations (8.164) and (8.165) yields

$$[ms^2 + (b_1 + b_2)s]V(s) = F(s) \tag{8.166}$$

and

$$(b_2 s + k)X(s) = b_2 s V_i(s) \tag{8.167}$$

Remember that when both subsystems are in contact, the velocity $V_i(s)$ equals the velocity $V(s)$. Considering this, solving equations (8.166) and (8.167) for $X(s)$ gives

$$\frac{X(s)}{F(s)} = \frac{(b_2/(k(b_1 + b_2)))}{(mb_2/k(b_1 + b_2))s^2 + (mk + b_1 b_2)/k(b_1 + b_2))s + 1} \tag{8.168}$$

This equation represents the transfer function of the second-order system when two first-order systems are initially in contact.

When we compare equation (8.168) with equation (8.163), they are completely different. This may be attributed to the fact that the first approach, which looks at the system as a collection of two separate first-order subsystems, may need to be corrected. This is because this result and this approach are usually not good enough to let two first-order subsystems come together to form a second-order system. The problem with the first approach is that specific physical assumptions are typically made when the two first-order models are analyzed as separated subsystems but are

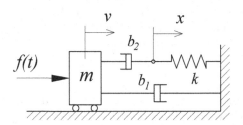

Figure 8.33 Two first-order subsystems

at least in part violated when the two systems are coupled. This problem happens with a link between two independently analyzed subsystems, not just the first-order case. This may be attributed to the fact that when the second subsystem uses power from the first subsystem, that was not specified in the first subsystem's model. So, when the second subsystem is connected, the transfer function indicated in equation (8.168) changes, and it is not just the addition of the two transfer functions shown on their own. This phenomenon is known as a loading effect; in some instances, it is insignificant enough to ignore, while in others, it is so significant, as in our case, that we need to reanalyze the whole system from the beginning rather than attempting to use models that have been established for each separated subsystem. Now, let us analyze the second-order system independently.

8.3.1.2   Second-order mechanical system formed independently

Second-order system models may arise when a complete system is mechanically studied separately. Figure 8.34 shows second-order mechanical systems, translational and rotational. Applying Newton's second law on both systems gives

a. For the translational system, shown in Figure 8.34a

$$m\frac{d^2x}{dt^2} + b\frac{dx}{dt} + kx = f(t) \tag{8.169}$$

b. For the rotational system, shown in Figure 8.34b

$$J\frac{d^2\theta}{dt^2} + b_t\frac{d\theta}{dt} + k_t\theta = T(t) \tag{8.170}$$

According to these two equations, the displacement x and the angular displacement θ are the independent coordinates of second-order systems, translational and rotational,

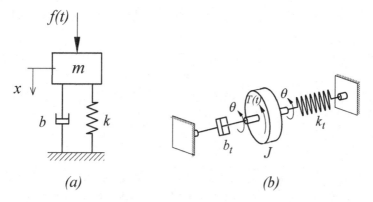

(a)   *(b)*

Figure 8.34   Mechanical second-order systems: (a) translational system; (b) rotational system

respectively. In the following analysis, we will only consider equations (8.169), representing the second-order translational system shown in Figure 8.34a. We prefer to use this configuration because it is widely considered the most straightforward system for imparting the fundamental principles of mechanical shocks and vibrations to students studying the critical subject of mechanical engineering. Considering this system and rewriting equation (8.169) as follows

$$\frac{d^2x}{dt^2} + \frac{b}{m}\frac{dx}{dt} + \frac{k}{m}x = \frac{1}{m}f(t) \tag{8.171}$$

Now, let us define

$$2\zeta\omega_n = \frac{b}{m}, \quad \omega_n^2 = \frac{k}{m} \text{ and } K = \frac{1}{m} \tag{8.172}$$

where ζ is the damping ratio, ω_n is the natural frequency, and K is the steady-state gain. Based on these relations, solving equation (8.172) for ζ and ω_n gives

$$\omega_n = \sqrt{\frac{k}{m}} \text{ and } \zeta = \frac{b}{2\sqrt{km}} \tag{8.173}$$

Rearrange this equation for the damping factor b as a function of ζ gives

$$b = 2\zeta\sqrt{km} \tag{8.174}$$

By substituting equation (8.173) in equation (8.171), the general standard form of the differential equation of a second-order system is given by

$$\frac{d^2x}{dt^2} + 2\zeta\omega_n\frac{dx}{dt} + \omega_n^2 x = Kf(t) \tag{8.175}$$

or,

$$\ddot{x} + 2\zeta\omega_n\dot{x} + \omega_n^2 x = Kf(t) \tag{8.176}$$

Taking the Laplace transform of equation (8.176), assuming zero initial conditions, gives

$$(s^2 + 2\zeta\omega_n s + \omega_n^2)X(s) = KF(s) \tag{8.177}$$

From this equation, we get

$$X(s) = \frac{K}{(s^2 + 2\zeta\omega_n s + \omega_n^2)}F(s) \tag{8.178}$$

This equation represents the transfer function of a second-order system applied to the input force $F(s)$. Usually, the general form of equation (8.178) is given by

$$X(s) = \frac{\omega_n^2}{(s^2 + 2\zeta\omega_n s + \omega_n^2)}F(s) \tag{8.179}$$

Based on this equation, it is necessary to obtain the roots of the characteristic equation $(s^2 + 2\zeta\omega_n s + \omega_n^2)$, which is the transfer function denominator, can be obtained using the following quadratic formula,

$$s_{1,2} = -\zeta\omega_n \pm \omega_n\sqrt{\zeta^2 - 1} \tag{8.180}$$

The roots of the complex quadratic formula indicated in this equation are classified as follows:

1. If $\zeta = 0$, the characteristic equation has two imaginary roots.

$$s_{1,2} = \pm i\omega_n$$

2. If $0 < \zeta < 1$, the characteristic equation has two complex conjugate roots,

$$s_1 = -\zeta\omega_n + i\omega_n \sqrt{1-\zeta^2} \text{ and } s_2 = -\zeta\omega_n - i\omega_n \sqrt{1-\zeta^2}.$$

3. If $\zeta = 1$, the characteristic equation has two real and equal roots (repeated roots),

$$s_{1,2} = -\omega_n.$$

4. If $\zeta > 1$, the characteristic equation has two real but not repeated roots,

$$s_1 = -\zeta\omega_n + \omega_n \sqrt{\zeta^2 - 1} \text{ and } s_2 = -\zeta\omega_n - \omega_n \sqrt{\zeta^2 - 1}$$

These roots can be characterized as follows:

1. *Undamped* (No damping): When $\zeta = 0$ ($b = 0$), it means that the damper is removed from the system, and the roots of the characteristic equation become $s_{1,2} = \pm i\omega_n$, which are purely imaginary. This implies that the system will be oscillatory, with ω_n as the natural frequency of the oscillations.

2. *Underdamped* : When $0 < \zeta < 1$, these oscillations will gradually die out by adding a damper to the system and gradually increasing ζ from 0 to 1 over the response time. In this case, the roots of the characteristic equation are $s_1 = -\zeta\omega_n + i\omega_n \sqrt{\zeta^2 - 1}$ and $s_2 = -\zeta\omega_n - i\omega_n \sqrt{\zeta^2 - 1}$ and the system is referred to as underdamped, exhibiting oscillations that eventually die out. Note that the damping ratio ζ never has a negative value.

3. *Critical damped* : When $\zeta = 1$, the characteristic equation is of two repeated roots, $s_1 = -\omega_n$ and $s_2 = -\omega_n$. This type of system is called critically damped, and the damping value results in this type of system is $b_c = 2\sqrt{km}$. The fact that $\zeta = b/b_c$ is the ratio of the actual damping to the critical value enables us to understand why ζ is referred to as the damping ratio and why it is defined as indicated in equation (8.173). Since we no longer have complex roots at this value, it is obvious that $\zeta = 1$ is unique in some system types.

4. *Overdamped* : When $\zeta > 1$, because they do not exhibit natural oscillations, systems with $\zeta > 1$ are over-damped systems. When ζ is extremely large and $\zeta \gg 1$ (b is very large), the damping effect considerably outweighs the inertia effect, implying that the system will behave quite closely to the first-order system. If inertia were completely neglected in the model, it would be first order with $\tau = b/k$. Since no natural system is free of inertia, analyses such as this help decide when inertia has a negligible effect and may be eliminated from the system model.

It is necessary to investigate how such systems respond to input forces: step, impulse, ramp, and sinusoidal. The frequency response, which is the system's response due to the sinusoidal input, will be discussed in Chapter 9.

8.3.2 SECOND-ORDER SYSTEM RESPONSE TO STEP INPUT

Consider the second-order transfer function indicated in equation (8.179) and assume that it is applied to a step input force $f(t) = A_i u(t)$, where the symbol $u(t)$ represents the unit step input. Taking the Laplace transform to the input function of the step input, $F(s) = A_i/s$, and inserting it in equation (8.179) gives

$$X(s) = \frac{A_i \omega_n^2}{s(s^2 + 2\zeta \omega_n s + \omega_n^2)} \tag{8.181}$$

This equation represents the transfer function of a second-order system applied to the step input force. Considering the classifications of the characteristic equation roots stated above, the second-order system step responses are obtained as follows:

1. *Undamped* (No damping $\zeta = 0$):
   Inserting $\zeta = 0$ in equation (8.181) gives

$$X(s) = \frac{A_i \omega_n^2}{s(s^2 + \omega_n^2)} \tag{8.182}$$

Applying partial fraction expansion to this equation yields

$$X(s) = \frac{A_i \omega_n^2}{s(s^2 + \omega_n^2)} = [\frac{A}{s} + \frac{Bs + C}{s^2 + \omega_n^2}] \tag{8.183}$$

Rearranging this equation gives

$$X(s) = \frac{A_i \omega_n^2}{s(s^2 + \omega_n^2)} = [\frac{A(s^2 + \omega_n^2) + s(Bs + C)}{s(s^2 + \omega_n^2)}] \tag{8.184}$$

Clearing fractions gives

$$A_i \omega_n^2 = (A \mid B)s^2 + Cs + A\omega_n^2 \tag{8.185}$$

Solving for A, B, and C as follows. Equating factors of both sides of equation (8.185) gives

$$s^2 : A + B = 0 \tag{8.186}$$

$$s : C = 0 \tag{8.187}$$

$$0 : A\omega_n^2 = A_i \omega_n^2 \text{ So } A = A_i \tag{8.188}$$

Substituting equation (8.188) in equation (8.186) gives

$$B = -A_i \tag{8.189}$$

Note that the value of C equals 0, as indicated in equation (8.187). So, substituting values of A, B and C in equation (8.183) gives

$$X(s) = A_i \frac{1}{s} - A_i \frac{s}{s^2 + \omega_n^2} \tag{8.190}$$

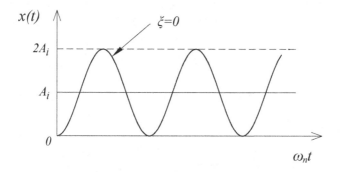

Figure 8.35 Generalized response of the second-order system applied to step input at $\zeta = 0$

Using Table 7.1, items (2) and (11), taking the inverse Laplace transform of equation (8.190) gives

$$x(t) = A_i(1 - \cos \omega_n t) \qquad (8.191)$$

This response equation represents the dynamic motion of a second-order system with no damping ($\zeta = 0$), applied to the step input force. It establishes that only the natural frequency ω_n governs the system response since the system has no damping, meaning that it does not include the exponential term $e^{-\zeta \omega_n t}$. As a result, the response would be a continuous signal with the same amplitude A_i as the input and the same natural frequency, as shown in Figure 8.35.

2. *Underdamping* $(0 < \zeta < 1)$:
Since $\zeta < 1$, ζ^2 is also less than 1, and so $1 - \zeta^2$ is always positive. In this case, the roots of the characteristic equation are complex conjugate as follows

$$s_{1,2} = -\zeta \omega_n \pm \omega_n \sqrt{(-1)(1 - \zeta^2)} = -\zeta \omega_n \pm i\omega_n \sqrt{1 - \zeta^2} \qquad (8.192)$$

So, rearrange the characteristic equation (denominator) of the transfer function indicated in equation (8.179) as follows:

$$(s^2 + 2\zeta \omega_n s + \omega_n^2) = (s + \zeta \omega_n)^2 + (\omega_n \sqrt{1 - \zeta^2})^2 = (s + a)^2 + b^2 \qquad (8.193)$$

where a and b are given by

$$a = \zeta \omega_n \text{ and } b = \omega_n \sqrt{1 - \zeta^2} \qquad (8.194)$$

Replacing equation (8.193) with the characteristic equation (denominator) of the transfer function indicated in equation (8.181) and rearranging the results gives

$$X(s) = \frac{A_i \omega_n^2}{s(s^2 + 2\zeta \omega_n s + \omega_n^2)} = \frac{A_i \omega_n^2}{s[(s + \zeta \omega_n)^2 + (\omega_n \sqrt{1 - \zeta^2})^2]} \qquad (8.195)$$

Applying partial fraction expansion to this equation gives

$$X(s) = \frac{A_i \omega_n^2}{s[(s+\zeta\omega_n)^2 + (\omega_n \sqrt{1-\zeta^2})^2]} = \frac{A}{s} + \frac{Bs+C}{(s+\zeta\omega_n)^2 + (\omega_n \sqrt{1-\zeta^2})^2}$$

(8.196)

Rearranging this equation gives

$$X(s) = \frac{A_i \omega_n^2}{s[(s+\zeta\omega_n)^2 + (\omega_n \sqrt{1-\zeta^2})^2]} =$$

$$\frac{A[(s+\zeta\omega_n)^2 + (\omega_n \sqrt{1-\zeta^2})^2] + (Bs+C)s}{s[(s+\zeta\omega_n)^2 + (\omega_n \sqrt{1-\zeta^2})^2]}$$

(8.197)

Clearing fractions gives

$$A_i \omega_n^2 = (A+B)s^2 + (2\zeta\omega_n A + C)s + \omega_n^2 A$$

(8.198)

Solving this equation for A, B, and C as follows. Equating like terms of equation (8.198) give

$$s^2 : A+B = 0$$

(8.199)

$$s : 2\zeta\omega_n A + C = 0$$

(8.200)

$$0 : \omega_n^2 A = A_i \omega_n^2 \text{ So } A = A_i$$

(8.201)

Substituting equation (8.201) in equation (8.200) gives

$$C = -2A_i \zeta \omega_n$$

(8.202)

From equation (8.199), we have $B = -A$, so using equation (8.201) gives

$$B = -A_i$$

(8.203)

Substituting values of A, B and C in equation (8.196) gives

$$X(s) = A_i \frac{1}{s} - A_i \frac{s}{(s+\zeta\omega_n)^2 + (\omega_n \sqrt{1-\zeta^2})^2} -$$

$$2\zeta\omega_n A_i \frac{1}{(s+\zeta\omega_n)^2 + (\omega_n \sqrt{1-\zeta^2})^2}$$

(8.204)

Using Table 7.1, take the inverse Laplace transform of equation (8.214) as follows:
- For the first term:
  Apply item (2) to the first term of equation (8.204) gives

$$x_1(t) = A_i$$

(8.205)

- For the second term:

We need to rewrite the second term of equation (8.204) so that we can use the Laplace transform table to transform it so it looks like this:

$$X_2(s) = -A_i \frac{s}{(s+\zeta\omega_n)^2 + (\omega_n\sqrt{1-\zeta^2})^2} = -A_i\left[\frac{s+a-a}{(s+a)^2+b^2}\right] =$$
$$-A_i\frac{s+a}{(s+a)^2+b^2} + A_i\frac{a}{b}\frac{b}{(s+a)^2+b^2} \tag{8.206}$$

where a and b are given by equation (8.194). Considering these, equation (8.206) becomes

$$X_2(s) = -A_i\frac{s+\zeta\omega_n}{(s+\zeta\omega_n)^2+(\omega_n\sqrt{1-\zeta^2})^2} +$$
$$A_i\frac{\omega_n\zeta}{\omega_n\sqrt{1-\zeta^2}}\frac{\omega_n\sqrt{1-\zeta^2}}{(s+\zeta\omega_n)^2+(\omega_n\sqrt{1-\zeta^2})^2} \tag{8.207}$$

Now, using Table 7.1 items (20) and (21), the inverse Laplace transform of equation (8.207) is given by

$$x_2(t) = -A_ie^{-\zeta\omega_n t}\cos(\omega_n\sqrt{1-\zeta^2}t) + \frac{A_i\zeta}{\sqrt{1-\zeta^2}}e^{-\zeta\omega_n t}\sin(\omega_n\sqrt{1-\zeta^2}t) \tag{8.208}$$

- For the third term:

Similarly, rewrite the third term of equation (8.204) so that we can use the inverse Laplace transform table item (22). Considering this, the third term of equation (8.204) becomes

$$X_3(s) = -2\zeta\omega_n A_i\frac{1}{b}\frac{b}{(s+\zeta\omega_n)^2+(\omega_n\sqrt{1-\zeta^2})^2} \tag{8.209}$$

where b is given by equation (8.194). Considering this, equation (8.209) becomes

$$X_3(s) = -2\zeta A_i\frac{1}{\sqrt{1-\zeta^2}}\frac{\omega_n\sqrt{1-\zeta^2}}{(s+\zeta\omega_n)^2+(\omega_n\sqrt{1-\zeta^2})^2} \tag{8.210}$$

Taking the inverse Laplace transform Table 7.1, item (20) gives

$$x_3(t) = \frac{-2\zeta A_i}{\sqrt{1-\zeta^2}}e^{-\zeta\omega_n t}\sin(\omega_n\sqrt{1-\zeta^2}t) \tag{8.211}$$

Now, combining equations (8.205), (8.208), and (8.211) gives

$$x(t) = x_1(t) + x_2(t) + x_3(t) = A_i - A_ie^{-\zeta\omega_n t}\cos(\omega_n\sqrt{1-\zeta^2}t) +$$
$$\frac{A_i\zeta}{\sqrt{1-\zeta^2}}e^{-\zeta\omega_n t}\sin(\omega_n\sqrt{1-\zeta^2}t) - \frac{2\zeta A_i}{\sqrt{1-\zeta^2}}e^{-\zeta\omega_n t}\sin(\omega_n\sqrt{1-\zeta^2}t) \tag{8.212}$$

Simplifying this equation gives

$$x(t) = A_i[1 - \frac{e^{-\zeta\omega_n t}}{\sqrt{1-\zeta^2}}\left\{\sqrt{1-\zeta^2}\cos(\omega_n\sqrt{1-\zeta^2}t) - \zeta\sin(\omega_n\sqrt{1-\zeta^2}t)\right\}]$$

(8.213)

If we consider $\sqrt{1-\zeta^2} = \sin\phi$, then ζ will be equal to $\cos\phi$. Considering this, equation (8.213) becomes

$$x(t) = A_i[1 - \frac{e^{-\zeta\omega_n t}}{\sqrt{1-\zeta^2}}\left\{\sin\phi\cos(\omega_n\sqrt{1-\zeta^2}t) - \cos\phi\,\zeta\sin(\omega_n\sqrt{1-\zeta^2}t)\right\}]$$

(8.214)

Now, we can rewrite this equation as follows

$$x(t) = A_i[1 - e^{-\zeta\omega_n t}\frac{\zeta}{\sqrt{1-\zeta^2}}\sin(\omega_n\sqrt{1-\zeta^2}t + \phi)]$$

(8.215)

where $\phi = \tan^{-1}(\frac{\sqrt{1-\zeta^2}}{\zeta})$. Equation (8.215) is the response equation representing the dynamic motion of the under-damped $(0 < \zeta < 1)$ second-order system applied to step input force. Note that the damping ratio never has negative values. So, equation (8.215) establishes that when ζ lies between 0 and 1, the system response exhibits damped oscillations (decreasing amplitude) since its response equation has the exponential term $e^{-\zeta\omega_n}$, as shown in Figure 8.36.

Note that if the transfer function of the system has the form indicated in equation (8.178), the under-damped response equation indicated in equation (8.208) becomes

$$x(t) = \frac{A_i K}{\omega_n^2}[1 - e^{-\zeta\omega_n t}\frac{\zeta}{\sqrt{1-\zeta^2}}\sin(\omega_n\sqrt{1-\zeta^2}t + \phi)]$$

(8.216)

3. *Critically damped* $(\zeta = 1)$:
   Inserting $\zeta = 1$ in equation (8.179) gives

$$X(s) = \frac{A_i\omega_n^2}{s(s^2 + 2\omega_n s + \omega_n^2)} = \frac{A_i\omega_n^2}{s(s + \omega_n)^2}$$

(8.217)

Applying partial fraction expansion to this equation gives

$$X(s) = \frac{A_i\omega_n^2}{s(s + \omega_n)^2} = \frac{A}{s} + \frac{B}{(s + \omega_n)} + \frac{C}{(s + \omega_n)^2}$$

(8.218)

Rearranging this equation gives

$$X(s) = \frac{A_i\omega_n^2}{s(s + \omega_n)^2} = \frac{A(s + \omega_n)^2 + Bs(s + \omega_n) + Cs}{s(s + \omega_n)^2}$$

(8.219)

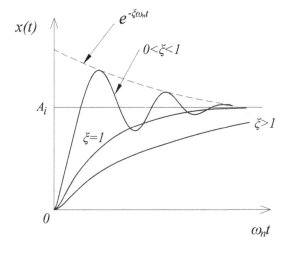

Figure 8.36 Generalized responses of the second-order system at different values of ζ

Clearing fractions gives

$$A_i\omega_n^2 = (A+B)s^2 + (2A\omega_n + B\omega_n + C)s + A\omega_n^2 \qquad (8.220)$$

Solving equation (8.220) for A, B, and C is as follows. Equating factors of both sides of this equation gives

$$s^2 : A + B = 0 \qquad (8.221)$$

$$s : 2A\omega_n + B\omega_n + C = 0 \qquad (8.222)$$

$$0 : A_i\omega_n^2 = A\omega_n^2 \ \text{ So } \ A = A_i \qquad (8.223)$$

From equation (8.221), we have $B = -A$, so

$$B = -A_i \qquad (8.224)$$

Substitute equations (8.223) and (8.224) in equation (8.222) as follows

$$2A_i\omega_n - A_i\omega_n + C = 0 \qquad (8.225)$$

Solving this equation for C, we get

$$C = -A_i\omega_n \qquad (8.226)$$

Substituting values of A, B, and C in equation (8.218) yields

$$X(s) = \frac{A_i\omega_n^2}{s(s+\omega_n)^2} = A_i\frac{1}{s} - A_i\frac{1}{s+\omega_n} - A_i\omega_n\frac{1}{(s+\omega_n)^2} \qquad (8.227)$$

Using Table 7.1, taking the inverse Laplace transform of equation (8.227), items (2), (6), and (7) gives

$$x(t) = A_i(1 - e^{-\omega_n t} - \omega_n t e^{-\omega_n t}) \qquad (8.228)$$

This response equation represents the dynamic motion of the critically damped ($\zeta = 1$) second-order system applied to the step input force. It establishes that when $\zeta = 1$, only the natural frequency ω_n governs the system response, as shown in Figure 8.36.

4. *Overdamping ($\zeta > 1$):*
   Since $\zeta > 1$, ζ^2 is also bigger than 1, and so $\zeta^2 - 1$ is always positive. In this case, the roots of the characteristic equation are real and distinct, as mentioned previously. In the same way, we can rearrange the characteristic equation of the transfer function indicated in equation (8.179) when $\zeta > 1$, as follows:

$$(s^2 + 2\zeta\omega_n s + \omega_n^2) = [s^2 + 2s(\zeta\omega_n) + (\zeta\omega_n)^2] + \omega_n^2 - (\zeta\omega_n)^2 =$$
$$= (s + \zeta\omega_n)^2 - \omega_n^2(\zeta^2 - 1) \tag{8.229}$$

Replacing this equation with the characteristic equation indicated in equation (8.179), the transfer function becomes

$$X(s) = \frac{A_i\omega_n^2}{s(s^2 + 2\zeta\omega_n s + \omega_n^2)} = \frac{A_i\omega_n^2}{s[(s + \zeta\omega_n)^2 - (\omega_n\sqrt{\zeta^2 - 1})^2]} \tag{8.230}$$

or

$$X(s) = \frac{A_i\omega_n^2}{s(s + \zeta\omega_n + \omega_n\sqrt{\zeta^2 - 1})(s + \zeta\omega_n - \omega_n\sqrt{\zeta^2 - 1})} \tag{8.231}$$

Applying partial fraction expansion to this equation gives

$$X(s) = \frac{A}{s} + \frac{B}{(s + \zeta\omega_n + \omega_n\sqrt{\zeta^2 - 1})} + \frac{C}{(s + \zeta\omega_n - \omega_n\sqrt{\zeta^2 - 1})} \tag{8.232}$$

Solving for A, B, and C as follows

$$A = \frac{A_i\omega_n^2}{(s + \zeta\omega_n + \omega_n\sqrt{\zeta^2 - 1})(s + \zeta\omega_n - \omega_n\sqrt{\zeta^2 - 1})}\bigg|_{s=0} = A_i \tag{8.233}$$

$$B = \frac{A_i\omega_n^2}{s(s + \zeta\omega_n - \omega_n\sqrt{\zeta^2 - 1})}\bigg|_{s=-\zeta\omega_n-\omega_n\sqrt{\zeta^2-1}} = \frac{A_i\omega_n}{2\sqrt{\zeta^2 - 1}} \tag{8.234}$$

$$C = \frac{A_i\omega_n^2}{s(s + \zeta\omega_n + \omega_n\sqrt{\zeta^2 - 1})}\bigg|_{s=-\zeta\omega_n+\omega_n\sqrt{\zeta^2-1}} = \frac{-A_i\omega_n}{2\sqrt{\zeta^2 - 1}} \tag{8.235}$$

Substituting equations (8.233), (8.234), and (235) in equation (8.232) gives

$$X(s) = A_i\frac{1}{s} + \frac{A_i\omega_n}{2\sqrt{\zeta^2 - 1}}\left(\frac{1}{(s + \zeta\omega_n + \omega_n\sqrt{\zeta^2 - 1}}\right) -$$
$$\frac{A_i\omega_n}{2\sqrt{\zeta^2 - 1}}\left(\frac{1}{(s + \zeta\omega_n - \omega_n\sqrt{\zeta^2 - 1}}\right) \tag{8.236}$$

Using Table 7.1, items (2) and (6), taking the inverse Laplace transform of this equation gives

$$x(t) = A_i[1 + \frac{\omega_n}{2(\sqrt{\zeta^2-1})}e^{-(\zeta\omega_n+\omega_n\sqrt{\zeta^2-1})t} - \frac{\omega_n}{2(\sqrt{\zeta^2-1})}e^{-(\zeta\omega_n-\omega_n\sqrt{\zeta^2-1})t}]$$

(8.237)

This response equation represents the dynamic motion of the over-damped, ($\zeta > 1$), second-order system applied to the step input force. Since it is over-damped, the step response of the second-order system will never reach the steady-state output value, as shown in Figure 8.36.

Finally, let us summarize the effect of different values of ζ discussed above on a second-order system applied to step input force.

1. Undamped (oscillatory system): A system is considered oscillatory or free vibration when $\zeta = 0$. In this case, the roots are given by

$$s_{1,2} = \pm i\omega_n$$

(8.238)

This equation indicates that the roots are complex with no real parts. They are located precisely on the imaginary axis of the s-plane, as shown in Figure 8.37c. There is no damping in this situation, implying that the system response would be a constant continuous-time signal, as shown in Figure 8.37d. So, the system is called marginally stable. The system will become unstable if the roots are moved to the right side of the s-plane, as shown in Figures 8.37a and 8.37b.

2. Under-damped system: A system is considered under-damped when the value of $\zeta < 1$ is less than one. In this case, the roots are given by

$$s_{1,2} = -\zeta\omega_n \pm i\omega_n\sqrt{1-\zeta^2}$$

(8.239)

This equation indicates that the roots are complex and the real parts are always negative. They are located on the left side of the complex region of the s-plane, as shown in 8.37e, and the system, in this case, is asymptotically stable, as shown in Figure 8.37f. Note that the value of the damping ratio ζ never has a negative value.

3. Critically damped system: A system is considered critically damped when $\zeta = 1$. In this case, the roots are given by

$$s_{1,2} = -\omega_n$$

(8.240)

This equation indicates that the roots are real, always repetitive, and negative. They are located on the left side of the real axis of the s-plane, as shown in Figure 8.38a. So, the system is stable, and there is no finite overshoot, as shown in figure Figure 8.38b.

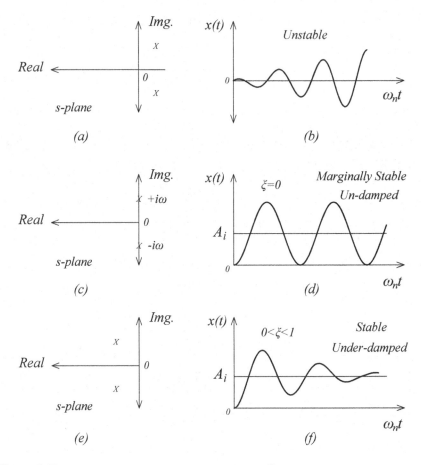

Figure 8.37 Second-order system at different values of ζ

4. Over-damped system: A system is considered over-damped when $\zeta > 1$. In this case, the roots are given by

$$s_{1,2} = -\zeta \omega_n \pm \omega_n \sqrt{\zeta^2 - 1} \qquad (8.241)$$

This equation indicates that the roots are real and distinct, and the real parts are always negative. They are located on the left side of the real axis of the s-plane, as shown in Figure 8.38c. So, the system is stable, and there is no finite overshoot, as shown in Figure 8.38d.

In general, equations (8.191), (8.215), (8.228), and (8.237) are the response equations representing the dynamic motion of a second-order system applied to step input, considering all damping conditions stated above. As a result, the system response to step input shows damped oscillations (decreasing amplitude) since its response equation contains the exponential component $e^{-\zeta \omega_n t}$. As a comparison, this damped

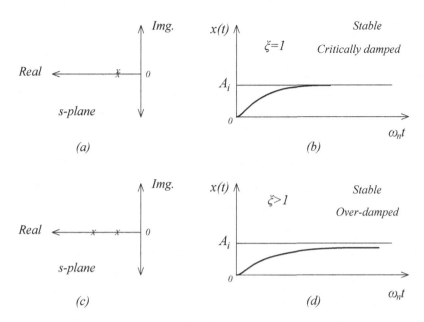

Figure 8.38 Second-order system at different values of ζ

oscillation component is presented only when $0 < \zeta < 1$, equation (8.215), but not in the other responses.

From the above analyses, we found that three parameters, K, ζ, and ω_n, are required to fully characterize the second-order system's dynamic motion. The following analyses will reveal the physical relevance of these parameters and the reasons for their definitions.

1. If we consider the standard form of the transfer function of the second-order system indicated in equation (8.178) and its under-damped response equation indicated in equation (8.216), the steady-state gain K has precisely the same meaning as in a first-order system, it is the steady-state output value generated for each unit of the input A_i. For all damping cases, the effect of K is always a straightforward scaling up or down of the response since it always appears as a multiplying factor on the input value of a transfer function. Therefore, K does not affect the speed of the response when we use the standard approach of defining it, which is the amount of time it takes to complete a specific proportion of the overall change while moving from one steady state to another.

2. The natural frequency ω_n is usually written as $\omega_n t$ in response equations, implying that it may quickly determine its effect on the system's dynamics. This is shown in Figure 8.36, when $0 < \zeta < 1$, which illustrates the motion as a damped oscillation with frequency ω_n, emphasizing the importance of the variable ω_n and

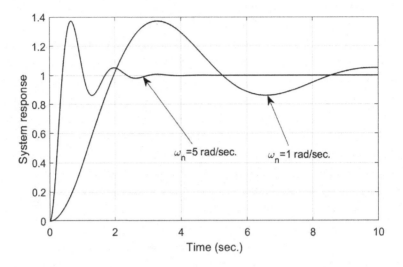

Figure 8.39 The step response for $\zeta = 0.3$ at $\omega_n = 1$ and $\omega_{=}5$

justifying its definition as indicated in the system responses equations. When K and ζ keep constant while ω_n is varied, we see that increasing ω_n results in a directly proportionate increase in the speed of the response, as shown clearly in Figure 8.39.

3. For every second-order system, regardless of how large or small the parameters K, ζ, and ω_n of A_i are, we may obtain a family of curves by plotting generalized and scaled step responses for various ζ values, as shown Figure 8.40. Because ζ enters the response equation in a complicated manner, we must depend on plotting specific numerical values based on the under-damped response equation (8.215), such as those shown in this Figure 8.40, to determine its effect.

4. As shown in Figure 8.40, the first overshoot is of practical importance and can be determined using basic calculus methods depending only on ζ, which will be indicated in equation (8.256).

5. The speed of the response is determined by ζ for a constant value of ω_n, as measured by the time it takes for the output to settle within a certain plus-minus tolerance around the final value ($\pm 2\%$) criterion. This concept of settling time, T_s indicated in equations (8.266) and (8.267) as the response speed indication, helps us find that $\zeta's$ is too small or large and take longer to fix, which should obtain its optimum value, which will be indicated in equation (8.267).

As a result of these analyses, we found that the value of K has no effect on response speed, whereas the other two factors, ω_n and ζ, both have an effect, but ω_n is the most significant, as clearly shown in Figure 8.39.

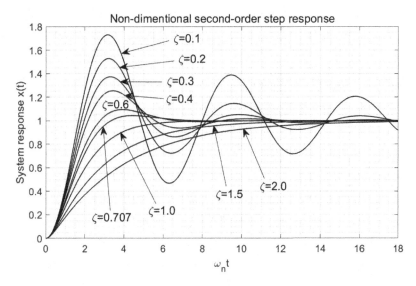

Figure 8.40 Generalized and scaled step responses of a second-order system for various values of the damping ratio

8.3.3 TRANSIENT RESPONSE SPECIFICATIONS OF SECOND-ORDER SYSTEMS

Because it is simple to generate, the performance of a second-order system can be expressed in terms of the time response to a step input function. The time response characteristics of under-damped second-order system conditions are shown in Figure 8.41. Many common terms for transient response characteristics can be found in this figure, as follows:

1. The delay time T_d is when the response reaches 50% of its final value (steady-state X_{ss} value) during its first oscillation cycle.
2. Peak time T_p is the time the response requires to reach its first peak (the peak of the first oscillation cycle). Let us determine the expression for the peak time. As per the definition, the response curve reaches its maximum value at the peak, as shown in Figure 8.41. Hence, at that point, we have

$$\frac{dx(t)}{dt} = 0 \tag{8.242}$$

So, taking the derivative of equation (8.215) yields

$$\frac{dx(t)}{dt} = A_i \left[-\frac{e^{-\zeta \omega_n t}}{\sqrt{1-\zeta^2}} \omega_n \sqrt{1-\zeta^2} \cos\left(\omega_n t \sqrt{1-\zeta^2} + \phi\right) \right.$$
$$\left. -\frac{(-\zeta \omega_n)e^{-\zeta \omega_n t}}{\sqrt{1-\zeta^2}} \sin\left(\omega_n t \sqrt{1-\zeta^2} + \phi\right) \right] \tag{8.243}$$

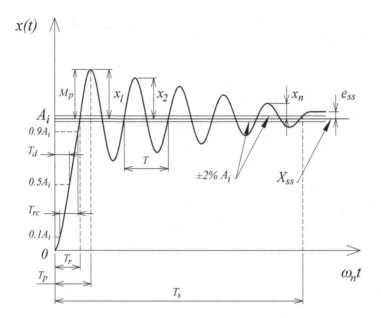

Figure 8.41 Under-damped second-order system step response

Inserting $t = T_p$ and putting $dx(t)/dt = 0$ yields

$$\frac{e^{-\zeta\omega_n T_p}}{\sqrt{1-\zeta^2}}[-\omega_n\sqrt{1-\zeta^2}\cos(\omega_n T_p\sqrt{1-\zeta^2}+\phi)$$

$$+\zeta\omega_n\sin(\omega_n T_p\sqrt{1-\zeta^2}+\phi)]=0 \qquad (8.244)$$

So,

$$\omega_n\sqrt{1-\zeta^2}\cos(\omega_n T_p\sqrt{1-\zeta^2}+\phi)=\zeta\omega_n\sin(\omega_n T_p\sqrt{1-\zeta^2}+\phi) \qquad (8.245)$$

Rearranging this equation gives

$$\frac{\zeta\omega_n\sin(\omega_n T_p\sqrt{1-\zeta^2}+\phi)}{\omega_n\sqrt{1-\zeta^2}\cos(\omega_n T_p\sqrt{1-\zeta^2}+\phi)}=1 \qquad (8.246)$$

So,

$$\tan(\omega_n T_p\sqrt{1-\zeta^2}+\phi)=\frac{\sqrt{1-\zeta^2}}{\zeta}=\tan\phi \qquad (8.247)$$

or,

$$\omega_n\sqrt{1-\zeta^2}T_p=n\pi \qquad (8.248)$$

where $n = 1, 2, 3, \ldots$ is the number of oscillation cycles of the response, as shown in Figure 8.42. The peak time occurs at the first peak $n = 1$. Considering this,

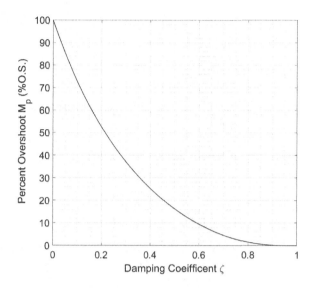

Figure 8.42 Percent overshoot versus damping ratio ζ for a second-order system

equation (8.248) becomes

$$T_p = \frac{\pi}{\omega_n \sqrt{1 - \zeta^2}} = \frac{\pi}{\omega_d} \qquad (8.249)$$

where ω_d is the damping frequency, which is equal to $\omega_d = \omega_n \sqrt{1 - \zeta^2}$.

3. Maximum peak (overshoot) M_p is a straight-way difference between the magnitude of the highest peak of the response and its steady state. It is sometimes referred to as %O.S. It is expressed in terms of the percentage of the steady-state value of the response. As the first peak of the response is usually the maximum in magnitude, the maximum overshoot M_p is simply the normalized difference between the response's first peak and steady-state value X_{ss}, as shown in Figure 8.41. As a result, the first overshoot is frequently of practical importance in underdamped systems. Let us determine the expression for the maximum overshoot M_p. Considering equation (8.215) and inserting $t = T_p$ gives

$$x(t)_{max} = A_i[1 - \frac{e^{-\zeta\omega_n T_p}}{\sqrt{1 - \zeta^2}} \sin(\omega_n T_p \sqrt{1 - \zeta^2} + \phi)] \qquad (8.250)$$

Substituting equation (8.249) in equation (8.250) gives

$$x(t)_{max} = A_i[1 - \frac{e^{-\zeta\omega_n \frac{\pi}{\omega_n \sqrt{1-\zeta^2}}}}{\sqrt{1 - \zeta^2}} \sin(\omega_n \frac{\pi}{\omega_n \sqrt{1-\zeta^2}} \sqrt{1 - \zeta^2} + \phi)] \qquad (8.251)$$

$$x(t)_{max} = A_i[1 - \frac{e^{\frac{-\pi\zeta}{\sqrt{1-\zeta^2}}}}{\sqrt{1 - \zeta^2}} \sin(\pi + \phi)] = A_i[1 - \frac{e^{\frac{-\pi\zeta}{\sqrt{1-\zeta^2}}}}{\sqrt{1 - \zeta^2}}(-\sin\phi)] \qquad (8.252)$$

Inserting $\sin\phi = \sqrt{1-\zeta^2}$ in this equation gives

$$x(t)_{max} = A_i[1 + \frac{e^{\frac{-\pi\zeta}{\sqrt{1-\zeta^2}}}}{\sqrt{1-\zeta^2}}\sqrt{1-\zeta^2}] = A_i[1 + e^{\frac{-\pi\zeta}{\sqrt{1-\zeta^2}}}] \qquad (8.253)$$

So, the peak amplitude M_{pp}, which is the difference between the $x(t)_{max}$ and the steady state value X_{ss} at peak time T_p, is given by

$$M_{pp} = x(t)_{max} - X_{ss} = x(t)_{max} - A_i = A_i(1 + e^{\frac{-\pi\zeta}{\sqrt{1-\zeta^2}}}) - A_i \qquad (8.254)$$

Rearranging this equation gives

$$M_{pp} = A_i e^{-\frac{\pi\zeta}{\sqrt{1-\zeta^2}}} = X_{ss}e^{-\frac{\pi\zeta}{\sqrt{1-\zeta^2}}} \qquad (8.255)$$

We must figure out the meaning of the peak amplitude and the maximum peak. The peak amplitude M_{pp} is the amplitude difference between the steady-state value X_{ss} and the maximum response point at the peak time T_p, as mentioned above. On the other hand, the maximum peak M_p is the percentage of peak amplitude M_{pp} over the steady state value X_{ss}. Considering these, the peak amplitude M_{pp} is given by equation (8.255), while the maximum peak M_p is given by

$$M_p = \frac{M_{pp}}{X_{ss}} = \frac{A_i e^{-\frac{\pi\zeta}{\sqrt{1-\zeta^2}}}}{A_i} = e^{-\frac{\pi\zeta}{\sqrt{1-\zeta^2}}} \qquad (8.256)$$

This equation indicates that the maximum peak M_p depends on only ζ. In this case, the percent overshoot $\%O.S.$ can be calculated as follows:

$$\%O.S. = M_p * 100\% = e^{-\frac{\pi\zeta}{\sqrt{1-\zeta^2}}} * 100\% \qquad (8.257)$$

Figure 8.42 shows the plot of M_p versus the damping ratio ζ. Equation (8.257) indicated that the percent overshoot $\%O.S.$ would decrease if the damping ratio ζ increased.

4. Rise time T_r is the time required to reach the final value of an under-damped time response signal during its first oscillation cycle. Let us determine an expression of the rise time T_r when the response is underdamped. As per definition, the magnitude of the response $x(t)$ at rise time is A_i, as shown in Figure 8.41. Considering equation (8.215) and inserting $x(t) = A_i$ yields

$$x(t) = A_i = A_i[1 - \frac{e^{-\zeta\omega_n t}}{\sqrt{1-\zeta^2}}\sin(\omega_n t\sqrt{1-\zeta^2}+\phi)] \qquad (8.258)$$

where $\phi = \tan^{-1}(\sqrt{1-\zeta^2}/\zeta)$. Rearranging equation (8.258) and inserting $t = T_r$ yields

$$A_i - A_i = \frac{e^{-\zeta\omega_n T_r}}{\sqrt{1-\zeta^2}}\sin(\omega_n T_r\sqrt{1-\zeta^2}+\phi) = 0 \qquad (8.259)$$

Rearranging this equation gives

$$\sin\left(\omega_n T_r \sqrt{1-\zeta^2}+\phi\right)=0 \tag{8.260}$$

So,

$$\omega_n T_r \sqrt{1-\zeta^2}+\phi=\pi \tag{8.261}$$

Solving this equation for T_r gives

$$T_r=\frac{\pi-\phi}{\omega_n\sqrt{1-\zeta^2}} \tag{8.262}$$

Inserting $\phi=\tan^{-1}(\sqrt{1-\zeta^s}/\zeta$, equation (8.262) becomes

$$T_r=\frac{\pi-\tan^{-1}\frac{\sqrt{1-\zeta^2}}{\zeta}}{\omega_n\sqrt{1-\zeta^2}} \tag{8.263}$$

If the response is critically or over-damped, peak time T_p is not defined, and in this case, the rise time is the amount of time it takes for the response to raise from 10% to 90% of its final value, as shown in Figure 8.41. Although it is difficult to acquire an accurate analytic expression for the rise time, we can use the following linear expression:

$$T_{rc}=\frac{2.16\zeta+0.6}{\omega_n} \tag{8.264}$$

where T_{rc} is the rise time when the response is critically or over-damped.

5. Settling time T_s is required for a response to become steady. It is the time the response requires to reach and remain steady within the specified range of $\pm2\%$ to $\pm5\%$ of its final value (usually considered $\pm2\%$ and called the 2% criterion). Note that ζ is the variable that determines the speed of response for a given ω_n value. This speed can be calculated as the time required for the output to settle within a defined plus-and-minus tolerance around the steady-state value X_{ss}. Using the concept of settling time as an indicator of the speed of the response, we see in Figure 8.40 that $\zeta's$ that are either too large or too small will take longer to settle, and as a result, an optimal value of ζ should exist. This is the case for bands in the $\pm2\%$ region. Since the response curve $x(t)$ is tangent to exponential curves, we can estimate the response time of a second-order system, such as the one shown in Figure 8.41, in terms of the settling time T_s, which is defined as follows:

$$T_s=4\tau \tag{8.265}$$

As shown in Figure 8.41, the response curve is tangent to envelope exponential $e^{-\zeta\omega_n t}$, with a time constant $\tau=1/(\zeta\omega_n)$. So,

$$T_s=4\tau=\frac{4}{\zeta\omega_n} \tag{8.266}$$

Since the response curve stays within the $\pm 2\%$ region of the total change for $T > T_s$, the settling time T_s can be considered an approximate system response time, as shown in Figure 8.41.

When the settling time is calculated based on the $\pm 5\%$ criterion, it is given by

$$T_s = 3\tau = \frac{3}{\zeta \omega_n} \tag{8.267}$$

6. Steady-state error e_{ss} is the difference between the actual output and the input at an infinite time range, as shown in Figure 8.41. Note that the steady-state value X_{ss} and the steady-state error e_{ss} for a second-order system have precisely the same meaning as in a first-order system. So, the second-order steady-state value X_{ss} can be determined using equation (8.17), while the steady-state error e_{ss} can be determined using equation (8.72).

Now, let us determine the damping ratio ζ when a step input force excites an under-damped second-order system. As shown in Figure 8.41, the time required to complete one oscillation, given by the time T and defined as the interval between two consecutive zero crossings, may be determined as follows:

$$T = \frac{2\pi}{\omega_d} = \frac{2\pi}{\omega_n \sqrt{1 - \zeta^2}} \tag{8.268}$$

To determine ζ, keeping track of the magnitude of amplitude decay per n cycles is necessary. Figure 8.41 shows that every peak will have zero velocity and some displacement value, which we may denote by x_n where $n = 1, 2, \ldots$ could be treated as a rebound response shape. Note that n cycles occur in a time nT. If we can measure x_1 and x_2 as indicated in Figure 8.41 for the first and second cycle intervals, ζ may be determined as follows:

$$\frac{x_1}{x_2} = \frac{e^{-\zeta \omega_n t_1}}{e^{-\zeta \omega_n (t_1 + T)}} = \frac{1}{e^{-\zeta \omega_n T}} = e^{\zeta \omega_n T} \tag{8.269}$$

The logarithm of the ratio of succeeding amplitudes is called logarithmic decrements δ. Thus

$$\delta = \log_e(\frac{x_1}{x_2}) = \ln(\frac{x_1}{x_2}) = \ln(e^{\zeta \omega_n T}) = \zeta \omega_n T \tag{8.270}$$

Substituting equation (8.270) in equation (8.272) gives

$$\delta = \frac{2\pi \zeta}{\sqrt{1 - \zeta^2}} \tag{8.271}$$

Solving this equation for ζ, yields

$$\zeta = \frac{\delta}{\sqrt{\delta^2 + 4\pi^2}} \tag{8.272}$$

Note that the value of δ is measured from the response curve.

Now, let us determine the natural frequency ω_n. To obtain the natural frequency ω_n, we must first get the damping frequency ω_d by measuring the transient period T as shown in Figure 8.41, which is given by

$$T = \frac{2\pi}{\omega_d} \quad \text{or} \quad \omega_d = \frac{2\pi}{T} \tag{8.273}$$

Assuming ζ has already been found, ω_n is computed from

$$\omega_n = \frac{\omega_d}{\sqrt{1-\zeta^2}} \tag{8.274}$$

It is worth noting that any natural oscillations observed in a second-order system always occur at frequency $\omega_n \sqrt{1-\zeta^2}$, not only ω_n.

8.3.4 SECOND-ORDER SYSTEM RESPONSE TO IMPULSE INPUT

Considering again the translational system shown in Figure 8.34a, which is applying to the impulse input $f(t) = I_i \delta(t)$, where $\delta(t)$ is the symbol of a unit impulse function, the transfer function of the system is given by

$$\ddot{x} + 2\zeta \omega_n \dot{x} + \omega_n^2 x = \omega_n^2 I_i \delta(t) \tag{8.275}$$

Assuming zero initial condition, taking the Laplace transform with zero initial conditions yields

$$(s^2 + 2\zeta \omega_n s + \omega_n^2)X(s) = I_i \omega_n^2 \tag{8.276}$$

Rearrange this equation yields

$$X(s) = \frac{I_i \omega_n^2}{(s^2 + 2\zeta \omega_n s + \omega_n^2)} \tag{8.277}$$

This equation represents the transfer function of a second-order system applied to impulse input force. Now, considering the classifications of characteristic equation roots stated previously in Section 8.3.1.2, the second-order system impulse response is obtained as follows.

a. *Undamped* (No damping $\zeta = 0$):
   Inserting $\zeta = 0$ in equation (8.277) yields

$$X(s) = \frac{I_i \omega_n^2}{(s^2 + \omega_n^2)} \tag{8.278}$$

When this transfer function was compared to item (10) of the Laplace transform Table 7.1, it was found to be the most useful. However, it needs to be rearranged for use. Thus, it appears that

$$X(s) = \frac{I_i \omega_n^2}{\omega_n} \frac{\omega_n}{(s^2 + \omega_n^2)} \tag{8.279}$$

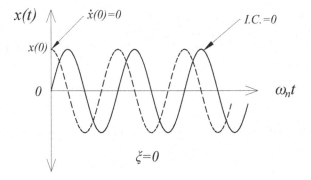

Figure 8.43 Second-order system response applied to impulse input at $\zeta = 0$ with and without I.C.

Now, we can extract the response equation directly from the table. So, taking the inverse Laplace transform, equation (8.279) becomes

$$x(t) = I_i \omega_n \sin(\omega_n t) \tag{8.280}$$

This response equation represents the dynamic motion of a second-order system with no damping ($\zeta = 0$) applied to the impulse input force. It establishes that only the natural frequency ω_n governs the system response since it has no damping. As a result, the response would be a continuous signal with the same amplitude and frequency as the input, as shown in Figure 8.43. This figure shows the response with and without the I.C. being considered.

b. *Underdamped* ($0 < \zeta < 1$):

As we discussed previously, when we study the system response applied to the step input, the roots of the characteristic equation of the under-damped second-order system are complex conjugates. In the same way, rearrange the characteristic equation of the transfer function indicated in equation (8.277) as follows

$$(s^2 + 2\zeta\omega_n s + \omega_n^2) = [s^2 + 2s(\zeta\omega_n) + (\zeta\omega_n)^2] + \omega_n^2 - (\zeta\omega_n)^2 =$$
$$= (s + \zeta\omega_n)^2 + \omega_n^2(1 - \zeta^2) = (s + \zeta\omega_n)^2 + (\omega_n\sqrt{1 - \zeta^2})^2 \tag{8.281}$$

Replacing this equation with the characteristic equation (denominator) of the transfer function indicated in equation (8.277) gives

$$X(s) = \frac{I_i \omega_n^2}{(s^2 + 2\zeta\omega_n s + \omega_n^2)} = \frac{I_i \omega_n^2}{(s + \zeta\omega_n)^2 + (\omega_n\sqrt{1 - \zeta^2})^2} \tag{8.282}$$

Applying partial fraction expansion to this equation yields

$$X(s) = \frac{I_i \omega_n^2}{(s + \zeta\omega_n)^2 + (\omega_n\sqrt{1 - \zeta^2})^2} = \frac{As + B}{(s + \zeta\omega_n)^2 + (\omega_n\sqrt{1 - \zeta^2})^2} \tag{8.283}$$

Solving this equation for A and B is as follows. Equating factors of both sides of equation (8.283) gives

$$s : A = 0 \tag{8.284}$$

$$0 : B = I_i \omega_n^2 \tag{8.285}$$

Substituting equations (8.289) and (285) in equation (8.283) gives

$$X(s) = \frac{I_i \omega_n^2}{(s + \zeta \omega_n)^2 + (\omega_n \sqrt{1 - \zeta^2})^2} \tag{8.286}$$

To use the inverse Laplace transform table item (10), we have to rearrange this equation to look like this:

$$X(s) = I_i \omega_n^2 \frac{1}{b} \frac{b}{(s + \zeta \omega_n)^2 + (\omega_n \sqrt{1 - \zeta^2})^2} \tag{8.287}$$

where b is equal to $\omega_n \sqrt{1 - \zeta^2}$. Considering this, equation (8.287) becomes

$$X(s) = \frac{I_i \omega_n^2}{\omega_n \sqrt{1 - \zeta^2}} \frac{\omega_n \sqrt{1 - \zeta^2}}{(s + \zeta \omega_n)^2 + (\omega_n \sqrt{1 - \zeta^2})^2} \tag{8.288}$$

Now, taking the inverse Laplace transform of this equation yields

$$x(t) = \frac{I_i \omega_n}{\sqrt{1 - \zeta^2}} e^{-\zeta \omega_n t} \sin(\omega_n \sqrt{1 - \zeta^2} t) \tag{8.289}$$

This response equation represents the dynamic motion of an under-damped ($0 < \zeta < 1$) second-order system applied to the impulse input force. It establishes that when ζ lies between 0 and 1, the system response exhibits damped oscillations (decreasing amplitude) since its response equation has an exponential term $e^{-\zeta \omega_n t}$, as shown in Figure 8.44. Since Table 7.1 item (10) is similar to the system transfer function indicated in equation (8.286), we may not need to carry out the analysis above (determining A and B), which means that we may extract the response equation directly from the table.

Note that if the transfer function of the system has the form indicated in equation (8.178), the under-damped response equation indicated in equation (8.289) becomes

$$x(t) = \frac{I_i K}{\omega_n \sqrt{1 - \zeta^2}} e^{-\zeta \omega_n t} \sin(\omega_n \sqrt{1 - \zeta^2} t) \tag{8.290}$$

c. *Critically damped* ($\zeta = 1$):
Inserting $\zeta = 1$ in equation (8.277) yields

$$X(s) = \frac{I_i \omega_n^2}{(s^2 + 2\omega_n s + \omega_n^2)} \tag{8.291}$$

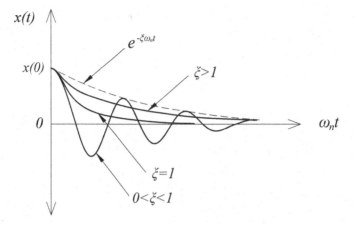

Figure 8.44   Second-order system applied to impulse input at different values of ζ considering the I.C.

or

$$X(s) = \frac{I_i \omega_n^2}{(s + \omega_n)^2} \qquad (8.292)$$

Since Table 7.1 item (7) is similar to this transfer function, we can extract the response equation directly from the table. So, taking the inverse Laplace transform of equation (8.191) yields

$$x(t) = I_i \omega_n^2 t e^{-\omega_n t} \qquad (8.293)$$

This response equation represents the dynamic motion of the critically damped second-order system applied to the impulse input force. It establishes that when $\zeta = 1$, only the natural frequency ω_n governs the system response. As can be seen from the corresponding system response equations, the oscillation component of the response is present when $0 < \zeta < 1$, equation (8.290), but not when $\zeta = 1$, equation (8.293). As a result, the critically damped system response to impulse input shows damped oscillations (decreasing amplitude) since its response equation contains the exponential component $e^{-\zeta \omega_n t}$, as shown in Figure 8.44.

d. *Overdamped* ($\zeta > 1$):
   In the same way, we can rearrange the characteristic equation of the transfer function equation (8.277) when $\zeta.1$, as follows:

$$(s^2 + 2\zeta \omega_n s + \omega_n^2) = [s^2 + 2s(\zeta \omega_n) + (\zeta \omega_n)^2] + \omega_n^2 - (\zeta \omega_n)^2 =$$
$$= (s + \zeta \omega_n)^2 - \omega_n^2(\zeta^2 - 1) = (s + \zeta \omega_n)^2 - (\omega_n(\sqrt{\zeta^2 - 1})^2 \qquad (8.294)$$

Replacing this equation with the characteristic equation of the transfer function indicated in equation (8.277) gives

$$X(s) = \frac{I_i \omega_n^2}{(s^2 + 2\zeta \omega_n s + \omega_n^2)} = \frac{I_i \omega_n^2}{[(s + \zeta \omega_n)^2 - (\omega_n \sqrt{\zeta^2 - 1})^2]} \qquad (8.295)$$

Rearranging this equation gives

$$X(s) = \frac{I_i\omega_n^2}{(s + [\zeta\omega_n + \omega_n\sqrt{\zeta^2 - 1}])(s + [\zeta\omega_n - \omega_n\sqrt{\zeta^2 - 1}])} \quad (8.296)$$

Applying partial fraction expansion to this equation yields

$$X(s) = \frac{B}{(s + [\zeta\omega_n + \omega_n\sqrt{\zeta^2 - 1}])} + \frac{C}{(s + [\zeta\omega_n - \omega_n\sqrt{\zeta^2 - 1}])} \quad (8.297)$$

Solving this equation for A, and B as follows

$$A = \frac{I_i\omega_n^2}{(s + [\zeta\omega_n - \omega_n\sqrt{\zeta^2 - 1}])}\bigg|_{s=-\zeta\omega_n - \omega_n\sqrt{\zeta^2 - 1)}} = \frac{-I_i\omega_n^2}{2\omega_n\sqrt{\zeta^2 - 1)}} \quad (8.298)$$

$$B = \frac{I_i\omega_n^2}{(s + [\zeta\omega_n + \omega_n\sqrt{\zeta^2 - 1}])}\bigg|_{s=-\zeta\omega_n + \omega_n\sqrt{\zeta^2 - 1)}} = \frac{I_i\omega_n^2}{2\omega_n(\sqrt{\zeta^2 - 1})} \quad (8.299)$$

Substituting equations (8.298) and (8.299) in equation (8.297) yields

$$X(s) = \frac{I_i\omega_n^2}{2\omega_n\sqrt{\zeta^2 - 1}}\left(\frac{1}{(s + \zeta\omega_n + \omega_n\sqrt{\zeta^2 - 1})}\right) - $$
$$\frac{I_i\omega_n^2}{2\omega_n\sqrt{\zeta^2 - 1}}\left(\frac{1}{(s + \zeta\omega_n - \omega_n\sqrt{\zeta^2 - 1})}\right) \quad (8.300)$$

Using Table 7.1 item (6), taking the inverse Laplace transform of equation (8.300) gives

$$x(t) = \frac{I_i\omega_n}{2\sqrt{\zeta^2 - 1}}\left(e^{-(\zeta\omega_n - \omega_n\sqrt{\zeta^2 - 1})t} - e^{-(\zeta\omega_n + \omega_n\sqrt{\zeta^2 - 1})t}\right) \quad (8.301)$$

This response equation represents the dynamic motion of the over-damped second-order system applied to the impulse input force. It establishes that when $\zeta > 1$, only the natural frequency ω_n governs the system response. Since it is overdamped, the impulse response of the second-order system will never reach step output in the steady state.

Now, let us determine the damping ratio ζ and natural frequency ω_n when an under-damped second-order system is excited by an impulse input force, as shown in Figure 8.45. Similarly, we will do the same calculation as provided in equations (8.269) to (8.274) when the step input is applied to the second-order system. As a result, the damping ratio ζ and natural frequency ω_n for impulse response is given by equations (8.272) and (8.274), respectively.

Finally, a generalized scaled family of impulse response curves is plotted as shown in Figure 8.46. From this figure, we observed that for every second-order system, regardless of how large or small the parameters K, ζ, and ω_n of I_i are, we may obtain a series of curves by plotting non-dimensional impulse responses for various ζ values.

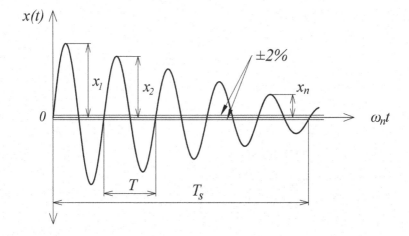

Figure 8.45 Impulse response of a second-order system

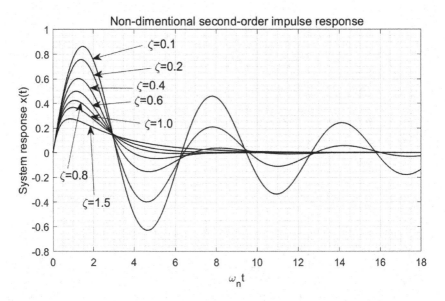

Figure 8.46 Generalized and scaled impulse response of second-order systems

8.3.5 SECOND-ORDER SYSTEM RESPONSE TO RAMP INPUT (LINEAR INPUT FUNCTION)

Considering again the translational system shown in Figure 8.34a and assuming that the ramp input force is $f(t) = R_i r(t)$, where $r(t) = t$, gives

$$\ddot{x} + 2\zeta\omega_n \dot{x} + \omega_n^2 x = \omega_n^2 R_i t \tag{8.302}$$

Assuming zero initial conditions and applying the Laplace transform to this equation yields

$$(s^2 + 2\zeta\omega_n s + \omega_n^2)X(s) = \frac{R_i}{s^2}\omega_n^2 \tag{8.303}$$

Rearranging this equation gives

$$X(s) = \frac{R_i\omega_n^2}{s^2(s^2 + 2\zeta\omega_n s + \omega_n^2)} \tag{8.304}$$

This equation represents the transfer function of a second-order system applied to ramp input force. Considering the classifications of the characteristic equation roots stated in Section 8.3.1.2, the second-order system responses due to ramp input force are obtained as follows.

a. *Undamped* (No damping $\zeta = 0$):
   Inserting $\zeta = 0$ in equation (8.304) yields

$$X(s) = \frac{R_i\omega_n^2}{s^2(s^2 + \omega_n^2)} \tag{8.305}$$

Applying partial fraction expansion to this equation yields

$$X(s) = \frac{R_i\omega_n^2}{s^2(s^2 + \omega_n^2)} = \frac{A}{s} + \frac{B}{s^2} + \frac{Cs+D}{(s^2 + \omega_n^2)} \tag{8.306}$$

Since we have repeated roots, cross-multiply equation (8.306), we get

$$s^2(s^2 + \omega_n^2)\frac{R_i\omega_n^2}{s^2(s^2 + \omega_n^2)} = s^2(s^2 + \omega_n^2)(\frac{A}{s} + \frac{B}{s^2} + \frac{Cs+D}{(s^2 + \omega_n^2)}) \tag{8.307}$$

Rearranging this equation gives

$$R_i\omega_n^2 = As(s^2 + \omega_n^2) + B(s^2 + \omega_n^2) + Cs^3 + Ds^2 \tag{8.308}$$

Clearing fractions yields

$$R_i\omega_n^2 = (A+C)s^3 + (B+D)s^2 + (A\omega_n^2)s + B\omega_n^2 \tag{8.309}$$

Solving this equation for A, B, C, and D as follows. Equating factors of both sides of equation (8.309) gives

$$s^3 : A+C = 0 \tag{8.310}$$
$$s^2 : B+D = 0 \tag{8.311}$$
$$s : A\omega_n^2 = 0 \text{ so } A = 0 \tag{8.312}$$
$$0 : B\omega_n^2 = R_i\omega_n^2 \text{ so } B = R_i \tag{8.313}$$

Substituting equation (8.312) in equation (8.310), we get $C = 0$. Substituting equation (8.313) in equation (8.311) gives

$$D = -R_i \tag{8.314}$$

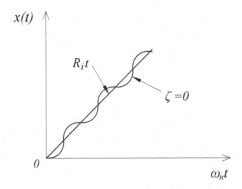

Figure 8.47 Second-order undamped system response applied to ramp input force

Substituting values of A, B, C, and D in equation (8.306) gives

$$X(s) = \frac{R_i}{s^2} - \frac{R_i}{(s^2 + \omega_n^2)} \tag{8.315}$$

Now, to be able to apply the inverse Laplace transform table, we need to rearrange the second term of equation (8.315) as follows:

$$X(s) = \frac{R_i}{s^2} - \frac{R_i}{\omega_n} \frac{\omega_n}{(s^2 + \omega_n^2)} \tag{8.316}$$

Using Table 7.1 items (3) and (10), taking the inverse Laplace transform of this equation, give

$$x(t) = R_i(t - \frac{1}{\omega_n} \sin \omega_n t) \tag{8.317}$$

This is the response equation representing the dynamic motion of a second-order system with no damping applied to the ramp input force. It establishes that only the natural frequency ω_n governs the system response since the system has no damping, meaning that it does not include the exponential term $e^{-\zeta \omega_n t}$. As a result, the response would be a continuous ramp signal of the same frequency and amplitude $(R_i t)$ as the input, as shown in Figure 8.47.

b. *Underdamped* $(0 < \zeta < 1)$:
Rearranging the characteristic equation of the transfer function indicated in equation (8.304) as follows:

$$(s^2 + 2\zeta \omega_n s + \omega_n^2) = [s^2 + 2s(\zeta \omega_n) + (\zeta \omega_n)^2] + \omega_n^2 - (\zeta \omega_n)^2 =$$
$$(s + \zeta \omega_n)^2 + \omega_n^2(1 - \zeta^2) = (s + \zeta \omega_n)^2 + (\omega_n \sqrt{(1 - \zeta^2)})^2 =$$
$$(s + a)^2 + b^2 \tag{8.318}$$

where $a = \zeta \omega_n$ and $b = \omega_n \sqrt{(1-\zeta^2)}$. Replacing equation (8.313) with the characteristic equation of the transfer function indicated in equation (8.304) gives

$$X(s) = \frac{R_i \omega_n^2}{s^2(s^2 + 2\zeta \omega_n s + \omega_n^2)} = \frac{R_i \omega_n^2}{s^2[(s + \zeta \omega_n)^2 + (\omega_n \sqrt{(1-\zeta^2)^2})]} \qquad (8.319)$$

Applying partial fraction expansion to this equation yields

$$X(s) = \frac{R_i \omega_n^2}{s^2[(s + \zeta \omega_n)^2 + (\omega_n \sqrt{(1-\zeta^2)^2})]} = \frac{A}{s} + \frac{B}{s^2} + \frac{Cs + D}{(s + \zeta \omega_n)^2 + (\omega_n \sqrt{(1-\zeta^2)^2}}$$
$$(8.320)$$

Since we have repeated roots, cross-multiplying equation (8.320) gives

$$s^2[(s + \zeta \omega_n)^2 + \omega_n^2(1-\zeta^2)] \frac{R_i \omega_n^2}{s^2[(s + \zeta \omega_n)^2 + (\omega_n \sqrt{(1-\zeta^2)^2})]} =$$
$$s^2[(s + \zeta \omega_n)^2 + \omega_n^2(1-\zeta^2)](\frac{A}{s} + \frac{B}{s^2} + \frac{Cs + D}{(s + \zeta \omega_n)^2 + (\omega_n \sqrt{(1-\zeta^2)^2}})$$
$$(8.321)$$

Rearranging this equation gives

$$R_i \omega_n^2 = As[(s + \zeta \omega_n)^2 + (\omega_n \sqrt{(1-\zeta^2)^2}] + B[(s + \zeta \omega_n)^2 +$$
$$(\omega_n \sqrt{(1-\zeta^2)^2}] + s^2(Cs + D) \qquad (8.322)$$

Clearing fractions yields

$$R_i \omega_n^2 = (A+C)s^3 + (2\omega_n A + B + D)s^2 + [\omega_n^2 A + (2\zeta \omega_n)B]s + \omega_n^2 B \qquad (8.323)$$

Solving this equation for A, B, C, and D as follows. Equating factors of both sides of equation (8.323) gives

$$s^3 : A + C = 0 \qquad (8.324)$$

$$s^2 : 2\zeta \omega_n A + B + D = 0 \qquad (8.325)$$

$$s : \omega_n^2 A + (2\zeta \omega_n)B = 0 \qquad (8.326)$$

$$0 : \omega_n^2 B = R_i \omega_n^2 \text{ so } B = R_i \qquad (8.327)$$

Substituting equation (8.327) in equation (8.326), we get

$$A = \frac{-2\zeta R_i}{\omega_n} \qquad (8.328)$$

From equation (8.324), we have $C = -A$, so

$$C = \frac{2\zeta R_i}{\omega_n} \qquad (8.329)$$

Substituting values of A and B indicated in equations (8.327) and (8.328) in equation (8.325) gives

$$D = R_i(4\zeta^2 - 1) \tag{8.330}$$

Substituting values of A, B, C and D in equation (8.320) and rearranging the results gives

$$X(s) = \frac{-2\zeta R_i}{\omega_n}\frac{1}{s} + \frac{R_i}{s^2} + \frac{2\zeta R_i}{\omega_n}\frac{s}{(s+\zeta\omega_n)^2 + (\omega_n\sqrt{1-\zeta^2})^2} +$$
$$R_i(4\zeta^2 - 1)\frac{1}{(s+\zeta\omega_n)^2 + (\omega_n\sqrt{1-\zeta^2})^2} \tag{8.331}$$

Using Table 7.1 , take the inverse Laplace transform of this equation as follows:
- For the first term:
  Applying item (2) to the first term of equation (8.331) gives

$$x_1(t) = \frac{-2\zeta R_i}{\omega_n} \tag{8.332}$$

- For the second term:
  Applying item (3) to the second term of equation (8.331) gives

$$x_2(t) = R_i t \tag{8.333}$$

- For the third term:
  We need to rewrite this term so that we can use the Laplace transform table, items (20) and (21), to transform it so it looks like this:

$$X_3(s) = \frac{2\zeta R_i}{\omega_n}\frac{s}{(s+\zeta\omega_n)^2 + (\omega_n\sqrt{1-\zeta^2})^2} = \frac{2\zeta R_i}{\omega_n}[\frac{s+a-a}{(s+a)^2+b^2}] =$$
$$\frac{2\zeta R_i}{\omega_n}\frac{s+a}{(s+a)^2+b^2} - \frac{2\zeta R_i}{\omega_n}\frac{a}{b}\frac{b}{(s+a)^2+b^2} \tag{8.334}$$

where $a = \zeta\omega_n$ and $b = \omega_n\sqrt{(1-\zeta^2)}$. Considering these, equation (8.334) becomes

$$X_3(s) = \frac{2\zeta R_i}{\omega_n}\frac{s+\zeta\omega_n}{(s+\zeta\omega_n)^2 + (\omega_n\sqrt{1-\zeta^2})^2} -$$
$$\frac{2\zeta R_i}{\omega_n}\frac{\zeta\omega_n}{\omega_n\sqrt{1-\zeta^2}}\frac{\omega_n\sqrt{1-\zeta^2}}{(s+\zeta\omega_n)^2 + (\omega_n\sqrt{1-\zeta^2})^2} \tag{8.335}$$

Using Table 7.1 items (20) and (21), the inverse Laplace transform of equation (8.335) is given by

$$x_3(t) = \frac{2\zeta R_i}{\omega_n}e^{-\zeta\omega_n t}\cos(\omega_n\sqrt{1-\zeta^2}t) -$$
$$\frac{2\zeta^2 R_{in}}{\omega_n\sqrt{1-\zeta^2}}e^{-\zeta\omega_n t}\sin(\omega_n\sqrt{1-\zeta^2}t) \tag{8.336}$$

- For the fourth term:

  Similarly, rewrite the fourth term using the inverse Laplace transform table, item (20). Considering this, the fourth term of equation (8.331) becomes

$$X_4(s) = R_i(4\zeta^2 - 1)\frac{1}{b}\frac{b}{(s + \zeta\omega_n)^2 + (\omega_n\sqrt{1-\zeta^2})^2} \tag{8.337}$$

where $b = \omega_n\sqrt{(1-\zeta^2)}$. Considering this, equation (8.337) becomes

$$X_4(s) = R_i(4\zeta^2 - 1)\frac{1}{\omega_n\sqrt{1-\zeta^2}}\frac{\omega_n\sqrt{1-\zeta^2}}{(s+\zeta\omega_n)^2 + (\omega_n\sqrt{1-\zeta^2})^2} \tag{8.338}$$

Using Table 7.1 item (20), the inverse Laplace transform of equation (8.338) gives

$$x_4(t) = \frac{R_i(4\zeta^2 - 1)}{\omega_n}\frac{1}{\sqrt{1-\zeta^2}}e^{-\zeta\omega_n t}\sin(\omega_n\sqrt{1-\zeta^2}t) \tag{8.339}$$

Now, combining equations (8.332), (8.333), (8.336) and (8.339) gives

$$x(t) = x_1(t) + x_2(t) + x_3(t) + x_4(t) = \frac{-2\zeta R_i}{\omega_n} + R_i t +$$

$$\frac{2\zeta R_i}{\omega_n}e^{-\zeta\omega_n t}\cos(\omega_n\sqrt{1-\zeta^2}t) - \frac{2\zeta^2 R_i}{\omega_n}\frac{1}{\sqrt{1-\zeta^2}}e^{-\zeta\omega_n t}\sin(\omega_n\sqrt{1-\zeta^2}t) +$$

$$\frac{R_i(4\zeta^2 - 1)}{\omega_n}\frac{1}{\sqrt{1-\zeta^2}}e^{-\zeta\omega_n t}\sin(\omega_n\sqrt{1-\zeta^2}t) \tag{8.340}$$

Simplifying this equation gives

$$x(t) = R_i[\frac{-2\zeta}{\omega_n} + t + \frac{e^{-\zeta\omega_n t}}{\omega_n}\{2\zeta\cos(\omega_n\sqrt{1-\zeta^2}t) + \frac{(2\zeta^2-1)}{\sqrt{1-\zeta^2}}\sin(\omega_n\sqrt{1-\zeta^2}t)\}] \tag{8.341}$$

We can rewrite this equation as follows

$$x(t) = R_i[\frac{-2\zeta}{\omega_n} + t + \frac{e^{-\zeta\omega_n t}}{\omega_n}\{2\zeta\cos(\omega_d t) + \frac{(2\zeta^2-1)}{\sqrt{1-\zeta^2}}\sin(\omega_d t)\}] \tag{8.342}$$

where ω_d is the damping frequency, which is equal to $\omega_n\sqrt{1-\zeta^2}$. Equation (8.342) is the response equation that represents the dynamic motion of an under-damped second-order system applied to the ramp input force. Since the values of the damping ratio, ζ, never have negative values, equation (8.342) establishes that when ζ lies between 0 and 1, the under-damped second-order system applied to the ramp input force exhibits damped oscillations since its response equation includes the exponential term $e^{-\zeta\omega_n t}$, as shown in Figure 8.48.

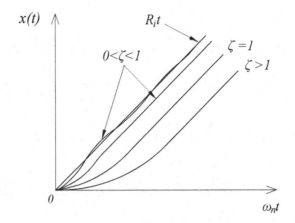

Figure 8.48 Generalized responses of a second-order system due to ramp input force at different values of damping ratio ζ

If the transfer function has the form indicated in equation (8.178), the response equation indicated in equation (8.342) becomes

$$x(t) = \frac{R_i K}{\omega_n^2} [\frac{-2\zeta}{\omega_n} + t + \frac{e^{-\zeta \omega_n t}}{\omega_n} \{2\zeta \cos(\omega_d t) + \frac{(2\zeta^2 - 1)}{\sqrt{1 - \zeta^2}} \sin(\omega_d t)\}] \quad (8.343)$$

c. *Critically damped ($\zeta = 1$):*
   Inserting $\zeta = 1$ in equation (8.304) gives

$$X(s) = \frac{R_i \omega_n^2}{s^2(s^2 + 2\omega_n s + \omega_n^2)} \quad (8.344)$$

or

$$X(s) = \frac{R_i \omega_n^2}{s^2(s + \omega_n)^2} \quad (8.345)$$

Applying partial fraction expansion to this equation yields

$$X(s) = \frac{R_i \omega_n^2}{s^2(s + \omega_n)^2} = \frac{A_1}{s} + \frac{A_2}{s^2} + \frac{B_1}{(s + \omega_n)} + \frac{B_2}{(s + \omega_n)^2} \quad (8.346)$$

Solving this equation for A_1, A_2, B_1, and B_2 as follows. Since we have repeated roots, cross-multiplying equation (8.346) gives

$$X(s) = s^2(s + \omega_n)^2 \frac{R_i \omega_n^2}{s^2(s + \omega_n^2)^2} = s^2(s + \omega_n)^2(\frac{A_1}{s} + \frac{A_2}{s^2} +$$

$$\frac{B_1}{(s + \omega_n)} + \frac{B_2}{(s + \omega_n)^2} \quad (8.347)$$

Rearranging this equation yields

$$R_i \omega_n^2 = A_1 s (s^2 + 2\omega_n s + \omega_n^2) + A_2 (s^2 + 2\omega_n s + \omega_n^2) +$$
$$B_1 s^2 (s + \omega_n) + B_2 s^2 \qquad (8.348)$$

Clearing fractions gives

$$R_i \omega_n^2 = (A_1 + B_1) s^3 + (2\omega_n A_1 + A_2 + \omega_n B_1 + B_2) s^2 +$$
$$(\omega_n^2 A_1 + 2\omega_n A_2) s + \omega_n^2 A_2 \qquad (8.349)$$

Equating factors of both sides of this equation gives

$$s^3 : A_1 + B_1 = 0 \qquad (8.350)$$

$$s^2 : 2\omega_n A_1 + A_2 + \omega_n B_1 + B_2 = 0 \qquad (8.351)$$

$$s : \omega_n^2 A_1 + 2\omega_n A_2 = 0 \qquad (8.352)$$

$$0 : \omega_n^2 A_2 = R_i \omega_n^2 \text{ so } A_2 = R_i \qquad (8.353)$$

Solving these equations for A_1, A_2, B_1, and B_2. as follows. Substituting equation (8.353) in equation (8.352), we get

$$A_1 = \frac{-2R_i}{\omega_n} \qquad (8.354)$$

From equation (8.350), we have $B_1 = -A_1$, so

$$B_1 = -A_1 = \frac{2R_i}{\omega_n} \qquad (8.355)$$

Substituting values of A_1, A_2 and B_1 in equation (8.351) and solving for B_2 gives

$$B_2 = R_i \qquad (8.356)$$

Now, substituting values of A_1, A_2, B_1 and B_2 in equation (8.346) gives

$$X(s) = \frac{R_i \omega_n^2}{s^2 (s + \omega_n)^2} = \frac{-2R_i}{\omega_n} \frac{1}{s} + R_i \frac{1}{s^2} + \frac{2R_i}{\omega_n} \frac{1}{(s + \omega_n)} + R_i \frac{1}{(s + \omega_n)^2} \qquad (8.357)$$

Using Table 7.1 items (2), (3), (6), and (7), taking the inverse Laplace transfer of this equation gives

$$x(t) = R_i \left(-\frac{2}{\omega_n} + t + \frac{2}{\omega_n} e^{-\omega_n t} + t e^{-\omega_n t} \right) \qquad (8.358)$$

This response equation represents the dynamic motion of a critically damped second-order system applied to the ramp input force. It establishes that only the natural frequency ω_n governs the system response when the system has $\zeta = 1$.

d. *Overdamped* ($\zeta > 1$):

Rearrange the characteristic equation of the transfer function indicated in equation (8.304), considering $\zeta > 1$, as follows:

$$(s^2 + 2\zeta \omega_n s + \omega_n^2) = [s^2 + 2s(\zeta \omega_n) + (\zeta \omega_n)^2] + \omega_n^2 - (\zeta \omega_n)^2 =$$
$$= (s + \zeta \omega_n)^2 - \omega_n^2(\zeta^2 - 1) = (s + \zeta \omega_n)^2 - (\omega_n \sqrt{\zeta^2 - 1})^2 \quad (8.359)$$

Replacing this equation with the characteristic equation of the transfer function indicated in equation (8.304) gives

$$X(s) = \frac{R_i \omega_n^2}{s^2(s^2 + 2\zeta \omega_n s + \omega_n^2)} = \frac{R_i \omega_n^2}{s^2[(s + \zeta \omega_n)^2 - (\omega_n \sqrt{\zeta^2 - 1})^2]} \quad (8.360)$$

Applying partial fraction expansion to this equation gives

$$X(s) = \frac{A}{s} + \frac{B}{s^2} + \frac{C}{(s + \zeta \omega_n + \omega_n \sqrt{\zeta^2 - 1})} + \frac{D}{(s + \zeta \omega_n - \omega_n \sqrt{\zeta^2 - 1})} \quad (8.361)$$

Solving this equation for A, B, C and D as follows

$$B = \frac{R_i \omega_n^2}{(s + \zeta \omega_n + \omega_n \sqrt{\zeta^2 - 1})(s + \zeta \omega_n - \omega_n \sqrt{\zeta^2 - 1})} \bigg|_{s=0} = R_i \quad (8.362)$$

$$C = \frac{R_i \omega_n^2}{s^2(s + \zeta \omega_n - \omega_n \sqrt{\zeta^2 - 1})} \bigg|_{s=-\zeta \omega_n - \omega_n \sqrt{\zeta^2 - 1}} =$$
$$\frac{R_i \omega_n^2}{(-\zeta \omega_n - \omega_n \sqrt{\zeta^2 - 1})^2[(-\zeta \omega_n - \omega_n \sqrt{\zeta^2 - 1}) + \zeta \omega_n - \omega_n \sqrt{\zeta^2 - 1}]} \quad (8.363)$$

Simplifying this equation gives

$$C = \frac{-R_i}{2\omega_n \sqrt{\zeta^2 - 1}(2\zeta - 1) + 4\zeta \omega_n(\zeta^2 - 1)} \quad (8.364)$$

and

$$D = \frac{R_i \omega_n^2}{s^2(s + \zeta \omega_n + \omega_n \sqrt{\zeta^2 - 1})} \bigg|_{s=-\zeta \omega_n + \omega_n \sqrt{\zeta^2 - 1}} =$$
$$\frac{R_i \omega_n^2}{(-\zeta \omega_n + \omega_n \sqrt{\zeta^2 - 1})^2[(-\zeta \omega_n + \omega_n \sqrt{\zeta^2 - 1}) + \zeta \omega_n + \omega_n \sqrt{\zeta^2 - 1}]} \quad (8.365)$$

Simplifying this equation gives

$$D = \frac{R_i}{2\omega_n \sqrt{\zeta^2 - 1}(2\zeta^2 - 1) - 4\zeta \omega_n(\zeta^2 - 1)} \quad (8.366)$$

Now, to determine A, cross-multiply both sides of equation (8.360) by s^2 as follows:

$$s^2 \frac{R_i \omega_n^2}{s^2(s + \zeta\omega_n + \omega_n\sqrt{\zeta^2 - 1})(s + \zeta\omega_n - \omega_n\sqrt{\zeta^2 - 1})} = sA + B +$$

$$\frac{Cs^2}{(s + \zeta\omega_n + \omega_n\sqrt{\zeta^2 - 1})} + \frac{Ds^2}{(s + \zeta\omega_n - \omega_n\sqrt{\zeta^2 - 1})} \quad (8.367)$$

Taking the derivative of this equation and evaluating it at $s = 0$ is as follows.
- For the left side:

$$\frac{d}{ds}\left(\frac{R_i \omega_n^2}{(s + \zeta\omega_n + \omega_n\sqrt{\zeta^2 - 1})(s + \zeta\omega_n - \omega_n\sqrt{\zeta^2 - 1})}\right)\bigg|_{s=0} =$$

$$\frac{0 - R_i\omega_n^2(2s + 2\zeta\omega_n)}{[(s + \zeta\omega_n + \omega_n\sqrt{\zeta^2 - 1})(s + \zeta\omega_n - \omega_n\sqrt{\zeta^2 - 1})]^2}\bigg)\bigg|_{s=0} =$$

$$\frac{-2R_i\zeta\omega_n^3}{\omega_n^4} = \frac{-2R_i\zeta}{\omega_n} \quad (8.368)$$

- For the right side:

$$\frac{d}{ds}\left(sA + B + \frac{Cs^2}{(s + \zeta\omega_n + \omega_n\sqrt{\zeta^2 - 1})} + \frac{Ds^2}{(s + \zeta\omega_n - \omega_n\sqrt{\zeta^2 - 1})}\right)\bigg|_{s=0} =$$

$$\left(A + 0 + \frac{(s + \zeta\omega_n + \omega_n\sqrt{\zeta^2 - 1})*2Cs - Cs^2*(\zeta\omega_n + \omega_n\sqrt{\zeta^2 - 1})}{(s + \zeta\omega_n + \omega_n\sqrt{\zeta^2 - 1})^2} + \right.$$

$$\left.\frac{(s + \zeta\omega_n - \omega_n\sqrt{\zeta^2 - 1})*2Ds - Ds^2*(\zeta\omega_n - \omega_n\sqrt{\zeta^2 - 1})}{(s + \zeta\omega_n - \omega_n\sqrt{\zeta^2 - 1})^2}\right)\bigg|_{s=0} =$$

$$A + 0 + 0 + 0 = A \quad (8.369)$$

So, equating equations (8.368) and (8.369) yields

$$A = \frac{-2R_i\zeta}{\omega_n} \quad (8.370)$$

Now, substituting equations (8.362), (8.364), (8.366), and (8.370) in equation (8.361) gives

$$X(s) = \frac{-2R_i\zeta}{\omega_n}\frac{1}{s} + \frac{R_i}{s^2} - \frac{R_i}{2\omega_n\sqrt{\zeta^2 - 1}(2\zeta^2 - 1) + 4\zeta\omega_n(\zeta^2 - 1)} \quad (8.371)$$

$$\left(\frac{1}{(s + \zeta\omega_n + \omega_n\sqrt{\zeta^2 - 1}}\right) + \frac{R_i}{2\omega_n\sqrt{\zeta^2 - 1}(2\zeta^2 - 1) - 4\zeta\omega_n(\zeta^2 - 1)}$$

$$\left(\frac{1}{(s + \zeta\omega_n - \omega_n\sqrt{\zeta^2 - 1}}\right) \quad (8.372)$$

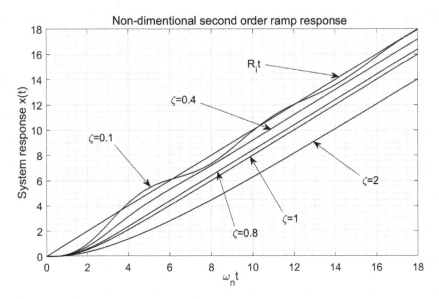

Figure 8.49 Generalized and scaled ramp responses of a second-order system at different values of the damping ratio ζ

From Table 7.1, taking the inverse Laplace transform of this equation, items (2), (3), (6) and (7), gives

$$x(t) = R_i \left(\frac{-2\zeta}{\omega_n} + t - \frac{1}{2\omega_n \sqrt{\zeta^2 - 1}(2\zeta^2 - 1) + 4\zeta\omega_n(\zeta^2 - 1)} e^{-(\zeta\omega_n + \omega_n \sqrt{\zeta^2 - 1})t} \right.$$
$$\left. + \frac{1}{2\omega_n \sqrt{\zeta^2 - 1}(2\zeta^2 - 1) - 4\zeta\omega_n(\zeta^2 - 1)} e^{-(\zeta\omega_n - \omega_n \sqrt{\zeta^2 - 1})t} \right) \qquad (8.373)$$

This response equation represents the dynamic motion of the over-damped second-order system, $(\zeta > 1)$, applied to the ramp input force.

Finally, a generalized scaled family of ramp response curves is plotted as shown in Figure 8.49. From this figure we observed that for every second-order system, regardless of how large or small the parameters K, ζ, and ω_n of I_i are, we may obtain a series of curves by plotting non-dimensional ramp responses for various ζ values.

8.3.6 SECOND-ORDER ELECTRICAL SYSTEMS

After reviewing the general properties of second-order mechanical systems and showing how they work well, let us focus on the electrical systems. Like mechanical systems, the second-order electrical system can be formed independently or by connecting two first-order electrical circuits, as shown in Figures 8.50 and 8.51.

8.3.6.1 Second-order electrical system formed independently

First, we examine the electrical circuit shown in Figure 8.50. We aim to obtain a differential equation representing the relationship between the input and the output voltages, $e_i(t)$ and $e_o(t)$, respectively. Assume the circuit is not linked to a load, meaning it has only one current i. Applying Kirchhoff's voltage loop law to the given circuit, we get

$$L\frac{di}{dt} + Ri + \frac{1}{C}\int i\,dt = e_i(t) \tag{8.374}$$

and the output voltage $e_o(t)$ is given by

$$e_o = \frac{1}{C}\int i\,dt \tag{8.375}$$

Taking the Laplace transform of equations (8.374) and (8.375) gives

$$LsI(s) + RI(s) + \frac{1}{Cs}I(s) = E_i(s) \tag{8.376}$$

and

$$E_o(s) = \frac{1}{Cs}I(s) \tag{8.377}$$

To obtain the output voltage $E_o(s)$, we must first determine the current $I(s)$. So, solving equation (8.376) for $I(s)$ gives

$$I(s) = \frac{CsE_i(s)}{(LCs^2 + CRs + 1)} \tag{8.378}$$

Substituting this equation in equation (8.377) and rearranging the result gives

$$\frac{E_i(s)}{E_o(s)} = \frac{1}{(LCs^2 + CRs + 1)} \tag{8.379}$$

Rearranging this equation gives

$$\frac{E_i(s)}{E_o(s)} = \frac{\frac{1}{LC}}{(s^2 + \frac{R}{C}s + \frac{1}{LC})} \tag{8.380}$$

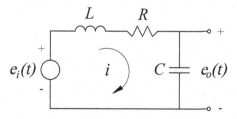

Figure 8.50 Second-order electrical system

This equation represents the transfer function of the electrical circuit shown in Figure 8.50. Compare this equation with the general form of the second-order transfer function as follows:

$$\frac{E_i(s)}{E_o(s)} = \frac{K\omega_n^2}{(s^2 + 2\zeta\omega_n s + \omega_n^2)} = \frac{\frac{1}{LC}}{(s^2 + \frac{R}{C}s + \frac{1}{LC})} \tag{8.381}$$

From this equation, we get:

$$K = 1 \text{ (volt/volt)}, \quad \omega_n = \sqrt{\frac{1}{LC}} \text{ (rad/sec)} \quad \text{and} \quad \zeta = \frac{R}{2\sqrt{\frac{C}{L}}} \tag{8.382}$$

8.3.6.2 Second-order electrical system formed from two first-order systems

Let us examine the electrical system shown in Figure 8.51. Remember that although currents in an electrical circuit are not necessary, we have to deal with them to achieve the desired voltage relationship. We always identify and label each of the various currents we are working on, even though knowing about these currents is not usually the end goal.

From Figure 8.51, we can identify three currents as follows: i_1 moves from a to b, i_2 moves from b to d, and i_3 moves from b to c. Remember that because there is no loading circuit, the current that flows from b to d is the same current from d to e, which is the current i_2. Note that the current i_3 equals $(i_1 - i_2)$. As a result, $(i_1 - i_2)$ rather than i_3 denotes the current flowing from b to c. This means there are only two unknown currents, i_1 and i_2. Apply Kirchhoff's voltage law; we get

$$R_1 i_1 + \frac{1}{C_1} \int (i_1 - i_2)\,dt = e_i(t) \tag{8.383}$$

$$R_2 i_2 + \frac{1}{C_1} \int (i_2 - i_1)\,dt \frac{1}{C_2} \int i_2\,dt = 0 \tag{8.384}$$

and

$$e_o(t) = \frac{1}{C_2} \int i_2\,dt \tag{8.385}$$

Assuming zero initial conditions, taking the Laplace transform of equations (8.383), (8.384) and (8.385) yields

$$(R_1 C_1 s + 1)I_1(s) - I_2(s) = (C_1 s)E_i(s) \tag{8.386}$$

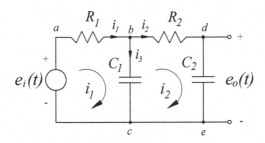

Figure 8.51 Second-order electrical system formed from two first-order electrical systems

$$(C_1 C_2 R_2 s + C_1 + C_2) I_2(s) - C_2 I_1(s) = 0 \tag{8.387}$$

and

$$E_o(s) = \frac{1}{Cs} I_2(s) \tag{8.388}$$

We could now easily solve for either of the two circuits. As indicated in equation (8.388), to determine $E_o(s)$, we must obtain the current $I_2(s)$. So, we use determinants of equations (8.386) and (8.387) as follows

$$I_2(s) = \frac{\begin{vmatrix} (R_1 C_1 s + 1) & C_1 s E_i(s) \\ -C_2 & 0 \end{vmatrix}}{\begin{vmatrix} (R_1 C_1 s + 1) & -1 \\ -C_2 & (C_1 C_2 R_2 s + C_1 + C_2) \end{vmatrix}} \tag{8.389}$$

Expanding the determinants and then cross-multiplying gives

$$I_2(s) = \frac{C_1 s E_i(s)}{(R_1 R_2 C_1 C_2) s^2 + (R_1 C_1 + R_1 C_2 + R_2 C_2) s + 1} \tag{8.390}$$

Substituting equation (8.390) in equation (8.388) and rearranging the results gives

$$\frac{E_o(s)}{E_i(s)} = \frac{1}{(R_1 R_2 C_1 C_2) s^2 + (R_1 C_1 + R_1 C_2 + R_2 C_2) s + 1} \tag{8.391}$$

Rearranging this equation gives

$$\frac{E_i(s)}{E_o(s)} = \frac{\frac{1}{(R_1 R_2 C_1 C_2)}}{s^2 + \left(\frac{R_1 C_1 + R_1 C_2 + R_2 C_2}{R_1 R_2 C_1 C_2}\right) s + \frac{1}{(R_1 R_2 C_1 C_2)}} \tag{8.392}$$

This equation represents the transfer function of the electrical circuit shown in Figure 8.51. Compare this equation with the general form of the second-order transfer function, as follows:

$$\frac{E_i(s)}{E_o(s)} = \frac{K \omega_n^2}{(s^2 + 2\zeta \omega_n s + \omega_n^2)} = \frac{\frac{1}{(R_1 R_2 C_1 C_2)}}{s^2 + \left(\frac{R_1 C_1 + R_1 C_2 + R_2 C_2}{R_1 R_2 C_1 C_2}\right) s + \frac{1}{(R_1 R_2 C_1 C_2)}} \tag{8.393}$$

From this equation, we get:

$$K = 1 \text{ (volt/volt)}, \quad \text{and} \quad \omega_n = \sqrt{\frac{1}{R_1 R_2 C_1 C_2}} \text{ (rad./sec)} \tag{8.394}$$

and

$$\zeta = \frac{C_1 R_1 + C_2 R_1 + C_2 R_2}{2} \sqrt{\frac{1}{R_1 R_2 C_1 C_2}} \tag{8.395}$$

Since equations (8.381) and (8.382), equations (8.393), (8.394), and (8.395) satisfy our description of a second-order electrical system, all standard responses from previous studies on the generalized second-order mechanical system discussed previously in Section 8.3.1 are readily available; there is no need for any extra analyses.

8.3.7 SECOND-ORDER LIQUID-LEVEL SYSTEMS

Figure 8.52 represents the consequence of combining two systems of the first-order of liquid-level systems. Neglect the liquid inertia inside the two tanks. During a period dt, applying the law of conservation of volume on the first tank gives

$$q_{in}\, dt - q_1 dt = A_1\, dh_1 \tag{8.396}$$

The flow rate q_1 passing through the valve of resistance R_{v_1} is given by

$$q_1 = \frac{\gamma(h_1 - h_2)}{R_{v_1}} \tag{8.397}$$

Substituting equation (8.397) in equation (8.396) gives

$$q_{in}\, dt - [\frac{\gamma(h_1 - h_2)}{R_{v_1}}]\, dt = A_1\, dh_1 \tag{8.398}$$

Similarly, for the second tank,

$$q_1\, dt - q_o\, dt = A_2\, dh_2 \tag{8.399}$$

The flow rate q_o passing through the valve of resistance R_{v_2} is given by

$$q_o = \frac{\gamma h_2}{R_{v_2}} \tag{8.400}$$

Substituting equations (8.397) and (8.400) in equation (8.399) gives

$$[\frac{\gamma(h_1 - h_2)}{R_{v_1}} - \frac{\gamma h_2}{R_{v_2}}]\, dt = A_2 dh_2 \tag{8.401}$$

Rearranging equations (8.398) and (8.401) yield

$$A_1 \frac{dh_1}{dt} + [\frac{\gamma(h_1 - h_2)}{R_{v_1}}] = q_{in} \tag{8.402}$$

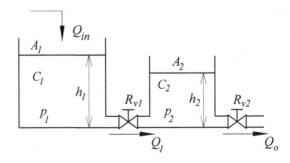

Figure 8.52 Second-order liquid-level system

$$A_2 \frac{dh_2}{dt} + \left[\frac{\gamma(h_1 - h_2)}{R_{v_1}} - \frac{\gamma h_2}{R_{v_2}}\right] = 0 \qquad (8.403)$$

We assume that the liquid resistances R_{v_1} and R_{v_2} are linear. Note that we can use the compaliaces C_1 and C_2 rather than A_1 and A_2, where $C_1 = (A_1/\gamma)$ and $C_1 = (A_1/\gamma)$. These compliances for both tanks can also be considered linearly. Using the assumed linear model and taking the Laplace transform of equations (8.402) and (8.403), we get:

$$A_1 s H_1(s) + \frac{\gamma}{R_{v_1}} H_1(s) - \frac{\gamma}{R_{v_1}} H_2(s) = Q_{in}(s) \qquad (8.404)$$

$$A_2 s H_2(s) + \frac{\gamma}{R_{v_1}} H_1(s) - \frac{\gamma}{R_{v_1}} H_2(s) - \frac{\gamma}{R_{v_2}} H_2(s) = 0 \qquad (8.405)$$

Rearranging these two equations yields

$$\left(\frac{A_1 R_{v_1}}{\gamma} s + 1\right) H_1(s) - H_2(s) = \frac{R_{v_1}}{\gamma} Q_{in}(s) \qquad (8.406)$$

$$\left[\left(\frac{A_2 R_{v_1} R_{v_2}}{\gamma(R_{v_1} + R_{v_2})}\right) s + 1\right)\right] H_2(s) - \left(\frac{R_{v_2}}{R_{v_1} + R_{v_2}}\right) H_1(s) = 0 \qquad (8.407)$$

We could now easily solve for either of the two tank levels. Our interest here is in $H_2(s)$, so use the determinants of equations (8.406) and (8.407) as follows:

$$H_2(s) = \frac{\begin{vmatrix} \left(\frac{R_{v_1} A_1}{\gamma} s + 1\right) & \frac{R_{v_1}}{\gamma} Q_{in}(s) \\ -\left(\frac{R_{v_2}}{R_{v_1} + R_{v_2}}\right) & 0 \end{vmatrix}}{\begin{vmatrix} \left(\frac{R_{v_1} A_1}{\gamma} s + 1\right) & -1 \\ -\left(\frac{R_{v_2}}{R_{v_1} + R_{v_2}}\right) & \left[\left(\frac{A_2 R_{v_1} R_{v_2}}{\gamma(R_{v_1} + R_{v_2})}\right) s + 1\right)\right] \end{vmatrix}}$$

Expanding the determinants and then cross-multiplying gives

$$\left[\left(\frac{R_{v_1} R_{v_2} A_1 A_2}{\gamma^2}\right) s^2 + \left(\frac{A_2 R_{v_1} + A_1 R_{v_2} + A_2 R_{v_2}}{\gamma}\right) s + 1\right] H_2(s) = \frac{R_{v_2}}{\gamma} Q_{in}(s) \qquad (8.408)$$

So,

$$\frac{H_2(s)}{Q_{in}(s)} = \frac{\frac{R_{v_2}}{\gamma}}{\left(\frac{R_{v_1} R_{v_2} A_1 A_2}{\gamma^2}\right) s^2 + \left(\frac{A_2 R_{v_1} + A_1 R_{v_2} + A_2 R_{v_2}}{\gamma}\right) s + 1} \qquad (8.409)$$

Rearranging this equation gives

$$\frac{H_2(s)}{Q_{in}(s)} = \frac{\frac{R_{v_2}}{\gamma}\left(\frac{\gamma^2}{R_{v_1} R_{v_2} A_1 A_2}\right)}{s^2 + \left(\frac{A_2 R_{v_1} + A_1 R_{v_2} + A_2 R_{v_2}}{\gamma}\right) \sqrt{\left(\frac{\gamma^2}{R_{v_1} R_{v_2} A_1 A_2}\right)} s + \left(\frac{\gamma^2}{R_{v_1} R_{v_2} A_1 A_2}\right)} \qquad (8.410)$$

This equation represented the transfer function of a second-order liquid-level system. Comparing equation (8.410) with the general form of the second-order transfer function gives

$$K = \frac{R_{v_2}}{\gamma} \; (\frac{m}{m^3/sec}) \quad \text{and} \quad \omega_n = \frac{\gamma}{\sqrt{R_{v_1}R_{v_2}A_1A_2}} \quad (rad./sec) \tag{8.411}$$

and

$$\zeta = \frac{A_1R_{v_1} + A_1R_{v_2} + A_2R_{v_2}}{2\sqrt{R_{v_1}R_{v_2}A_1A_2}} \tag{8.412}$$

From equation (8.412), we can show that ζ cannot be less than 1, meaning that oscillatory behavior is impossible. This is because we ignored any inertial effects that may have been present in the liquid inside both tanks. Oscillating behavior may be caused by the inertance (liquid inertia) in the pipe that connects the two tanks through the valve of resistance R_{v_1}. However, this would only be possible if this resistance was low, like what would happen when a liquid with a low viscosity rubbed against the pipe walls.

Since equations (8.410), (8.411), and (8.412) satisfy our description of the given liquid-level system, all standard responses from previous studies on the generalized second-order mechanical system discussed previously in Section 8.3.1 are readily available; there is no need for any extra analyses.

8.3.8 SECOND-ORDER THERMAL SYSTEMS

Oscillation is theoretically impossible for thermal systems to participate in since thermal systems only have one storage element, thermal capacitance. This implies that a single thermal system will be first-order because it has no inertia element. This is in contrast to mechanical, electrical, and liquid systems, which may exhibit natural oscillations due to the presence of two energy storage elements: inertia and capacitance. On the other hand, the roots of the characteristic equation will be precise every time, so there is no chance that the system will participate in any oscillations. We can obtain a system equation of any arbitrarily higher order if we utilize heat resistance and capacitance numbers. As a result, a high-order thermal system is composed of multiple systems of a first-order system. When the heater's energy storage cannot be neglected, we consider the system shown in Figure 8.53. This is because the energy storage capacity of the heater is essential for successful operation. When the product of the temperature increases and the thermal capacitance of the heater reaches a particular threshold, the energy storage capacity of this heater can become significant (in comparison to that of the liquid inside the tank). A situation similar to this could occur if the resistance or capacitance of the heater is significantly higher than usual. To take into consideration the effects described above, we will need to develop conservation of energy equations for both the heater and the liquid mass shown in Figure 8.53 as follows:

$$q_i \, dt - U_hA_h(T_h - T_l) \, dt = M_hc_h \, dT_h \tag{8.413}$$

and

$$U_hA_h(T_h - T_l)dt - U_wA_w(T_l - T_i) \, dt = M_lc_l \, dT_l \tag{8.414}$$

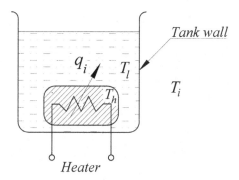

Figure 8.53 Second-order thermal system

where

- A_h=surface area of the heater.
- A_w=surface area of the tank.
- c_l=specific heat of the heater material.
- c_h=specific heat of the heater material.
- M_l=liquid mass.
- M_h=heater mass.
- T_h=heater temperature.
- T_i=atmospheric temperature.
- T_l=liquid temperature.
- U_h=overall heat transfer coefficient between the heater and the liquid.
- U_w=overall heat transfer coefficient between the liquid and environment.

Rearranging equations (8.413) and (8.414) gives

$$M_h c_h \frac{dT_h}{dt} + U_h A_h T_h - U_h A_h T_l = q_i \qquad (8.415)$$

$$M_l C_l \frac{dT_l}{dt} + U_w A_w T_l + U_h A_h T_l - U_h A_h T_h = U_w A_w T_i \qquad (8.416)$$

Taking the Laplace transform of equations (8.415) and (8.416) gives

$$(M_h c_h s + U_h A_h) T_h(s) - U_h A_h T_l(s) = Q_i(s) \qquad (8.417)$$

$$(M_l C_l s + U_w A_w) T_l(s) + U_h A_h T_l(s) - U_h A_h T_h(s) = U_w A_w T_i(s) \qquad (8.418)$$

Rearranging these two equations gives

$$(\frac{M_h c_h}{U_h A_h} s + 1) T_h(s) - T_l(s) = \frac{1}{U_h A_h} Q_i(s) \qquad (8.419)$$

$$(-\frac{M_l C_l}{U_w A_w} s + 1 + \frac{U_h A_h}{U_w A_w}) T_l(s) - \frac{U_h A_h}{U_w A_w} T_h(s) = T_i(s) \qquad (8.420)$$

These two simultaneous equations may be rewritten as

$$(\tau_h s + 1)T_h(s) - T_l(s) = \frac{1}{U_h A_h}Q_i(s) \qquad (8.421)$$

$$(\tau_l s + 1 + \frac{U_h A_h}{U_w A_w})T_l(s) - (\frac{M_h c_h}{U_w A_w})T_h = T_i(s) \qquad (8.422)$$

where

$$\tau_h = \frac{M_h c_h}{U_h A_h} \quad \text{and} \quad \tau_l = \frac{M_l C_l}{U_w A_w} \qquad (8.423)$$

Our interest here is in $T_l(s)$, so using determinants of equations (8.421) and (8.422) gives

$$T_l(s) = \frac{\begin{vmatrix} (\tau_h s + 1) & \frac{q_i}{U_h A_h} \\ (-\frac{U_h A_h}{U_w A_w}) & T_i(s) \end{vmatrix}}{\begin{vmatrix} (\tau_h s + 1) & -1 \\ (-\frac{U_h A_h}{U_w A_w}) & (\tau_l s + 1 + (\frac{U_h A_h}{U_w A_w})) \end{vmatrix}} \qquad (8.424)$$

Expanding the determinants and then cross-multiplying gives

$$[\tau_l \tau_h s^2 + (\tau_l + \tau_h + \frac{U_h A_h}{U_w A_w})s + 1]T_l(s) = (\tau_h s + 1)T_i(s) + (\frac{1}{U_w A_w})Q_i(s) \qquad (8.425)$$

Rearranging this equation gives

$$T_l(s) = \frac{(\tau_h s + 1)T_i(s) + (\frac{1}{U_w A_w})Q_i(s)}{(\tau_l \tau_h s^2 + (\tau_l + \tau_h + \frac{U_h A_h}{U_w A_w})s + 1)} \qquad (8.426)$$

Comparing this equation to the general form of the second-order system gives

$$T_l(s) = \frac{K}{(\frac{s^2}{\omega_n^2} + (\frac{2\zeta}{\omega_n})s + 1)} = \frac{(\tau_h s + 1)T_i(s) + (\frac{1}{U_w A_w})Q_i(s)}{(\tau_l \tau_h s^2 + (\tau_l + \tau_h + \frac{U_h A_h}{U_w A_w})s + 1)} \qquad (8.427)$$

From this equation, we get:

$$K = \frac{1}{U_w A_w} \quad (\frac{°C}{\text{Joules/sec}}) \qquad (8.428)$$

$$\omega_n = \sqrt{\frac{1}{\tau_l \tau_h}} \quad (\text{rad./sec}) \qquad (8.429)$$

and

$$\zeta = \frac{\tau_l + \tau_h + \frac{M_h c_h}{U_w A_w}}{2\sqrt{\tau_l \tau_h}} \qquad (8.430)$$

When considering thermal systems, it is essential to know that they are considered overdamped, meaning that the damping coefficient must be $\zeta \geq 1$. Equation (8.428) shows that even though responding to the outside temperature T_i has numerator dynamics and necessitates a different solution, the response to the heating rate is our typical second-order form.

Since equations (8.427) to (8.430) satisfy our description of a second-order thermal system, all standard responses from previous studies on the generalized second-order mechanical system discussed previously in Section 8.3.1 are readily available; there is no need for any extra analyses.

Finally, remember that all differential equations that describe the dynamics of systems (mechanical, electrical, liquid, and thermal) will lead to transfer functions after applying the Laplace transform. In other words, a transfer function represents one of these systems. Therefore, we will utilize different transfer functions rather than systems in the following examples to demonstrate the previous analyses. This will allow us to understand how to get the response equations, plot the output responses, and obtain their characteristic definitions for different input-forcing functions.

Example (8.6):

Consider the following transfer function:

$$X(s) = \frac{10}{(s^2 + 5s + 50)}$$

For a unit-step input, determine,

1. The damping ratio ζ, natural frequency ω_n, and damping frequency ω_d.
2. Steady-state value X_{ss} and the steady-state error e_{ss}.
3. Peak time T_p, Maximum peak M_p, percent overshoot %O.S., rise time T_r and settling time T_s.
4. The response equation $x(t)$.
5. Use MATLAB to generate and plot the response equation $x(t)$. Verify the results obtained in steps 1 to 4 (use the *stepinfo* command to generate the response characteristics values).
6. Use Simulink to plot the response $x(t)$.

Solution:

1. Damping ratio ζ, natural frequency ω_n and damping frequency ω_d:
   Comparing the characteristic equation of the given transfer function with the general characteristic equation of the second-order system gives

$$s^2 + 5s + 50 = s^2 + 2\zeta\omega_n s + \omega_n^2 \qquad (8.431)$$

Equating like terms gives

$$2\zeta\omega_n = 5 \quad \text{and} \quad \omega^2 = 50 \qquad (8.432)$$

Solving this equation for ω_n and ζ gives

$$\omega_n = 7.071 \text{ (rad/sec)} \quad \text{and} \quad \zeta = 0.3536 < 1 \text{ (under damped)} \qquad (8.433)$$

The damping frequency ω_d can be calculated as follows:

$$\omega_d = \omega_n \sqrt{1 - \zeta^2} = 7.071 \sqrt{1 - (0.3536)^2} = 6.614 \quad \text{(rad/sec)} \qquad (8.434)$$

2. Steady-state value X_{ss} and steady-state error e_{ss}.
   The steady-state value X_{ss} can be calculated as follows:

$$X_{ss} = \lim_{s \to 0} s U(s) X(s) = \lim_{s \to 0} s \frac{1}{s} \left(\frac{10}{s^2 + 5s + 50} \right) = 0.2 \qquad (8.435)$$

The steady-state error e_{ss} can be calculated as follows:

$$e_{ss} = \lim_{s \to 0} s * U(s)(1 - X(s)) = \lim_{s \to 0} s \frac{1}{s} \left(1 - \frac{10}{s^2 + 5s + 50} \right) = 0.8 \qquad (8.436)$$

3. Peak time T_p, maximum peak M_p, percent overshoot %O.S., rise time T_r and settling time T_s:
   - The peak time T_p can be calculated as follows:

$$T_p = \frac{\pi}{\omega_n \sqrt{1 - \zeta^2}} = \frac{\pi}{\omega_d} = \frac{\pi}{6.614} = 0.475 \text{ (sec)} \qquad (8.437)$$

- The maximum peak M_p can be calculated as follows:

$$M_p = e^{-\pi\zeta/\sqrt{1-\zeta^2}} = e^{-\pi*0.3536/\sqrt{1-(0.3536)^2}} = 0.305 \qquad (8.438)$$

- Percent overshoot %O.S. can be calculated as follows:

$$\%O.S. = M_p * 100 = 30.5\% \qquad (8.439)$$

We can calculate the peak amplitude M_{pp}, which is the difference between the steady-state value X_{ss} and the peak value of the response at the peak time T_p, as follows:

$$M_{pp} = \%O.S. * X_{ss} = 0.305 * 0.2 = 0.061 \qquad (8.440)$$

Now, the value of the maximum response starting from 0 to the peak point at the peak time can be calculated as follows:

$$M_{pmax} = X_{ss} + M_{pp} = 0.2 + 0.061 = 0.261 \qquad (8.441)$$

- Since the system is underdamped, using the formula indicated in equation (8.263), which is the time needed to raise the response to reach the steady state value, the rise time T_r can be calculated as follows:

$$T_r = \frac{\pi - \tan^{-1} \frac{\sqrt{1-\zeta^2}}{\zeta}}{\omega_n \sqrt{1 - \zeta^2}} = \frac{\pi - \tan^{-1} \frac{\sqrt{1-(0.3536)^2}}{0.3536}}{7.071 \sqrt{1 - (0.3536)^2}} = 0.292 \text{ (sec)} \qquad (8.442)$$

We can also determine the rise time based on the formula indicated in equation (8.264), which is the time needed to raise the response from 10% to 90% of the steady state value, as follows:

$$T_{rc} = \frac{2.16\zeta + 0.6}{\omega_n} = \frac{2.16 * 0.3536 + 0.6}{7.071} = 0.193 \text{ (sec)} \qquad (8.443)$$

- The settling time T_s can be calculated as follows:

$$T_s = \frac{4}{\zeta \omega_n} = \frac{4}{0.3536 * 7.071} = 1.6 \text{ (sec)} \qquad (8.444)$$

4. Response equation $x(t)$:

To determine the response equation $x(t)$ due to unit step input, rewrite the given transfer function as follows

$$X(s) = \frac{10}{s(s^2 + 5s + 50)} \tag{8.445}$$

Applying partial fraction expansion to this equation gives

$$X(s) = \frac{10}{s(s^2 + 5s + 50)} = \frac{A}{s} + \frac{Bs + C}{(s^2 + 5s + 50)} \tag{8.446}$$

Clearing fractions gives

$$X(s) = \frac{10}{s(s^2 + 5s + 50)} = \frac{(A + B)s^2 + (5A + C)s + 50A}{s(s^2 + 5s + 50)} \tag{8.447}$$

Equating like terms gives

$$s^2 : A + B = 0 \tag{8.448}$$

$$s : 5A + C = 0 \tag{8.449}$$

$$0 : 50A = 10 \tag{8.450}$$

Solving these equations for A, B and C gives

$$A = \frac{1}{5}, \quad B = \frac{-1}{5} \quad \text{and} \quad C = 0 \tag{8.451}$$

Substituting these values in equation (8.446) gives

$$X(s) = \frac{1}{5}\frac{1}{s} - \frac{1}{5}\frac{s}{(s^2 + 5s + 50)} \tag{8.452}$$

Since the given transfer function is under-damped, the roots of the characteristic equation, $(s^2 + 5s + 50)$, are complex conjugate, which are given by

$$s_{1,2} = -2.5 \pm 6.61i \tag{8.453}$$

So, rewrite equation (8.452) as follows:

$$X(s) = \frac{1}{5}\frac{1}{s} - \frac{1}{5}\frac{s}{(s + 2.5)^2 + (6.61)^2} \tag{8.454}$$

To apply the inverse Laplace transform, we must rearrange this equation to be able to use Table 7.1, items (20) and (21), as follows:

$$X(s) = \frac{1}{5}\frac{1}{s} - \frac{1}{5}\frac{s + 2.5 - 2.5}{(s + 2.5)^2 + (6.61)^2} = \frac{1}{5}\frac{1}{s} - \frac{1}{5}\frac{s + 2.5}{(s + 2.5)^2 + (6.61)^2} +$$
$$\frac{1}{5}\frac{2.5}{6.61}\frac{6.61}{(s + 2.5)^2 + (6.61)^2} \tag{8.455}$$

Now, taking the inverse Laplace transform of this equation gives

$$x(t) = \frac{1}{5} - \frac{1}{5}e^{-2.5t}\left(\cos(6.61t) + \frac{2.5}{6.61}\sin(6.61t)\right) \tag{8.456}$$

This is the response equation of the given transfer function applied to the unit step input.

5. Generate $x(t)$ using MATLAB:
   Figure 8.54 shows the MATLAB live script file for generating the response equation and response plot. It also generates the response characteristics values using

```
% Generating the time response of a second-order system applied to a unit step input.
% Write the system transfer function.
xs=tf([10],[1 5 50]);
% Use syms command to create the symbolic expression of the given transfer function
% applied to the unit step input.
syms s; xss=(10)/(s^3+5*s^2+50*s);
% Use ilaplace command to generate the response equation x(t).
xt=ilaplace(xss)
```

xt =

$$\frac{1}{5} - \frac{e^{-\frac{5t}{2}}\left(\cos\left(\frac{5\sqrt{7}\,t}{2}\right) + \frac{\sqrt{7}\sin\left(\frac{5\sqrt{7}\,t}{2}\right)}{7}\right)}{5}$$

```
% Use step or stepplot command to plot the system response to a unit step input.
fig_raw_resp = figure('position', [0, 0, 700, 350]);
step(xs);title('Unit step response'); xlabel('Time'); ylabel('Step response x(t)')
```

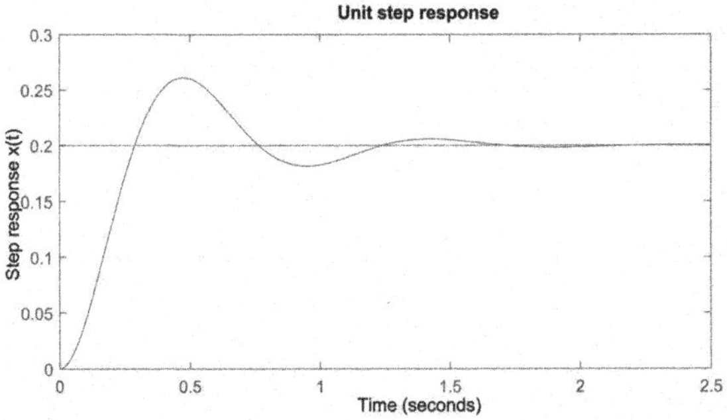

```
% Use stepinfo command to generate the response characteristics values
stepinfo(xs)
```

ans = struct with fields:
         RiseTime: 0.1977
    TransientTime: 1.5484
     SettlingTime: 1.5484
      SettlingMin: 0.1814
      SettlingMax: 0.2610
        Overshoot: 30.4890
       Undershoot: 0
             Peak: 0.2610
         PeakTime: 0.4789

Figure 8.54 MATLAB live script file of transfer function $X(s) = \frac{10}{(s^2+5s+50)}$

the command *stepinfo*. Note that the script provides a definition for each step it contains. Figure 8.55 shows the same plot, viewing the values of the response characteristics in datatips.

When comparing the calculated values of T_p, M_P, %O.S., T_r, and T_s with the corresponding generated ones shown in Figure 8.54 and the data tips shown in Figure 8.55, we found that they are approximately the same. It is essential to know that the rise time calculated in MATLAB is based on the time required to increase the response from 10% to 90% of the steady state value, even when the system is underdamped.

6. Generate the response plot using Simulink.
   The Simulink transfer function block diagram and the generated response are shown in Figures 8.56 and 8.57, respectively. When comparing Figure 8.57, the Simulink response, to Figure 8.55, the MATLAB response, they are the same.

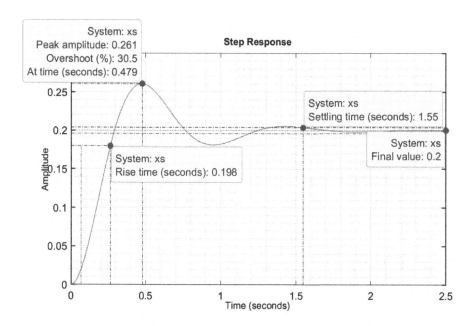

Figure 8.55 Unit step response of the given transfer function generated from MATLAB

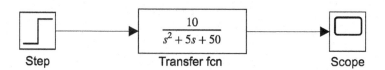

Figure 8.56 Simulink transfer function block diagram

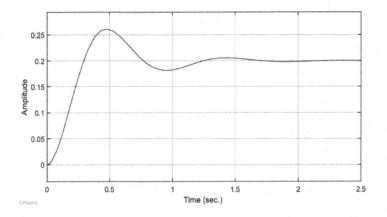

Figure 8.57 Unit step response of the given transfer function generated from Simulink

Example (8.7):

Consider the following transfer function:

$$X(s) = \frac{s+4}{(s^2 + 4s + 4)}$$

For a unit impulse input, determine

1. The damping ratio ζ, natural frequency ω_n, damping frequency ω_d.
2. The response equation $x(t)$.
3. The steady-state value X_{ss} and the steady-state error e_{ss}.
4. Use MATLAB to generate and plot the response equation (verify the results obtained in step 2).

Solution:

1. Damping ratio ζ, natural frequency ω_n, damping frequency ω_d.
    Comparing the characteristic equation of the given transfer function with the general characteristic equation of the second-order system gives

$$s^2 + 4s + 4 = s^2 + 2\zeta\omega_n s + \omega_n^2 \tag{8.457}$$

Equating like terms gives

$$2\zeta\omega_n = 4 \quad \text{and} \quad \omega^2 = 4 \tag{8.458}$$

Solving this equation for ω_n and ζ gives

$$\omega_n = 2 \ \ (\text{rad/sec}) \quad \text{and} \quad \zeta = 1 \ (\text{critically damped}) \tag{8.459}$$

Since the system is critically damped, there is no damping frequency ω_d.

2. Response equation $x(t)$:

To determine the response equation $x(t)$ due to unit impulse input, rewrite the given transfer function as follows

$$X(s) = \frac{s+4}{(s^2+4s+4)} = \frac{s+4}{(s+2)^2} \tag{8.460}$$

Applying partial fraction expansion to this equation gives

$$X(s) = \frac{s+4}{(s+2)^2} = \frac{A}{(s+2)} + \frac{B}{(s+2)^2} \tag{8.461}$$

Clearing fractions gives

$$X(s) = \frac{s+4}{(s+2)^2} = \frac{A(s+2)+B}{(s+2)^2} = \frac{As+(2A+B)}{(s+2)^2} \tag{8.462}$$

Equating like terms and solving for A and B gives

$$A = 1, \quad \text{and} \quad B = 2 \tag{8.463}$$

Substituting these values in equation (8.461) gives

$$X(s) = \frac{1}{(s+2)} + \frac{2}{(s+2)^2} \tag{8.464}$$

Using Table 7.1, items (6) and (7), taking the inverse Laplace transform to equation (8.464) gives

$$x(t) = e^{-2t} + 2te^{-2t} \tag{8.465}$$

This is the response equation of the given transfer function applied to the unit impulse input.

3. Steady-state value X_{ss} and the steady-state error e_{ss}:

The steady-state value X_{ss} can be calculated as follows,

$$X_{ss} = \lim_{s \to 0} s\, U(s)\, X(s) = \lim_{s \to 0} s\left(\frac{s+4}{(s^2+4s+4)}\right) = 0 \tag{8.466}$$

The steady-state error e_{ss} can be calculated as follows:

$$e_{ss} = \lim_{s \to 0} s * U(s)(1 - X(s)) = \lim_{s \to 0} s\left(1 - \frac{s+4}{(s^2+4s+4)}\right) = 0 \tag{8.467}$$

```
% Generating the time response of a second-order system applied to unit impulse input.
% Write the system transfer function.
xs=tf([1 4],[1 4 4]);
% Use syms command to create the symbolic expression of the given transfer function
% applied to the unit impulse input.
syms s;
xss=(s+4)/(s^2+4*s+4);
% Use ilaplace command to generate the response equation x(t) to unit
% impulse input
xt=ilaplace(xss)
```

$xt = e^{-2t} + 2te^{-2t}$

```
% Use impulse command to plot the system response to unit impulse input.
fig_raw_resp = figure('position', [0, 0, 700, 400]);
impulse(xs);title('Unit impulse response'); xlabel('Time');
ylabel('Impulse response x(t)')
```

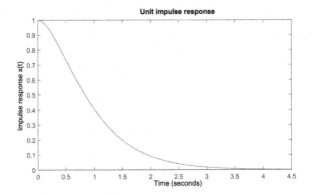

Figure 8.58   MATLAB live script file of the transfer function $X(s) = \frac{s+4}{(s^2+4s+4)}$

4. Generate $x(t)$ using MATLAB:
   Figure 8.58 shows the MATLAB live script file for generating the response equation and the response plot for a unit impulse input. Note that the script provides a definition for each step it contains. Comparing equation (8.465) with the corresponding generated one shown in Figure 8.58, they are the same.

Example (8.8):
Consider the following transfer function:

$$X(s) = \frac{25}{(s^2 + 4s + 25)}$$

For a unit ramp input, find:

1. The damping ratio ζ, natural frequency ω_n, and damping frequency ω_d.
2. The steady-state value X_{ss} and steady-state error e_{ss}.
3. The response equation $x(t)$.
4. Use MATLAB to generate and plot the response equation $x(t)$.

Solution:

1. Damping ratio ζ, natural frequency ω_n, and damping frequency ω_d:

   Comparing the characteristic equation of the given transfer function with the general characteristic equation of the second-order system gives

   $$s^2 + 4s + 25 = s^2 + 2\zeta\omega_n s + \omega_n^2 \qquad (8.468)$$

   Equating like terms gives

   $$2\zeta\omega_n = 4 \quad \text{and} \quad \omega^2 = 25 \qquad (8.469)$$

   Solving this equation for ω_n and ζ gives

   $$\omega_n = 5 \quad (\text{rad/sec}) \quad \text{and} \quad \zeta = 0.4 < 1 \ (\text{under damped}) \qquad (8.470)$$

   The damping frequency ω_d is calculated as follow:

   $$\omega_d = \omega_n \sqrt{1 - \zeta^2} = 5\sqrt{1 - (0.4)^2} = 4.58 \quad (\text{rad/sec}) \qquad (8.471)$$

2. The steady-state value X_{ss} and the steady-state error e_{ss}:

   The steady-state value X_{ss} can be calculated as follows,

   $$X_{ss} = \lim_{s \to 0} s\, U(s)X(s) = \lim_{s \to 0} s\, \frac{1}{s^2}\left(\frac{25}{(s^2 + 4s + 25)}\right) = \infty \qquad (8.472)$$

   The steady-state error e_{ss} can be calculated as follows:

   $$e_{ss} = \lim_{s \to 0} s\, U(s)(1 - X(s)) = \lim_{s \to 0} s\frac{1}{s^2}\left(1 - \frac{25}{(s^2 + 4s + 25)}\right) = \infty \qquad (8.473)$$

3. Response equation $x(t)$:
   To determine the response equation $x(t)$ due to unit ramp input, rewrite the given transfer function as follows

   $$X(s) = \frac{25}{s^2(s^2 + 4s + 25)} \qquad (8.474)$$

   Applying partial fraction expansion gives

   $$X(s) = \frac{25}{s^2(s^2 + 4s + 25)} = \frac{A}{s} + \frac{B}{s^2} + \frac{Cs + D}{s^2 + 4s + 25} \qquad (8.475)$$

   Clearing fractions gives

   $$X(s) = \frac{s + 25}{s^2(s^2 + 4s + 25)} = \frac{(A+C)s^3 + (4A + B + D)s^2 + (25A + 4)s + 25B}{s^2(s^2 + 4s + 25)} \qquad (8.476)$$

Solving for A, B, and C as follows. Equating factors of both sides of equation (8.475) gives

$$s^3 : A + C = 0 \tag{8.477}$$

$$s^2 : 4A + B + D = 0 \tag{8.478}$$

$$s : 25A + 4B = 0 \tag{8.479}$$

$$0 : 25B = 25 \tag{8.480}$$

Solving these equations gives

$$A = \frac{-4}{25}, \quad B = 1, \quad C = \frac{4}{25}, \quad \text{and} \quad D = \frac{-9}{25} \tag{8.481}$$

Substituting these values in equation (8.475) and rearranging the results gives

$$X(s) = \frac{-4}{25}\frac{1}{s} + \frac{1}{s^2} + \frac{4}{25}\frac{s}{(s^2 + 4s + 25)} - \frac{9}{25}\frac{1}{(s^2 + 4s + 25)} \tag{8.482}$$

Since the given transfer function is under-damped, the roots of the characteristic equation are complex conjugates, which are given by

$$s_{1,2} = -2 \pm \sqrt{21}i \tag{8.483}$$

Considering this, equation (8.482) becomes

$$X(s) = -\frac{4}{25}\frac{1}{s} + \frac{1}{s^2} + \frac{4}{25}\frac{s}{(s+2)^2 + (\sqrt{21})^2} - \frac{9}{25}\frac{1}{(s+2)^2 + (\sqrt{21})^2} \tag{8.484}$$

To use the inverse Laplace transform Table 7.1, items (2), (3), (20), and (21), rewrite this equation as follows:

$$X(s) = -\frac{4}{25}\frac{1}{s} + \frac{1}{s^2} + \frac{4}{25}\frac{s+2-2}{(s+2)^2 + (\sqrt{21})^2} - \frac{9}{25}\frac{1}{\sqrt{21}}\frac{\sqrt{21}}{(s+2)^2 + (\sqrt{21})^2} \tag{8.485}$$

Rearranging this equation gives

$$X(s) = -\frac{4}{25}\frac{1}{s} + \frac{1}{s^2} + \frac{4}{25}\frac{s+2}{(s+2)^2 + (\sqrt{21})^2} - \frac{8}{25}\frac{1}{\sqrt{21}}\frac{\sqrt{21}}{(s+2)^2 + (\sqrt{21})^2} -$$

$$\frac{9}{25}\frac{1}{\sqrt{21}}\frac{\sqrt{21}}{(s+2)^2 + (\sqrt{21})^2} \tag{8.486}$$

Taking the inverse Laplace transform of equation (8.486) gives

$$x(t) = -\frac{4}{25} + t + e^{-2t}\left(\frac{4}{25}\cos(\sqrt{21}t) - \frac{17}{25\sqrt{21}}\sin(\sqrt{21}t)\right) \tag{8.487}$$

This is the response equation of the given transfer function due to a unit impulse input.

4. Generate and plot the response equation $x(t)$ using MATLAB:

Figure 8.59 shows the MATLAB live script file generating the response equation $x(t)$ and its plot to a unit ramp input. The script provides a definition for each step it contains.

```
% Generating the time response of a second-order system applied to unit ramp input.
% Write the system transfer function.
xs=tf([25],[1 4 25]);
% Use syms command to create the symbolic expression of the given transfer function
% applied to the unit ramp input.
syms s;
xss=(25)/(s^4+4*s^3+25*s^2);
% Use ilaplace command to generate the response equation x(t).
xt=ilaplace(xss)
```

xt =

$$t + \frac{4\,e^{-2t}\left(\cos(\sqrt{21}\,t) - \dfrac{17\,\sqrt{21}\,\sin(\sqrt{21}\,t)}{84}\right)}{25} - \frac{4}{25}$$

```
% Define the unit ramp function.
u=t;
% Define the location of the plot.
fig_raw_resp = figure('position', [0, 0, 700, 350]);
% Use lsim command to generate the plot of the response to the unit ramp input.
[x,t]=lsim(xs,t,u); plot(t,x);hold on; plot(t,u);
title('Unit Ramp response'); xlabel('Time (sec)'); ylabel('Ramp response x(t)')
```

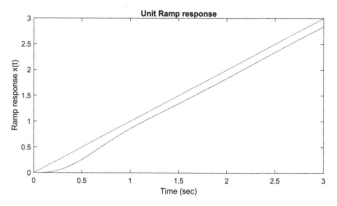

Figure 8.59 MATLAB live script file for the transfer function $X(s) = \frac{25}{(s^2+4s+25)}$

8.3.9   BOND GRAPH SIMULATION OF THE SYSTEM RESPONSE

There are many types of software for obtaining the differential equations and generating the system response through Bond Graphs. One software, 20-sim, was developed at the University of Twente, Netherlands. The bond numbers, power directions, causalities, and differential equations can be automatically assigned. The user has to give the parameter values and the initial conditions. The program then solves the differential equations and plots the responses. Many other analysis and control functions can also be performed using 20-sim software.

Example (8.9):

Consider the system shown in Figure 3.38; its Augmented Bond Graph is shown in Figure 8.62. Their equations of motion are indicated in equations (3.107) and (3.108), which are

$$m_1\ddot{x}_1 + b(\dot{x}_1 - \dot{x}_2) + +b_1\dot{x}_1 + k(x_1 - x_2) + k_1x_1 = F(t) \qquad (8.488)$$

and

$$m_2\ddot{x}_2 + b(\dot{x}_2 - \dot{x}_1) + b_2\dot{x}_2 + k(x_2 - x_1) + k_2 x_2 = 0 \tag{8.489}$$

Considering the following parameter values; $m_1 = 4$ (kg), $m_2 = 2$ (kg), $b = 10$ (N sec/m), $b_1 = b_2 = 5$ (N sec/m), $k = 100$ (N m), $k_1 = k_2 = 50$ (N.m) and $F(t) = 50\,u(t)$ (N) (step input).

1. Create the Simulink block diagram and plot the responses $x_1(t)$ and $x_2(t)$.
2. Create the 20-sim Bond Graph model and plot the responses $x_1(t)$ and $x_2(t)$. Compare the results obtained with the Simulink-generated results.

Solution:
1. Simulink model:
The Simulink block diagram of equations (8.488) and (4.489) is shown in Figure 8.60. The procedure for generating a Simulink block diagram for coupled equations is illustrated in this figure. Insert the parameter values into the default values of the Simulink blocks, and then run the model. The generated plots of the system responses $x_1(t)$ and $x_2(t)$ are shown in Figure 8.61.

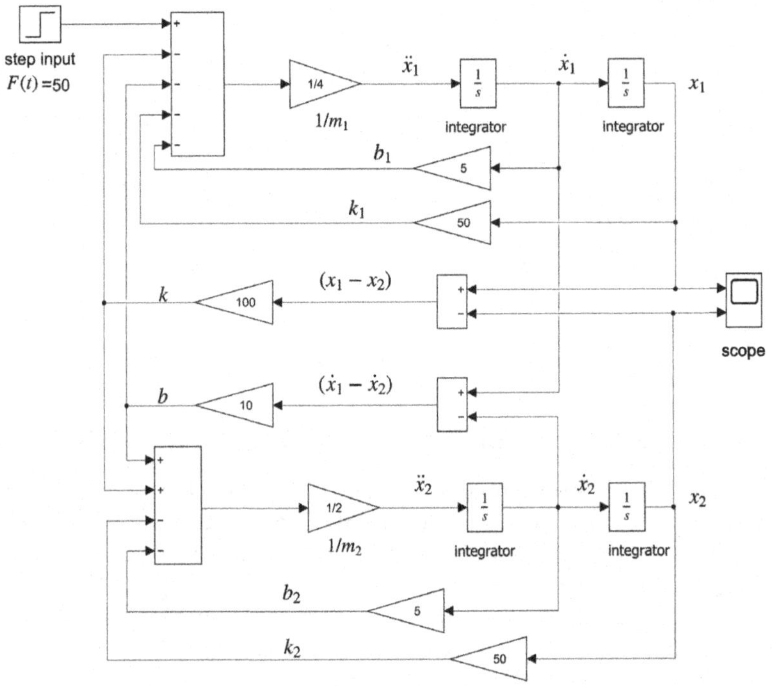

- To create the Simulink model, rearrange equation (8.488) and (8.489) as follows:

$$\ddot{x}_1 = \frac{1}{m_1}[F(t) - b(\dot{x}_1 - \dot{x}_2) - b_1\dot{x}_1 - k(x_1 - x_2) - k_1 x_1]$$

$$\ddot{x}_2 = \frac{1}{m_2}[-b(\dot{x}_2 - \dot{x}_1) - b_2\dot{x}_2 - k(x_2 - x_1) - k_2 x_2]$$

- Use the Simulink Library Browser to build the Simulink block digram shown above see Appendix (2).

Figure 8.60 Simulink block diagram of equations (8.488) and (8.489)

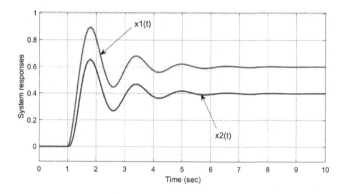

Figure 8.61   Generated responses from Simulink block diagram

2. 20-sim Bond Graph model:

The 20-sim model of the Augmented Bond Graph model of the system shown in Figure 8.62 should be created in the same manner as shown in Figure 8.63. Insert the parameter values, and then run the model. To supply the step input to the model, we must use a *MSE* element to control the start time and step values. To generate the responses $x_1(t)$ and $x_2(t)$, we created two integrators, one for each 1-junction, as shown in Figure 8.63. Both are utilized to integrate the outputs of the two 1-junctions of common flow (velocity) v_1 and v_2, which are the responses $x_1(t)$ and $x_2(t)$. Comparing Figure 8.64, 20-sim results, with Figure 8.61, they are the same.

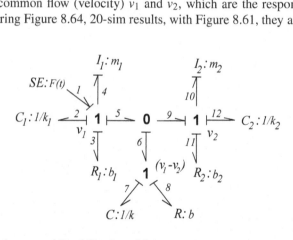

Figure 8.62   Augmented Bond Graph model

Example (8.10):

Consider the system shown in Figure 4.56; its Augmented Bond Graph is shown in Figure 8.62. Their equations of motion are indicated in equations (4.264) and (4.265), which are

$$L_a \frac{di_a}{dt} + R_a i_a + k_m \dot{\theta} = E(t) \tag{8.490}$$

and

$$J_l \ddot{\theta} + b_l \dot{\theta} = k_m i_a \tag{8.491}$$

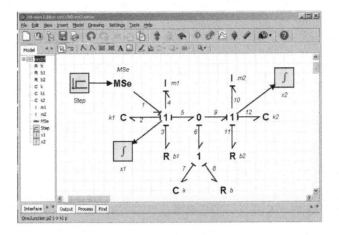

Figure 8.63 20-sim bond graph model

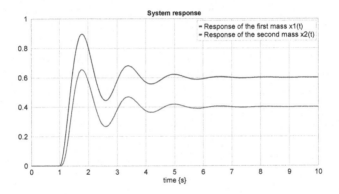

Figure 8.64 Generated responses from 20-sim Bond Graph model

Considering the following parameter values: $R_a = 0.07$ (Ohms), $L_a = 0.17$ (H), $J_m = 0.17$ (N m^2), $J_l = 0.1$ (N m^2), $b_l = 0.02$ (N m/sec) and $F(t) = 5$ (Volt.) (step input).

1. Create the Simulink block diagram and plot the motor speed $\dot{\theta}(t)$ and the motor current $i_a(t)$.
2. Consider the Augmented Bond Graph model shown in Figure 8.65; create the 20-sim Bond Graph model and generate the motor speed $\dot{\theta}(t)$ and motor current $i_a(t)$. Compare the results with the Simulink results obtained in step (1).

Solution:

1. Simulink model:
   The Simulink block diagram of equations (8.490) and (4.491) is shown in Figure 8.66. The procedure for generating a Simulink block diagram for coupled equations is illustrated in this figure. Insert the parameter values into the default values

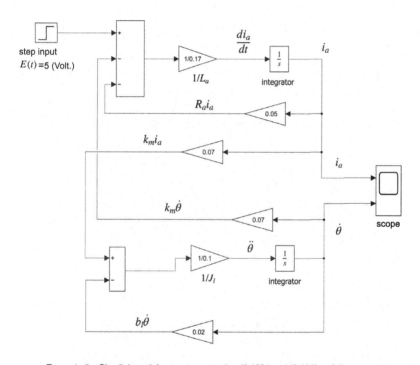

Figure 8.65 Augmented Bond Graph model

- To create the Simulink model, rearrange equation (8.490) and (8.491) as follows:

$$\frac{di}{dt} = \frac{1}{L_a}[E(t) - R_a i_a - k_m \dot{\theta}]$$

$$\ddot{\theta} = \frac{1}{J_l}[k_m i_a - b_l \dot{\theta}]$$

- Use the Simulink Library Browser to build the Simulink block digram shown above, see Appendix (2).

Figure 8.66 Simulink block diagram model of equations (8.490) and (8.491)

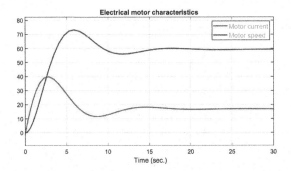

Figure 8.67   Generated responses from Simulink model

of the Simulink blocks, and then run the model. The generated plots of the motor speed $\dot{\theta}$ and the current i_a are shown in Figure 8.67.

2. 20-sim Bond Graph model:

The 20-sim model of the Augmented Bond Graph model of the system shown in Figure 8.68 should be created in the same manner as shown in Figure 8.65. Insert the parameter values, and then run the model. The generated results of motor speed $\dot{\theta}(t)$ and motor current $i_a(t)$, are shown in Figure 8.67. Comparing Figure 8.67, 20-sim results, with Figure 8.64, they are the same.

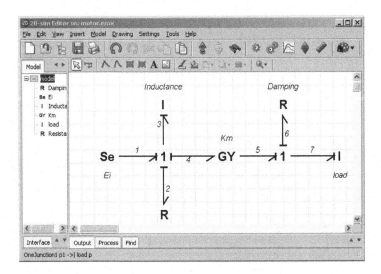

Figure 8.68   20-sim Bond Graph model

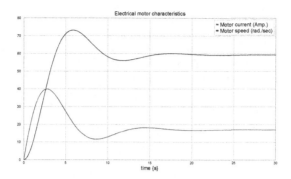

Figure 8.69   Generated responses from 20-sim Bond Graph model

8.4   PROBLEMS

Problem (8.1):
For each differential equation, obtain the system gain K and time constant τ.

1. $8\dot{x} + 7x = 0$ if $x(0) = 3$
2. $6\dot{x} + 5x = 15$ if $x(0) = 3$
3. $13\dot{x} + 3x = 0$ if $x(0) = -2$
4. $7\dot{x} - 5x = 0$ if $x(0) = 9$

Problem (8.2):
For each of the following differential equations, determine

a. The transfer function $X(s)$.
b. The steady-state gain K and the time constant τ.
c. The response equation $x(t)$ and plot it over a time interval ranging from $t = 0$ to 5τ.
d. The steady-state value X_{ss} and the steady-state error e_{ss}.
e. Use MATLAB to generate the plot of $x(t)$.

Note that $\delta(t)$, $u(t)$, and $r(t)$ are the unit impulse, the unit step, and the unit ramp input functions, respectively.

1. $10\dot{x} + 5x = 25\,\delta(t)$ if $x(0) = 0$
2. $6\dot{x} + 12x = 3\,\delta(t)$ if $x(0) = 1$
3. $12\dot{x} - 6x = 18\,\delta(t)$ if $x(0) = -2$
4. $10\dot{x} + 5x = 25\,u(t)$ if $x(0) = 0$
5. $6\dot{x} + 12x = 3\,u(t)$ if $x(0) = 1$
6. $13\dot{x} - 6x = 18\,u(t)$ if $x(0) = -2$
7. $10\dot{x} + 5x = 25\,r(t)$ if $x(0) = 0$
8. $6\dot{x} + 12x = 3\,r(t)$ if $x(0) = 1$
9. $13\dot{x} - 6x = 18\,r(t)$ if $x(0) = -2$

Problem (8.3):
If applicable, determine the damping coefficient ζ, the time constant τ, the natural frequency ω_n, and the damping frequency ω_d for the following roots and find the corresponding characteristics equation.

1. $s = -2 \pm 6i$
2. $s = -1 \pm 5i$
3. $s = -5, -5$
4. $s = -2, -3 \pm i$
5. $s = -3, -2 \pm 2i$

Problem (8.4):
For each of the following differential equations, determine

a. The damping ratio ζ, natural frequency ω_n, and damping frequency ω_d.
b. Steady-state value X_{ss} and the steady-state error e_{ss}.
c. Peak time T_p, maximum peak M_p, percent overshoot %O.S., rise time T_r and settling time T_s.
d. The transfer function $X(s)$.
e. The response equation $x(t)$ and plot it over a time interval ranging from $t = 0$ to 5τ.
f. Use MATLAB to generate the plot of $x(t)$.
g. Use Simulink to generate the $x(t)$ plot and verify the MATLAB results obtained in step (f).

Note that $\delta(t)$, $u(t)$, and $r(t)$ are the unit impulse, the unit step, and the unit ramp input functions, respectively.

1. $2\ddot{x} + 2\dot{x} + 20x = 4\,\delta(t)$ if $x(0) = 0$ and $\dot{x}(0) = 4$
2. $3\ddot{x} + 21\dot{x} + 75 = 7\,\delta(t)$ if $x(0) = 2$ and $\dot{x}(0) = 6$
3. $\ddot{x} + 8\dot{x} + 40x = 5\,\delta(t)$ if $x(0) = 5$ and $\dot{x}(0) = 0$
4. $2\ddot{x} + 20\dot{x} + 2x = 4\,u(t)$ if $x(0) - 0$ and $\dot{x}(0) = 4$
5. $3\ddot{x} + 21\dot{x} + 75x = 7\,u(t)$ if $x(0) = 2$ and $\dot{x}(0) = 6$
6. $\ddot{x} + 8\dot{x} + 40x = 5\,u(t)$ if $x(0) = 5$ and $\dot{x}(0) = 0$
7. $2\ddot{x} + 2\dot{x} + 20x = 4t\,r(t)$ if $x(0) = 0$ and $\dot{x}(0) = 4$
8. $3\ddot{x} + 21\dot{x} + 75x = 7t\,r(t)$ if $x(0) = 2$ and $\dot{x}(0) = 6$
9. $\ddot{x} + 8\dot{x} + 40x = 5t\,r(t)$ if $x(0) = 5$ and $\dot{x}(0) = 0$

Problem (8.5):
Figure 8.70 shows the response of a system to step input. Determine,

a. The damping ratio ζ, natural frequency ω_n, and damping frequency ω_d.
b. Peak time T_p, maximum peak M_p, percent overshoot %O.S., rise time T_r and settling time T_s.
c. The transfer function $X(s)$.

Problem (8.6):
Considering the following differential equation, which describes a certain second-order system, estimate the value of damping factor b so that it meets the following specifications:

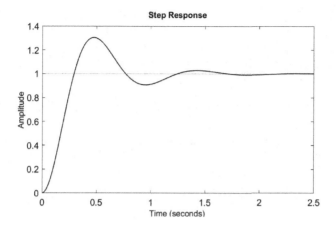

Figure 8.70   Second-order step response

a. The maximum percent overshoot is $O.S. \leq 15\%$.
b. The rise time $T_r \leq 4$

$$20\ddot{x} + b\dot{x} + 5x = u(t)$$

Problem (8.7):
Figure 8.71 shows the response of a second-order system to step input of $f(t) = 5u(t)$. The equation of motion is

$$m\ddot{x} + b\dot{x} + kx = f(t)$$

Estimate the values of m, b, and k.

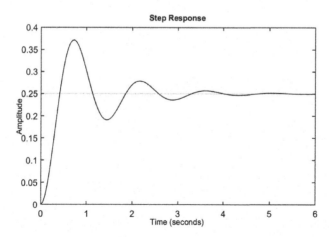

Figure 8.71   Second-order step response

Problem (8.8):

Consider the system shown in Figure 3.45, presented in Example (3.5), and consider the following parameter values: $m_1 = 2$ (Kg), $m_2 = 5$ (kg), $b = 10$ (N-m/sec), $k = 100$ (N/m) and $f(t) = 20$ (N) (impulse input). Create 20-sim Bond Graph models and generate the displacements $x_1(t)$ and $x_2(t)$ considering the following two cases

1. When m_2 moves in the same direction as the mass m_1, (use the Augmented Bond Graph model shown in Figure 3.46).
2. When m_2 moves in the opposite direction as the mass m_1, (use the Augmented Bond Graph model shown in Figure 3.47).
3. Compare the results and discuss the effect of changing the movement direction of the mass m_2 on the system's motion.

9 Frequency Response

9.1 INTRODUCTION

What is the Frequency Response, and how does it perform? It measures the system response to sinusoidal inputs of varying frequencies. Here, we look for the sinusoidal steady-state achieved after the transient has passed out. In other words, we look for the forced part of the system response rather than nature. When the system's input is sinusoidal, the steady-state output of the system response will be sinusoidal, of the same frequency but possibly different magnitude and phase. So, if the input to a linear system is a sinusoidal function with magnitude F_i and frequency ω,

$$f(t) = F_i \sin(\omega t + \phi_i) \tag{9.1}$$

then the steady-state output of the system response will be sinusoidal of the same frequency ω, but with a different magnitude and phase as follows

$$x(t) = X_o \sin(\omega t + \phi_o) \tag{9.2}$$

where X_o is the magnitude of the output, while ϕ_i and ϕ_o are phase angles of the input and output sinusoidal functions, respectively. Usually, the sinusoidal inputs are assumed to be of zero phases ($\phi_i = 0$), implying that the phase angle of the system response is equal to the phase angle of the output ϕ_o, as shown in Figure 9.1. Equation (9.2) indicates that if we can be fined X_o and ϕ_o as a function of the frequency ω, we will know how the system response $x(t)$, due to a sinusoidal input, behaves.

9.2 FREQUENCY RESPONSE CONCEPT

Considering the system shown in Figure 9.2, the equation representing its dynamic motion due to a sinusoidal input is given by

$$m\ddot{x} + b\dot{x} + kx = F_i \sin(\omega t + \phi_i) \tag{9.3}$$

First, consider a convenient method to construct a sinusoidal function using a complex number algebra to specify the frequency response described below in polar form

$$a + ib = M \angle \phi \quad \text{and} \quad M = |(a + ib)| = \sqrt{a^2 + b^2}, \quad \phi = -\tan^{-1}\frac{b}{a} \tag{9.4}$$

where M is the magnitude and ϕ is the phase angle of the complex number. Equation (9.4) indicates that the complex number's magnitude and phase represent the sinusoidal function's magnitude and phase. Therefore, applying equation (9.4), the input $f(t) = F_i \sin(\omega t + \phi_i)$ and the output $x(t) = X_o \sin(\omega t + \phi_o)$ can be represented by

DOI: 10.1201/9781032685656-9

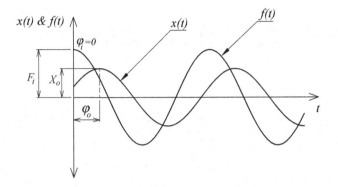

Figure 9.1 Input and output sinusoidal signals

complex numbers as $f(t) = F_i \angle \phi_i(\omega)$ and $x(t) = X_o(\omega) \angle \phi_o(\omega)$, respectively. Assuming that the system itself, $m\ddot{x} + b\dot{x} + kx$, is also represented by $M(\omega) \angle \phi(\omega)$ as shown in Figure 9.3, the steady-state output of the system response is determined by multiplying the complex number representation of the input by the complex number representation of the system. Thus, the steady-state output of the system is given by

$$X_o(\omega) \angle \phi_o(\omega) = F_i M(\omega) \angle [\phi_i(\omega) + \phi(\omega)] \tag{9.5}$$

From this equation, we found that the magnitude of the system is given by

$$M(\omega) = \frac{X_o(\omega)}{F_i} \tag{9.6}$$

and phase angle is given by

$$\phi(\omega) = \phi_o(\omega) - \phi_i(\omega) \tag{9.7}$$

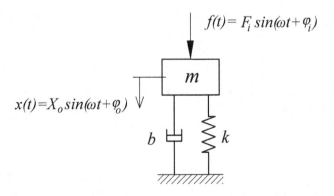

Figure 9.2 Second-order system

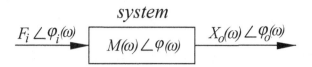

Figure 9.3 System definition

Assuming that the sinusoidal input is of zero phases, $\phi_i(\omega) = 0$, equation (9.7) becomes

$$\phi(\omega) = \phi_o(\omega) \tag{9.8}$$

Equations (9.6) and (9.8) indicate that the combination of magnitudes and phase angles of the input and the output is referred to as the frequency response of a dynamic system. In this case, the magnitude of the system response $M(\omega)$ will be defined as the ratio of the output magnitude $X_o(\omega)$ to the input magnitude F_i as indicated in equation (9.6). In contrast, the phase angle of the system $\phi(\omega)$ will be defined as the phase angle of the output $\phi_o(\omega)$ as indicated in equation (9.8), considering that the phase angle of the input is always assumed to be zero. It also notes that the magnitude and phase of the system are functions of the frequency ω and only apply to the system's steady-state sinusoidal response.

Since we are looking for the forced steady-state output of the system response, we can do that in two ways: either from the particular solution part of the system response equation or directly from its transfer function.

9.2.1 PARTICULAR SOLUTION

9.2.1.1 First-order system

Considering the first-order system shown in Figure 9.4, the equation representing its dynamic motion due to sinusoidal input is given by

$$b\dot{x} + kx = F_i \sin(\omega t) \tag{9.9}$$

Recall the general form of this equation as follows

$$\tau \dot{x} + x = K F_i \sin(\omega t) \tag{9.10}$$

Consider the particular solution part of equation (9.10) in the following form,

$$x(t) = x_p(t) = A \sin(\omega t) + B \cos(\omega t) \tag{9.11}$$

Taking the derivative of this equation gives

$$\dot{x}(t) = \dot{x}_p(t) = A\omega \cos(\omega t) - B\omega \sin(\omega t) \tag{9.12}$$

Substituting equations (9.11) and (9.12) in equation (9.10) gives

$$\tau[A\omega \cos(\omega t) - B\omega \sin(\omega t)] + [A \sin(\omega t) + B \cos(\omega t)] = K F_i \sin(\omega t) \tag{9.13}$$

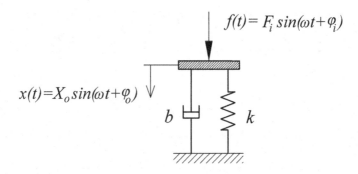

Figure 9.4 First-order system

Rearranging this equation yields

$$(\tau A \omega + B)\cos(\omega t) + (-\tau B \omega + A)\sin(\omega t) = K F_i \sin(\omega t) \tag{9.14}$$

From this equation, we get,

$$\tau A \omega + B = 0 \tag{9.15}$$

$$-\tau B \omega + A = K F_i \tag{9.16}$$

Solving equations (9.15) and (9.16) for A and B gives

$$A = \frac{K F_i}{\tau^2 \omega^2 + 1} \tag{9.17}$$

$$B = \frac{-\tau \omega K F_i}{\tau^2 \omega^2 + 1} \tag{9.18}$$

Substituting equations (9.17) and (9.18) in equation (9.11) gives

$$x(t) = \left(\frac{K F_i}{\tau^2 \omega^2 + 1}\right)\sin \omega t - \left(\frac{\tau \omega K F_i}{\tau^2 \omega^2 + 1}\right)\cos \omega t \tag{9.19}$$

This equation may be simplified as follows,

$$A \sin \omega t + B \cos \omega t = \sqrt{A^2 + B^2}\, \sin\left(\omega t + \tan^{-1}\frac{B}{A}\right) \tag{9.20}$$

Substituting values of A and B, equations (9.17) and (9.18), in equation (9.20) gives

$$x(t) = \sqrt{\left(\frac{K F_i}{\tau^2 \omega^2 + 1}\right)^2 + \left(\frac{-\tau \omega K F_i}{\tau^2 \omega^2 + 1}\right)^2}\, \sin\left(\omega t + \tan^{-1}\left(\frac{\frac{-\tau \omega K F_i}{\tau^2 \omega^2 + 1}}{\frac{K F_i}{\tau^2 \omega^2 + 1}}\right)\right) \tag{9.21}$$

Rearranging this equation yields

$$x(t) = \left(\frac{K F_i}{\sqrt{\tau^2 \omega^2 + 1}}\right)\sin\left(\omega t + \tan^{-1}(-\tau \omega)\right) \tag{9.22}$$

This equation represents the forced steady-state response of a first-order system applied to a sinusoidal input, obtained from the particular solution part of the response equation applied to a sinusoidal input. Now, let us obtain the system response in magnitude-phase form, which is given by

$$X(\omega) = X_o(\omega) \angle \phi_o(\omega) = \sqrt{A^2 + B^2} \angle \tan^{-1} \frac{B}{A} \tag{9.23}$$

Substituting values of A and B, equations (9.17) and (9.18), in equation (9.23), gives

$$X(\omega) = X_o(\omega) \angle \phi_o(\omega) = \sqrt{\left(\frac{KF_i}{\tau^2 \omega^2 + 1}\right)^2 + \left(\frac{-\tau \omega KF_i}{\tau^2 \omega^2 + 1}\right)^2} \angle \tan^{-1} \left(\frac{\frac{-\tau \omega KF_i}{\tau^2 \omega^2 + 1}}{\frac{KF_i}{\tau^2 \omega^2 + 1}}\right) \tag{9.24}$$

Rearrange this equation yields

$$X(\omega) = \left(\frac{KF_i}{\sqrt{\tau^2 \omega^2 + 1}}\right) \angle \tan^{-1}(-\tau \omega) \tag{9.25}$$

This equation represents the magnitude-phase form of the frequency response of a first-order system obtained from the particular solution part of the system equation applied to a sinusoidal input.

9.2.1.2 Second-order system

Consider the system shown in Figure 9.2, and recall the general form of the equation that describes its dynamic motion as follows:

$$\ddot{x} + 2\zeta \omega_n \dot{x} + \omega_n^2 x = KF_i \sin(\omega t) \tag{9.26}$$

Consider the particular solution part of this equation in the following form,

$$x(t) = x_p(t) = A \sin(\omega t) + B \cos(\omega t) \tag{9.27}$$

Taking the first derivative of this equation gives

$$\dot{x}(t) = \dot{x}_p(t) = A \omega \cos(\omega t) - B \omega \sin(\omega t) \tag{9.28}$$

Taking the second derivative gives

$$\ddot{x}(t) = \ddot{x}_p(t) = -A \omega^2 \sin(\omega t) - B \omega^2 \cos(\omega t) \tag{9.29}$$

Substituting equations (9.28) and (9.29) in equation (9.26) gives

$$[-A \omega^2 \sin(\omega t) - B \omega^2 \cos(\omega t)] + 2\zeta \omega_n [A \omega \cos(\omega t) - B \omega \sin(\omega t)] +$$
$$\omega_n^2 [A \sin(\omega t) + B \cos(\omega t) = K F_i \sin(\omega t) \tag{9.30}$$

Rearranging this equation yields

$$[-A \omega^2 - 2\zeta \omega_n \omega B + \omega_n^2 A] \sin(\omega t) + [-B \omega^2 + 2\zeta \omega_n \omega A + \omega_n^2 B] \cos(\omega t)$$
$$= KF_i \sin(\omega t) \tag{9.31}$$

From this equation, we get

$$-A\omega^2 - 2\zeta\omega_n\omega B + \omega_n^2 A = KF_i \tag{9.32}$$

and

$$-B\omega^2 + 2\zeta\omega_n\omega A + \omega_n^2 B = 0 \tag{9.33}$$

Solving equations (9.32) and (9.33) for A and B gives

$$A = \frac{KF_i(\omega_n^2 - \omega^2)}{(\omega_n^2 - \omega^2)^2 - (2\zeta\omega_n\omega)^2} \tag{9.34}$$

$$B = \frac{KF_i(2\zeta\omega_n\omega)}{(\omega_n^2 - \omega^2)^2 - (2\zeta\omega_n\omega)^2} \tag{9.35}$$

Substituting equations (9.34) and (9.35) in equation (9.27) and solving for $x(t)$ gives

$$x(t) = \left(\frac{KF_i(\omega_n^2 - \omega^2)}{(\omega_n^2 - \omega^2)^2 - (2\zeta\omega_n\omega)^2}\right)\sin(\omega t) - \left(\frac{KF_i(2\zeta\omega_n\omega)}{(\omega_n^2 - \omega^2)^2 - (2\zeta\omega_n\omega)^2}\right)\cos(\omega t) \tag{9.36}$$

This equation may be simplified as follows

$$x(t) = \sqrt{\left(\frac{KF_i(\omega_n^2 - \omega^2)}{(\omega_n^2 - \omega^2)^2 - (2\zeta\omega_n\omega)^2}\right)^2 + \left(\frac{KF_i(2\zeta\omega_n\omega)}{(\omega_n^2 - \omega^2)^2 - (2\zeta\omega_n\omega)^2}\right)^2} \sin(\omega t + \phi) \tag{9.37}$$

and

$$\phi = \tan^{-1}\left(\frac{B}{A}\right) = \tan^{-1}\left(\frac{\frac{KF_i(2\zeta\omega_n\omega)}{(\omega_n^2 - \omega^2)^2 - (2\zeta\omega_n\omega)^2}}{\frac{KF_i(\omega_n^2 - \omega^2)}{(\omega_n^2 - \omega^2)^2 - (2\zeta\omega_n\omega)^2}}\right) = \tan^{-1}\left(\frac{2\zeta\omega_n\omega}{(\omega_n^2 - \omega^2)}\right) \tag{9.38}$$

Equations (9.37) and (9.38) represent the forced steady-state response of a second-order system applied to a sinusoidal input. We can obtain the system response in the magnitude-phase form as follows.

$$X(\omega) = \sqrt{\left(\frac{KF_i(\omega_n^2 - \omega^2)}{(\omega_n^2 - \omega^2)^2 - (2\zeta\omega_n\omega)^2}\right)^2 + \left(\frac{KF_i(2\zeta\omega_n\omega)}{(\omega_n^2 - \omega^2)^2 - (2\zeta\omega_n\omega)^2}\right)^2} \angle\phi \tag{9.39}$$

and

$$\phi = \tan^{-1}\left(\frac{2\zeta\omega_n\omega}{(\omega_n^2 - \omega^2)}\right) \tag{9.40}$$

Rearrange equations (9.39) and (9.40) yields

$$X(\omega) = X_o(\omega)\angle\phi_o(\omega) = \left(\frac{KF_i[(\omega_n^2 - \omega^2) - (2\zeta\omega_n\omega)]}{(\omega_n^2 - \omega^2)^2 - (2\zeta\omega_n\omega)^2}\right)\angle\tan^{-1}\left(\frac{2\zeta\omega_n\omega}{(\omega_n^2 - \omega^2)}\right) \tag{9.41}$$

This equation represents the magnitude-phase form of the frequency response of a second-order system obtained from the particular solution part of the system response applied to a sinusoidal input. Now, let us obtain the frequency response from the transfer function.

9.2.2 TRANSFER FUNCTION

Whenever we speak of transfer functions and frequency response, we deal with the forced response of a system. Since we are looking for the forced steady-state output of the system response, we will solve for the forced response portion of the output, which will allow us to determine the system's frequency response. First, we must distinguish between the forced and transient solutions by performing the partial fraction expansion of the system transfer function in the manner described below.

$$X(s) = \text{Partial fraction expansion terms of the system (Transient terms)}+$$

$$\text{Partial Fraction Expansion terms of the input (Forced terms)} \qquad (9.42)$$

Taking the Laplace transform of the input sinusoidal function, $f(t) = F_i \sin(\omega t)$, gives

$$F(s) = \frac{\omega F_i}{s^2 + \omega^2} \qquad (9.43)$$

Taking the partial fraction expansion of this equation gives

$$F(s) = \frac{\omega F_i}{s^2 + \omega^2} = \left(\frac{C_1 s + C_2}{s^2 + \omega^2} \right) \qquad (9.44)$$

Now, recall the general transfer function of a dynamic system and consider equation (9.42); the partial fraction expansion of the system transfer function is given by

$$X(s) = \left(\frac{k_1}{s + p_1} + \frac{k_2}{s + p_2} + \cdots + \frac{k_n}{s + p_n} \right) + \left(\frac{C_1 s + C_2}{s^2 + \omega^2} \right) \qquad (9.45)$$

where p_n are assumed to be distinct poles. Taking the inverse Laplace transform of equation (9.45) gives

$$x(t) = \left(k_1 e^{-p_1 t} + k_2 e^{-p_2 t} + \cdots + k_n e^{-p_n t} \right) + \mathscr{L}^{-1} \left(\frac{C_1 s + C_2}{s^2 + \omega^2} \right) \qquad (9.46)$$

where k_n, C_1 and C_2 are constants. Note that if the system is stable, all p_n have positive nonzero real parts, implying that each exponential term, indicated in equation (9.46), decays to zero as the time interval approaches ∞. In other words, once the system reaches a steady state ($t \rightarrow \infty$), all transient terms approach zero, and only the forced terms are considered. Considering this fact and insert $s = i\omega$, equation (9.46) becomes

$$x(t) = \mathscr{L}^{-1} \left(\frac{C_1 s + C_2}{s^2 + \omega^2} \right) = \mathscr{L}^{-1} \left(\frac{C_1 s + C_2}{(s + i\omega)(s - i\omega)} \right) \qquad (9.47)$$

In this case, in the limit for $x(t)$ when the time $t \rightarrow \infty$, we can obtain the steady state output as follows

$$x(t) = \mathscr{L}^{-1} \left(\frac{C_1 s + C_2}{(s + i\omega)(s - i\omega)} \right) = \frac{1}{\omega} |\omega F_i X(i\omega)| \sin(\omega t + \phi) \qquad (9.48)$$

So,

$$x(t) = F_i |X(i\omega)| \sin(\omega t + \phi) \tag{9.49}$$

where ϕ is given by

$$\phi(\omega) = \angle X(i\omega) \tag{9.50}$$

Equations (9.49) and (9.50) indicate that if we want to get the frequency response of a system from the transfer function, we must first insert $s = i\omega$ in $X(s)$ and then obtain the magnitude and phase.

9.2.2.1 First-order system

Recall the general form of the first-order system, equation (9.10), and applying Laplace transform gives

$$X(s) = \frac{K}{\tau s + 1} \frac{\omega F_i}{(s^2 + \omega^2)} \tag{9.51}$$

Now, we will solve for the forced response portion of the output, allowing us to determine the system's frequency response. To get the frequency response from the transfer function, insert $s = (i\omega)$ as stated above and convert it to magnitude-phase form. First, we must distinguish between the transient and forced solutions by performing a partial fraction expansion.

$$X(s) = \text{Partial Expansion terms of the system (Transient terms)} +$$

$$\left(\frac{C_1}{(s + i\omega)} + \frac{C_2}{(s - i\omega)} \right) \text{ (Forced terms)} \tag{9.52}$$

As discussed previously, the partial expansion terms of the transfer function provide exponential terms that decay as the system reaches a steady state response. So, the constants C_1 and C_2 can be determine as follows

$$C_1 = \frac{\omega}{(s - i\omega)} \frac{KF_i}{\tau s + 1} \bigg|_{s = -i\omega} \tag{9.53}$$

$$C_1 = \frac{\omega}{(-i\omega - i\omega)} \frac{KF_i}{\tau(-i\omega) + 1} \tag{9.54}$$

and,

$$C_2 = \frac{\omega}{(s + i\omega)} \frac{KF_i}{\tau s + 1} \bigg|_{s = i\omega} \tag{9.55}$$

$$C_2 = \frac{\omega}{(i\omega + i\omega)} \frac{KF_i}{\tau(i\omega) + 1} \tag{9.56}$$

Substituting equations (9.54) and (9.56) in equation (9.49) and neglecting the transient terms, the magnitude is given by

$$|X(i\omega)| = \frac{\sqrt{K^2 F_i^2}}{\sqrt{(\tau\omega)^2 + (1)^2}} = \frac{KF_i}{\sqrt{(\tau\omega)^2 + 1}} \tag{9.57}$$

and the phase angle is given by

$$\angle X(i\,\omega) = \frac{\angle KF_i}{\angle \tan^{-1}((\tau\omega)^2 + 1)} = \frac{\angle 0^o}{\angle \tan^{-1}(\tau\omega)} = \angle(0^o - \tan^{-1}(\tau\omega)) \quad (9.58)$$

or

$$\angle X(i\,\omega) = \tan^{-1}(-\tau\omega) \quad (9.59)$$

Substituting equations (9.57) and (9.59) in equation (9.23) gives

$$X(i\,\omega) = \frac{KF_i}{\sqrt{\tau^2\,\omega^2 + 1}}\angle \tan^{-1}(-\tau\omega) \quad (9.60)$$

Sometimes, the magnitude is represented by a magnitude ratio, which is the ratio of the output $X(\omega)$ to the input F_i: $M(\omega) = |X(i\omega)/F_i|$. So, as a magnitude ratio equation (9.60) becomes

$$M(\omega) = |X(i\omega)/F_i| = \frac{K}{\sqrt{\tau^2\,\omega^2 + 1}}\angle \tan^{-1}(-\tau\omega) \quad (9.61)$$

This equation represents the magnitude-phase form of the frequency response obtained from the transfer function for a first-order system applied to a sinusoidal input. Some remarks on equation (9.61) are outlined below,

1. The first-order systems have a small magnitude for high-frequency input and a larger magnitude for low-frequency input.
2. Whether a frequency is considered high or low for the system response depends on the time constant τ.
3. A high frequency for a system with large τ may be a relatively low frequency for a small τ system.

9.2.2.2   Second-order system

Considering the second-order system shown in Figure 9.2 and assuming that the input is a sinusoidal function, the equation representing the dynamic motion of the system is given by

$$\ddot{x} + 2\,\zeta\,\omega_n\dot{x} + \omega^2 x = KF_i\sin(\omega t) \quad (9.62)$$

Apply the Laplace transform to equation (9.62) and set $s = (i\omega)$ as follows. Note that the symbol i is equal to $\sqrt{-1}$.

$$X(i\omega) = \frac{KF_i}{(i\omega)^2 + 2\,\zeta\,\omega_n(i\omega) + \omega_n^2} \frac{\omega}{((i\omega)^2 + \omega^2)} \quad (9.63)$$

The magnitude of the response $|X(i\omega)|$ can be calculated as follows

$$|X(i\omega)| = \frac{\sqrt{(KF_i)^2}}{\sqrt{(\omega_n^2 - \omega^2)^2 + (2\zeta\,\omega_n\omega)^2}} = \frac{KF_i}{\sqrt{(\omega_n^2 - \omega^2)^2 + (2\zeta\,\omega_n\omega)^2}} \quad (9.64)$$

and the phase angle $\angle X(i\omega)$ is calculated as follows

$$\angle X(i\omega) = \frac{\angle 0^o}{\angle \tan^{-1}(\frac{2\zeta\omega_n\omega}{\omega_n^2-\omega^2})} = \angle \tan^{-1} -(\frac{2\zeta\omega\omega_n}{\omega_n^2-\omega^2}) \qquad (9.65)$$

Then, the total response is given by

$$|X(i\omega)| = \frac{F_iK\angle 0^0}{\sqrt{(\omega_n^2-\omega^2)^2+(2\zeta\omega\omega_n)^2}\angle\tan^{-1}-(\frac{2\zeta\omega\omega_n}{\omega_n^2-\omega^2})} \qquad (9.66)$$

Rearrange this equation yields

$$|X(i\omega)| = \frac{F_iK}{\sqrt{(\omega_n^2-\omega^2)^2+(2\zeta\omega\omega_n)^2}}\angle-\tan^{-1}(\frac{2\zeta\omega\omega_n}{\omega_n^2-\omega^2}) \qquad (9.67)$$

or

$$|X(i\omega)| = F_iK[(\omega_n^2-\omega^2)^2+(2\zeta\omega\omega_n)^2]^{-1/2}\angle-\tan^{-1}(\frac{2\zeta\omega\omega_n}{\omega_n^2-\omega^2}) \qquad (9.68)$$

As a magnitude ratio, this equation becomes

$$M(\omega) = |X(i\omega)/F_i| = K[(\omega_n^2-\omega^2)^2+(2\zeta\omega\omega_n)^2]^{-1/2}\angle-\tan^{-1}(\frac{2\zeta\omega\omega_n}{\omega_n^2-\omega^2}) \quad (9.69)$$

This equation represents the magnitude-phase form of the frequency response of a second-order system obtained from the transfer function. This equation indicated that when the frequency ω is equal to the natural frequency of the system ω_n, the amplitude of the response is dependent on the value of the damping ratio ζ. This situation will be further investigated later in detail.

9.3 MAGNITUDE-PHASE FORM OF GENERAL TRANSFER FUNCTION

Consider the general transfer function of a dynamic system as follows

$$X(s) = \frac{K(s+z_1)(s+z_2)[s^2+2\zeta_{z_1}\omega_{z_1}s+\omega_{z_1}^2]...}{s^k(s+p_1)(s+p_2)[s^2+2\zeta_{p_1}\omega_{p_1}s+\omega_{p_1}^2]...} \qquad (9.70)$$

The factors of the transfer function numerator are known as *Zeros*, $(z_1,z_2,\omega_{z_1},...)$, while the factors of the denominator (characteristic equation) are known as *Poles*, $(p_1,p_2,\omega_{p_1},...)$. Note that the number of zeros is always less than the poles. Inserting $s=i\omega$ in equation (9.70) gives

$$X(i\omega) = \frac{K(i\omega+z_1)(i\omega+z_2)[(i\omega)^2+2\zeta_{z_1}\omega_{z_1}(i\omega)+\omega_{z_1}^2]...}{(i\omega)^k(i\omega+p_1)(i\omega+p_2)[(i\omega)^2+2\zeta_{p_1}\omega_{p_1}(i\omega)+\omega_{p_1}^2]...} \qquad (9.71)$$

Now, let us estimate analytical expressions for the magnitude and phase of the frequency response from the general transfer function. Rearrange equation (9.71) so that each zero and pole appear to take on their normalized form as follows.

$$X(i\omega) = \frac{K(z_1 z_2 \omega_z ...)(\frac{i\omega}{z_1}+1)(\frac{i\omega}{z_2}+1)((\frac{i\omega}{\omega_{z_1}})^2 + 2\zeta_{z_1}(\frac{i\omega}{\omega_{z_1}})+1)\cdots}{(p_1 p_2 \omega_{p_1}...)(i\omega)^k(\frac{i\omega}{\omega_{p_1}}+1)(\frac{i\omega}{p_2}+1)((\frac{i\omega}{\omega_{p_1}})^2 + 2\zeta_{p_1}(\frac{i\omega}{\omega_{p_1}})+1)\cdots} \tag{9.72}$$

or,

$$X(i\omega) = \frac{K_t(\frac{i\omega}{z_1}+1)(\frac{i\omega}{z_2}+1)((\frac{i\omega}{\omega_{z_1}})^2 + 2\zeta_{z_1}(\frac{i\omega}{\omega_{z_1}})+1)\cdots}{(i\omega)^k(\frac{i\omega}{p_1}+1)(\frac{i\omega}{p_2}+1)((\frac{i\omega}{\omega_{p_1}})^2 + 2\zeta_{p_1}(\frac{i\omega}{\omega_{p_1}})+1)\cdots} \tag{9.73}$$

where K_t is a constant and it is given by

$$K_t = \left(\frac{K z_1 z_2 \omega_{z_1} \cdots}{p_1 p_2 \omega_{p_1} \cdots}\right) \tag{9.74}$$

Applying equation (9.23), the magnitude-phase expression of the general transfer function indicated in equation (9.73) is given by

$$|X(i\omega)|\angle X(i\omega) = \frac{K_t \angle \alpha_{K_t} * |(\frac{i\omega}{z_1}+1)|\angle \alpha_1 * |(\frac{i\omega}{z_2}+1)|\angle \alpha_2 * |((\frac{i\omega}{\omega_{z_1}})^2 + 2\zeta_{z_1}(\frac{i\omega}{\omega_{z_1}})+1)|\angle \alpha_{z_1}\cdots}{|(i\omega)^k|\angle \phi_k * |(\frac{i\omega}{p_1}+1)|\angle \phi_1 * |(\frac{i\omega}{p_2}+1)|\angle \phi_2 * |((\frac{i\omega}{\omega_{p_1}})^2 + 2\zeta_{p_1}(\frac{i\omega}{\omega_{p_1}})+1)|\angle \phi_{p_1}\cdots} \tag{9.75}$$

Rearranging this equation yields

$$|X(i\omega)|\angle X(i\omega) = \frac{K_t|(\frac{i\omega}{z_1}+1)||(\frac{i\omega}{z_2}+1)||((\frac{i\omega}{\omega_z})^2 + 2\zeta_{z_1}(\frac{i\omega}{\omega_z})+1)|\cdots}{|(i\omega)^k||(\frac{i\omega}{p_1}+1)||(\frac{i\omega}{p_2}+1)||((\frac{i\omega}{\omega_p})^2 + 2\zeta_{p_1}(\frac{i\omega}{\omega_p})+1)|\cdots}$$

$$* \frac{\angle(\alpha_{K_t} + \alpha_1 + \alpha_2 + \alpha_{z_1}...)}{\angle(\phi_k + \phi_1 + \phi_2 + \phi_{p_1}\cdots)} \tag{9.76}$$

where the magnitude is given by

$$M(i\omega) = |X(i\omega)| = \frac{K_t|(\frac{i\omega}{z_1}+1)||(\frac{i\omega}{z_2}+1)||[(\frac{i\omega}{\omega_{z_1}})^2 + 2\zeta_{z_1}(\frac{i\omega}{\omega_{z_1}})+1]|\cdots}{|(i\omega)^k||(\frac{i\omega}{p_1}+1)||(\frac{i\omega}{p_2}+1)||[(\frac{i\omega}{\omega_{p_1}})^2 + 2\zeta_{p_1}(\frac{i\omega}{\omega_{p_1}})+1]|\cdots} \tag{9.77}$$

or

$$M(i\omega) = K_t + |(\frac{i\omega}{z_1}+1)| + |(\frac{i\omega}{z_2}+1)| + |[(\frac{i\omega}{\omega_{z_1}})^2 + 2\zeta_{z_1}(\frac{i\omega}{\omega_{z_1}})+1]| + \cdots$$
$$- |(i\omega)^k| - |(\frac{i\omega}{p_1}+1)| - |(\frac{i\omega}{p_2}+1)| - |[(\frac{i\omega}{\omega_{p_1}})^2 + 2\zeta_{p_1}(\frac{i\omega}{\omega_{p_1}})+1]| - \cdots \tag{9.78}$$

From this equation, we get

$$M(\omega) = K_t + \sqrt{(\frac{\omega}{z_1})^2 + 1^2} + \sqrt{(\frac{\omega}{z_2})^2 + 1^2} + \sqrt{[1^2 - (\frac{\omega}{\omega_{z_1}})^2]^2 + [2\zeta_{z_1}(\frac{\omega}{\omega_{z_1}})]^2} + \cdots$$
$$- k\sqrt{(\omega)^2} - \sqrt{(\frac{\omega}{p_1})^2 + 1^2} - \sqrt{(\frac{\omega}{p_2})^2 + 1^2} - \sqrt{[1^2 - (\frac{\omega}{\omega_{p_1}})^2]^2 + [2\zeta_{p_1}(\frac{\omega}{\omega_{p_1}})]^2} - \cdots \tag{9.79}$$

From equation (9.76), the phase angle is given by

$$\phi = \angle X(i\omega) = \frac{\angle(\alpha_{K_t} + \alpha_1 + \alpha_2 + \alpha_{z_1}\ldots)}{\angle(\phi_k + \phi_1 + \phi_2 + \phi_{p_1}\ldots)} \qquad (9.80)$$

Rearranging this equation yields

$$\phi = \angle \alpha_{K_t} + \angle \alpha_1 + \angle \alpha_2 + \angle \alpha_{z_1} + \cdots - \angle \phi_k - \angle \phi_1 - \angle \phi_2 + \angle \phi_{p_1} - \cdots \qquad (9.81)$$

Applying equation (9.38), equation (9.81), becomes

$$
\phi(\omega) = \angle \tan^{-1} K_t + \angle \tan^{-1}\left(\frac{\omega/z_1}{1}\right) + \angle \tan^{-1}\left(\frac{\omega/z_2}{1}\right) + \tan^{-1}\left(\frac{2\zeta_{z_1}\left(\frac{\omega}{\omega_{z_1}}\right)}{1 - \left(\frac{\omega}{\omega_{z_1}}\right)^2}\right)\ldots
$$

$$
- \angle \tan^{-1}\left(\frac{\omega/p_1}{1}\right) - \angle \tan^{-1}\left(\frac{\omega/p_2}{1}\right) - \tan^{-1}\left(\frac{2\zeta_{p_1}\left(\frac{\omega}{\omega_{p_1}}\right)}{1 - \left(\frac{\omega}{\omega_{p_1}}\right)^2}\right) - \cdots
$$

$$(9.82)$$

Equations (9.78) and (80) represent the magnitude and phase forms of the frequency response of the general transfer function of a dynamic system. These equations indicated that if we know zeros and poles of $X(s)$, we can be fined the magnitude $M(\omega) = |X(i\omega)|$ and the phase angle $\phi(\omega) = \angle X(i\omega)$.

Now, the question is, how do we obtain $M(\omega)$ and $\phi(\omega)$ from the transfer function? To do that, the above calculations for $M(\omega)$ and $\phi(\omega)$ should be performed as ω changing from 0 to ∞ and then plotting the frequency response. Because this is too difficult, we will use the frequency response plotting techniques developed by Bode. The Bode plot is a semi-log plot with a logarithmic x-axis coordinate for the frequency ω and linear rectangular y-axis coordinates for magnitude and phase. Later on, when we look at the Bode plot, we will find it much easier to understand the frequency response than the calculation!

9.4 BODE PLOTTING TECHNIQUES

The logarithmic of the magnitude is expressed in terms of the logarithm to the base 10 of a unit defined in decibels dB; hence, it is given by

$$\text{Logarthmic gain} = 20\log_{10}|X(i\omega)| = 20\log_{10}M(\omega) \qquad (9.83)$$

where $M(\omega) = |X(i\omega)|$ and is a constant for each frequency. Note that the following analysis uses the symbol log to mean $\log_{10}$. The Bode plot has two plots: one for the magnitude and another for the phase angle. The magnitude plot shows first, followed by the phase plot, and both are a function of frequency ω.

This section will demonstrate Bode plot techniques and explain how to get an analytical formula: insert $s = i\omega$ into $X(s)$ and solve the system's frequency response. We will demonstrate how to estimate the magnitude, and phase plots are drawn rapidly using straight lines (asymptotes) approximation. In the following analysis, we will call straight lines approximation asymptotes. Also, a low frequency is near 0, and a high frequency is when the frequency approaches ∞. Note that when the transfer function is factorizing in normalized form as indicated in equation (9.73), it will take on a Bode transfer function with a Bode gain K_t. When plotting the frequency response of this transfer function, it is essential to note that after normalization, each factor in the numerator and denominator will have the same low-frequency asymptote. This common low-frequency asymptote makes combining the zero and the pole factors easier to obtain the Bode plot, which will be shown later. Now, convert the magnitude response to decibels (dB) as follows. Using equation (9.79) gives

$$20 \log M(\omega) = 20 \log K_t + 20 \log[(\frac{\omega}{z_1})^2 + 1^2]^{\frac{1}{2}} + 20 \log[(\frac{\omega}{z_2})^2 + 1^2]^{\frac{1}{2}}$$

$$+ 20 \log \left([1 - (\frac{\omega}{\omega_{z_1}})^2]^2 + [2\zeta_{z_1}(\frac{\omega}{\omega_{z_1}})]^2 \right)^{\frac{1}{2}} \cdots - k * 20 \log(\omega) - 20 \log[(\frac{\omega}{p_1})^2 + 1^2]^{\frac{1}{2}}$$

$$- 20 \log[(\frac{\omega}{p_2})^2 + 1^2]^{\frac{1}{2}} - 20 \log \left([1 - (\frac{\omega}{\omega_{p_1}})^2]^2 + [2\zeta_{p_1}(\frac{\omega}{\omega_{p_1}})]^2 \right)^{\frac{1}{2}} \cdots \qquad (9.84)$$

This equation represents the analytical expression of the magnitude response of the transfer function in dB, while equation (9.82) represents the analytical expression of the phase response. Equation (9.84) indicates that the product of the magnitudes of the frequency responses of each pole and zero factor yields the magnitude of the system's frequency response. This means we can figure out how much response we have by knowing the magnitude response of each pole and zero factors.

Since the Bode plot can be approximated as a sequence of asymptotes, it is feasible to simplify the sketching of these plots. In other words, working with the logarithm of the magnitude can make the process easier because the magnitude response of zeros factors would be added together, and the magnitude response of poles factors would be subtracted together rather than being multiplied or divided as indicated in equations (9.82) and (9.84). This means that if we know the magnitude of each term, the algebraic addition or subtraction will produce the full magnitude of the transfer function in dB. Furthermore, if we could create an approximation of each factor consisting of asymptotes, would the graphical addition or subtraction of terms be significantly simplified, as we will show later.

Now, let us demonstrate how to approximate the frequency response of zeros and poles terms using asymptote approximation and how to combine these responses to sketch the composite frequency response of the system, meaning by plotting the frequency response using the Bode plot technique.

Generally, any transfer function can be factorized, in normalized form, to be composed of the following seven types of terms.

1. constant K_t
2. zeros at the origin, $(i\omega)^m$.
3. poles at the origin $\frac{1}{(i\omega)^n} = (i\omega)^{-n}$.
4. real zero $(\frac{i\omega}{z} + 1)$.

5. real pole $\frac{1}{(\frac{i\omega}{p}+1)} = (\frac{i\omega}{p}+1)^{-1}$.

6. complex conjugate zeros $((\frac{i\omega}{\omega_n})^2 + 2\zeta(\frac{i\omega}{\omega_n}) + 1)$.

7. complex conjugate poles $\frac{1}{((\frac{i\omega}{\omega_n})^2+2\zeta(\frac{i\omega}{\omega_n})+1)} = ((\frac{i\omega}{\omega_n})^2 + 2\zeta(\frac{i\omega}{\omega_n}) + 1)^{-1}$.

In the following analysis, we will demonstrate how to plot the magnitude and phase for each type of these terms, and at the end, we will explain how to generate a composite plot for a general transfer function. The following figures illustrate the reasoning behind these plots.

1. Constant, K_t:
- **Magnitude plot:** The magnitude of the constant K_t in dB is given by

$$20\log K_t = \text{Constant} \tag{9.85}$$

It is represented by a horizontal asymptote starting from the value obtained from equation (9.85) until it approaches ∞, as shown in Figure 9.5a.

- **Phase plot:** The phase angle of the constant K_t is given by

$$\phi(\omega) = \angle\tan^{-1}(K_t) = 0 \tag{9.86}$$

It is represented by a horizontal asymptote starting from 0 until it approaches ∞, as shown in Figure 9.5a. Note that if the value of K_t is negative, the magnitude gain remains $20\log K_t$, but the phase angle accounts for the negative sign equals -180°.

2. Zeros at the origin, $(i\omega)^m$:
Consider a zero at the origin, for which we want to sketch separate magnitude and phase response plots. The magnitude of zeros at the origin can be calculated as follows. Using equation (9.87), gives

$$M(\omega) = |(i\omega)^m| \tag{9.87}$$

and the phase angle is given by

$$\phi(\omega) = m\angle\tan^{-1}(\omega) \tag{9.88}$$

- **Magnitude plot:** The magnitude response of zeros at the origin in dB is given by

$$20\log M(\omega) = 20\log|(i\omega)^m| = m*20\log\omega \tag{9.89}$$

It is represented by an asymptote extending the frequency, which varies from 0 to infinite and has a slope of +20 dB/decade or +6 dB/octave, crossing the 0 dB line at $\omega = z$, as shown in Figure 9.5a. Note that the *decade* is the interval between two frequencies with a ratio of 10, while the *octave* is the interval between two frequencies with a ratio of 2, as shown in Figure 9.5a. Since both slopes are equivalent, we will refer to the following analysis as dB/decade. Note that the slope of the asymptote of

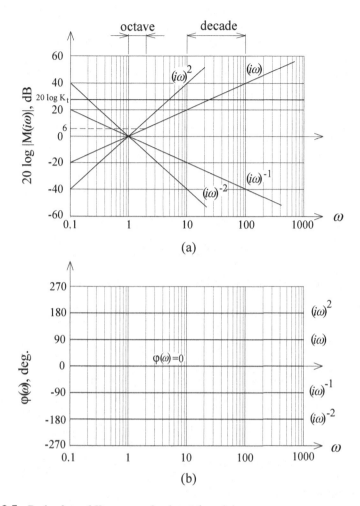

Figure 9.5 Bode plots of K_t, zeros and poles at the origin

zero at the origin depends on how many zeros are at the origin. So, in the case of multiple zeros at the origin, the slope of the asymptote is $20m$ dB/decade, as indicated in equation (9.89). For example, if we have two zeros at the origin ($m = 2$), the asymptote will start from -40 dB at the low-frequency point, passing through $\omega = 1$ until it reaches ∞ with a slope +40 dB/decade, as shown in Figure 9.5a. It is important to note that the starting point of the zero at the origin is -20 dB if the low-frequency point is $\omega = 0.1$ and then intersects the 0 dB line at $\omega = z$, whereas it is -40 dB if the low-frequency point is $\omega = 0.01$ and then intersects the 0 dB line at $\omega = z$.

- Phase plot: Using equation (9.88), the phase angle of zeros at the origin is given by

$$\phi(\omega) = m\angle\tan^{-1}(\omega) = +90^o m \tag{9.90}$$

It is represented by a horizontal asymptote starting from $+90^o$ at the low-frequency point and continuing to ∞. The value of the phase angle depends on how many zeros are at the origin, meaning that if we have two zeros at the origin, $(m = 2)$, the phase angle will be a horizontal asymptote starting from $+180^o$ at the low-frequency point and continuing to ∞ as indicated in equation (9.90) and shown in Figure 9.5b.

3. Poles at the origin, $1/(i\omega)^n$:
Consider the pole at the origin, for which we want to sketch separate magnitude and phase response plots. The magnitude of poles at the origin is given by

$$M(\omega) = |\frac{1}{(i\omega)^n}| = |(i\omega)^{-n}| \tag{9.91}$$

and the phase angle is given by

$$\phi(\omega) = \angle\tan^{-1}(\omega)^{-n} = -n\angle\tan^{-1}(\omega) \tag{9.92}$$

- **Magnitude plot:** The magnitude response of poles at the origin in dB is given by

$$20\log|\frac{1}{(i\omega)^n}| = 20\log 1 - n*20\log\omega = -n*20\log\omega \tag{9.93}$$

It is represented by an asymptote extending the frequency from 0 to infinite and has a slope of -20 dB/decade, crossing the 0 dB line at $\omega = p$, as shown in Figure 9.5a. In the case of multiple poles at the origin, the slopes of the asymptotes are $-20n$ dB/decade, as indicated in equation (9.93). For example, if we have two poles at the origin $(n = 2)$, the asymptote will start from +40 dB at low frequency, passing through $\omega = 1$ until it approaches ∞ with a slope of -40 dB/decade as indicated in equation (9.94) and shown in Figure 9.5a. Like the zero at the origin, the starting point of the pole at the origin is +20 dB if the low-frequency point is $\omega = 0.1$ and then intersects the 0 dB line at $\omega = 1$, whereas it is +40 dB if the low-frequency point is $\omega = 0.01$ then intersects the 0 dB line at $\omega = 1$.

- **Phase plot:** The phase angle of poles at the origin is given by

$$\phi(\omega) = -n\angle - \tan^{-1}(\omega) = -n*-90^o \tag{9.94}$$

It is represented by a horizontal asymptote starting from -90° at low frequency until it reaches ∞. The value of the phase angle depends on how many poles are, meaning that if we have two poles at the origin, $(n = 2)$, the phase angle will be a horizontal asymptote starting from -180° until it approaches ∞, as shown in Figure 9.5b. Note that the asymptotes of zeros and poles at the origin are called low-frequency asymptotes.

4. Real zero, $\left(\frac{i\omega}{z} + 1\right) = (i\omega\tau + 1)$

It is worth noting that the real zero represents a first-order system nominator dynamic with $\tau = (1/z)$. Consider a real zero, for which we want to sketch separate magnitude and phase response plots. The magnitude of a real zero is given by

$$M(\omega) = |(\frac{i\omega}{z}) + 1| = \sqrt{(\frac{\omega}{z})^2 + 1^2} = [(\frac{\omega}{z})^2 + 1^2]^{\frac{1}{2}} \qquad (9.95)$$

and the phase angle is given by

$$\phi(\omega) = \angle\tan^{-1}(\frac{\omega}{z} + 1) = \angle\tan^{-1}(\frac{\omega/z}{1}) = \angle\tan^{-1}(\frac{\omega}{z}) \qquad (9.96)$$

- **Magnitude plot** The magnitude response of a real zero in dB can be calculated as follows. Using equation (9.95) gives
- For $\omega \ll z$:

$$20\log M(\omega) = 20\log[(\frac{\omega}{z})^2 + 1^2)]^{\frac{1}{2}} = 10\log[(\frac{\omega}{z})^2 + 1^2] = 10\log 1 = 0 \qquad (9.97)$$

It is represented by an asymptote starting from 0 at a low frequency and continuing horizontally until it reaches $\omega = z$, as shown in Figure 9.6a.

- When $\omega = z$, breakpoint frequency:

It is represented by an asymptote starting from 0 at a low frequency and continuing horizontally until it reaches $\omega = z$, as shown in Figure 9.6a. At this frequency, the low-frequency asymptote is equal to the high-frequency asymptote, and the transition point between them, known as the breakpoint frequency (corner frequency), occurs. Also, at this frequency, the slope of the asymptote will begin to change, indicating the beginning of the transition.

- For $\omega \gg z$:

$$20\log M(\omega) = 10\log[(\frac{\omega}{z})^2 + 1^2] = 20\log(\frac{\omega}{z}) \qquad (9.98)$$

The slope of the asymptote representing the real zero at $\omega \gg z$ can be determined using this equation. A decade between two frequencies ω_1 and ω_2 will be used to calculate it, with $\omega_2 = 10\omega_1$. The difference in the logarithmic increases over a decade of the frequency for $\omega \gg z$ can be calculated as follows. Using equation (9.98) gives

$$20\log M(\omega_2) - 20\log M(\omega_1) = 20\log(\omega_2/z) - 20\log(\omega_1/z) =$$
$$20\log(\omega_2 z/\omega_1 z) = 20\log(10\omega_1 z/\omega_1 z) = 20\log(10) = +20\,\text{dB} \qquad (9.99)$$

The slope of the asymptote for the real zero is +20 dB/decade for $\omega \gg z$. So, for $\omega \gg z$, the real zero is represented by an asymptote starting at the breakpoint frequency and then extends upward until it reaches infinite frequency, passing through a magnitude of +20 dB one decade above $\omega = z$, as shown in Figure 9.6a.

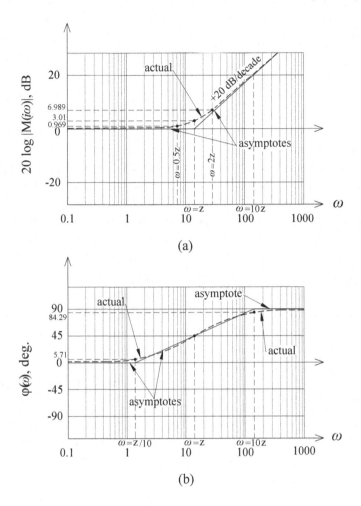

Figure 9.6 The asymptote and the actual curve of a real zero

To summarize, the magnitude of the real zero is represented by a horizontal asymptote starting at 0 dB at the low-frequency point and continuing until it reaches the breakpoint frequency, $\omega = z$. The slope changes upward by +20 dB/decade to the infinite frequency at this frequency. Now, let us draw the exact (actual) magnitude curve of the real zero. This can be accomplished directly using a computer program such as Matlab to solve equation (9.95) for various frequency ω values. Some of the results and the corresponding values obtained using the asymptote approximation for comparison are presented in Table 9.1. Alternatively, to sketch the actual magnitude quickly, instead of using the computer program, we will use equation (9.99) to apply it to three different frequencies, including the breakpoint frequency, one below and one above within a decade interval, for example, $\omega = 0.5z$ and $\omega = 2z$. So, using equation (9.95) gives

- At $\omega = 0.5z$:

$$20\log M(0.5z) = 10\log[(\frac{\omega}{z})^2 + 1^2] = 10\log[(\frac{0.5z}{z})^2 + 1^2)] =$$
$$10\log[(0.5)^2 + 1^2] = 0.969 \text{ dB} \qquad (9.100)$$

- At breakpoint frequency, $\omega = z$:

$$20\log M(z) = 10\log[(\frac{\omega}{z})^2 + 1^2] = 10\log(2) = 3.01\,\text{dB} \qquad (9.101)$$

- At $\omega = 2z$:

$$20\log M(2z) = 10\log[(\frac{\omega}{z})^2 + 1^2] = 10\log[(\frac{2z}{z})^2 + 1^2] =$$
$$10\log[(2)^2 + 1^2] = 10\log(5) = 6.989 \text{ dB} \qquad (9.102)$$

Because the values of the actual magnitude have approximately the same values as the asymptotic approximations at low and high frequencies regardless of the value of the frequency as indicated in Table 9.1, and the results obtained in equations (9.100), (9.101), and (9.102), the actual magnitude of the real zero is sketched (dotted curve) as shown in Figure 9.6a. Now, let us check out the maximum difference in magnitude response between the actual and the asymptotic approximation shown in this figure. As a result, the real zero has a disparity in magnitude response of no more than 3.01 dB at the breakpoint frequency, $\omega = z$.

- **Phase plot:** The phase angle of a real zero can be determined as follows. First, it is worth noting that the slope of the asymptote in the magnitude plot of the real zero starts to change at the breakpoint frequency, meaning that it was zero below this point (at low frequencies). As we will see below, the phase angle of the real zero has a specific value at the breakpoint frequency, meaning that the slope of the phase asymptote will begin below the breakpoint frequency. Thus, the slope of the phase asymptote must begin one decade below the breakpoint frequency, $\omega = z/10$, with zero phases and end one decade above the breakpoint frequency, $\omega = 10z$. Now, let us turn to the phase plot of a real zero, which can be drawn as follows. Using equation (9.96) gives

Table 9.1

Comparison between the asymptote and actual magnitude values of a real zero

ω	0.01z	0.1z	0.5z	z	5z	10z	50z	100z	500z	1000z
Asymptote (dB)	0	0	0	0	14.12	20	33.97	40	53.98	60
Actual (dB)	0	0	0.969	3.01	14.15	20.04	33.98	40	53.98	60

- For $\omega \ll z$:

$$\phi(\omega) = \angle\tan^{-1}(\frac{\omega}{z}) \approx \angle\tan^{-1}(0) = 0^o \qquad (9.103)$$

- When $\omega = z$, breakpoint frequency:

$$\phi(\omega) = \angle\tan^{-1}(\frac{\omega}{z}) = \angle\tan^{-1}(1) = +45^o \qquad (9.104)$$

- For $\omega \gg z$:

$$\phi(\omega) = \angle\tan^{-1}(\frac{\omega}{z}) \approx +90^o \qquad (9.105)$$

It is represented by an asymptote starting horizontally from 0 at the low-frequency point until it reaches $\omega = z/10$, then extends upward until it reaches $\omega = 10z$, passing through $+45^o$ at the breakpoint frequency $\omega = z$, where it levels off at $+90^o$, as shown in Figure 9.6b. At this frequency, the asymptote continues horizontally from the leveled-off phase, $+90^o$ at $\omega = 10z$, until it approaches ∞.

Now, let us draw the actual phase curve of the real zero. This can be accomplished directly using the Matlab program to solve equation (9.96) for various frequency ω values. Some of the results and the corresponding values obtained using the asymptote approximation for comparison are presented in Table 9.2. Alternatively, to sketch the actual phase of the real zero quickly rather than using a computer program, we must first establish the phases at specific frequencies. So, let us determine the actual phase values at $\omega = 0.01$, $\omega = 0.1z$, $\omega = 10z$, and $\omega = 1000z$ as follows. Using equation (9.96) gives

- At $\omega = 0.01$:

$$\phi(\omega) = -\angle\tan^{-1}(\frac{0.01z}{z}) = -\angle\tan^{-1}(0.01) = +0.573^o \qquad (9.106)$$

- At $\omega = z/10$:

$$\phi(\omega) = \angle\tan^{-1}(\frac{z}{10z}) = \angle\tan^{-1}(0.1) = +5.71^o \qquad (9.107)$$

- At $\omega = 10z$:

$$\phi(\omega) = \angle\tan^{-1}(\frac{10z}{z}) = \angle\tan^{-1}(10) = +84.29^o \qquad (9.108)$$

Table 9.2

Comparison between the asymptote and actual phase values of a real zero

ω	$0.01z$	$0.1z$	$0.5z$	z	$5z$	$10z$	$50z$	$100z$	$500z$	$1000z$
Asymptote (deg.)	0	0	31.5	45	76.5	90	90	90	90	90
Actual (deg.)	0.57	5.71	26.6	45	78.69	84.29	88.85	89.42	89.88	89.94

- At $\omega = 1000$:

$$\phi(\omega) = -\angle\tan^{-1}(\frac{1000z}{z}) = -\angle\tan^{-1}(1000) = +89.94^o \qquad (9.109)$$

Using the phase values obtained from equations (9.106) to (9.109), we can quickly sketch the actual phase response instead of the computer program. Equations (9.107) indicated that the difference between the actual and the asymptotic phases at $\omega = 10z$ is $\phi(10z) = +90^o - (+84.29^o) = +5.71^o$. Now, let us check out the maximum difference in phase response between the actual and the asymptotic approximation shown in Figure 9.6b. As a result, the maximum difference was found to be equally $+5.71^o$ at both $\omega = z/10$ and $\omega = 10z$, implying that the real zero has a disparity in phase response of no more than 5.71^o.

Note that all previous analyses were carried out on a real zero in its normalized form, implying that we can apply them to any real zero.

4. Real pole, $\frac{1}{(i\omega\tau+1)}$:

It is worth noting that the real pole represents a first-order dynamic system with $\tau = (1/p)$. Consider a real pole for which we want to sketch separate magnitude and phase response plots. The magnitude of a real pole is given by

$$M(\omega) = |\frac{1}{(\frac{i\omega}{p})+1}| = \frac{\sqrt{1^2}}{\sqrt{(\frac{\omega}{p})^2 + 1^2}} = [(\frac{\omega}{p})^2 + 1^2]^{\frac{-1}{2}} \qquad (9.110)$$

and the phase angle is given by

$$\phi(\omega) = \angle\tan^{-1}\frac{1}{(\frac{\omega}{p}+1)} = -\angle\tan^{-1}(\frac{\omega}{p}+1) = -\angle\tan^{-1}(\frac{\omega/p}{1}) = -\angle\tan^{-1}(\frac{\omega}{p}) \qquad (9.111)$$

- **Magnitude plot:** The magnitude response of a real pole in dB can be calculated as follows. Using equation (9.110) gives

- For $\omega \ll p$:

$$20\log M(\omega) = 20\log[(\frac{\omega}{p})^2 + 1^2)]^{\frac{-1}{2}} = -10\log[(\frac{\omega}{p})^2 + 1^2] = -10\log 1 = 0 \qquad (9.112)$$

It is represented by an asymptote starting from 0 at the low frequency and continuing horizontally until it reaches $\omega = p$, as shown in Figure 9.7a.

- When $\omega = p$, breakpoint frequency:
Like real zero, at this frequency, the low-frequency asymptote is equal to the high-frequency asymptote, implying the transition point between them occurs. At this frequency, the slope of the asymptote will begin to change, indicating the beginning of the transition.

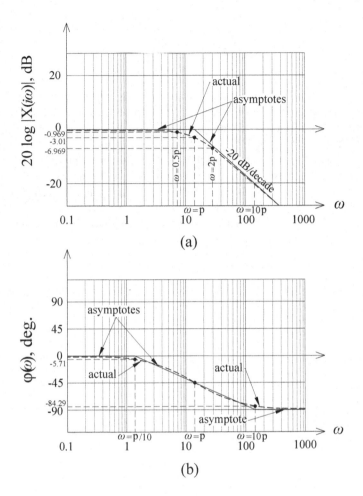

Figure 9.7 The asymptote and the actual curve of a real pole

- For $\omega \gg p$:

$$20\log M(\omega) = -10\log[(\frac{\omega}{p})^2 + 1^2] = -20\log(\frac{\omega}{p}) \qquad (9.113)$$

The slope of the asymptote representing the real pole at $\omega \gg p$ can be determined using equation (9.110). Like real zero, we will use a decade between two frequencies, ω_1 and ω_2 to calculate it, with $\omega_2 = 10\omega_1$. Using equation (9.110) gives

$$20\log M(\omega_2) - 20\log M(\omega_1) = -20\log(\omega_2/p) - (-20\log(\omega_1/p)) =$$
$$20\log(\omega_1 p/\omega_2 p) = 20\log(\omega_1 p/10\omega_1 p) = 20\log(0.1) = -20\,\text{dB} \quad (9.114)$$

That is, the slope of the asymptote for the real pole is -20 dB/decade for $\omega \gg p$. So, for $\omega \gg p$, the real pole is represented by an asymptote starting at the breakpoint frequency and then extends downward until it reaches infinite frequency, passing through a magnitude of -20 dB one decade above $\omega = p$, as shown in Figure 9.7a. To summarize, the magnitude of the real pole is represented by a horizontal asymptote beginning at 0 dB at the low-frequency point and continuing until it reaches the breakpoint frequency $\omega = p$. The slope changes downward at this frequency from -20 dB/decade to an infinite frequency.

Now, let us draw the actual magnitude curve of the real pole. This can be accomplished directly using the Matlab program to solve equation (9.110) for various frequency ω values. Some of the results and the corresponding values obtained using the asymptote approximation for comparison are presented in Table 9.3. Alternatively, to sketch the actual magnitude quickly, instead of using the computer program, we will use equation (9.110) to apply it to three different frequencies, including the breakpoint frequency, one below and one above within a decade interval, for example, $\omega = 0.5p$ and $\omega = 2p$. So, using equation (9.110) gives

- At $\omega = 0.5p$:

$$20\log M(\omega) = -10\log[(\frac{\omega}{p})^2 + 1^2] = -10\log[(\frac{0.5p}{p})^2 + 1^2)] =$$
$$-10\log[(0.5)^2 + 1^2] = -10\log(1.25) = -0.969 \text{ dB} \qquad (9.115)$$

- At breakpoint frequency, $\omega = p$:

$$20\log M(\omega) = -10\log[(\frac{\omega}{p})^2 + 1^2] = -10\log(2) = -3.01 \text{ dB} \qquad (9.116)$$

- At $\omega = 2p$:

$$20\log M(\omega) = -10\log[(\frac{\omega}{p})^2 + 1^2] = -10\log[(\frac{2p}{p})^2 + 1^2] =$$
$$-10\log[(2)^2 + 1^2] = -10\log(5) = -6.989 \text{ dB} \qquad (9.117)$$

Because the values of the actual magnitude have approximately the same values as the asymptotic approximations at low and high frequencies regardless of the value of the frequency, as indicated in Table 9.3, and the results obtained in equations (9.115), (9.116), and (9.117), the actual magnitude is sketched (dotted curve) as shown in Figure 9.7a.

Now, let us check out the maximum difference in magnitude response between the actual and the asymptotic approximation shown in Figure 9.7a. As a result, the real pole has a disparity in magnitude response of no more than -3.01 dB at the breakpoint frequency.

- **Phase plot:** The phase angle of a real pole can be determined as follows. Using equation (9.111) gives
- For $\omega << p$:

$$\phi(\omega) = -\angle\tan^{-1}(\frac{\omega}{p}) \approx -\angle\tan^{-1}(0) \approx 0^o \qquad (9.118)$$

Table 9.3

Comparison between the asymptote and actual magnitude values of a real pole

ω	$0.01p$	$0.1p$	$0.5p$	p	$5p$	$10p$	$50p$	$100p$	$500p$	$1000p$
Asymptote (dB)	0	0	0	0	-14.12	-20	-33.97	- 40	-53.98	-60
Actual (dB)	0	0	-0.969	- 3.01	-14.15	-20.04	-33.98	-40	-53.98	-60

- When $\omega = p$, breakpoint frequency:

$$\phi(\omega) = -\angle\tan^{-1}(\frac{\omega}{p}) = -\angle\tan^{-1}(1) = -45^o \qquad (9.119)$$

- For $\omega >> p$:

$$\phi(\omega) = -\angle\tan^{-1}(\frac{\omega}{p}) \approx -90^o \qquad (9.120)$$

It is represented by an asymptote starting from 0 at the low-frequency point and continuing horizontal until it reaches $\omega = p/10$, then extends downward until it reaches $\omega = 10p$, passing through -45° at the breakpoint frequency $\omega = p$, where it levels off at -90° at $\omega = 10p$, as shown in Figure 9.7b. At this frequency, the asymptote continues horizontally from the leveled-off phase -90° at $\omega = 10p$ until it approaches ∞.

Now, let us draw the actual phase curve of the real pole. This can be accomplished directly using the Matlab program to solve equation (9.111) for various frequency ω values, as shown in Figure 9.7b (dotted line). Some of the results and the values obtained using the asymptote approximation for comparison are presented in Table 9.4. Alternatively, to sketch the actual phase of the real pole quickly rather than using a computer program, we must establish the phases at some specific frequencies. So, let us determine the actual phase values at $\omega = 0.1$, $\omega = p/10$, $\omega = 10p$, and $\omega = 1000p$ as follows. Using equation (9.111) gives

- At $\omega = 0.01p$:

$$\phi(\omega) = -\angle\tan^{-1}(\frac{0.01p}{p}) = -\angle\tan^{-1}(0.01) = -0.573^o \qquad (9.121)$$

- At $\omega = p/10$:

$$\phi(\omega) = -\angle\tan^{-1}(\frac{p}{10p}) = -\angle\tan^{-1}(0.1) = -5.71^o \qquad (9.122)$$

- At $\omega = 10p$:

$$\phi(\omega) = -\angle\tan^{-1}(\frac{10p}{p}) = -\angle\tan^{-1}(10) = -84.29^o \qquad (9.123)$$

Table 9.4

Comparison between the asymptote and actual phase values of a real pole

ω	$0.01p$	$0.1p$	$0.5p$	p	$5p$	$10p$	$50p$	$100p$	$500p$	$1000p$
Asymptote (deg.)	0	0	-31.5	-45	-76.5	-90	-90	- 90	- 90	-90
Actual (deg.)	-0.57	-5.71	-26.6	-45	-78.69	-84.29	-88.85	- 89.42	-89.88	-89.94

- At $\omega = 1000p$:

$$\phi(\omega) = -\angle\tan^{-1}(\frac{1000p}{p}) = -\angle\tan^{-1}(1000) = -89.94^o \qquad (9.124)$$

Using the values of phases obtained from equations (9.121) to (9.124), we can quickly sketch the actual phase response instead of the computer program, as shown in Figure 9.7b (dotted line). Equations (9.122) indicated that the difference between the actual and the asymptotic phases at $\omega = 10p$ is $\phi = -90^o - (-84.29^o) = -5.71^o$. Now, let us check out the difference in phase response between the actual phase and the asymptotic approximation shown in Figure 9.7b. As a result, the difference is found to be equal at both $\omega = p/10$ and $\omega = 10p$, equal to -5.71^o, implying that the real pole has a disparity in phase response of no more than 5.71^o. Note that all previous analyses are carried out on a real pole in its normalized form, implying that we can apply them to any real pole.

6. Complex conjugate zeros, $((\frac{i\omega}{\omega_n})^2 + 2\zeta(\frac{i\omega}{\omega_n}) + 1)$:
It is worth noting that the complex conjugate zeros represent a second-order nominator dynamic. Consider a pair of complex conjugate zeros, for which we want to sketch separate magnitude and phase response plots. The quadratic factor of a pair of complex conjugate zeros can be written in normalized form as

$$(\frac{i\omega}{\omega_n})^2 + 2\zeta(\frac{i\omega}{\omega_n}) + 1 = \left(-(\frac{\omega}{\omega_n})^2 + 2\zeta(\frac{i\omega}{\omega_n}) + 1\right) = \left((1 - (\frac{\omega}{\omega_n})^2) + 2\zeta(\frac{i\omega}{\omega_n})\right)$$
$$(9.125)$$

Then, the magnitude is given by

$$M(\omega) = \left\|\left((1 - (\frac{\omega}{\omega_n})^2) + 2\zeta(\frac{i\omega}{\omega_n})\right)\right\| = \sqrt{(1 - (\frac{\omega}{\omega_n})^2)^2 + (2\zeta(\frac{\omega}{\omega_n}))^2} =$$
$$\left((1 - (\frac{\omega}{\omega_n})^2)^2 + (2\zeta(\frac{\omega}{\omega_n}))^2\right)^{\frac{1}{2}} \qquad (9.126)$$

and the phase angle is given by

$$\phi(\omega) = \angle\tan^{-1}\left(\frac{2\zeta(\frac{\omega}{\omega_n})}{1 - (\frac{\omega}{\omega_n})^2}\right) \qquad (9.127)$$

where ω_n is the natural frequency of the system.

- **Magnitude plot:** The magnitude response of complex conjugate zeros in dB is given by

$$
\begin{aligned}
20\log M &= 20\log\left(\left(1-\left(\frac{\omega}{\omega_n}\right)^2\right)^2 + \left(2\zeta\left(\frac{\omega}{\omega_n}\right)\right)^2\right)^{\frac{1}{2}} \\
&= 10\log\left(\left(1-\left(\frac{\omega}{\omega_n}\right)^2\right)^2 + \left(2\zeta\left(\frac{\omega}{\omega_n}\right)\right)^2\right) \\
&= 20\log\left(1-\left(\frac{\omega}{\omega_n}\right)^2\right) + 20\log\left(2\zeta\left(\frac{\omega}{\omega_n}\right)\right) \tag{9.128}
\end{aligned}
$$

- For $\omega \ll \omega_n$:

$$
20\log M = 20\log(1) = 0 \tag{9.129}
$$

It is represented by an asymptote starting from 0 dB at the low-frequency point and continuing horizontally until it reaches $\omega = \omega_n$, as shown in Figure 9.8a.

- When $\omega = \omega_n$:

$$
20\log M(\omega) = 20\log(2\zeta) \tag{9.130}
$$

This is called corner frequency (breakpoint frequency), $\omega = \omega_n$, the transition point between low- and high-frequency asymptotes. At this frequency, the slope of the asymptote will begin to change, indicating the beginning of the transition. Equation (9.130) indicates that the actual magnitude of the complex conjugate zeros at the corner frequency depends on the value of the damping ratio ζ.

- For $\omega \gg \omega_n$:

$$
20\log M(\omega) = 10\log\left(-\left(\frac{\omega}{\omega_n}\right)^2\right)^2 + 10\log\left(2\zeta\left(\frac{\omega}{\omega_n}\right)\right)^2 = 40\log\left(\frac{\omega}{\omega_n}\right) +
$$
$$
20\log\left(2\zeta\left(\frac{\omega}{\omega_n}\right)\right) \tag{9.131}
$$

The slope of the asymptote representing the complex zero when $\omega \gg \omega_n$ can be determined using equation (9.131). A decade between two frequencies ω_1 and ω_2 will be used to calculate it, with $\omega_2 = 10\omega_1$. The difference in the logarithmic magnitude changes over a decade of the frequency for $\omega \gg \omega_n$ for a complex conjugate zero can be estimated as follows. Considering that the asymptote is unaffected by ζ, we can ignore the term in equation (9.131) that contains ζ and assume that the frequency ω_1 equals ω_n, meaning that $\omega_2 = 10\omega_n$. Using equation (9.131) and considering the above assumptions, the slope of the asymptote of the complex conjugate zero is given by

$$
20\log M(\omega_2) - 20\log M(\omega_1) = 40\log\left(\frac{\omega_2}{\omega_n}\right) - 40\log\left(\frac{\omega_1}{\omega_n}\right) =
$$
$$
40\log\left(\frac{10\omega_n}{\omega_n}\right) - 40\log\left(\frac{\omega_n}{\omega_n}\right) = 40\log(10) - 40\log(1) = +40\,\mathrm{dB} \tag{9.132}
$$

Table 9.5

Comparison between the asymptote and actual magnitude values of a complex zero

(ω/ω_n)	0.1	0.3	0.5	0.8	1.0	2.0	4.0	6.0	8.0	10.0
Asymptote (dB)	0	0	0	0	0	12.04	24.08	31.12	36.12	40
Actual (dB)	0.09	0.75	1.94	4.3	6.02	13.93	24.61	31.36	36.26	40.09

This equation indicated that the slope of the asymptote of the complex zero is +40 dB/decade for $\omega >> \omega_n$. So, for $\omega >> \omega_n$, the complex zero is represented by an asymptote starting at the corner frequency and then extends upward to infinite frequency, passing through a magnitude of +40 dB one decade above $\omega = \omega_n$, as shown in Figure 9.8a. In other words, it is represented by an asymptote starting at the corner frequency, $\omega = \omega_n$, until it approaches ∞ with a slope of +40 dB.

Now, let us draw the actual magnitude curve of the complex zero compared with its asymptotes. We may accomplish this directly by solving equation (9.128) in the Matlab program for various frequency ω/ω_n values. Since the magnitude of the complex zero is affected by the damping ratio, the magnitude response indicated in equation (9.128) is carried out at different values of the damping ratio of $1 < \zeta < 0$, and the results obtained are plotted in Figure 9.9a. The actual magnitude curve of the complex zero carried out at $\zeta = 1$ is plotted in Figure 9.8a, and some results are presented in Table 9.5 to compare with the corresponding asymptote values.

- **Phase plot:** The phase angle of complex conjugate zeros is determined as follows. Using equation (9.127) gives,

- For $\omega << \omega_n$:

$$\phi(\omega) = \angle \tan^{-1}\left(\frac{2\zeta * 0}{1 - 0}\right) \approx \angle \tan^{-1}(0) \approx 0 \qquad (9.133)$$

- When $\omega = \omega_n$:

$$\phi(\omega) = \angle \tan^{-1}\left(\frac{2\zeta * 1}{1 - 1}\right) = \angle \tan^{-1}(\infty) = +90^o \qquad (9.134)$$

- For $\omega >> \omega_n$:

$$\phi(\omega) = \angle \tan^{-1}\left(\frac{2\zeta(\frac{\omega}{\omega_n})}{(1 - (\frac{\omega}{\omega_n})^2)}\right) \approx 2 * +90 \approx +180^o \qquad (9.135)$$

Equation (9.135) indicates that the phase response of the complex pole approaches $+180^o$ at high frequencies, regardless of the values of the damping ratios. It is represented by an asymptote starting from 0 at the low frequency and continuing horizontally until it reaches $\omega = \omega_n/10$, then extending upward, passing through $+90^o$ at

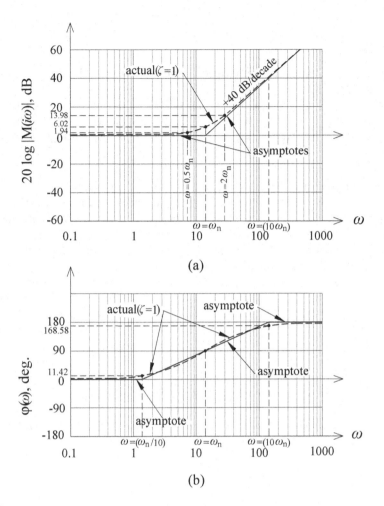

Figure 9.8 The asymptotes and actual curve of a complex zero

$\omega = \omega_n$ until it reaches $\omega = 10\omega_n$, where it levels off at $+180^o$. At this frequency, the asymptote continues horizontally from the leveled-off $+180^o$ until it approaches ∞.

Now, let us draw the actual phase curve of the complex zero compared with its asymptotes. We may accomplish this directly by solving equation (9.127) in the Matlab program for various frequency ω/ω_n values. The magnitude response indicated in equation (9.127) is carried out at different values of the damping ratio of $1 < \zeta < 0$, and the results obtained are plotted in Figure 9.9b. The actual phase response of the complex zero carried out at $\zeta = 1$ is plotted in Figure 9.8b, and some results are presented in Table 9.6 to compare with the corresponding asymptote values.

Note that the results shown in Figure 9.9 have been plotted against (ω/ω_n) so that we can be applied to any ω_n.

7. Complex conjugate poles, $1/[(\frac{i\omega}{\omega_n})^2 + 2\zeta(\frac{i\omega}{\omega_n}) + 1]$:
Note that the complex conjugate poles represent a second-order system. Consider a pair of complex conjugate poles for which we want to sketch separate magnitude and

Table 9.6

Comparison between the asymptote and actual phase values of a complex zero

(ω/ω_n)	0.1	0.3	0.5	0.8	1.0	2.0	4.0	6.0	8.0	10.0
Asymptote (deg.)	0	42.94	62.91	81.28	90	117.09	144.18	160.03	171.28	180
Actual (deg.)	11.42	33.4	53.13	77.32	90	126.87	151.93	161.08	165.75	168.58

phase response plots. The quadratic factor of a pair of complex conjugate poles can be written in normalized form as

$$\frac{1}{(\frac{i\omega}{\omega_n})^2 + 2\zeta(\frac{i\omega}{\omega_n}) + 1} = \left(-(\frac{\omega}{\omega_n})^2 + 2\zeta(\frac{i\omega}{\omega_n}) + 1 \right)^{-1} = \left((1 - (\frac{\omega}{\omega_n})^2) + 2\zeta(\frac{i\omega}{\omega_n}) \right)^{-1}$$

(9.136)

Then, the magnitude is given by

$$M = \left| \left((1 - (\frac{\omega}{\omega_n})^2) + (2\zeta(\frac{i\omega}{\omega_n})) \right)^{-1} \right| = \sqrt{\left((1 - (\frac{\omega}{\omega_n})^2)^2 + (2\zeta(\frac{\omega}{\omega_n}))^2 \right)^{-1}} =$$

$$\left((1 - (\frac{\omega}{\omega_n})^2)^2 + (2\zeta(\frac{\omega}{\omega_n}))^2 \right)^{-\frac{1}{2}}$$

(9.137)

and the phase angle is given by

$$\phi(\omega) = -\angle \tan^{-1} \left(\frac{2\zeta(\frac{\omega}{\omega_n})}{1 - (\frac{\omega}{\omega_n})^2} \right)$$

(9.138)

- **Magnitude plot:** The magnitude response of complex conjugate poles in dB is given by

$$20\log M(\omega) = 20\log \left((1 - (\frac{\omega}{\omega_n})^2)^2 + (2\zeta(\frac{\omega}{\omega_n}))^2 \right)^{-\frac{1}{2}}$$

$$= -10\log \left((1 - (\frac{\omega}{\omega_n})^2)^2 + (2\zeta(\frac{\omega}{\omega_n}))^2 \right)$$

$$= -20\log(1 - (\frac{\omega}{\omega_n})^2) - 20\log(2\zeta(\frac{\omega}{\omega_n}))$$

(9.139)

- For $\omega << \omega_n$:

$$20\log M(\omega) = -20\log(1) = 0$$

(9.140)

It is represented by an asymptote starting from 0 dB at a low-frequency point and continuing horizontally until it reaches the corner frequency $\omega = \omega_n$, as

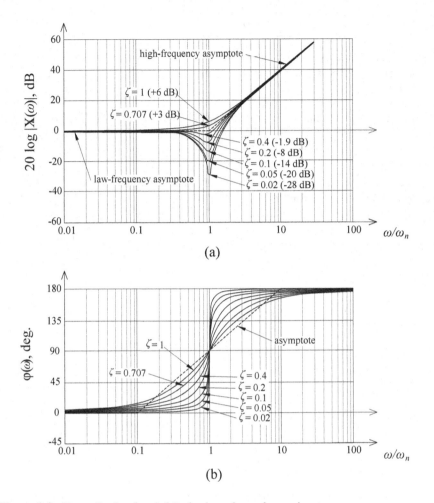

Figure 9.9 Normalized and scaled Bode plots of complex conjugate zeros

shown in Figure 9.10a.

- When $\omega = \omega_n$, corner frequency:

$$20\log M(\omega) = -20\log(2\zeta) \tag{9.141}$$

This equation indicated that at the corner frequency, $\omega = \omega_n$, the magnitude depends on the value of ζ.

- For $\omega \gg \omega_n$:

$$20\log M(\omega) = -20\log(-(\frac{\omega}{\omega_n}))^2 - 20\log(2\zeta(\frac{\omega}{\omega_n})) = -40\log(\frac{\omega}{\omega_n}) - 20\log(2\zeta(\frac{\omega}{\omega_n})) \tag{9.142}$$

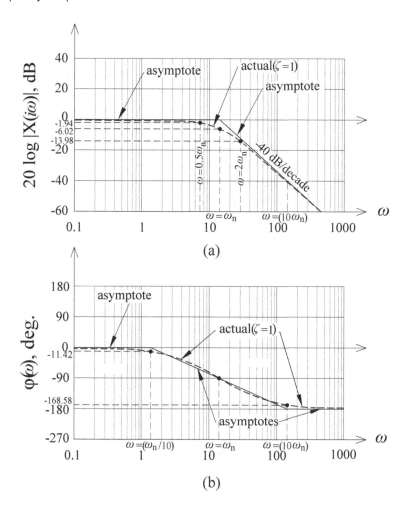

Figure 9.10 The asymptotes and actual curve of a complex pole

The difference in the logarithmic magnitude changes over a decade of the frequency when $\omega \gg \omega_n$ for a complex conjugate pole can be estimated as follows. Considering that the asymptote is unaffected by ζ, we can ignore the term in equation (9.139) that contains ζ and assume that the frequency ω_1 equals ω_n, meaning that $\omega_2 = 10\omega_n$. Using equation (9.139) and considering the above assumptions, the slope of the asymptote of the complex conjugate pole is given by

$$-20\log M(\omega_2) - 20\log M(\omega_1) = -40\log(\frac{\omega_2}{\omega_n}) - 40\log(\frac{\omega_1}{\omega_n}) =$$

$$-40\log(\frac{10\omega_n}{\omega_n}) - 40\log(\frac{\omega_n}{\omega_n}) = -40\log(10) - 40\log(1) = -40\,\text{dB} \qquad (9.143)$$

This equation indicated that the slope of the asymptote of the complex pole is -40 dB/decade for $\omega \gg \omega_n$. So, for $\omega \gg \omega_n$, the complex pole is represented by an asymptote starting at the corner frequency and then extends upward until it reaches

Table 9.7

Comparison between the asymptote and actual magnitude values of a complex pole

(ω/ω_n)	0.1	0.3	0.5	0.8	1.0	2.0	4.0	6.0	8.0	10.0
Asymptote (dB)	0	0	0	0	0	-12.04	-24.08	-31.12	-36.12	-40
Actual (dB)	-0.09	-0.75	-1.94	-4.3	-6.02	-13.93	-24.61	-31.36	-36.26	-40.09

infinite frequency, passing through a magnitude of -40 dB one decade above $\omega = \omega_n$, as shown in Figure 9.10a. In other words, it is represented by an asymptote starting at the corner frequency, $\omega = \omega_n$, until it approaches ∞ with a slope of -40 dB.

Now, let us draw the actual magnitude curve of the complex pole compared with its asymptotes. We may accomplish this directly by solving equation (9.139) in the Matlab program for various frequency ω/ω_n values. Since the phase of the complex zero is affected by the damping ratio, the magnitude response indicated in equation (9.139) is carried out at different values of the damping ratio of $1 < \zeta < 0$, and the results obtained are plotted in Figure 9.11a. The actual magnitude response of the complex pole carried out at $\zeta = 1$ is plotted in Figure 9.10a, and some results are presented in Table 9.7 to compare with the corresponding asymptotes values.

- Phase plot: The phase angle of complex conjugate poles is determined as follows. Using equation (9.138) gives

- For $\omega << \omega_n$:

$$\phi(\omega) = -\angle\tan^{-1}(\frac{2\zeta * 0}{1-0}) = -\angle\tan^{-1}(0) \approx 0 \tag{9.144}$$

- When $\omega = \omega_n$:

$$\phi(\omega) = -\angle\tan^{-1}(\frac{2\zeta * 1}{1-1}) = -\angle\tan^{-1}(\infty) = -90^o \tag{9.145}$$

- For $\omega >> \omega_n$:

$$\phi(\omega) = -\angle\tan^{-1}(\frac{\omega}{\omega_n})^2 \approx 2 * -90 \approx -180^o \tag{9.146}$$

It is represented by an asymptote starting from 0 at low frequency and continuing horizontally until it reaches $\omega = \omega_n/10$, then extending downward, passing through -90° at $\omega = \omega_n$ until it reaches $\omega = 10\omega_n$, where it levels off at -180°. At this frequency, the asymptote continues horizontally from the leveled-off -180° at $\omega = 10\omega_n$ until it approaches ∞.

Now, let us draw the actual phase curve of the complex pole compared with its asymptotes. We may accomplish this directly by solving equation (9.138) in the Matlab program for various frequency ω/ω_n values. Since the phase of the complex

Table 9.8

Comparison between the asymptote and actual phase values of a complex pole

(ω/ω_n)	0.1	0.3	0.5	0.8	1.0	2.0	4.0	6.0	8.0	10.0
Asymptote (deg.)	0	-42.94	-62.91	-81.28	-90	-117.09	-144.18	-160.03	-171.28	-180
Actual (deg.)	-11.42	-33.4	-53.13	-77.32	-90	-126.87	-151.93	-161.08	-165.75	-168.58

zero is affected by the damping ratio, the magnitude response indicated in equation (9.138) is carried out at different values of the damping ratio of $1 < \zeta < 0$, and the results obtained are plotted in Figure 9.11b. The actual phase response of the complex pole carried out at $\zeta = 1$ is plotted in Figure 9.10b, and some results are presented in Table 9.8 to compare with the corresponding asymptotes values.

Note that the results shown in Figure 9.11 have been plotted against (ω/ω_n) so that we can be applied to any ω_n.

Figure 9.11 shows the Bode plot of the magnitude and phase angle of a pair of complex conjugate poles for various values of the damping ratio ζ. From this figure, we can observe the following

- When $\omega = \omega_n$, $(\omega/\omega_n = 1)$, the actual magnitude of $\zeta < 0.707$ is located near the 0 dB line, and the magnitude response does not have a peak value.
- The magnitude value approaches ∞ for the damping ratio $\zeta = 0$ (undamped system) as the input frequency ω approaches the natural frequency ω_n.
- The frequency response reaches its maximum peak M_p at the resonant frequency ω_r. In other words, as the damping ratio approaches zero, the resonant frequency ω_r approaches the natural frequency ω_n, implying that the magnitude response has reached its maximum value. This means that when the damping ratio ζ approaches 0, a slight change in the input force might result in dangerously large oscillations. This *resonance* phenomenon is explained by the fact that when $\omega = \omega_r = \omega_n$, the input force is applied in exact synchronism with the natural movements of the system, thus increasing its magnitude in the same way.

Now, let us determine the resonant frequency ω_r. To accomplish this, take the derivative of equation (9.137) for ω/ω_n and set it equal to zero. As a result, it is given by

$$\omega_r = \omega_n \sqrt{1 - 2\zeta^2} \qquad \zeta < 0.707 \qquad (9.147)$$

Remember that the damped natural frequency of a pair of complex conjugate poles is given by

$$\omega_d = \omega_n \sqrt{1 - \zeta^2} \qquad \zeta < 0.707 \qquad (9.148)$$

The maximum peak M_p of the magnitude of a pair of complex conjugate poles can be obtained as follows. Considering equation (9.137) and setting $\omega_n = \omega_r$ gives

$$M_p = |M(\omega_r)| = \frac{1}{2\zeta\sqrt{1 - \zeta^2}} \qquad \zeta < 0.707 \qquad (9.149)$$

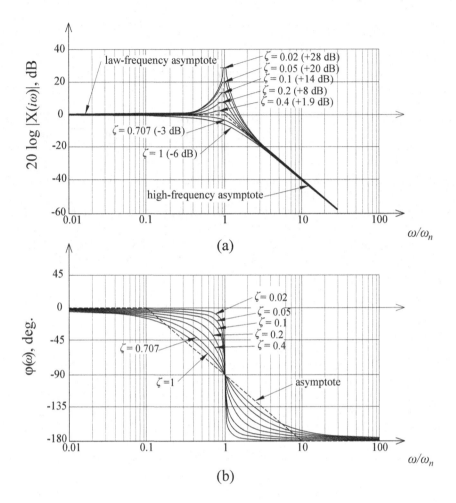

Figure 9.11 Normalized and scaled Bode plots of complex conjugate poles

Note that the value of the maximum peak M_p that is obtained from this equation is given a magnitude ratio. This value is also called the resonance magnification factor. As shown in Figure 9.11a, a resonant peak of the magnitude response exists when $\zeta < 0.707$, meaning that equations (9.147), (9.148), and (9.149) should be applied only for $\zeta < 0.707$. Some remarks on these equations can be observed as follows.

- At $\zeta < 0.707$, the magnitude of the output will be magnified by a resonant magnification factor M_p because the input force is applied statically ($\omega = 0$). For example, as indicated in equation (9.149), if $\zeta = 0.01$, a given input force can create 50 times the amount of the output motion it would produce statically at resonance.
- For the damping ratio $\zeta > 0.707$, the magnitude ratio decreases steadily over time to zero as ω approaches infinity; consequently, there is no magnification.

Furthermore, if $\zeta > 0.707$, the frequency of the peak response shifts further below ω_n, which is not the same as the frequency of damped natural oscillations.
- It is worth noting that for step inputs, overshoots occur if $\zeta < 1$, whereas for sinusoidal inputs, a resonant peak occurs when $\zeta < 0.707$. In practice, any real system must have some damping, so the behavior for $\zeta = 0$ is somewhat educational; however, real systems with ζ as low as 0.01 are also not rare.

Now, let us turn to the Bode plot. It is important to remember that we can always factor any given transfer function $X(s)$ into a form composed of the seven terms discussed previously. We can plot the composite magnitude plot of $M(i\omega)$ by adding plots of its factors on a semi-log scale. Adding the log of the magnitude of the individual terms gives the log of the magnitude of the composite transfer function. Similarly, the phase $\phi(\omega)$ of the transfer function is provided by adding phases of individual poles and zeros. These two remarks are the basis for Bode's plotting technique. We should follow the following outline steps to generate the composite sketch of the Bode plot.

1. Put the transfer function in Bode form. Identify terms according to the seven types stated previously.
2. Plot the magnitude associated with $(i\omega)^m$ and $(i\omega)^{-n}$ terms (low frequency asymptotes).
3. Plot the magnitude for each individual term of zeros and poles on a semi-log scale.
4. Add an individual magnitude plot to generate a composite magnitude curve—work from low to high frequency. Here, we must add distances, not magnitude values.
5. Plot the phase associated with $(i\omega)^m$ and $(i\omega)^{-n}$ terms (low frequency asymptotes).
6. Sketch approximate phase curve by stepping $\pm 90^o$ or $\pm 180^o$ at breakpoint frequency—work from low to high frequency. Here, we must add distances, not phase angle values.
7. Draw the transition asymptotes
8. Sketch the composite curves of the magnitude and phase separately.

Finally, it is worth noting that a system having n-degrees of freedom (n^{th} pairs of complex conjugate poles) will have n^{th} natural frequencies.

Composite Bode plot: It is important to note that the composite magnitude and phase plots are sketching plots (drawing by hand), which are approximate representations of the actual plots intended to be used. There are, however, reasons to create a methodology for manually sketching Bode plots. Drawing the plots by hand helps to understand how poles and zeros affect the shape of the frequency response plots. In addition to that, by simply observing the system's transfer function (poles and zeros), we may estimate how it will behave in the frequency domain. If we know the shape of the transfer function we need, we can generate it using our knowledge of Bode plots. On the other hand, a computer program, such as Matlab, will generate the Bode plot immediately by carrying out an extensive computation, varying the frequency from low to high using the given transfer function accessible through the program's toolboxes.

Therefore, following the creation of these sketching plots, to ensure that they accurately represent the system's frequency response, it is helpful to compare them with their exact values at the breakpoint frequencies, especially when the system has complex roots (resonance frequencies) and, of course, at any other frequency. The exact magnitude and phase values from the system transfer function can be calculated using equations (9.82) and (9.84). As a comparison, if the difference between the exact and sketching findings is significant, we must return to the Bode plot and determine the cause of the disparity between the results to represent the system's frequency response for stability analysis accurately.

9.5   SYSTEM CHARACTERISTICS USING BODE PLOTS

In this section, we determine gain and phase margins and the gain range required for stability. Understanding gain and phase margins is essential to grasping Bode plots and their implications. The question is, what do the terms gain and phase margins mean, and how do they affect the stability of a dynamic system? To understand that, consider them to be safety margins for you. When walking next to a cliff, a positive space or margin of safety between yourself and a potentially catastrophic event should be desirable. This intuition will assist you in keeping track of how gain and phase margins are defined so that positive margins indicate the presence of a safety margin for you before instability. On the other hand, negative margins indicate issues with instability if it happen.

Let us now define the gain and the phase margins using the Bode plot. The following symbols are used for evaluating them.

- ω_{mcu} = magnitude crossover frequency of the unstable stable response.
- ω_{mcm} = magnitude crossover frequency of the marginally stable response.
- ω_{mcs} = magnitude crossover frequency of the stable response.
- G_{un} = gain of the unstable response.
- G_{ms} = gain of the marginally stable response.
- G_{st} = gain of the stable response.
- ϕ_{un} = phase lag of the unstable response.
- ϕ_{ms} = phase lag of the marginally stable response.
- ϕ_{st} = phase lag of the stable response.

9.5.1   GAIN MARGIN

The gain margin is defined as the value of the gain that may increase without causing the system to become unstable. When the gain increases to a specific threshold, the system becomes marginally stable; nonetheless, at this point, any more gain variation may result in an unstable system. The gain margin is determined using the phase plot to identify the frequency ω_{pc}, called phase crossover frequency, at which the phase plot intersects the phase angle -180° line. At this frequency, we look at the magnitude plot to find the gain value necessary to move down the magnitude plot to 0 dB, where the system is marginally stable. It is worth noting that when the magnitude is equal to 0 dB, the system output and input amplitudes (magnitude) are identical. The gain margin can be expressed as follow.

$$G_m = 0 - G \quad \text{dB} \tag{9.150}$$

where G is the gain; it is the magnitude value in dB that can be found by reading the vertical distance between the magnitude plot and the 0 dB line at the phase crossover frequency ω_{pc}. Note that the gain G and the gain margin G_m are different parameters. When measured in decibels, the gain margin G_m is the negative of the gain G, which is apparent when examining the gain margin formula indicated in equation (9.150). Let us now determine the gain margin G_m for a particular system. Assuming we have the Bode plot shown in Figure 9.12. From the magnitude plot shown in Figure 9.12a, we found that the value of G is approximately equal to $G_{un} = +16$. We can calculate the gain margin G_m using equation (9.150) as follows.

$$G_m = 0 - G_{un} = 0 - (+16) = -16 \quad \text{dB} \tag{9.151}$$

Because the value of the gain margin is negative, as indicated in this equation, the system is unstable. Since G_m is expressed in decibels, we may find it helpful to convert between decibels and magnitude as a ratio, M_r. To do so, we can use the following analysis:

$$G_m = 20 \log M_r \tag{9.152}$$

or,

$$M_r = 10^{(G_m/20)} \tag{9.153}$$

So,

$$G_m = \frac{1}{M_r} \tag{9.154}$$

We can see from this equation that the gain margin is equal to the reciprocal of the magnitude ratio M_r if we want to be not in dB. So, as indicated in equation (9.154), decreasing the magnitude ratio M_r leads to increasing the gain margin G_m, implying that the system will be more stable. As a result, the greater the gain margin G_m, the more stable the system, and vice versa.

9.5.2   PHASE MARGIN

The phase margin is the difference between the phase plot and the -180° line when the magnitude plot intersects with the 0 dB line. It is found using the magnitude plot to identify the magnitude crossover frequency ω_{mc}; it is also called the gain crossover frequency, at which the magnitude plot intersects with the 0 dB line. At this frequency, on the phase plot, read the vertical distance ϕ between the phase plot and the -180° line, as shown in Figure 9.10b. This value is called phase lag, a number less than 0°. The phase margin is expressed in degrees, and it is given by

$$\phi_m = \phi - (-180^o) \tag{9.155}$$

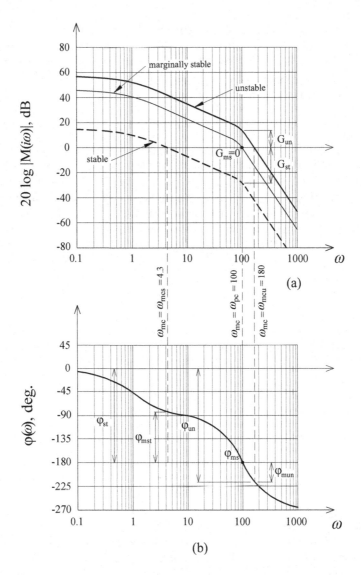

Figure 9.12 Gain margin and phase margin

This equation indicated that the phase margin ϕ_m and the phase lag ϕ are different parameters. In our example shown in Figure 9.12b, it is found that the phase lag ϕ is approximately equal to $\phi_{un} = -215^o$; hence, using equation (9.155) gives

$$\phi_m = \phi_{mun} = \phi_{un} - (-180^o) = -215^o - (-180^o) = -35^o \qquad (9.156)$$

Since the phase margin value is negative, the system is unstable. So, the larger the phase margin, the higher the system's stability, and vice versa.

9.5.3 SYSTEM STABILITY

As stated previously, determining stability for a linear dynamic system is straight-forward; all roots of the characteristic equation (transfer function denominator) must have negative real parts to stabilize the system. The system's stability is satisfied for the frequency response if the gain margin G_m is greater than 0 dB. When the gain margin equals 0 dB, the system will be marginally stable; when it is less than 0 dB, it will be unstable. Therefore, for a dynamic system to be stable, we must determine the value of the system gain K, ensuring that the gain margin is greater than 0 dB. So, changing the system gain leads the magnitude plot of the unstable system to move by a distance of $\leq G_{un}$, causing the magnitude crossover frequency ω_{mc} to shift, as shown in Figure 9.12a. We can see from this figure that changing the system gain K that achieves system stability causes the magnitude plot to move, whereas the phase plot does not shift. This is because the magnitude plot's location is primarily deter-mined by the logarithmic magnitude of the Bode transfer function gain K_t, a function of the system gain K, as indicated in equation (9.74). The phase angle of the bode gain K_t, on the other hand, is always equal to 0 even if it is changed since it is a constant. As a result, changing the system gain K does not affect the location of the phase plot. In addition, changing the system gain results in a change in the phase margin since the magnitude crossover frequency ω_{mc} for the stable system, which is the crossing of the magnitude plot with the 0 dB line, is already changed, as shown clearly in Figure 9.12a.

Return to our example, at the phase crossover frequency $\omega_{pc} = 100$, which is the intersection point of the phase plot with the -180° line, the gain margin G_{un} is -16 dB as obtained in equation (9.151), indicating that the system is unstable, as stated previously. For the system to be marginally stable, G_{ms} must equal 0 dB, as shown in Figure 9.12a. Consequently, move the unstable response downward until $G_{un} = G_{ms} = 0$, at which both magnitude and phase crossover frequencies are equal, as shown in Figure 9.12. It is essential to decrease the value of the gain G_{un} to be less than 0 dB for the system to be stable, meaning that its value must be negative. As shown in Figure 9.12a, the gain G_{st} value must be negative. Note that to avoid the system becoming unstable, we must adjust the magnitude plot to where the gain G_{un} is equal to 0 dB at the phase crossover frequency ω_{pc}; otherwise, the system will remain unstable. To ensure the system will not be unstable and consistently stable, we must determine the value of the system gain K required to achieve stability.

Note that, in some systems, the phase plot does not intersect with the -180° line. Thus, there is no phase crossover frequency ω_{pc}, implying that the system has an infinite gain margin, and in this case, it is unconditionally stable. This means the sys-tem will always be stable no matter how much we increase the gain G. On the other hand, when the magnitude plot does not intersect the 0 dB line, there is no magnitude crossover frequency ω_{mc}, meaning that the system is either unconditionally unstable or unconditionally stable, depending on whether the magnitude plot location is in the positive or negative dB zone, respectively. It should be noted that neither the gain margin nor the phase margin alone is an adequate indicator of the system's stability, and both should be considered when determining the system's stability.

To summarize, the stability of the dynamic system based on the relation between phase crossover frequency and magnitude crossover frequency is listed below.

- If the phase crossover frequency ω_{pc} is greater than the magnitude crossover frequency ω_{mc}, the system is stable.
- If the phase crossover frequency ω_{pc} is equal to the magnitude crossover frequency ω_{mc}, the system is marginally stable.
- If the phase crossover frequency ω_{pc} is less than the magnitude crossover frequency ω_{mc}, the system is unstable.

The stability of the dynamic system based on the relationship between the gain margin and phase margin is listed below.

- If both the gain margin G_m and the phase margin ϕ_m are positive, the system is stable.
- The system is marginally stable if both the gain margin G_m and the phase margin ϕ_m are equal to zero.
- If the gain margin G_m or the phase margin ϕ_m is negative, the system is unstable.

Let us now explain how the Bode plot depicts the frequency response of a particular transfer function. The following two examples will cover most of the seven previously described terms.

Example (9.1):
For the following transfer function,

$$X(s) = \frac{K(s+0.5)}{s(s+10)(s+50)}$$

where $K = 2000$.

a. Sketch the Bode plot and verify the magnitude and phase values at the breakpoint frequencies with their corresponding exact values.
b. Determine the gain margin G_m, phase margin ϕ_m, and system stability. Use MAT-LAB to verify the results.

Solution:
a. Bode plot:
Convert the given transfer function to the Bode transfer function as follows.

$$X(s) = \frac{2000*0.5(\frac{s}{0.5}+1)}{10*50\,s(\frac{s}{10}+1)(\frac{s}{50}+1)} = \frac{K_t(\frac{s}{0.5}+1)}{s(\frac{s}{10}+1)(\frac{s}{50}+1)} \qquad (9.157)$$

where,

$$K_t = \frac{K*0.5}{10*50} = \frac{2000*0.5}{10*50} = 2 \qquad (9.158)$$

The factors of the Bode transfer function are listed below in ascending order of occurrence as frequency increases.

1. A constant gain $K_t = 2$.
2. A pole at the origin.
3. A real zero at $\omega = 0.5$.
4. A real pole at $\omega = 10$.
5. A real pole at $\omega = 50$.

- **Magnitude plot:** We begin by plotting the logarithmic magnitude of the gain K_t and the zero and each pole factor as follows, see Figure 9.13.

1. The logarithmic magnitude of the Bode gain K_t is given by

$$20\log 2 = 6.02 \quad \text{dB} \tag{9.159}$$

It is represented by a horizontal asymptote starting from 6.02 dB at the low-frequency point $\omega = 0.01$ until it approaches ∞.
2. The asymptotic magnitude of the pole at the origin starts from +40 dB at the low-frequency point $\omega = 0.01$ and extends downward to the infinite frequency with a slope of -20 dB/decade, crossing the 0 dB line at $\omega = 1$.
3. The asymptotic magnitude of the real zero at $\omega = 0.5$ starts at 0 dB at the low-frequency point $\omega = 0.01$ and continues until it reaches the breakpoint frequency $\omega = 0.5$ and then extends upward to the infinite frequency with a slope of +20 dB/decade.
4. The asymptotic magnitude of the real pole at $\omega = 10$ starts at 0 dB at the low-frequency point $\omega = 0.01$ and continues until it reaches the breakpoint frequency $\omega = 10$ and then extends downward to the infinite frequency with a slope of -20 dB/decade.
5. The asymptotic magnitude of the real pole at $\omega = 50$ starts at 0 dB at the low-frequency point $\omega = 0.01$ and continues until it reaches the breakpoint frequency $\omega = 50$ and then extends downward to the infinite frequency with a slope of -20 dB/decade.

Figure 9.13a shows that each magnitude asymptote is plotted with its actual curve (dotted line). To sketch the composite magnitude plot, insert its entire asymptote first. The entire asymptotic magnitude can be obtained directly by summing the asymptotes of the factor of each pole and zero from low to high frequency. Since the given transfer function contains one real zero at a breakpoint frequency of $\omega = 0.5$ and two real poles at breakpoint frequencies of $\omega = 10$ and $\omega = 50$, the entire asymptotic magnitude plot must include these breakpoint frequencies. Remember that we must add distances rather than magnitude values here, except for the starting point at the low-frequency point $\omega = 0.01$. That is, the starting point of the entire magnitude asymptote at the low-frequency point $\omega = 0.01$ equals 46.02 dB, the sum of the logarithmic magnitude of the gain K_t, 6.02 dB, and the value of the starting point of the pole at the origin is 40 dB at the low-frequency point $\omega = 0.01$. Thus, the slope is -20 dB/decade due to the pole at the origin and continues until it reaches the first breakpoint frequency at $\omega = 0.5$. At this frequency, the slope becomes 0 dB/decade

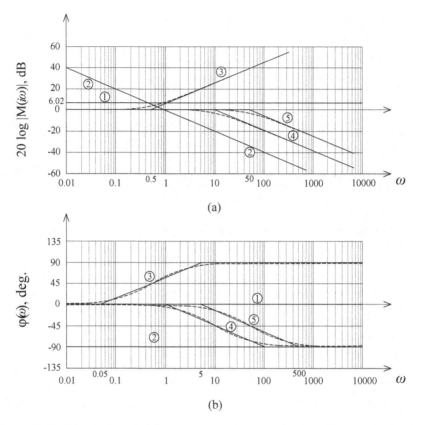

Figure 9.13 Magnitude and phase asymptotes of zeros and poles of the transfer function
$X(s) = \frac{2000(s+0.5)}{s(s+10)(s+50)}$

due to the existence of the real zero (the sum of the slopes of the pole at origin -20 dB and the real zero $+20$ dB and then continues until it reaches the second break-point frequency at $\omega = 10$. At this frequency, the slope becomes -20 dB/decade due to the existence of the first real pole and continues until it reaches the third break-point frequency $\omega = 50$. At this frequency, the slope changes to -40 dB/decade due to the existence of the second real pole. Using these comments as a guide, sketch the composite magnitude plot, shown in Figure 9.14a by a bold solid line.

- Phase plot: We will first draw the asymptotic phase of each factor in the given transfer function, see Figure 9.13b.

1. The asymptotic phase of the Bode gain $K_t = 6.02$ is represented by a horizontal asymptote starting from $0°$, at the low-frequency point $\omega = 0.01$, and continuing until it approaches ∞.
2. The pole at the origin makes a constant phase angle of $-90°$. It is represented by a horizontal asymptote that spans the frequency range from the low-frequency point $\omega = 0.01$ to approaching ∞.
3. The asymptotic phase of the real zero at $\omega = 0.5$ starts from $0°$ at the low-frequency point $\omega = 0.01$ and continues until it reaches $\omega = 0.5/10$. At this

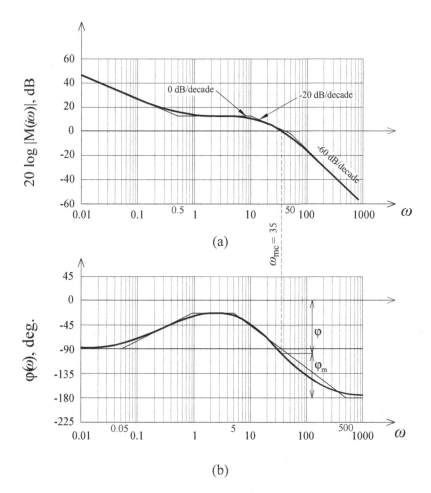

(a)

(b)

Figure 9.14 Composite Bode plots of the transfer function $X(s) = \frac{2000(s+0.5)}{s(s+10)(s+50)}$

frequency, it shifts upward until it levels off at $+90^o$ at $\omega = 0.5 * 10$ passes through $+45^o$ at breakpoint frequency $\omega = 0.5$ and then continues horizontally from the leveled-off angle $+90^o$ until it approaches ∞.

4. The asymptotic phase of the first real pole at $\omega = 10$ starts from 0° at the low-frequency point $\omega = 0.01$ and continues until it reaches $\omega = 10/10$. At this frequency, it shifts downward until it levels off at -90° at $\omega = 10 * 10$ passes through -45° at breakpoint frequency $\omega = 10$, and then continues horizontally from the leveled-off angle -90° until it approaches ∞.

5. The asymptotic phase of the second real pole at $\omega = 50$ starts from 0° at the low-frequency point $\omega = 0.01$ and continues until it reaches $\omega = 50/10$. At this frequency, it shifts downward until it levels off at -90° at $\omega = 50 * 10$ passes through -45° at breakpoint frequency $\omega = 50$, and then continues horizontally from the leveled-off angle -90° until it approaches ∞.

Figure 9.13b shows the asymptotic phases with their actual curves (dotted lines). Because the three poles will eventually make a phase angle of -270° at high frequencies, whereas the zero makes a phase angle of $+90^o$, a phase angle scale ranging from $+90^o$ to -270° will almost be sufficient. It is important to note that the final value of the phase angle of the composite phase plot must be level off -180°, which is the sum of -270° (three poles) and $+90^o$ (one zero). We will keep this comment in mind when sketching the composite phase plot. Now, let us draw the composite phase response. To accomplish this, insert its entire asymptote first. The entire asymptotic phase can be obtained directly by summing the asymptotes of the factor of each pole and zero from low to high frequency, as shown in Figure 9.14b. That is due to the existence of a pole at the origin; the entire asymptotic phase must begin at -90° and then continue until it reaches $\omega = 0.5/10$. At this frequency, the phase shifts upward due to the existence of the real zero at $\omega = 0.5$ until it reaches $\omega = 10/10$. Due to the existence of the first real pole at $\omega = 10/10$, the phase becomes 0 and continues until it reaches $\omega = 50/10$. At this frequency, the phase shifts downward due to the existence of the second real pole at $\omega = 50$ until it levels off at a phase of -180° at $\omega = 50*10$, which is the final value of the phase angle as stated above. Remember that, here, we must add distances rather than phase values. Now, sketch the composite phase curve, shown in Figure 9.14b, by a bold solid line.

Remember that both the composite magnitude and phase plots are sketching plots, so we need to look at them and figure out what they imply. This can be verified by comparing the exact magnitude and phase values at the breakpoint frequencies 0.5, 10, and 50 with the corresponding values obtained from the composite plots at the same frequencies, i.e., whether the composite magnitude and phase sketching plots represented the frequency response of the given transfer function adequately. The exact values of the magnitude and phase at any frequency for the given transfer function are provided by

$$M(\omega) = 20\log K_t + 10\log((2\omega)^2 + 1) - 20\log\omega - 10\log((0.1\omega)^2 + 1)$$
$$- 10\log((0.02\omega)^2 + 1) \tag{9.160}$$

and

$$\phi(\omega) = \tan^{-1} K_t + \tan^{-1}(2\omega) - \tan^{-1}(\omega) - \tan^{-1}(0.1\omega) - \tan^{-1}(0.02\omega) \tag{9.161}$$

Now, calculate the exact magnitude and phase values at breakpoint frequencies as follows. Using equations (9.160) and (9.161) gives
- At first breakpoint frequency, $\omega = 0.5$:

$$M(0.5) = 6.02 + 10\log((2*0.5)^2 + 1) - 20\log(0.5) - 10\log((0.1*0.5)^2 + 1)$$
$$- 10\log((0.02*0.5)^2 + 1) = 6.02 + 3.01 - (-6.02) - 0.01 - 4.34*10^{-4}$$
$$= 15.04 \quad \text{dB} \tag{9.162}$$

and

$$\phi(0.5) = 0^o + \tan^{-1}(2*0.5) - 90^o - \tan^{-1}(0.1*0.5) - \tan^{-1}(0.02*0.5) =$$
$$0^o + 45^o - 90^o - 2.86^o - 0.57^o = -48.43^o \tag{9.163}$$

- At second breakpoint frequency $\omega = 10$:

$$M(10) = 6.02 + 10\log((2*10)^2 + 1) - 20\log(10) - 10\log((0.1*10)^2 + 1)$$
$$- 10\log((0.02*10)^2 + 1) = 6.02 + 26.03 - 20 - 3.01 - 0.17 = 8.87 \quad \text{dB}$$
$$(9.164)$$

and

$$\phi(10) = 0^o + \tan^{-1}(2*10) - 90^o - \tan^{-1}(0.1*10) - \tan^{-1}(0.02*10) =$$
$$0^o + 87.13^o - 90^o - 45^o - 11.31^o = -59.18^o \tag{9.165}$$

- At third breakpoints frequency $\omega = 50$:

$$M(50) = 6.02 + 10\log((2*50)^2 + 1) - 20\log(50) - 10\log((0.1*50)^2 + 1)$$
$$- 10\log((0.02*50)^2 + 1) = 6.02 + 40 - 33.98 - 14.15 - 3.01 = -5.12 \quad \text{dB}$$
$$(9.166)$$

and

$$\phi(50) = 0^o + \tan^{-1}(2*50) - 90^o - \tan^{-1}(0.1*50) - \tan^{-1}(0.02*50) =$$
$$0^o + 89.42^o - 90^o - 78.69^o - 45^o = -124.27^o \tag{9.167}$$

To make the results obtained from the composite plot more trustworthy, let us verify them at two more frequencies, one near the low-frequency point, for example, $\omega = 0.05$, and the other near the high-frequency end, for example, $\omega = 1000$.

- At $\omega = 0.05$:

$$M(0.05) = 6.02 + 10\log((2*0.05)^2 + 1) - 20\log(0.05) - 10\log((0.1*0.05)^2 + 1)$$
$$- 10\log((0.02*0.05)^2 + 1) = 6.02 + 0.043 - (-26.02)$$
$$- 1.08*10^{-4} - 4.34*10^{-6} = +32.08 \quad \text{dB} \tag{9.168}$$

and

$$\phi(0.05) = 0^o + \tan^{-1}(2*0.05) - 90^o - \tan^{-1}(0.1*0.05) - \tan^{-1}(0.02*0.05) =$$
$$0^o + 5.71^o - 90^o - 0.28^o - 0.057^o = -84.62^o \tag{9.169}$$

- At $\omega = 1000$:

$$M(1000) = 6.02 + 10\log((2*1000)^2 + 1) - 20\log(1000)$$
$$- 10\log((0.1*1000)^2 + 1) - 10\log((0.02*1000)^2 + 1)$$
$$= 6.02 + 66.02 - 60 - 40 - 26.03 = -53.99 \quad \text{dB} \tag{9.170}$$

Table 9.9

Exact and composite values of the frequency response of the transfer function $X(s) = \frac{2000(s+0.5)}{s(s+10)(s+50)}$

ω	M (dB)		Phase ϕ	
	exact	composite	exact	composite
0.05	32.08	32	-84.62^o	-82°
0.5	15.04	17	-48.43^o	-45°
10	8.87	9	-59.18^o	-55°
50	-5.12	-6	-124.27^o	-120°
1000	-53.99	-52	-176.58^o	-175°

---

and

$$\phi(1000) = 0^o + \tan^{-1}(2*1000) - 90^o - \tan^{-1}(0.1*1000) - \tan^{-1}(0.02*1000) =$$
$$0^o + 89.97^o - 90^o - 89.42^o - 87.13^o = -176.58^o \qquad (9.171)$$

Let us now obtain the magnitude and phase values from their composite plot shown in Figure 9.14 at the system frequencies $\omega = 0.05, 0.5, 3, 10$, and 1000. Remember that the values obtained from the composite magnitude and phase plot are approximate, and keep this in mind when evaluating them. The results obtained and their exact values calculated above for comparison are given in Table 9.9. As a comparison, it was found that the disparity between the results is small enough to be negligible. As a result, we may utilize the sketching composite Bode plot of the given transfer function, shown in Figure 9.14 to study system stability accurately.

b. Gain margin, phase margin, and system stability: Figure 9.14b shows that the phase plot does not intersect with the -180° line. Thus, there is no phase crossover frequency in the system, implying that the system has an infinite gain margin and is unconditionally stable, as stated previously. Because both gain and phase margins should indicate system stability, not only one of them, let us check the phase margin to ensure the system is stable. The magnitude plot intersects with the 0 dB line at a magnitude crossover frequency approximately equal to $\omega_{mc} = 35$, as shown in Figure 9.14a. Look at the phase plot at this frequency and find the phase lag ϕ, is approximately equal to -105°. So, the phase margin can be calculated as follows,

$$\phi_m = \phi - (-180^o) = -105^o - (-180^o) = +75^o \qquad (9.172)$$

This equation indicates that the phase margin is positive, ensuring the system is stable. Figure 9.15 shows the Matlab live script file and the generated results. Because the magnitude crossover frequency ω_{cm} and phase margin ϕ_m values, which are $\omega_{mc} = 35$ and $\phi_m = 75^o$, obtained from the composite plot shown in Figure 9.14 are approximated values, it was found that there is a slight difference between them and the corresponding generated one from MATLAB, $\omega_{mc} = 32.133$ and $\phi_m = 73.667^o$.

```
% Bode plots and system characteristics, Example (9-1).
% Write the given transfer function.
n=2000*[1 0.5];d1=conv([1 10],[1 50]);d=conv([1 0],d1);
xs=tf(n,d);grid on
% Use the allmargin command to generate the system characteristics.
s=allmargin(xs)
```

```
s = struct with fields:
   GainMargin: Inf
  GMFrequency: Inf
  PhaseMargin: 73.6673
  PMFrequency: 32.1334
  DelayMargin: 0.0400
  DMFrequency: 32.1334
       Stable: 1
```

```
% or we can use the following.
[Gm,Pm,Wcp,Wcg]=margin(xs)
```

```
Gm = Inf
Pm = 73.6673
Wcp = Inf
Wcg = 32.1334
```

```
% Use the bode or bodeplot command to plot the response.
bode(xs);
```

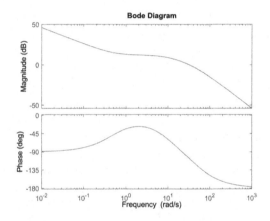

Figure 9.15 MATLAB live script file generating the Bode plots of the transfer function
$X(s) = \frac{2000(s+0.5)}{s(s+10)(s+50)}$ and system characteristics

Example (9.2):
For the following transfer function, draw the Bode plots. From the plot,

$$X(s) = \frac{2000(s+0.5)}{s(s+50)(s^2+s+100)}$$

a. Determine and verify the magnitude and phase values at the breakpoint frequencies with their corresponding exact values.
b. Determine the magnitude and phase crossover frequencies. Determine the gain margin G_m, phase margin ϕ_m, and maximum peak M_p and check the system's stability.
c. Using Matlab, generate the Bode plots and verify the values of the gain margin, phase margin, and maximum peak determined in step (b).

Solution:
a- Bode plot:
Convert the given transfer function to the Bode transfer function as follows

$$X(s) = \frac{2000 * 0.5(\frac{s}{0.5}+1)}{50 * 100\, s(\frac{s}{50}+1)((\frac{s}{10})^2 + \frac{1}{10}s + 1)} = \frac{0.2(\frac{s}{0.5}+1)}{s(\frac{s}{50}+1)((\frac{s}{10})^2 + \frac{1}{10}s + 1)} \quad (9.173)$$

Let us first determine the complex pole's natural frequency and damping ratio. Comparing factors of the complex pole with factors of the general form of the second-order system is as follows:

$$(s^2 + s + 100) = (s^2 + 2\zeta\omega_n s + \omega_n^2) \quad (9.174)$$

Equalizing factors of both sides of this equation yields

$$\omega_n^2 = 100 \quad \text{and} \quad 2\zeta\omega_n = 1 \quad (9.175)$$

Solving this equation for ω_n and ζ gives

$$\omega_n = 10 \quad \text{and} \quad \zeta = 0.05 \quad (9.176)$$

The factors of the transfer function are listed below in ascending order of occurrence as frequency increases.

1. A constant gain $K_t = 0.2$.
2. A pole at the origin.
3. A real zero at $\omega = 0.5$.
4. A pair of complex poles at $\omega_n = 10$.
5. A real pole at $\omega = 50$.

- **Magnitude plot:** The transfer function contains one real zero with a breakpoint frequency of $\omega = 0.5$, a pair of complex conjugate poles with a natural frequency of $\omega_n = 10$, and a real pole at breakpoint frequency at $\omega_n = 50$. As a result, the entire asymptotic magnitude plot must include breakpoint frequencies at 0.5, 10, and 50. Let us now draw the composite magnitude plot. First, we will plot the asymptotic magnitude of the gain K_t and each factor of poles and zero, as shown in Figure 9.16a.

1. The logarithmic magnitude of the Bode gain $K_t = 0.2$ in dB is given by

$$20\log 0.2 = -13.98 \quad \text{dB} \quad (9.177)$$

   It is represented by a horizontal asymptote starting from -13.98 dB at the low-frequency point $\omega = 0.01$ until it approaches ∞.
2. The asymptotic magnitude of the pole at the origin starts from +40 dB at the low-frequency point $\omega = 0.01$ and extends downward to the infinite frequency with a slope of -20 dB/decade, crossing the 0 dB line at $\omega = 1$.
3. The asymptotic magnitude of the real zero at $\omega = 0.5$ starts from 0 dB at $\omega = 0.01$ and continues until it reaches the breakpoint frequency $\omega = 0.5$ and then extends upward to the infinite frequency with a slope of +20 dB/decade.

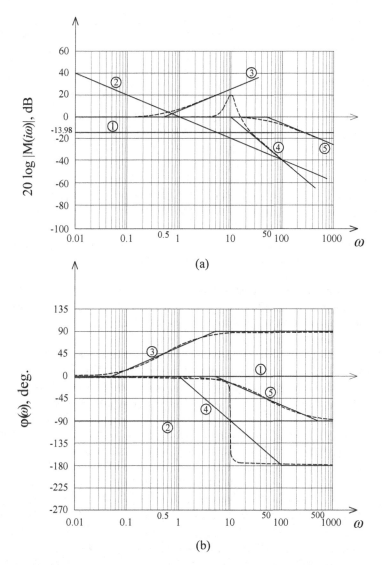

Figure 9.16 Magnitude and phase asymptotes of the zeros and poles of the transfer function
$X(s) = \frac{2000(s+0.5)}{s(s+50)(s^2+s+100)}$

4. The asymptotic magnitude of the complex poles starts at 0 dB at the low frequency
$\omega = 0.01$. It continues until it reaches its natural frequency $\omega = \omega_n = 10$ and
then extends downward to the infinite frequency with a slope of -40 dB/decade.
Referring to Figure 9.11, the maximum peak of the complex poles of a damping
ratio $\zeta = 0.05$ at the resonance frequency $\omega_r = \omega_n = 10$ is determined and found
to be 20 dB.

5. The asymptotic magnitude of the real pole at $\omega = 50$ starts at 0 dB at the low
frequency and continues until it reaches the breakpoint frequency $\omega = 50$ and
then extends downward to the infinite frequency with a slope of -20 dB/decade.

Figure 9.16a shows that each asymptote is plotted with its actual curve (dotted line). To obtain the composite magnitude plot, insert its entire asymptotes first. The entire asymptotic magnitude can be directly plotted by summing the asymptotes due to the factor of each pole and zero from low to high frequency. Thus, due to the pole at the origin, the slope is -20 dB/decade starting from approximately 26.02 dB at the low-frequency point $\omega = 0.01$, which is the sum of +40 dB and -13.98 dB, intersecting the -13.98 dB line at $\omega = 1$ until it reaches $\omega = 0.5$. At this frequency, the slope changes from -20 dB/decade to 0 dB/decade due to the existence of the real zero at $\omega = 0.5$ and continues until it reaches $\omega_n = 10$. At this frequency, the slope changes to -40 dB/decade due to the existence of the complex pole at $\omega = 10$ until it reaches $\omega = 50$. Finally, at $\omega = 50$, the slope becomes -60 dB/decade due to the existence of the real pole. Using these comments as a guide, sketch the composite magnitude curve, shown by a bold solid line in Figure 9.17a.

- Phase plot: To draw the composite phase plot, we will first draw the asymptotic phase of each factor in the provided transfer function, see Figure 9.16b.

1. The phase angle of the Bode gain K_t is equal to 0°.
2. The pole at the origin makes a constant phase angle of -90°. It is represented by a horizontal asymptote that spans the frequency range from the low-frequency point to infinity.
3. The asymptotic phase of the real zero at $\omega = 0.5$ starts from the 0° phase at the low frequency $\omega = 0.01$ and continues until it reaches $\omega = 0.5/10$. At this frequency, it shifts upward until it levels off at $+90^o$ at $\omega = 0.5 * 10$ passes through $+45^o$ at $\omega = 0.5$ and continues horizontally from the leveled-off angle +90 until it approaches ∞.
4. The asymptotic phase of the complex pole at $\omega = 10$ starts from the 0° phase at the low frequency $\omega = 0.01$ and continues until it reaches $\omega = 10/10$. At this frequency, it shifts downward until it levels off at -180° at $\omega = 10 * 10$ passes through -90° at $\omega = 10$ and continues horizontally from the leveled-off angle -180 until it approaches ∞. The actual phase plot of the complex poles at the resonance frequency, $\omega_r = \omega_n = 10$, is determined at damping ratio $\zeta = 0.05$ using Figure 9.11.
5. The asymptotic phase of the real pole at $\omega = 50$ starts from the 0° phase at the low frequency and continues until it reaches $\omega = 50/10$. At this frequency, it shifts downward until it levels off at -90° at $\omega = 50 * 10$, passes through -45° at $\omega = 50$, and continues horizontally from the leveled-off angle -90 until it approaches ∞.

Figure 9.16b shows the asymptotic phases with their actual curves (dotted lines). So, the phase angle of the gain K_t equals 0°. The pole at the origin makes -90° spans the frequency range from the low-frequency point until it approaches ∞, while the real zero and the real pole make $+90^o$ and -90°, respectively, at high frequencies. The complex pole makes a phase angle of -180° at high frequencies. Therefore, a phase angle scale ranging from $+90^o$ (one zero) to -360° (three poles) will almost be sufficient. Note that the final value of the phase angle should be -270°, which is

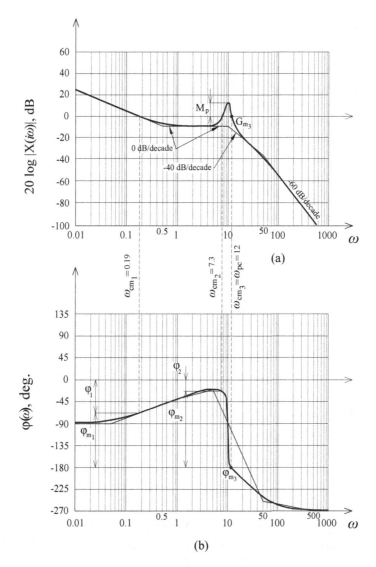

Figure 9.17 Composite Bode plots of the transfer function $X(s) = \frac{2000(s+0.5)}{s(s+50)(s^2+s+100)}$

the sum of -360° (three poles) and $+90^o$ (one zero). We will sketch the composite magnitude and phase while keeping with the above guidelines.

To obtain the composite phase curve, insert its entire asymptotic plot first. Remember that, here, we must add distances rather than phase values. Due to the existence of a pole at the origin, the asymptotic phase must begin at -90° and then continue until it reaches $\omega = 0.5/10$. At this frequency, the phase shifts upward due to the existence of the real zero until it reaches $\omega = 10/10$. Due to the existence of the complex pole at $\omega = 10/10$, the phase angle changes until it reaches $\omega = 50/10$.

At this frequency, the phase shifts downward due to the existence of the second real pole until it reaches $\omega = 50 * 10$. At this frequency, the phase angle level off at $-270°$, which is the final value of the phase plot as stated above. Now, sketch the composite phase plot, shown in Figure 9.17b, by a bold solid line. We will now check whether the composite magnitude and phase plots appropriately represent the frequency response of the given transfer function. We will evaluate this by comparing the composite magnitude and phase values at the breakpoint frequencies 0.5, 10, and 50 with the corresponding exact values. The exact values of the magnitude and phase at any frequency for the given transfer function are provided by

$$M(\omega) = 20\log 0.2 + 10\log((2\omega)^2 + 1) - 10\log \omega - 20\log((0.02\omega)^2 + 1)$$
$$- 10\log\left([1 - (\frac{\omega}{\omega_n})^2]^2 + [2\zeta(\frac{\omega}{\omega_n})]^2\right) \tag{9.178}$$

and

$$\phi(\omega) = 0° + \tan^{-1}(2\omega) - 90° - \tan^{-1}(0.02\omega) - \tan^{-1}\left(\frac{2\zeta(\frac{\omega}{\omega_n})}{1 - (\frac{\omega}{\omega_n})^2}\right) \tag{9.179}$$

Now, calculate the exact magnitude and phase values at breakpoint frequencies as follows. Using equations (9.178) and (9.179) gives

- At first breakpoint frequency, $\omega = 0.5$:

$$M(0.5) = -13.98 + 10\log((2*0.5)^2 + 1) - 20\log(0.5) - 10\log((0.02*0.5)^2 + 1)$$
$$- 10\log\left([1 - (\frac{0.5}{10})^2]^2 + [2*0.05*(\frac{0.5}{10})]^2\right) = -13.98 + 3.01$$
$$- 4.34 * 10^{-4} - 0.08 = -4.94 \quad \text{dB} \tag{9.180}$$

and

$$\phi(0.5) = 0° + \tan^{-1}(2*0.5) - 90° - \tan^{-1}(0.02*0.5) - \tan^{-1}\left(\frac{2*0.05*(\frac{0.5}{10})}{1 - (\frac{0.5}{10})^2}\right) =$$
$$0° + 45° - 90° - 0.57° - 0.29° = -45.86° \tag{9.181}$$

- At the second breakpoint frequency (natural or resonance frequency), $\omega_r = 10$:

$$M(10) = -13.98 + 10\log((2*10)^2 + 1) - 20\log(10) - 10\log((0.02*10)^2 + 1)$$
$$- 10\log\left([1 - (\frac{10}{10})^2]^2 + [2*0.05*(\frac{10}{10})]^2\right) = -13.98 + 26.03$$
$$- 20 - 0.17 + 20 = 11.88 \quad \text{dB} \tag{9.182}$$

and

$$\phi(10) = 0° + \tan^{-1}(2*10) - 90° - \tan^{-1}(0.02*10) - \tan^{-1}\left(\frac{2*0.05*(\frac{10}{10})}{1 - (\frac{10}{10})^2}\right) =$$
$$0° + 87.13° - 90° - 11.31° - 90° = -104.18° \tag{9.183}$$

- At third breakpoint frequency $\omega = 50$:

$$M(50) = -13.98 + 10\log((2*50)^2 + 1) - 20\log(50) - 10\log((0.02*50)^2 + 1)$$
$$- 10\log\left([1 - (\frac{50}{10})^2]^2 + [2*0.05*(\frac{50}{10})]^2\right) = -13.98 + 40$$
$$- 33.98 - 0.17 - 27.6 = -35.65 \quad \text{dB} \tag{9.184}$$

and

$$\phi(50) = 0^o + \tan^{-1}(2*50) - 90^o - \tan^{-1}(0.02*50) - \tan^{-1}\left(\frac{2*0.05*(\frac{50}{10})}{1 - (\frac{50}{10})^2}\right)$$
$$= 0^o + 89.42^o - 90^o - 45^o - 1.14^o = -46.72^o + (-180^o) = -226.72^o \tag{9.185}$$

Now, let us verify the results at two more frequencies, one near the low-frequency point, for example, $\omega = 0.05$, and the other at the frequency end, for example, $\omega = 100$. - At $\omega = 0.05$:

$$M(0.05) = -13.98 + 10\log((2*0.05)^2 + 1) - 20\log(0.05) - 10\log((0.02*0.05)^2 + 1)$$
$$- 10\log\left([1 - (\frac{0.05}{10})^2]^2 + [2*0.05*(\frac{0.05}{10})]^2\right) = -13.98 + 0.043 - (-26.02)$$
$$- 4.34 * 10^{-4} = 12.04 \quad \text{dB} \tag{9.186}$$

and

$$\phi(0.05) = 0^o + \tan^{-1}(2*0.05) - 90^o - \tan^{-1}(0.02*0.05) - \tan^{-1}\left(\frac{2*0.05*(\frac{0.05}{10})}{1 - (\frac{0.05}{10})^2}\right)$$
$$= 0^o + 5.71^o - 90^o - 0.05 - (-0.028^o) = -84.31^o \tag{9.187}$$

- At $\omega = 100$:

$$M(100) = -13.98 + 10\log((2*100)^2 + 1) - 20\log(100) - 10\log((0.02*100)^2 + 1)$$
$$- 10\log\left([1 - (\frac{100}{10})^2]^2 + [2*0.02(\frac{100}{10})]^2\right) = -13.98 + 46.02$$
$$- 40 - 6.99 - 40.08 = -55.03 \quad \text{dB} \tag{9.188}$$

and

$$\phi(100) = 0^o + \tan^{-1}(2*100) - 90 - \tan^{-1}(0.02*100) - \tan^{-1}\left(\frac{2*0.05*(\frac{100}{10})}{1 - (\frac{100}{10})^2}\right)$$
$$0^o + 89.71^o - 90^o - 63.43^o - 0.57^o = -63.15 + (-180^o) = -243.15^o \tag{9.189}$$

Table 9.10

Exact and composite values of the gain and phase margins of the transfer function $X(s) = \frac{2000(s+0.5)}{s(s+50)(s^2+s+100)}$

	M (dB)		Phase ϕ	
ω	exact	composite	exact	composite
0.05	12.04	11	-84.31^o	-85°
0.5	-4.94	-4	-45.86^o	-48°
10	-11.88	-11	-104.18^o	-108°
50	-35.65	-40	-226.72^o	-230°
100	-55.03	-57	-243.15^o	-240°

Note that because the phase response has already been shifted by -180° due to the effect of the complex pole, the exact values of phases above the resonance frequency $\omega_r = 10$ must be added to a phase value of -180°, as indicated in equations (9.185) and (9.189). However, the complex pole does not affect the phase response below the resonance frequency.

Let us now obtain the magnitude and phase values from their sketching composite plot, shown in Figure 9.17, at the system breakpoints frequencies $\omega = 0.5$, $\omega_r = 10$, and $\omega = 50$ and the frequencies $\omega = 0.05$ and $\omega = 100$. Since the values obtained from the composite magnitude and phase plots are approximate, keep this in mind while evaluating them. The results obtained and the corresponding exact values calculated above for comparison are given in Table (9.10). As a comparison, it was found that the disparity between the results was small enough to be negligible. As a result, we may utilize the sketching composite Bode plot of the given transfer function, shown in Figure 9.17 to study system stability accurately.

b. Gain margin G_m, phase margin ϕ_m, maximum peak M_p and system stability: As shown in Figure 9.17a, the magnitude plot intersects the 0 dB line at three different magnitude crossover frequencies: approximately $\omega_{cm_1} = 0.19$, $\omega_{mc_2} = 7.3$, and $\omega_{mc_3} = 12$. On the other hand, the phase plot intersects the -180° line at a phase crossover frequency of approximately $\omega_{pc} = 12$. Remember that the gain margin occurs only when the phase plot intersects with the -180° line. Now, let us determine the gain margin G_m and phase margin ϕ_m and check the system stability at these frequencies.

- At the first magnitude crossover frequency, $\omega_{mc_1} = 0.19$:
  Look at the phase plot shown in Figure 9.17b; at this frequency, the phase lag is approximately equal to $\phi_1 = -67^o$. So, the phase margin is given by

$$\phi_{m_1} = \phi_1 - (-180) = -67^o + 180^o = +113^o \qquad (9.190)$$

This equation indicates that the phase margin has a positive value when $\omega_{mc_1} = 0.19$, meaning the system is stable at this frequency.

- At the second magnitude crossover frequency, $\omega_{mc_2} = 7.3$:
At this frequency, the phase lag is approximately equal to $\phi = -23^o$. So, the phase margin is given by

$$\phi_{m_2} = \phi_2 - (-180) = -23^o + 180^o = +157^o \qquad (9.191)$$

This equation indicates that the phase margin has a positive value at $\omega_{mc_2} = 7.3$, meaning the system is stable at this frequency.

- At the third magnitude crossover frequency, $\omega_{mc_3} = 12$:
At this frequency, the magnitude crossover frequency ω_{mc_3} equals the phase crossover frequency ω_{pc}, meaning the gain margin G_{m_3} and phase margin ϕ_{m_3} are equal to 0 as shown in Figure 9.17. As a result, the system is marginally stable at this frequency.

Equations (9.190) and (9.191) indicate that the system is stable at the first two crossover frequencies, $\omega_{mc_1} = 0.19$ and $\omega_{mc_2} = 7.3$, while is marginally stable at the third, $\omega_{mc_3} = 12$.

In general, the Bode plot method is insufficient to examine the stability of a system in a situation in which there is more than one magnitude crossover frequency. In this case, we must analyze the system's stability using methods different from those covered in this course, such as the root locus technique or the Nyquist criterion.

To determine the maximum peak M_p, looking at the magnitude plot shown in Figure 9.17a, we found that it occurs at the resonance frequency $\omega_r = \omega_n = 10$ and is approximately equal to $+13$ dB.

c. Bode plot through Matlab: The Matlab live script file and the generated results are shown in Figure 9.18, while the generated Bode plot is shown in Figure 9.19. Since Matlab generates both the gain margin and the maximum peak as magnitude ratios, we need to convert them to logarithmic magnitude in dB to validate the corresponding ones performed in step (b). We can accomplish this by adding codes to convert the magnitude ratios to dB, as shown in Figure 9.18. The Matlab results and the corresponding results obtained from the composite Bode plot performed in step (b) for comparison are given in Table 9.11. In comparison, the disparities between the

---

Table 9.11

Comparison between system characteristics obtained from composite Bode plot and that generated from MATLAB for the transfer function $X(s) = \frac{2000(s+0.5)}{s(s+50)(s^2+s+100)}$

Bode plot	ω_{mc_1}	ω_{mc_2}	ω_{mc_3}	ω_{pc}	ϕ_{m_1}	ϕ_{m_2}	ϕ_{m_3}	G_m	M_p
Composite	0.19	7.3	12	12	113°	157°	0	0	13
Matlab	0.2183	7.8217	11.7124	11.92	113.215^o	156.059^o	1.856^o	1.0168	11.893

```
% Bode plots and system characteristics, Example (9-2).
% Write the given transfer function.
n=2000*[1 0.5];d1=conv([1 50],[1 1 100]);d=conv([1 0],d1);
xs=tf(n,d);
% Use the allmargin command to generate the system characteristics.
s=allmargin(xs)
```

```
s = struct with fields:
    GainMargin: 1.1242
   GMFrequency: 11.9216
   PhaseMargin: [113.2151 156.0598 1.8567]
   PMFrequency: [0.2183 7.8217 11.7124]
   DelayMargin: [9.0498 0.3482 0.0028]
   DMFrequency: [0.2183 7.8217 11.7124]
        Stable: 1
```

```
% or we can use the following.
[Gm,Pm,Wcp,Wcg]=margin(xs)
```

```
Gm = 1.1242
Pm = 1.8567
Wcp = 11.9216
Wcg = 11.7124
```

```
% calculate the gain margin in dB.
Gm_dB = 20*log10(Gm)
```

```
Gm_dB = 1.0168
```

```
% Use the following command to obtain the maximum peak.
gpeak = getPeakGain(xs,1e-2,[8,12])
```

```
gpeak = 3.9326
```

```
% calculate the maximum peak in dB.
gpeak_dB = 20*log10(gpeak)
```

```
gpeak_dB = 11.8935
```

```
% Use the bode or bodeplot command to plot the frequency response.
bodeplot(xs);
```

Figure 9.18 MATLAB live script file of the transfer function $X(s) = \frac{2000(s+0.5)}{s(s+50)(s^2+0.4s+100)}$ and the generated results of the system characteristics

results are relatively small to ignore, implying that the stability analyses performed in step (b) from the composite Bode plot shown in Figure 9.17 are acceptable.

Finally, as a result of the previous analysis, we noted that the behavior of a dynamic system in the frequency domain becomes apparent when plotting the Bode plot by factorizing the overall transfer function and using a plotting scheme with a logarithmic scale.

The following are summarized outlines of criteria for drawing Bode plots and evaluating the system stability.

- **Gain Margin:** The greater the gain margin, the greater the system's stability. The gain can be changed without causing the system to become unstable. It is commonly expressed in decibels (dB).

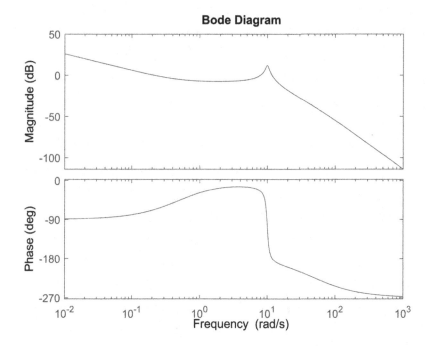

Figure 9.19 Bode plots of the transfer function $X(s) = \frac{2000(s+0.5)}{s(s+50)(s^2+s+100)}$ generated from the MATLAB

- **Phase Margin:** The greater the phase margin, the more stable the system. It is the phase of the system that can be changed without causing the system to become unstable. It is usually expressed in degree.
- **Magnitude Crossover Frequency:** It is the frequency at which the magnitude plot crosses the 0 dB line.
- **Phase Crossover Frequency:** It is the frequency at which the phase plot intersects with the -180° line.
- **Breakpoint point frequency (Corner Frequency):** When the two asymptotes meet or break, this is known as the "Break frequency" or the "Corner frequency."
- **Resonant Frequency:** The frequency value at which the complex pole or complex zero has a peak value is known as the resonant frequency.

9.6 ESTIMATING THE TRANSFER FUNCTION FROM EXPERIMENTAL BODE PLOT

The internal configuration of the system may be locked down, or specific component settings may be unknown. In this case, it is possible to experimentally identify the system's frequency response from input to output, which is then used to estimate the transfer function. In this section, we will demonstrate how to estimate the transfer function of a Bode plot that has already been provided. Once the system's frequency response has been determined, its transfer function can be roughly estimated using the slopes of the asymptotes and break frequencies. We will attempt to superimpose the entire plot of asymptotes that matches the Bode plot provided. The provided Bode plot is a composite one, clearly demonstrated in examples (9.1) and (9.2), representing a known transfer function. The following steps are a guide to making the transfer function estimation much more accessible.

1. The pole-zero configuration of the system can be estimated by looking at both magnitude and phase plots. When determining the type of system (poles and zeros at the origin), look at the initial slope on the magnitude plot. We may determine the difference between the number of poles and zeros by looking at the phase deviations, as discussed in examples (9.1) and (9.2).
2. Examine the magnitude and phase curves to see whether any locations clearly show first- or second-order poles or zeros.
3. Examine the magnitude plot to determine whether or not it contains any obvious peaking or depressions that indicate an underdamped second-order pole or zero, respectively.
4. In general, the slopes of the asymptotes of the magnitude curves must be only $0, \pm 20, \pm 40, \pm 60$, or $\pm 80...$; this indicates that no asymptote has a slope between these values. So, estimate the break frequencies by superimposing appropriate ± 20 or ± 40 dB/decade lines on the magnitude curve or $\pm 45^o$ or $\pm 90^o$ deg./decade lines on the phase curves. This step is necessary to find the locations at break frequencies and then the locations of the poles and zeros. Use the standard curves shown in Figures 9.9 and 9.11 to determine the damping ratio ζ for the second-order poles and zeros.
5. Create a transfer function with a unity gain for each pole and zero we identified.
6. The difference between the number of poles and the number of zeros can be found by specifying the starting angle on the phase plot and then determining the angle at which it is leveled off.

As a comparison, we will find that the approaches to estimating the transfer function from the Bode plot frequency response can potentially be more accurate than estimates from the transient response discussed in Chapter 8.

Example (9.3):
Find the transfer function of the system whose Bode plot is shown in Figure 9.20.

Solution:
To estimate the transfer function of the Bode plot, we need to follow the guidelines provided above. To begin, we will try to superimpose all asymptotes that are consistent with the provided Bode plot.

1. Look at the magnitude plot shown in Figure 9.20a; we will see that it begins with a line of a particular slope beginning at the low-frequency point $\omega = 0.01$. When we superimposed a parallel asymptote to this slope, we found that the slope is of

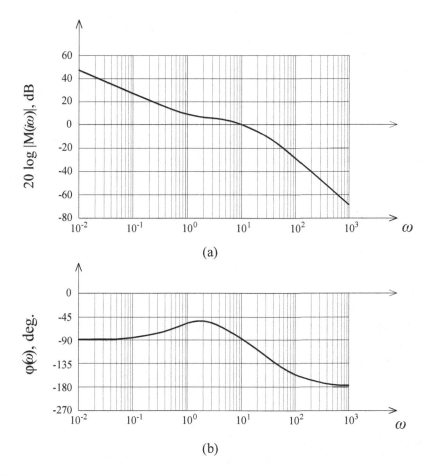

Figure 9.20 Bode plot of a system with unknown transfer function

-20 dB/decade, which implies that the system has a pole at the origin. With unity gain, it is given by

$$X_1(s) = \frac{1}{s} \qquad (9.192)$$

Look at the phase plot shown in Figure 9.20b; it begins at -90°, ensuring the system has a pole at the origin.

2. This asymptote reached the first break frequency at $\omega = 1$; at this point, the slope of the asymptote must be 0 dB/decade. If we attempt to draw an asymptote starting from the first break frequency $\omega = 1$ with a slope other than 0 dB/decade (-20 dB/decade or -40 dB/decade) and close to the magnitude curve, we will fail. As mentioned above, The slopes of the asymptotes must be 0, ± 20, ± 40, ± 60, or ± 80 dB/decade,..., and should be very close to the magnitude curves. Therefore, the slope of the asymptote starting at the break frequency $\omega = 1$ should be 0 dB/decade, as shown in Figure 9.21a. When the slope of the asymptote changes

at a specific break frequency, it is reasonable to assume that the pole, or zero, is located at this frequency. When the asymptote slope is changed from having a slope of -20 dB/decade to a slope of 0 dB/decade, a real zero is indicated at this frequency. Note that this slope is equal to the sum of the slope of the previous pole at the origin, which is -20 dB/decade, and the slope of the real zero, which is +20 dB/decade. With unity gain, this real zero is given by:

$$X_2(s) = (s+1) \qquad (9.193)$$

3. Extend the 0 dB/decade asymptote until it reaches the second break frequency at $\omega = 5$. At this frequency, we found that the slope of an asymptote downward and tangent to the magnitude plot is -20 dB/decade, implying that the system has a real pole at $\omega = 5$ since the slope goes from 0 dB/decade to -20 dB/decade, as shown in Figure 9.21a. With unity gain, it is given by

$$X_3(s) = \frac{5}{s+5} \qquad (9.194)$$

4. Extend this asymptote until it reaches the third break frequency at $\omega = 40$. At this frequency, draw an asymptote downward and attach it to the magnitude curve at high frequency. We found that its slope is -40 dB/decade, which implies that the system has a real pole at $\omega = 40$. This slope is equal to the sum of the slope of the second pole, which is -20 dB/decade, and the slope of the third real pole, which is also -20 dB/decade. With unity gain, this real pole is given by

$$X_4(s) = \frac{40}{s+40} \qquad (9.195)$$

5. Back to the starting point of the magnitude plot at low-frequency $\omega = 0.01$, we found that it is of +46 dB. According to the definition of the asymptote of the pole at the origin, the slope must begin at +40 dB at a low frequency of $\omega = 0.01$. Given the starting point value, +46 dB, which is more than 40 dB, the transfer function has a Bode gain K_t, which may be calculated as follows. The logarithmic magnitude of the Bode gain K_t is given by

$$20\log K_t = (46-40) = 6 \qquad (9.196)$$

Solving for K_t gives

$$K_t = 10^{(6/20)} = 10^{0.3} = 1.99 \qquad (9.197)$$

6. Now that we have one real zero and three real poles, the poles are greater than the zeros by two. This may be demonstrated by looking at the phase plot shown in Figure 9.21b; we discovered that it began at -90° and leveled off at -180°, representing the two real poles, which is the difference between the number of the poles and the number of the zeros.

According to the above calculations, the estimated transfer function of the given system is given by

$$X(s) = K_t X_1(s) X_2(s) X_3(s) X_4(s) = \frac{1.99 * 1 * 5 * 40(s+1)}{s(s+5)(s+40)} = \frac{398(s+1)}{s(s+5)(s+40)}$$
$$(9.198)$$

Let us now use Matlab to verify this result. To accomplish this, we must provide the scale from Matlab for the Bode plot we have been given. Figure 9.22 shows the generated Bode plot from Matlab. Comparing the given Bode plot shown in Figure 9.21 with the Bode plot generated from Matlab shown in Figure 9.22, we found both are approximately the same.

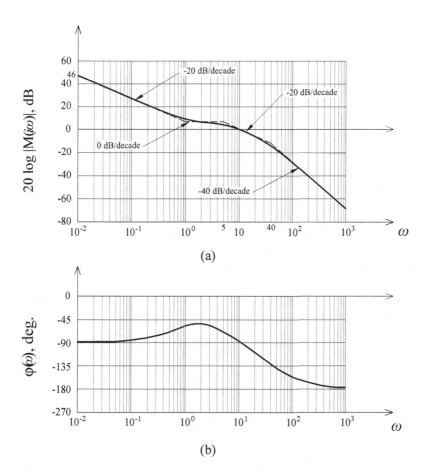

Figure 9.21 Bode plot of the given system with the asymptotes

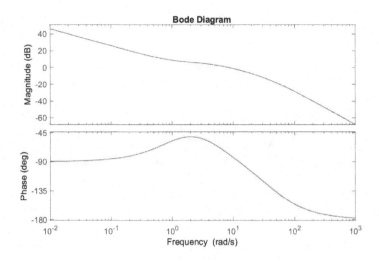

Figure 9.22 Bode plot of the estimated transfer function generated from MATLAB

9.7 PROBLEMS

Problem (9.1):

For each of the following transfer functions,

a. Set $s = i\omega$, put the transfer function into Bode form, and determine the Bode gain K_t.

b. Sketch the asymptotes of each zero and pole in the magnitude and phase frequency responses of $X(s)$. Define the breakpoints frequencies. Note that the magnitude must be in dB and the phase angle in degrees.

c. Sketch the entire asymptote for the magnitude and phase and draw their composite Bode plots.

d. Verify the magnitude and phase values of the composite plot with their exact values at the breakpoints frequencies.

$$1. \quad X(s) = \frac{1}{s(s+1)(s+8)}$$

$$2. \quad X(s) = \frac{1}{s(s^2+3s+10)}$$

$$3. \quad X(s) = \frac{(s+2)}{s(s^2+2s+6)}$$

$$4. \quad X(s) = \frac{1}{(s+2)(s^2+3s+8)}$$

$$5. \quad X(s) = \frac{(s+3)}{s(s+1)(s+5)(s+8)}$$

$$6. \quad X(s) = \frac{s^2+8s+10}{s(s+10)(s^2+6s+5)}$$

Problem (9.2):
For the following transfer functions,

a. Draw the Bode plot and clearly label all corner (breakpoint) frequencies, slopes, and magnitude values.
b. Verify the magnitude and phase values of the composite plot with their corresponding exact values at the corner frequencies.
c. Determine magnitude crossover ω_{mc} and phase crossover ω_{pc} frequencies.
d. Determine gain margin G_m, phase margin ϕ_m, and maximum peak M_p and verify the results using MATLAB.
e. Check the system's stability. If the system is unstable, determine the range of system gain K that makes the system first marginally stable and then choose the value of the gain K that makes the system stable and draw the Bode plot of the stable response.

$$1. \quad X(s) = \frac{1+0.4s}{s^2}$$

$$2. \quad X(s) = \frac{-100(s-1)}{s(s+1)(s+8)}$$

$$3. \quad X(s) = \frac{30(s+2)}{s(s+0.5)(s^2+3.2s+49)}$$

Problem (9.3):
Sketch the Bode plot (Composite Bode plot) of the frequency response for the following transfer functions. After completing the composite plot,

a. Verify the magnitude and phase values at breakpoint frequencies with their exact values.
b. Using Matlab and verify the composite Bode plot (turn in the composite plot and Matlab results on the same scales).
c. Determine magnitude crossover frequency ω_{mc}, phase crossover frequency ω_{pc}, gain margin G_m, phase margin ϕ_m, and maximum peak M_p. Verify the results with their corresponding generated from Matlab.
d. Check the system's stability. If the system is unstable, determine the range of system gain K that makes the system first marginally stable and then choose the value of the gain K that makes the system stable and sketch the Bode plot of the stable response. Verify the stable response with the corresponding generated from Matlab (turn in the composite plot and Matlab results on the same scales).

$$1. \quad X(s) = \frac{1}{(0.25s+1)(3s+1)}$$

$$2. \quad X(s) = \frac{(s+4)}{(s^2+3s+4)}$$

$$3. \quad X(s) = \frac{10(s+8)}{s(s+2)(s+4)}$$

Problem (9.4):

Sketch the asymptotes of the Bode plot magnitude and phase of the following transfer functions and the transition at the second-order breakpoint based on the value of the damping ratio. After completing the hand sketches (composite plot), verify the magnitude and phase values with their exact values at breakpoint frequencies. Draw the Bode plot using Matlab and verify the composite Body plot (turn in the composite plot and Matlab results on the same scales).

$$1. \quad X(s) = \frac{2000}{s(s+200)}$$

$$2. \quad X(s) = \frac{1}{s(s^2+3s+8)}$$

$$3. \quad X(s) = \frac{(s^2+2s+4)}{s(s^2+3s+10)}$$

$$4. \quad X(s) = \frac{1}{s(s+1)(0.02s+1)}$$

Problem (9.5):

For the following transfer functions, sketch the Bode plot and determine the magnitude and phase crossover frequencies ω_{mc} and ω_{pc}. Determine gain margin G_m, phase margin ϕ_m, and response peak M_p. Check the system's stability. If the system is unstable, determine the range of system gain K that causes the system to become marginally stable, then select the value of gain K that causes the system to become stable and draw the Bode plot of the stable response.

$$1. \quad X(s) = \frac{1}{s(0.5s+1)(2s+1)}$$

$$2. \quad X(s) = \frac{0.5}{s(0.1s+1)(0.001s^2+0.002s+1)}$$

$$3. \quad X(s) = \frac{80(s+2)}{s(s^2+6s+25)}$$

Problem (9.6):

Sketch the Bode plot for the following transfer function.

$$X(s) = \frac{80(s+1)}{s(s+10)(s+100)}$$

Follow the steps below:

a. Write the transfer function in the Bode transfer function form.
b. Identify and label each as part of the transfer function.
c. For each constituent part of the transfer function, find the break frequency.
d. Neatly draw the Bode diagram for each constituent part using different labels or colors.

e. Draw the overall (composite plot) Bode plot by adding up the results from step (d) and highlighting any corrections applied to the composite plot.

Problem (9.7):

Given

$$X(s) = \frac{160(\frac{s}{10}+1)}{s(\frac{s}{1.75}+1)(\frac{s}{60}+1)}$$

a. Draw Bode plot.
b. Determine the magnitude crossover frequency ω_{mc}, the phase crossover frequency ω_{pc}, the gain margin G_m, and the phase margin ϕ_m.
c. Check the system's stability. If the system is unstable, determine the range of system gain K that makes the system first marginally stable, then choose the value of the gain K that makes the system stable, and draw the Bode plot of the stable response.

Problem (9.8):

Draw the Bode plot for a system whose transfer function is

$$X(s) = \frac{2(s/10+1)}{s^2(5s+1)((s/9)^2+(s/9)+1)}$$

Determine the gain margin and phase margin and check the stability.

Problem (9.9):

Draw the Bode plot for a system whose transfer function is

$$X(s) = \frac{K(1+s/5)}{s(1+s/2)(1+s/10)}$$

Here, $K = 10.5$. Demonstrate that the system crossover frequency is 5 rad/sec and phase margin ϕ_m is around 40°.

Problem (9.10):

For the following transfer function

$$X(s) = \frac{8(1+0.4s)}{s(1+2s)(0.04s^2+0.24s+1)}$$

a. Draw the Bode plot and determine the gain margin G_m phase margin ϕ_m and maximum peak M_p.
b. Check the system's stability. If the system is unstable, determine the range of system gain K that makes the system first marginally stable and then choose the value of the gain K that makes the system stable and draw the Bode plot of the stable response.

Problem (9.11):

The transfer function of a certain system is given by

$$X(s) = \frac{K}{s(s+1)(s+12)}$$

Draw the magnitude and phase angle plots for the a value of K such that the system becomes marginally stable.

Problem (9.12):
A certain system is given by

$$X(s) = \frac{K}{s(s+1)(0.1s+1)}$$

Draw the Bode plot and determine the value of K to have

a. The gain margin of G_m approximately equal to 20 dB.
b. The phase margin of ϕ_m approximately equal to 50°.

Problem (9.13):
For the following transfer function, use Matlab to create the Bode plot for $K = 1$, determine from the plot what values of K the system will be stable, and justify the answer.

$$X(s) = \frac{K}{s(s+1)(s+\sqrt{8})}$$

Problem (9.14):
Sketch the magnitude and phase plots for the following transfer function and determine the system gain K for the magnitude crossover frequency to be 5 rad/sec.

$$X(s) = \frac{Ks^2}{(0.2s+1)(0.02s+1)}$$

Problem (9.15):
The asymptotic magnitude of a transfer function is shown in Figure 9.23. Slopes are given in dB/decade. Sketch the corresponding asymptotic phase angle curves and estimate the transfer function of this system.

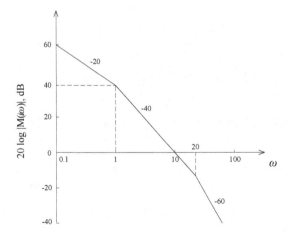

Figure 9.23 Asymptotes of the magnitude plot of a system

Problem (9.16):
For the given Bode plot shown in Figure 9.24, determine

1. The magnitude crossover frequencies.
2. The gain margin G_m.
3. The phase crossover frequencies.
4. The phase margin ϕ_m.
4. The maximum peak M_p.
5. The system transfer function.
6. Check the system's stability.
7. Use MATLAB to verify the results obtained in steps (1) to (5).

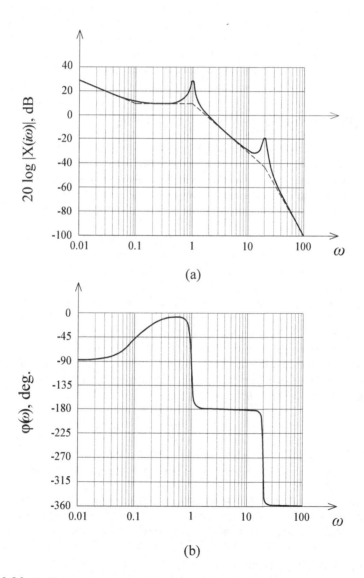

(a)

(b)

Figure 9.24 Bode plot of a system with unknown transfer function

10 Nonlinear Dynamic Systems

10.1 INTRODUCTION

Many engineering books focus only on linear systems because of their relative simplicity analysis. However, it is not easy to come across a truly linear system in either nature or engineering. In the earlier chapters, when we were studying the generation of system equations, we saw that even relatively simple systems might create nonlinear differential equations. Where the derived equations were linear, the modeling process used approximations. For example, we thought that springs and other mechanical parts that cause friction would have linear properties, which made their nonlinear properties easier to understand.

10.2 LINEAR AND NONLINEAR SYSTEMS

Generally, a set of linear equations is a local linear approximation of a nonlinear system near the equilibrium point, as stated previously. This is true whether the system is linear or nonlinear. Let us look at an example to illustrate this point. We will derive the equation of motion for the simple pendulum system shown in Figure 10.1, assuming that the angular displacement θ cannot be considered small. Applying Newton's law gives

$$J\ddot{\theta} + mgx = 0 \tag{10.1}$$

where

$$x = l\sin\theta \tag{10.2}$$

Substituting this equation in equation (10.1) gives

$$J\ddot{\theta} + mg(l\sin\theta) = 0 \tag{10.3}$$

This equation represents the nonlinear differential equation of motion of the pendulum system shown in Figure 10.1. If we consider the angular displacement θ is small, meaning that, $\sin\theta \approx \theta$, equation (10.3) becomes

$$J\ddot{\theta} + mgl\theta = 0 \tag{10.4}$$

Although most high school students have encountered equation (10.4) while studying dynamics, they must be aware that it represents a local linear equation (10.3) in nonlinear form.

In engineering, linear system representation is still widely used because, in general, the nominal operating point of any system is located at an equilibrium point. If perturbations are small, then the linear approximation gives a simple, workable

DOI: 10.1201/9781032685656-10

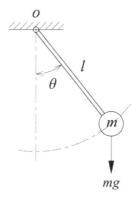

Figure 10.1 Simple pendulum system

model of the dynamical system. This is why linear system representation is still widely used in engineering processes. Nevertheless, if a system has strong disturbances or perturbations, the system's variables may move very far from the equilibrium point, and the linear approximation no longer remains valid. The following three examples demonstrate how to derive the equations of motion of nonlinear systems.

Example (10.1):
For the system shown in Figure 10.2, assuming that the angular displacement θ cannot be considered small,

1. Draw the free-body diagram and derive the equation(s) of motion.
2. Using SIMULINK, generate the transient response of the angular displacement $\theta(t)$, considering the following parameter values $m = 5$ kg, $k = 200$ N m, $l = 0.5$ m and I.C. $\theta(0) = 0.1$ rad and $\dot{\theta}(0) = 0$ rad/sec.

Solution:
Figure 10.2 shows a mass m fixed at the end of a rod that can rotate about the point o. When the massless pulley moves, the gravitational force pulls the mass m down while the sliding spring pulls it up in the opposite direction.

1. Free-body diagram and equations of motion:
The free-body diagram of forces acting on the mass m is shown in Figure 10.3. Assuming no friction, applying Newton's law on the system's motion gives

$$\sum M_o = J\ddot{\theta} \tag{10.5}$$

As shown in Figure 10.3, taking the moment about point o gives

$$-(ky)x + (-mg)x = J\ddot{\theta} \tag{10.6}$$

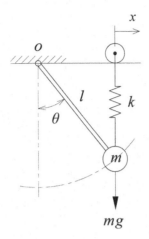

Figure 10.2 Nonlinear mechanical system

where
$$x = l\sin\theta, \quad y = (l - l\cos\theta) \quad \text{and} \quad J = ml^2 \tag{10.7}$$

Substituting equation (10.7) in equation (10.6) and rearranging the results gives

$$(ml^2)\ddot{\theta} + k(l - l\cos\theta)(l\sin\theta) + mg(l\sin\theta) = 0 \tag{10.8}$$

This equation represents the nonlinear differential equation of motion of the given system.

2. SIMULINK block diagram model:
Using the SIMULINK program, we can represent equation (10.8) by creating the SIMULINK block diagram shown in Figure 10.4. Figure 10.5 shows the transient response of the angular displacement $\theta(t)$ generated by the SIMULINK model when the parameters $m = 5$ kg, $k = 200$ N m, $l = 0.5$ m, $\theta(0) = 0.1$ rad and $\dot{\theta}(0) = 0$ rad/sec

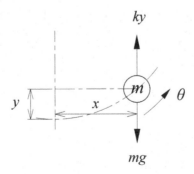

Figure 10.3 Free body diagram

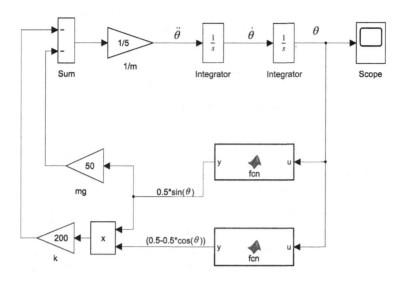

Figure 10.4   SIMULINK block diagram of equation (10.8)

are set to their respective default values in the SIMULINK block diagram shown in Figure 10.4. Because we assume no friction in the system, the response appears to be oscillatory, as shown in Figure 10.5.

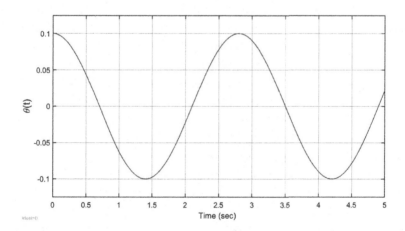

Figure 10.5   The angular displacement $\theta(t)$ generated by SIMULINK

Example (10.2):
A quarter-car model of a vehicle suspension with two degrees of freedom subjected to a bumpy road profile is shown in Figure 10.6. The profile of the bump surface is

represented by a half-sin curve given by

$$y_i(t) = y_o \sin \omega t \qquad (10.9)$$

where ω is the excitation frequency. Assuming that the force exerted by the vehicle damper has a nonlinear function of a form of $F_d = b_V \dot{z}^2$, where $\dot{z}$ is the velocity of the damper, draw the free-body diagram and derive the equation(s) of motion.

Solution:
The model consists of two masses, m_V, corresponding to a quarter of a vehicle mass and a wheel mass m_w. The displacements x_V and x_w are the positions of masses m_V and m_w concerning their static equilibrium positions (i.e., where the springs balance the gravity forces). The suspension linking the vehicle body to the wheel consists of a spring of stiffness k_V parallel with a shock absorber of damping coefficient b_V, as shown in Figure 10.6. The free-body diagram of the forces acting on both masses when the suspension comes into contact with the bumpy road is shown in Figure 10.7. Meanwhile, the damper is modeled as a nonlinear component with a damping force of a nonlinear form of $F_d = b_V \dot{z}^2$, and $\dot{z}$ is the relative velocity of the damper, which is equal to $(\dot{x}_w - \dot{x}_V)$, as shown in Figure 10.7. The input to the suspension is the road surface displacement $y_i(t)$, which depends on the bumpy road profile, which takes the form of a half-sine curve, as indicated in equation (10.9). Note that the excitation frequency ω is due to the speed of the vehicle v and the length l of the half-sine excitation curve on the road, which is equal to $\omega = (\pi v/l)$. Assume the vehicle is moving at a constant velocity v when it comes into contact with the bump. Applying Newton's law on both masses gives

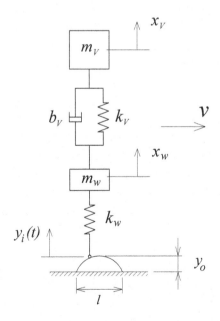

Figure 10.6 A quarter-car active suspension system

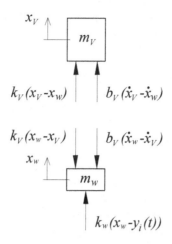

Figure 10.7   Free-body diagram of the acting forces in a quarter-car suspension system

- For the wheel mass m_w:
Note that the relative displacement of the wheel spring of stiffness k_w is equal to $(x_w - y_i(t))$, as shown in Figure 10.7. Considering this, the equation of motion of the wheel mass m_w is given by

$$m_w \ddot{x}_w + k_w(x_w - y_i(t)) + b_V(\dot{x}_w - \dot{x}_V)^2 + k_V(x_w - x_V) = 0 \qquad (10.10)$$

Substituting equation (10.9) in equation (10.10) and insert $\omega = (\pi v / l)$ and rearranging the results gives

$$m_w \ddot{x}_w + b_V(\dot{x}_w - \dot{x}_V)^2 + k_V(x_w - x_V) + k_w x_w = k_w y_o \sin(\frac{\pi v}{l})t \qquad (10.11)$$

- For the vehicle mass m_V:

$$m_V \ddot{x}_V + b_V(\dot{x}_V - \dot{x}_w)^2 + k_V(x_V - x_w) = 0 \qquad (10.12)$$

Equations (10.11) and (10.12) are the nonlinear differential equations of motion of the quarter-car suspension system.

Example (10.3):
Using the Lagrange method, derive the nonlinear differential equations of motion for the inverted pendulum gantry system of a massless rod of length l shown in Figure 10.8.

Solution:
Let the generalized coordinates be x and θ, as shown in Figure 10.9. When the system is at rest, the angle θ corresponds to a completely vertical, upward-pointing pendulum rod of zero value. When facing the system, we assumed that the rotation of the

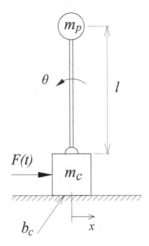

Figure 10.8 Inverted pendulum gantry system

pendulum rod that is considered to be in a positive direction is counter-clockwise (CCW), as shown in Figure 10.9. In addition, the positive direction of the cart displacement x_c is to the right, as indicated by the Cartesian frame of coordinates (x, y) shown in Figure 10.9. This figure shows that when the cart moves a displacement x_c, the displacements x_p and y_p of the bob mass m_p are given by

$$x_p = x_c - l\sin\theta \quad \text{and} \quad y_p = l\cos\theta \qquad (10.13)$$

Taking the time derivative of this equation gives

$$\dot{x}_p = \dot{x}_c - l\cos\theta\,\dot{\theta} \quad \text{and} \quad \dot{y}_p = -l\sin\theta\,\dot{\theta} \qquad (10.14)$$

Let us first determine the total kinetic energy of the system. It is the sum of the translational and rotational kinetic energies arising from the cart mass m_c and pendulum bob mass m_p. Note that the cart mass considered only translational kinetic energy, while the bob mass considered both translational and rotational kinetic energies. So, the translational kinetic energy of the cart mass m_c is given by

$$T_c = \frac{1}{2}m_c\dot{x}_c^2 \qquad (10.15)$$

and the translational kinetic energy of the bob mass m_p can be expressed as a function of its velocity components as follows

$$T_{pt} = \frac{1}{2}m_p\left(\sqrt{\dot{x}_p^2 + \dot{y}_p^2}\right)^2 \qquad (10.16)$$

Substituting equation (10.14) in equation (10.16) and rearranging the results gives

$$T_{pt} = \frac{1}{2}m_p[(\dot{x}_c - l\cos\theta\,\dot{\theta})^2 + (-l\sin\theta\dot{\theta})^2] \qquad (10.17)$$

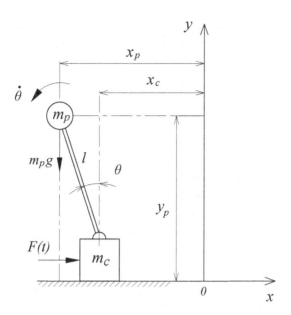

Figure 10.9 Motion of the inverted pendulum gantry system

The rotational kinetic energy of the bob mass m_p is given by

$$T_{pr} = \frac{1}{2} J_p \dot{\theta}^2 \qquad (10.18)$$

Thus, the total kinetic energy of the system is the sum of the three individual kinetic energies indicated in equations (10.15), (10.17), and (10.18). It is given by

$$T = \frac{1}{2} m_c \dot{x}_c^2 + \frac{1}{2} m_p [(\dot{x}_c - l \cos \theta \, \dot{\theta})^2 + (-l \sin \theta \dot{\theta})^2] + \frac{1}{2} J_p \dot{\theta}^2 \qquad (10.19)$$

This equation indicated that the total kinetic energy regarding the generalized coordinates and their time derivative could be expressed.

Since the system does not include a spring, the potential energy is only due to gravity. It is usually caused by the mass's vertical movement from rest. So, the total potential energy in the system is due to the gravitational potential energy of the bob mass m_p, which is given by

$$U = m_p g y_p \qquad (10.20)$$

Using equation (10.13), equation (10.20) becomes

$$U = m_p g l \cos \theta \qquad (10.21)$$

This equation indicated that potential energy can only be expressed in terms of generalized coordinates. So, the Lagrange function in x_c and θ coordinates is given by

$$L = T - U = \frac{1}{2} m_c \dot{x}_c^2 + \frac{1}{2} m_p [(\dot{x}_c - l \cos \theta \, \dot{\theta})^2 + (-l \sin \theta \dot{\theta})^2] + \frac{1}{2} J_p \dot{\theta}^2 - m_p g l \cos \theta$$

$$(10.22)$$

Considering the friction of damping factor b_c that is provided to the cart, as shown in Figure 10.8, the dissipation (non-conservative) energy is given by Rayleigh potential as,

$$D_c = -\frac{1}{2} b_c \dot{x}_c^2 \qquad (10.23)$$

Taking the derivative of this equation a follows

$$\frac{\partial D_c}{\partial x_c} = -b_c \dot{x}_c \qquad (10.24)$$

Let us now derive Lagrange equations of motion for the system. Applying the Lagrange equation indicated in equation (6.22) on equation (10.24) gives

$$\frac{d}{dt}\left(\frac{\partial L}{\partial \dot{x}_c}\right) - \frac{\partial L}{\partial x_c} = \frac{d}{dt}\left[m_c \dot{x}_c + m_p(\dot{x}_c - l\cos\theta \, \dot{\theta})\right] = Q_c \qquad (10.25)$$

and

$$\frac{d}{dt}\left(\frac{\partial L}{\partial \dot{\theta}}\right) - \frac{\partial L}{\partial \theta} = \frac{d}{dt}\left[m_p(\dot{x}_c - l\cos\theta \, \dot{\theta}) - m_p l \sin\theta \dot{\theta}\right] + J_p \dot{\theta}] - \frac{\partial}{\partial \theta}(m_p g \cos\theta) = Q_p \qquad (10.26)$$

Since the external force $F(t)$ and friction forces are applied only to the cart, so the generalized forces of the system are given by

$$Q_c = F(t) - b_c \dot{x}_c \quad \text{and} \quad Q_p = 0 \qquad (10.27)$$

Substituting equation (10.27) in equations (10.25) and (10.26) and carrying out the partial derivative $(\partial/\partial\theta)$ presented in equation (10.26) gives

- For the generalized coordinate x_c, we have

$$\frac{d}{dt}\left[m_c \dot{x}_c + m_p \dot{x}_c - m_p l \cos\theta \, \dot{\theta}\right] = F(t) - b_c \dot{x}_c \qquad (10.28)$$

- For the generalized coordinate θ, we have

$$\frac{d}{dt}\left[m_p(\dot{x}_c - l\cos\theta \, \dot{\theta}) - m_p l \sin\theta\dot{\theta} + J_p\dot{\theta}\right] + m_p g \, l \sin\theta = 0 \qquad (10.29)$$

Carrying out the derivative terms presented in equations (10.28) and (10.29) and rearranging the results give

$$(m_c + m_p)\ddot{x}_c - (m_p l \cos\theta)\ddot{\theta} + m_p l \sin\theta \dot{\theta}^2 + b_c \dot{x}_c = F(t) \qquad (10.30)$$

and

$$(J_p + m_p l^2)\ddot{\theta} - (m_p l \cos\theta)\ddot{x}_c + m_p l \sin\theta \dot{x}_c \dot{\theta} + m_p g l \sin\theta = 0 \qquad (10.31)$$

Equations (10.30) and (10.31) are the nonlinear equations of motion derived from the Lagrange method for the inverted pendulum gantry system shown in Figure 10.8.

10.3   BOND GRAPH MODULATED ELEMENTS

Providing the modeling process with a graphical representation of the causality (the input-output relationships) in the operating platform is one of the distinguishing characteristics of the Bond-Graph modeling technique. When the modeler understands causality, he can recognize algebraic loops and implicit differential equations earlier in finding a solution. In addition, it clarifies the effect of the model's features being added or removed. Even if some variables cannot be easily stated in equation form, the modeler can still develop simulation programs for nonlinear systems using the information about causal relationships. The generated program is similar to a collection of equations, but rather than including algebraic expressions, it uses subroutines. The modeler can create programs utilizing sets of variables rather than individual variables for numerous complicated systems by using vector bond graphs to express causal connections among extensive subsystems. In general, the elements of the Bond Graph that comprise it are allowed to be functions of another parameter. This section will explain the nonlinear power conservation of the Bond Graph elements, also known as modulated Bond Graph elements.

10.3.1   ACTIVE BOND

Bonds are characterized by their ability to explain how different physical variables interact. As a result, these variables will influence each other. For example, an electromotor's load will affect how it behaves dynamically. Even though a direction and causality have been established, this return action variable remains in a Bond Graph using the active bond. As shown in Figure 10.10, the active bond is represented by a line with a full arrow indicating the flow of information. Now that we have established that, let us show how the active bond works with the modulated Bond Graph elements.

10.3.2   MODULATED BOND GRAPH ELEMENTS *MC* AND *MR*

We always thought that springs and frictional damping would have linear properties for a mechanical system to be linear; this simplified their nonlinear properties. If we cannot consider these assumptions, the spring and the damper will be nonlinear elements. In this case, these elements are called modulated Bond Graph elements. This means another parameter, say x, which is the output of another element, affects how these elements work when the Bond Graph model is being simulated. The active bond and the conventional symbols for a modulated capacitance (nonlinear spring) *MC* and modulated resistance (nonlinear damper) *MR* conservation are shown in Figures 10.10a and 10.10b. In this case, the variable parameter x is the output of another element.

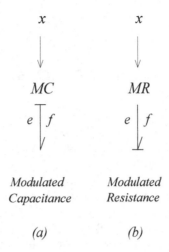

Figure 10.10   Modulated capacitance MC and resistance MR elements

10.3.3   MODULATED 1-PORT ELEMENTS MSE AND MSF (MODULATED SOURCES)

Figures 10.11a and 10.11b show the active bond and the conventional symbols for conserving the modulated source of flow MSF and the modulated source of effort MSE, respectively. The output, whether an amount of effort or a flow, is proportional to the variation provided by the input signal $F(t)$ or $v(t)$, which is often a function of time.

10.3.4   MODULATED 2-PORTS ELEMENTS MTF AND MGY

The TF and the GY are representations of linear power-conserving elements with fixed ratios, and their detailed definitions can be found at the beginning of Chapter 3. Here, we will discuss the nonlinear power conservation of the TF and GY, known as the modulated transformer MTF and modulated gyrator MGY, respectively. The active bond and the conventional symbols for modulated transformers MTF and modulated gyrators MGY are shown in Figures 10.12a and 10.12b, respectively. Their constitutive relations are given by

$$F(t) \longrightarrow MSE \xrightarrow{\ e\ }_{f}$$ $$v(t) \longrightarrow MSF \vdash\!\!\xrightarrow{\ e\ }_{f}$$

Modulated source of effort   *Modulated source of flow*

(a)   (b)

Figure 10.11   Modulated 1-port MSE and MSF elements

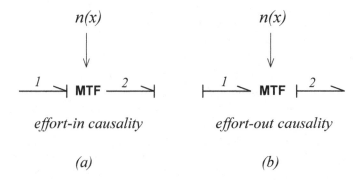

effort-in causality   *effort-out causality*

(a)   *(b)*

Figure 10.12   Modulated 2-ports MTF element

1. Modulated Transformers MTF:

$$f_1 = n(x)f_2 \quad \text{and} \quad n(x)e_1 = e_2 \tag{10.32}$$

2. Modulated Gyrators MGY:

$$m(x)f_1 = e_2 \quad \text{and} \quad e_1 = m(x)f_2 \tag{10.33}$$

As shown in Figures 10.12a and 10.12b, when it comes to the modulated transformer MTF, an effort-in causality on the incoming bond will result in an effort-out causality on the outgoing bond and vice versa. Similarly, as shown in Figures 10.13a and 10.13b for the modulated gyrator MGY, an effort-in causality on the incoming bond will result in an effort-in causality on the outgoing bond, and vice versa. It should be noted that regardless of what the values of $m(x)$ or $n(x)$ may be, the power $e_1 f_1$ for both elements is always equal to the power $e_2 f_2$. In addition, the values of the parameters $m(x)$ and $n(x)$ are represented as shifting through the use of the active bond rather than a power bond. As a result, one of the distinguishing features of these elements with modulated parameters is that the parameters' values $m(x)$ and $n(x)$ can shift without a corresponding change in the power flow.

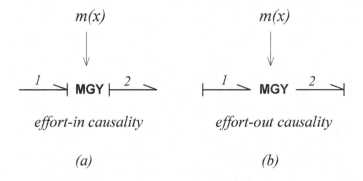

effort-in causality   *effort-out causality*

(a)   *(b)*

Figure 10.13   Modulated 2-ports MGY element

Example (10.4):
For the system shown in Figure 10.14, assuming that the angular displacement θ cannot be considered small,

1. Create the preliminary nonlinear Bond Graph and simplify it.
2. Create the Augmented nonlinear Bond Graph.
3. Derive the nonlinear equation(s) of motion.
4. Check if Newton's law is embedded in the Augmented Bond Graph model.
5. Build the Augmented nonlinear Bond Graph model using the 20-sim program and generate the transient response of the angular displacement $\theta(t)$, considering the following parameter values $m = 5\,\text{kg}$, $k = 200\,\text{N/m}$, $l = 0.5\,\text{m}$ and I.C. $\theta(0) = 0.1$ rad and $\dot{\theta}(0) = 0$ rad/sec.

Solution:
As shown in Figure 10.14, the mass m rotates with an angular velocity $\dot{\theta}$, gravity acts with a velocity $\dot{y}$, and the spring position depends on both $\dot{y}$ and the velocity of the massless pulley in the y-direction, which is the same as the ground speed v_g (to be eliminated). Therefore, three 1-junctions are created for $\dot{\theta}$, $\dot{y}$ and v_g as shown in Figure 10. The moment of inertia of the rotating mass is represented by an inertia element J connected to the 1-junction of a common flow $\dot{\theta}$. The gravity force, the external force that acts on the mass m once it starts to move, is represented by a source of effort SE connected to the 1-junction of a common flow $\dot{y}$. Since the spring share the same effort (force), a capacitance element C, represents the spring stiffness k is connected to a 0-junction positioned between the two 1-junctions of the common flows of $\dot{y}$ and v_g, as shown in Figure 10.15. Note that the vertical displacement y of the mass equals zero when the system is at rest ($t = 0$). It is generated when the mass starts to rotate an angular displacement θ, and a resulting geometric relationship between θ and y is given by

$$y = (l - l\cos\theta) \tag{10.34}$$

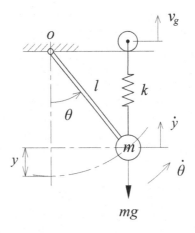

Figure 10.14 Nonlinear mechanical system

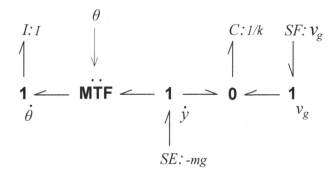

Figure 10.15 Preliminary nonlinear Bond Graph model of the given system

Taking the derivative of this equation gives

$$\dot{y} = (l \sin \theta)\, \dot{\theta} \tag{10.35}$$

Since in the linear transformer TF, we assumed that the angular displacement θ is small, the velocity $\dot{y}$ is determined by the constant value of the angular velocity $\dot{\theta}$. To determine the transformation ratio $n(\theta)$ (modulus) between $\dot{\theta}$ and $\dot{y}$, rearrange equation (10.35) as follows

$$n(\theta) = \frac{\dot{\theta}}{\dot{y}} = \left(\frac{1}{l \sin \theta}\right) \tag{10.36}$$

As indicated in this equation, the modulus $n(\theta)$ is no longer a constant but a function of θ. As a result, we create a modulated transformer MTF with a modulus $n(\theta)$ positioned between the two 1-junctions of the common flows $\dot{\theta}$ and $\dot{y}$, represented by an active bond of the variable θ as shown in Figure 10.16.

At this point, the preliminary Bond Graph model of the given system is created, as shown in Figure 10.16. The half-arrows indicating the positive directions of power flow are selected, as shown in the same figure. Remember that sources supply power while elements receive it. To simplify this Bond Graph model, all connections that hold the ground speed v_g with their associated bonds must be removed, and then any 2-port 1-junctions or 0-junctions with through power flow should be swapped with single bonds, as shown in Figure 10.16. As a result, the 1-junction connected to the source of flow SF, the 0-junction connected to the capacitance C, and the 1-junction connected to inertia element J are removed. The resulting simplified nonlinear Bond Graph model of the given system is shown in Figure 10.17.

Now, let us assign the causality of the Bond Graph model. Since only one source of effort SE exists, the causal stroke marks should be located on its bond at the junction side, as shown in Figure 10.17. Considering the causality restrictions in Section 3.3, apply the causal stork marks to all elements and junctions. It is essential to remember that as we work through assigning causality, we must consider the integral causality that applies to all associated elements and bonds. We are finally numbing

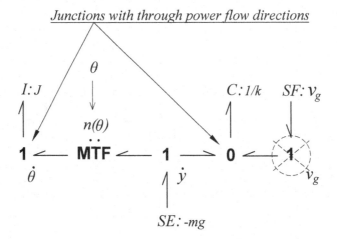

Figure 10.16 Simplified nonlinear Bond Graph model of the given system

all bonds to get the Augmented nonlinear Bond Graph model of the system shown in Figure 10.17. Note that the power flow direction and causality of the bond (2) are imposed on the element side of the *MTF*, which implies that the transformer will transform $\dot{y}$ to $\dot{\theta}$ with a transformation ratio of $(1/n(\theta))$, as shown in Figure 10.17. This figure shows two integral causality storage elements, J and C, represented by bonds 1 and 3.

To derive the nonlinear differential equations of motion of the given system, we must first define the key variables of the model: Input, State variables, and Co-energy (power) variables and their constitutive relations.

- Key variables:
- Inputs: $e_4 = SE = -mg$.
- State variables: p_1 and q_3 .
- Co-energy (Power) variables: f_1 and e_3.

Figure 10.17 Augmented nonlinear Bond Graph model of the given system

- Constitutive relations:
Write equations describing the relationship between co-energy variables and state variables for each energy storage element, as follows:

$$f_1 = \frac{1}{J}p_1 \quad \text{and} \quad e_3 = \frac{1}{C}q_3 \tag{10.37}$$

Now that we have the connections of flow and effort variables of the model junctions shown in Figure 10.17, let us derive the equations that reflect them.

- As shown in Figure 10.17, since the effort exerted on one side of the modulated transformer element, MTF relates to the effort on the other side and vice versa, the flow and effort variables are given by

$$f_1 = n(\theta)f_2 \quad \text{or} \quad n(\theta) = \frac{f_1}{f_2} \tag{10.38}$$

and

$$n(\theta)e_1 = e_2 \quad \text{or} \quad e_1 = \dot{p}_1 = \frac{1}{n(\theta)}e_2 \tag{10.39}$$

where $\dot{p}_1$ the momentum change of the rotating mass J. Equations (10.38) and (10.39) indicate that the power balance $(e_1 f_1 = e_2 f_2)$ at both sides of the modulated transformer is satisfied. Note that the transformer modulus $n(\theta)$ can be obtained through equation (10.36).
- The flow variables of the 1-junction are given by

$$f_2 = f_3 = f_4 \tag{10.40}$$

and the effort variables are given by

$$e_4 = e_2 + e_3 \tag{10.41}$$

Now, let us derive the nonlinear differential equations of motion of the given system. To determine the momentum change $\dot{p}_1$ of the rotating mass, considering equation (10.39), we must first determine the effort e_2. Equation (10.41) indicates that the effort e_2 equals $(e_4 - e_3)$. Considering this and insert $e_4 = SE$, equation (10.39) becomes

$$e_1 = \dot{p}_1 = \frac{1}{n(\theta)}(e_4 - e_3) = \frac{1}{n(\theta)}(SE - e_3) \tag{10.42}$$

Using the constitutive relations indicated in equation (10.37), equation (10.42) becomes

$$\dot{p}_1 = \frac{1}{n(\theta)}\left(SE - \frac{1}{C}q_3\right) \tag{10.43}$$

Rearranging this equation gives

$$\dot{p}_1 + \frac{1}{C}\frac{1}{n(\theta)}q_3 = \frac{1}{n(\theta)}SE \tag{10.44}$$

This equation is the nonlinear differential equation of motion of the given system.

Now, we will check if Newton's law of motion is embedded in the Augmented nonlinear Bond Graph model shown in Figure 10.17. The following analyses are what we get when using the constitutive relations indicated in equation (10.37).

- The momentum of the rotating mass m is given by

$$p_1 = Jf_1 = J\dot{\theta} \tag{10.45}$$

Taking the derivative of this equation gives

$$\dot{p}_1 = J\ddot{\theta} \tag{10.46}$$

- The state variable q_3 is equal to the displacement y of the spring which is given by

$$q_3 = (l - l\cos\theta) \tag{10.47}$$

Substituting equations (10.46) and (10.47) in equation (10.44) and inserting $SE = -mg$ and $1/C = k$ gives

$$J\ddot{\theta} + k(l - l\cos\theta)(l\sin\theta) = -mg(l\sin\theta) \tag{10.48}$$

Rearranging this equation and inserting $J = ml^2$ gives

$$ml^2\ddot{\theta} + k(l - l\cos\theta)(l\sin\theta) + mg(l\sin\theta) = 0 \tag{10.49}$$

By comparing equation (10.49) with equation (10.8), we found that they are identical, implying that Newton's law is satisfied in the Augmented nonlinear Bond Graph model of the given system shown in Figure 10.17.

- 20-sim Bond Graph model:
The 20-sim program package allows the construction and simulation of nonlinear models as Bond Graph structures. Based on the Augmented Bond Graph of the given system shown in Figure 10.17, create the corresponding 20-sim Bond Graph model as shown in Figure 10.18. The block diagram in this figure is created using the 20-sim blocks to calculate the modulus $n(\theta)$ of the modulated transformer MTF. It is processed and fed back into the Bond Graph at the MTF. Figure 10.19 shows the transient response of the angular displacement $\theta(t)$ generated by the 20-sim model when the parameters $m = 5\,\text{kg}$, $k = 200\,\text{N/m}$, and $l = 1\,\text{m}$, $\theta(0) = 0.1$ rad and $\dot{\theta}(0) = 0$ rad/sec are set to their respective default values in the 20-sim diagram shown in Figure 10.18. Comparing these results with those generated using SIMULINK shown in Figure 10.5, we found that they are the same.

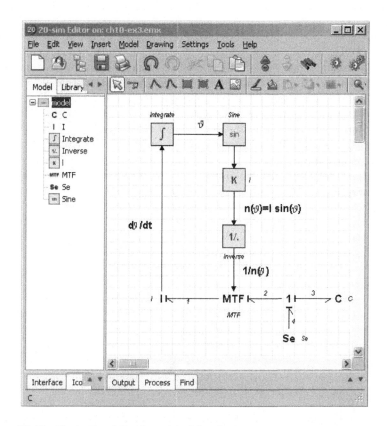

Figure 10.18   20-sim Bond Graph model of the given system

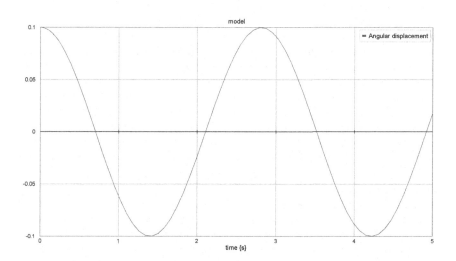

Figure 10.19   The angular displacement $\theta(t)$ generated by 20-sim

10.4 PROBLEMS

Problem (10.1):
For the crankshaft system shown in Figure 10.20, assuming that both rods are massless and the angular displacements θ_1 and θ_2 cannot be considered small,

a. Derive the nonlinear equations of motion.
b. Create the augmented nonlinear Bond Graph model (add the half-arrow positive power flow directions, assign causalities, and number all bonds). Check the derivative causality.
c. Derive the nonlinear equations of motion (write the key variables and their constitutive relations). Verify the results with step (a).

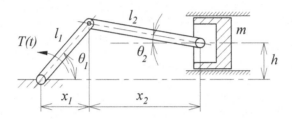

Figure 10.20   Mechanical system

Problem (10.2):
For the system shown in Figure 10.21, assuming that both rods are massless and the angular displacements θ_1 and θ_2 cannot be considered small, derive the nonlinear equations of motion.

Problem (10.3):
For the system shown in Figure 10.22, assuming that both rods are massless, and the angular displacements θ_1 and θ_2 can not be considered small,

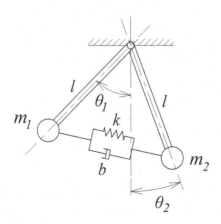

Figure 10.21   Mechanical system

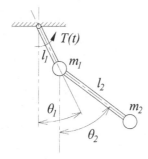

Figure 10.22 Mechanical system

a. Derive the nonlinear equations of motion of the given system.
b. Create the augmented nonlinear Bond Graph model (add the half-arrow positive power flow directions, assign causalities, and number all bonds). Check the derivative causality.
c. Derive the nonlinear equations of motion (write the key variables and their constitutive relations). Compare the results with step (a).
d. Use SIMULINK and 20-sim to generate the plots of the system responses $\theta_1(t)$ and $\theta_2(t)$ considering the values of the following parameters: $m_1 = 0.25$ kg, $m_2 = 0.5$ kg, $l_1 = 0.2$ m, $l_2 = 0.4$ m, and $T(t) = 10u(t)$ N/m (step input). Verify the results (to accomplish this, turn in the SIMULINK and 20-sim plots on the same scales).

Problem (10.4):
For the system shown in Figure 10.23, assume the lever rods are massless, and the angular displacements θ_r and θ_p cannot be considered small. Note that the mass m moves only in the x direction. Considering the values of the following parameters: $m = 0.125$ kg, $r = 0.2$ m, $l = 0.35$ m, and $T(t) = 5u(t)$ N m (step input),

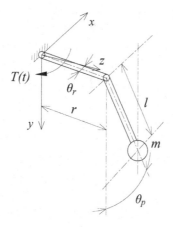

Figure 10.23 Mechanical system

a. Derive the nonlinear equations of motion describing the dynamics of the angular displacements $\theta_r(t)$ and $\theta_p(t)$.
b. Using SIMULINK, create the SIMULINK block diagram of the derived equations and generate the plots of responses $\theta_r(t)$ and $\theta_p(t)$.
c. Create the augmented nonlinear Bond Graph model (add the half-arrow positive power flow directions, assign causalities, and number all bonds). Check the derivative causality.
d. Derive the nonlinear equations of motion (write the key variables and their constitutive relations). Compare the results with step (a).
e. Using 20-sim, generate the plot of the responses $\theta_r(t)$ and $\theta_p(t)$. Verify the results with step (b) (to accomplish this, turn in the SIMULINK and 20-sim plots on the same scales).

Problem (10.5):
For the system shown in Figure 10.24, assuming the angular displacements θ cannot be considered small,

a. Derive the nonlinear equations of motion.
b. Create the augmented nonlinear Bond Graph model (add the half-arrow positive power flow directions, assign causalities, and number all bonds). Check the derivative causality.
c. Derive the nonlinear equations of motion (write the key variables and their constitutive relations). Compare the results with step (a).
d. Choose real values of a car parameters: M, l_f, l_r, b_f, k_f, b_r, k_r, v_f, and v_r. Using SIMULINK and 20-sim, generate the plots of the car responses $v_a(t)$ and $v_b(t)$. Verify the results (to accomplish this, turn in the SIMULINK and 20-sim plots on the same scales).

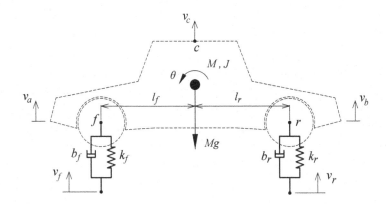

Figure 10.24 Car suspension system

Index

Printed in the United States
by Baker & Taylor Publisher Services